Braun-Brumfield
AUTHORIZED
ITEM

AUTHORIZED BY
NOT FOR SALE

ENVIRONMENTAL IMPACTS OF MINING
Monitoring, Restoration, and Control

ENVIRONMENTAL IMPACTS OF MINING
Monitoring, Restoration, and Control

M. SENGUPTA

LEWIS PUBLISHERS
Boca Raton   Ann Arbor   London   Tokyo

Library of Congress Cataloging-in-Publication Data

Sengupta, M.
  Environmental impacts of mining: monitoring, restoration, and control / Mritunjoy Sengupta.
    p. cm.
  Includes bibliographical references and index.
  ISBN 0-87371-441-5
  1. Mineral industries—Environmental aspects. I. Title.
TD195.M5S46   1993
622—dc20   92-43805
                                                                          CIP

COPYRIGHT © 1993 by LEWIS PUBLISHERS
ALL RIGHTS RESERVED

This book represents information obtained from authentic and highly regarded sources. Reprinted material is quoted with permission, and sources are indicated. A wide variety of references are listed. Every reasonable effort has been made to give reliable data and information, but the author and the publisher cannot assume responsibility for the validity of all materials or for the consequences of their use.

Neither this book nor any part may be reproduced or transmitted in any form or by any means, electronic or mechanical, including photocopying, microfilming, and recording, or by any information storage and retrieval system, without permission in writing from the publisher.

Direct all inquiries to CRC Press, Inc., 2000 Corporate Blvd., N.W., Boca Raton, Florida 33431.

PRINTED IN THE UNITED STATES OF AMERICA
  2 3 4 5 6 7 8 9 0
Printed on acid-free paper

Nupur

Shyam Sundar

Preface

This book has been written to serve as an introductory text to the state of the art on environmental impacts of mining and their control. It endeavors to fulfill the needs of students and professional engineers concerned with the environmental problems created by mining operations. The area of environmental impacts of mining has not received comprehensive treatment in other texts. The need for such a contribution is understood by the widespread scattering of literature in government reports, journals, and conference proceedings.

The author is grateful to the Society of Mining, Metallurgy and Exploration and to the Southern Illinois University Press for their kind permission to reproduce materials from their publications.

M. Sengupta
Fairbanks, Alaska
May 1992

The Author

Dr. Mritunjoy Sengupta received his Ph.D. from the Colorado School of Mines. He also holds an E.M. in Mining Engineering and an M.S. in Industrial Engineering from the School of Engineering and Applied Science, Columbia University, New York. He received a B.S. degree in Mining Engineering from the Indian School of Mines. He is a registered professional engineer in Colorado, Idaho, New Mexico, and Alaska.

Dr. Sengupta has 10 years of professional experience in the U.S. mining industry. His former employers include AMAX Inc., Continental Oil Company, Texas Gulf Inc., Morrison-Knudsen Company, United Nuclear Corporation, and Hawley Coal Mining Company. Dr. Sengupta has served as a scientist with the Central Mining Research Station of the Government of India and as a United Nations expert to the Government of India.

Contents

CHAPTER 1
Mining and the Environment   1
    Introduction   1
    Uniqueness of Mining   2
    Environmental Effects of Surface Mining   4
    Water Pollution   17
    Land Use   18
    Subsidence of Mined Land   26
    Environmental Audits   29
        Types of Environmental Audits   31
    References   33

CHAPTER 2
Surface Coal Mining with Reclamation   35
    Dragline Operation   35
        Keycutting and Layered Cutting   37
            Panel (Pit) Width   37
            Extended Bench   39
    Dragline Stripping Procedures   39
        Single Seam Stripping with Nonselective
          Spoil Placement   40
        Single Machine Subsystems   40
            Shallow Overburden, One Lift, One Pass   40
        Box Pits   45
            End Cut   46

      Side Cut   46
      Rehandle (End Cut)   46
      Rehandle (Borrow Pit)   49
   Moderate Overburden Depth, Single Lift, Single Pass   51
   Moderately Deep Overburden, Two Lifts, One Pass   53
   Deep Overburden, Split Bench Mining   54
   Deep Overburden, Multiple Lifts, One Pass   58
   Two-Pass Extended Bench Method   62
   Pullback Method   63
Terrace Mining   67
Tandem Machine Systems   67
   Tandem Dozer/Dragline Stripping   67
   Single-Seam Stripping with Selective Spoil Placement   70
   Moderate Overburden Depth, One Lift, Single Pass   70
   Moderate Overburden Depth, Two Lifts, One Pass   71
   Deep Overburden, Two Lifts, One Pass   71
   Multiple Seam Stripping with Draglines   72
   Single Dragline, Two Seams, Nonselective
     Spoil Placement   72
      Moderate Overburden and Interburden Depths   72
   The Extended Bench Method for Two Seams   73
   The One-Pass Extended Bench Method for
     Two-Seam Stripping   74
   The Two-Pass Spoil Side Method   77
   Elevated Bench Method   78
   Tandem Machine, Two-Seam Condition, Nonselective
     Spoil Placement   80
   The Tandem Shovel-Dragline System   80
   Dozer-Dragline System   81
   Tandem Machine-Multiple Dragline   81
   Selective Spoil Placement in Two-Seam Condition   81
   Stripping of Unstable Burden Material   83
Multiple Seam Systems   85
Horseshoe Mining Sequence   89
Steep Slope Mining   91
  General Sequence of Mining and
    Reclamation Operations   91
     One-Cut, Single-Seam Conventional Contour Mining   92
   Other Conventional Contour Mining Situations   96
   Haulback Mining Methods   96
     Single-Cut Haulback Mining of Single Seams   96
References   101

CHAPTER 3
Reclamation and Revegetation of Mined Land   103
   Introduction   103
   Reclamation of Surface Mined Land in Australia   105
      Material Characterization   106
      Landform Design   107
      Use of Topsoil   108
      Reclamation Procedures   109
      Revegetation Methods   110
      Tree Planting   111
      Pasture Management   111
   Revegetation of a Surface Mined Land in Montana   112
   Factors Affecting Natural Revegetation of Coal
     Mine Spoil Banks in Ohio   113
   Vegetation Development on Old and Abandoned
     Lead and Zinc Mine   115
   Revegetation at the Usibelli Coal Mine, Alaska   117
   References   119

CHAPTER 4
The Acid Mine Drainage Problem from Coal Mines   121
   Introduction   121
   Chemistry of Formation   122
      The Role of Bacteria   124
   Conventional Neutralization Process Using Lime   126
      High-Density Sludge Process   128
      Other Treatment Processes   128
   Chemical Treatment   129
      Lime   129
      Limestone   132
      Caustic Soda   135
   Iron Oxidation   135
      Aeration Systems   137
      Biological Oxidation   140
         Oxidation Rate   142
   Sludge Dewatering and Disposal   143
   Reverse Osmosis   151
   Ion Exchange   155
      Sul-Bisul Process   156
      Modified Desal Process   158
      Two-Resin Process   158

Chemical Softening   160
    Lime-Soda Process   160
      Alumina-Lime-Soda Process   161
  Bactericides in AMD Control   162
  Determination of Acid-Generating Potential   163
  References   165

CHAPTER 5
Acid Rock Drainage and Metal Migration   167
  Introduction   167
  The Acid-Generation Process   168
  Sulfide Minerals   169
  Chemical and Biological Reactions Related to Acid Generation   171
    Metal Leaching and Migration Processes   175
    Prediction of Acid Drainage   176
    Static Tests   178
    Kinetic Tests   180
  Control of Acid Generation   182
    Available Control Measures   183
  Conditioning of Tailings/Waste Rock   184
  Waste Segregation and Blending   185
  Bactericides   186
  Base Additives   186
  Covers and Seals to Control Acid Generation   190
    Soil Covers   190
    Synthetic Membrane Covers   191
    Water Cover   192
    Saturated Soil or Bog   192
  Subaqueous Deposition   192
    Disposal into Manmade Impoundments   193
    Disposal into Flooded Mine Workings   194
    Lake Disposal   194
    Marine Disposal   195
    Migration Control of ARD   195
      Diversion of Surface Water   197
      Underground Mines   197
      Open Pits   197
      Waste Rock Dumps and Spoil Piles   198
      Tailings Deposits   198
      Stockpiles and Spent Heap-Leach Piles   198
    Groundwater Interception   198

        Underground Mines   199
        Open Pits   199
        Waste Rock Dumps and Spoil Piles   200
        Tailings Deposits   200
        Stockpiles and Spent Heap-Leach Piles   200
    Covers and Seals to Control Infiltration   200
        Soil Covers   200
        Synthetic Covers   209
        Placement of Covers   212
    Waste Rock and Tailings Placement Methods   214
Monitoring   215
    Specific Monitoring Programs for
      Each Mine Component   217
        Environmental Monitoring of Open Pits   218
        Environmental Monitoring of Underground Workings   219
        Environmental Monitoring of Waste Rock Dumps, Ore
          Stockpiles, and Heap-Leach Sites   221
        Environmental Monitoring of Tailings Impoundments   222
        Environmental Monitoring of Quarries   223
        Environmental Monitoring of Haul Roads   224
    Impact of an Abandoned Mine on Water Quality   224
    Hydrologic Solution to Acid Mine Drainage   227
    Water Resource Problems in a Lead Belt   228
    Environmental Control Measures after the Closure of a Lead-Zinc
      Mine in Greenland   232
    Mine Environmental Rehabilitation   238
    Designing Closure of an Open Pit Mine in Canada   246
    Metal Contents and Treatment of Mine Water   249
        Water Types and Contents   250
    References   259

CHAPTER 6
Hydrologic Impact   261
    Introduction   261
    Hydrologic Impact of Phosphate Mining   263
    Hydrologic Impact of Phosphate Gypsum Disposal Areas in
      Central Florida   269
    Hydrologic Effect of Subsurface Coal Mining in the
      Appalachian Region   274
    Effects of Longwall Mining on Hydrology   276
    References   281

CHAPTER 7
Erosion and Sediment Control   283
    Preliminary Site Evaluation   284
        Land Type   285
        Soil and Rock   285
        Streams   285
        Floodplains   286
        Impoundments   286
        Groundwater Conditions   286
        Vegetative Cover   287
    Planning   287
        Preliminary Site Investigation   287
        Preliminary Design   287
        Subsurface Investigations   288
        Final Design   290
    Formulation of an Erosion and Sediment Control Plan   296
        Operation   297
        Maintenance   298
    Sedimentation Control in a Surface Coal Mine   300
    Surface Mine Sedimentation Control   307
    Surface Mine Drainage Control   317
    References   323

CHAPTER 8
Wetlands   325
    Introduction   325
    Constructed Wetlands   328
    Constructed Wetlands for Mine Drainage Treatment   335
        Metal Removal in Constructed Wetlands   342
        Site Selection   347
        Performance Expectations   349
        Hydraulic Design and Control Structures   351
    Substrate   354
        Substrate Evaluation for AMD Systems   354
        Vegetation   361
        Water and Soil Parameters Affecting Growth of Cattails   361
            Where Cattails Grow   363
    Effects of Cattails (*Typha*) on Metal Removal   368
    Metal Retention Capacity of Wetlands for Treatment of
      Acid Mine Drainage   374
    Role of *Sphagnum* Plants in Iron Uptake   376

Iron and Manganese Removal in a *Typha*-Dominated Wetland 381
The Role of Algae in the Treatment of Acid Mine Drainage 386
Constructed Wetlands for Acid Drainage Control in the
   Tennessee Valley 390
Windsor Coal Company Wetland 396
The Tracy Wetland 399
Wetland Treatment in Metal Mining 405
   Big Five Tunnel Experimental Wetland, Colorado 407
   Nickel and Copper Removal by a Natural Wetland 413
References 422

CHAPTER 9
Blasting 425

CHAPTER 10
Mining Subsidence 431
Introduction 431
Subsidence Investigations 434
Structural Damage 436
Damage Criteria 437
Remedial Measures 438
References 440

CHAPTER 11
Postmining Land Use 441
Introduction 441
Appalachian Region Case Study 445
   Evaluation of Alternatives 446
      Economic Evaluation 447
      Environmental Evaluation 447
      Social Impact Evaluation 447
   Selected Alternatives 447
Midwest Case Study 448
   Evaluation of Alternatives 449
      Economic Evaluation 449
      Environmental Evaluation 450
      Social Impact Evaluation 450
   Selected Alternative 450
References 451

CHAPTER 12
Environmental Effects of Gold Heap-Leaching Operations   453
    Introduction   453
    Hazard Identification   454
    Exposure Assessment   454
        Groundwater Pathway   459
        Surface Water Pathway   462
    Toxicity Assessment   466
    Risk Characterization   467
        Decommissioning of Heap-Leach Facilities —
          Industry Experience   468
            Borealis Mine, Echo Bay Minerals
              Company, Hawthorne, NV   468
            Barrick Goldstrike Mines, Inc., Carlin, NV   470
            Fondaway Mine, Tenneco Minerals   471
            Gilt Edge Mine, Deadwood, SD   472
            Annie Creek Mine, Lead, SD   472
            Golden Maple Gilt-Edge Mine   473
            Previous Experience at ZMI   474
    References   477

INDEX   479

CHAPTER 1

Mining and the Environment

INTRODUCTION

Mining operations have been seen by environmentalists and conservationists alike as causing problems. Undoubtedly, the operations of metal and coal producers have caused varying degrees of environmental damage in mining areas, which are often located in remote regions. In the urban, suburban, and rural settings of agricultural communities, the operators of rock quarries, gravel pits, and certain industrial mines have been considered the more visible and significant offenders. Much of the concern has been focused on the concurrent and subsequent physical and aesthetic effects that their operations have had on the land — as a basic resource. Mining is only a temporary occupier of the land surface and, hence, is of a transient nature. Although active mines at any particular time are not as widespread as other land uses, they dramatically change the landscape and tend to leave evidence of their past use. Thus, results of abandonment or closure become most conspicuous to the general public.

There have been continuous confrontations between citizen groups, governmental agencies, and members of the mining industry. The degree of conflict and its nature usually depended on the current land use and the estimated consequences of proposed disturbances. The conflicts centered on the following issues:

- destruction of the landscape
- degradation of the visual environment
- disturbance of watercourses
- destruction of agricultural and forest lands

- damage to recreational lands
- noise pollution
- dust
- truck traffic
- sedimentation and erosion
- land subsidence
- vibration from blasting and air blasts

Environmental conscience has developed dramatically and led to widespread public opinion that governments at all levels should be able to control the depletion of natural resources and excessive environmental damage.

UNIQUENESS OF MINING

Mineral deposits have fixed locations, so mining activities, unlike renewable resource activities (such as fishing, agriculture, and forestry), are not subject to rational selection or advanced planning. Due to unique physical conditions associated with their location, there is no choice about the characteristics of their ecological setting, the biological and chemical characteristics, mineral composition, or grade of ore in question. All of these factors influence the ultimate design, layout, and size of the operation, as well as the basic environmental problems and the potential longer range of regional impacts. The nature of the ecological setting also determines other land uses or activities that would be affected by the proposed mining.

Mines have a finite life. Because of the nonrenewable nature of mineral deposits, mining is only a temporary land user. However, in some situations, ore reserves are so great that mining activities appear to be permanent fixtures in the life of the region inhabitants.

Mines are usually located in a setting of relatively unspoiled nature. The contrast between the mine itself, its dump, mill, and newly constructed tower and the wooded valley or otherwise unscarred mountainside is always there for all to see. A generation ago, this isolated outpost of industry, winning wealth from untapped nature, was looked on as a symbol of man's ingenuity and as a proud demonstration of progress toward an ever expanding better future. Now it is looked on by some, perhaps by an increasing number, as a forerunner of the destruction of the environment which supports us and of which we are a part.

In many instances, the original mine is the very reason for the existence of a town. Other means of economic support are generated. Often, because of the isolated nature of these mining towns in forested and lake regions, alternative activities grow in forestry, recreation, and tourism. The latter two usually rely on increased access by a traveling public whose desire for a relatively clear, unspoiled environment contributes to changing viewpoints and increased opposition to mining activities.

The major difference between alternative land uses and mining in isolated forest, barren tundra, or alpine areas is that most alternative land uses are related to renewable resources perceived as less damaging than mining. Mining activities are also related to coal, sand, gravel, stone, and potash deposits underlying prime agricultural lands or bordering expanding urban centers. In these instances, the problem of land allocation is further compounded by political, social, economic, and environmental considerations.

How can the mineral deposits be extracted in the short term without permanently altering the land values for post-mining uses? Alternative land uses with a measurable economic value can present considerable competition for mineral-rich lands or can lead to conflicts. In some cases, no matter how economically valuable a mineral deposit is, the sociopolitical and environmental factors in opposition are so strong that no new mine development will take place.

Another important aspect of mine development is time lag. After initial discovery and evaluation of a potential mine, years of development and construction pass before a mine begins production. Time can have an important effect on the ultimate impact of mining operations on the environment. Unforeseen physical and chemical changes to the environment can emerge at any stage of the life of the mine or even long after closure, despite precautions.

In the recent past, the legislation of new environmental controls and resource management procedures has resulted in a number of significant changes in the traditional approach to both mining and resource development. These developments include the following:

- environmental impact assessment and public inquiries
- conditions for "permit" approval
- resource management and land-use planning
- land reclamation and rehabilitation

These control measures have created a lengthy and complicated development process in the life of a future mine.

With a growing practice for careful and formalized proposals for all major resource developments, the use of environmental impact assessment procedures as a guide to the development and management of a new project has spread widely. Throughout the development and operation stage of a mine, two parallel processes function: (1) the engineering design, layout, and technological requirements normally associated with mining and (2) an environmental program that meets all existing regulations and standards for air, water, and land quality. A new mine development process requires collection of environmental baseline data, an environmental impact assessment, and, sometimes, a socioeconomic impact assessment. A series of formal submissions, public hearings, and applications may follow. In most instances, additional constraints designed to reduce the environmental impacts are imposed.

Concern is growing within the industry that the increasing complexity of regulations, policies, and guidelines designed to protect and conserve our resources has led to considerable delay in the development of mines. A fundamental question has been how to balance the requirements of society for continued economic growth with its desire to preserve the environmental quality of the land resource base. This illustrates the intimate relationship between the exploitation and use of mineral resources and the consequences for the environment.

The environment is integrated, and its components are linked by dynamic processes. We cannot use or affect any part without affecting some other parts. No matter how beneficial for our own desired purposes the principal intended results of our activities may be, our actions are bound to cause effects additional to the principal effects we have in mind. Such additional or unintended effects may not be to our advantage, and then we have the problem of controlling the environmental consequences of our own environmental control.

ENVIRONMENTAL EFFECTS OF SURFACE MINING

Potentially, many adverse environmental impacts result from area surface mining of coal if no mitigating measures (reclamation practices) are used.[1] Such measures are used with varying degrees of effectiveness throughout the United States.

A summary of types of environmental impacts, causal factors, and mitigating measures is shown in Table 1.1. As shown in that table, air quality, on and off the mine site, may be affected in several ways. Fugitive dust from a variety of sources — coal haul roads, unvegetated spoil surfaces, topsoil stockpiles, and coal stockpiles — is a potential problem in areas of low rainfall, high winds, and erodable soils. Dust levels are not dependent upon the mining method to any great extent. Watering of haul roads and stockpiles is a commonly used dust-suppression measure.

Overburden blasting may adversely affect the environment in several ways. Two of these, noise and air shock, are loosely classified as air impacts. Blasting noise and air shock can be troublesome if people live within a radius of several miles (km) of the blasting site. The magnitude of the problem is dependent primarily upon the depth and type of overburden being blasted, the powder factor, the amount of explosive detonated at a given instant, the population density in the vicinity of the blasting site, and the times of day during which blasting takes place. Common practice is to use millisecond delays between rows of blast holes in a given blasting pattern in order to reduce the amount of explosive charge detonated at any given instant. Reduction of the powder factor, that is, use of less explosive per cubic yard of overburden, and restriction of blasting to daylight hours are additional mitigating measures which are sometimes used.

Ground vibration due to overburden blasting is another potential problem where people live near the mine site. The practices used to reduce blasting noise and air shock also reduce ground vibration levels.

The quality and quantity of surface water and groundwater, both on and off the mine site, can be adversely affected if effective reclamation practices are not used.

Table 1.1. Summary of Possible Environmental Impacts, Causal Factors, and Possible Mitigating Measures: Area Mining

Environment	Environmental Impact Category	Specific Environmental Impact	Uncontrollable Causal Factors	Controllable Causal Factors	Possible Mitigating Measures
Air	Air quality	Fugitive dual	Precipitation (lack of) Wind Soil types	Coal haul road surfaces Haulage road surface areas Surface material (soil) Vegetative density on mined areas	Dust control (watering) on coal haul roads Revegetation of mined areas
	Noise levels	Blasting noise	Overburden characteristics	Spacing and of overburden blast holes Blasting sequence	Delaying shots: hole-to-hole or row-to-row Decking (delaying shots vertically within holes) Use of less explosive charge Use of blasting machines
	Air pressure				Same as above
Surface water	Physical quality	Air shock Erosion and sedimentation	Same as above Precipitation Natural topography Natural drainage patterns Natural vegetative density	Same as above Length and slope of spoil surfaces Water intercepted by open cut Type of "soil" on mined areas Vegetative density on mined areas	Reduction of spoil grades Reduction of length of unbroken spoil slopes Diversion of surface drainage around active mining areas Settlement of suspended solid prior to discharge to natural drainageways Restoration of approximate original drainage patterns Revegetation of mined areas
	Chemical quality	Acid or mineralized surface water	Precipitation Overburden geochemistry Overburden stratigraphy	Inversion of overburden materials on spoil piles Water intercepted by open cut Change in permeability (spoil) Length and slope of spoil surface	Identification and selective placement of undesirable overburden materials Drainage diversion Spreading of topsoil on spoil surfaces Grading of spoils soon after placement Chemical water treatment

ENVIRONMENTAL IMPACTS OF MINING

Table 1.1 (Continued). Summary of Possible Environmental Impacts, Causal Factors, and Possible Mitigating Measures: Area Mining

Environmental Impact Environment	Specific Environmental Category	Uncontrollable Causal Impact	Controllable Causal Factors	Possible Mitigating Factors	Measures
Ground-water	Quantity	Drawdown	Natural height of water table; Rates and directions of natural groundwater flow	Interception of groundwater by open cut	Reduce length of open cut; Deepen wells on properties adjoining the mine site
		Altered flow rates	Overburden characteristics; Aquifer characteristics	Replacement of coal seam aquifer by spoil material; Differences in percolation rates for overburden and spoil; Open final cut (can be beneficial)	Selective placement of overburden materials; Spreading of topsoil on spoil surfaces; Backfilling of final cut
	Chemical quality	Acid or mineralized groundwater	Precipitation; Nature of aquifer; Overburden geochemistry	Same as above	Same as above
Land-use potential	Topography	Major changes in topography (more rugged)	Natural topography; Spoil swell factor; Coal seam thickness; Natural repose angle of spoil	Method of spoil placement; Removal of coal (thick coal only)	Spoil grading (during or after mining); Backfilling of final cut
	Drainage	Disruption of natural drainage	Natural drainage patterns; Overburden geochemistry; Precipitation	Method of spoil placement	Selective placement of overburden materials; Spoil grading
	Vegetation	Removal of native vegetation	Native vegetation	Overburden removal	Revegetation

Surface texture	Rocks on spoil surface	Overburden characteristics	Method of spoil placement	Selective placement of overburden materials
Appearance	Changed appearance	Natural topography Natural vegetation	Overburden removal Method of spoil placement	Spreading of topsoil on spoil surfaces Spoil grading Revegetation Backfilling of final cut

From Cook, F. Evaluation of Current Surface Coal Mining Overburden Handling Techniques and Reclamation Practices, U.S. Department of Commerce, National Technical Information Service, PB-264-111 (1976), pp. 60–123.

Sedimentation of surface waters may result from several factors in combination. In sidecast area mining, the outslope side of the box-cut spoil pile drains externally. Since, immediately after placement of the box-cut spoils the outslope is steep and unvegetated, erosion and transport of sediment from the outslope to external drainage ways may occur, particularly if rainfall is frequent and intense and spoil surface materials are erodable. Procedures that may be used to prevent these kinds of problems include reduction of the outslope angle by grading, construction of terraces or contour ditches on the outslope, mulching and revegetation of the outslope, and placement of sediment basins in drainage ways below the box-cut spoil pile.

Where the sidecast method of spoil placement is used, erosion and sedimentation from spoil areas in the second and subsequent cuts during mining generally are not a problem. This is because, before extensive grading of the spoil piles, most drainage in the areas disturbed by mining is internal. That is, surface runoff from spoil piles does not enter external drainage systems unless water is pumped from the pit into those systems. At large-area mining operations where 10 to 50 cuts may be made, the outslopes of box-cut spoils, which drain externally, constitute a relatively small percentage of the total mining disturbance.

As mining progresses, sidecast spoils are graded to make the topography unsuitable for the planned post-mining land use. In practice, this generally means restoration of approximate original contours. First, in the standard procedure used for grading of sidecast spoils, dozers are used to construct roads or flat working areas along the ridge lines of the spoil piles to be graded. Next, working from these roads, spoil is pushed sideways or in a herringbone pattern by the dozers into the vee's between adjacent spoil piles.

As large areas are graded, erosion and transport of sediment to external drainage systems become a potential problem. The magnitude of the problem is dependent upon the lengths and steepnesses of slopes on graded areas, the frequency and intensity of rainfall, the erodability of spoil surface materials, and the types and density of vegetative cover on reclaimed areas. The erosion and sediment control procedures that may be used on these areas are identical to those described above in discussing box-cut spoils. These include reduction through grading, terracing, contour ditching, mulching, revegetation, and use of sediment basins.

Diversion of surface water around the active pit and interception of surface water and groundwater by the pit are further potential sources of sedimentation. Drainage diversion systems, which are used primarily for production reasons to keep the pits from flooding, are of two basic types. The first type is diversion or rerouting of a major drainage way, such as a perennial stream crossed by the pit. Standard practice in such cases is to divert the stream around the mining area by constructing a new stream channel. The tendency is to straighten and channelize the stream, thereby increasing the hydraulic gradient and possibly causing headcutting in alluvial material. In order to prevent headcutting and the associated sedimentation, the gradients of diverted streams must be carefully controlled.

A more common kind of drainage diversion used at area mines involves construction of ditches, usually near the perimeter of the area to be mined, to divert surface runoff around the pit and generally around reclaimed areas as well. These ditches discharge directly onto undisturbed land or into sediment basins. If the hydraulic gradient in the ditches is too great or the ditches are not adequately vegetated, sediment loads in surface water carried by the ditches may increase over natural levels. Additionally, unless some means for dissipation of water energy at the outlet of the ditch is provided, erosion of the land surface near the outlet and sedimentation may occur. After all mining and reclamation have been completed, the drainage diversion ditches may be removed, or they may be left in place permanently. The choice, among other factors, depends upon prevailing reclamation regulations.

Some water will enter the pit as direct rainfall, surface runoff from the highwall side of the pit, groundwater seepage from the highwall, surface runoff down the inclines, and seepage from the spoil. Water that collects in the pit will usually become heavily sedimented, even if sediment loads in the pit were low prior to the interception of water. One reason is transit of coal haulage trucks through water that has collected in the pit.

Unless the pit is self-draining, water will be pumped from the pit. If the quality of the pit water is acceptable, the water may be discharged directly onto undisturbed or reclaimed land. More typically, however, pit water is discharged into a diversion ditch, the outlet of which is at or near a sediment basin or basins. Alternately, the pit discharge may be piped all the way to the sediment basin from the pit via plastic tubing.

The chemical quality of surface waters may also be adversely affected if preventive measures are not used. There are two possible causes of chemical pollution of surface water. One, just discussed, is interception of surface and groundwater by the pit. Where the coal, overburden, spoil, or incline surfaces are high in pyrite or marcasite — as is generally the case when the coal being mined is medium to high in sulfur — or where there are soluble minerals or trace elements in the overburden, spoil, coal, or pit floor, then collection of water in the pit may lead to chemical changes in that water. These changes may result from oxidation of pyritic materials (thereby increasing the acidity of the water) or from dissolution of soluble salts in the spoil, overburden, or coal (thereby increasing levels of dissolved solids in the water). The oxidation process requires that both air and water come in contact with pyritic materials, whereas air is not required for dissolution of soluble salts.

Several measures are used to prevent or minimize oxidation and dissolution. The standard ones are to minimize the amount of water that enters the pit, using the procedures described above, and where acid water is a problem, to chemically treat the water in the pit or in an out-of-pit treatment facility prior to its discharge to offsite drainage ways. An additional procedure, discussed below, is to identify and selectively place toxic or otherwise undesirable materials during placement of spoil.

Runoff from graded or ungraded spoil surfaces may be chemically altered if the surface spoil materials are pyritic or high in soluble minerals or trace elements. This can be a problem where undesirable overburden materials are located stratigraphically close to and above the coal seam or seams being mined and the sidecast method of spoil placement is used.

A characteristic of this method, in the absence of special handling procedures, is that overburden materials are inverted on the spoil pile. That is, the overburden material stratigraphically just above the coal seam being mined is placed on the top of the spoil piles. If this material is undesirable, as is often the case, the chemical quality of surface runoff and water that percolate into the spoil may be altered.

The problem can be particularly acute where two or more coal seams are mined in one pit using the sidecast method of spoil placement. In the absence of special spoil handling procedures, the interburden separating two coal seams (which usually has undesirable chemical or physical characteristics) is placed on the tops of spoil piles. A procedure frequently used to alleviate this problem, with varying degrees of effectiveness, is to bury the undesirable materials during the overburden removal and spoil placement process.

An additional technique — used in part to bury undesirable materials, but primarily to aid revegetation of graded lands — is to salvage topsoiling material prior to overburden removal and then to spread that material over spoil surfaces soon after grading of the spoil piles. Experience indicates that covering acid-producing spoil with as little as 2 ft (0.6 m) of clean fill is generally sufficient to greatly inhibit acid production. Erosion on topsoiled spoils must, of course, be controlled, or the topsoiling material may be washed away, exposing the underlying toxic material. As mentioned above, revegetation is a major erosion-control practice.

Whether or not topsoil is used where undesirable spoil materials are a potential problem, it is important to grade and revegetate spoil soon after placement. Several factors affect when these procedures may be used. The most important of these are the proximity of grading activities to the active pit, as well as the width and length of the pit itself.

If mining takes place below the water table, groundwater will be intercepted by the open cut, pumped out, or lost by evaporation, and the water table will be lowered in the mining and adjacent areas. This could result in loss of head or dewatering of wells within a radius of several miles (km) of the pit. Where this situation occurs, responsible coal companies generally remedy it by deepening wells or hauling water to affected residents for as long as groundwater levels are affected by mining. Generally, after mining, the water table will rise close to its original level.

After mining and reclamation have been completed, groundwater quantity can still be affected, although not necessarily adversely. If the mine is located in a groundwater recharge area — that is, an area in which surface waters percolate into the groundwater system — the recharge characteristics may be affected because the coal seam has been removed or because the spoil has different permeability characteristics than the overburden. Suppose, for example, that the coal seam itself was an aquifer. Sometimes, water enters or recharges a coal seam aquifer where the seam outcrops. If so, removal of the cropline coal and replacement of coal with spoil may increase or decrease recharge rates. Alternatively, the aquifer may be recharged from above by percolation of surface water or by groundwater from higher-lying aquifers. Replacement of overburden with spoil that has different permeability characteristics from the original overburden might also affect groundwater recharge rates.

Rates of groundwater flow (gallons per minute or liters per minute) may be altered if spoil materials have different physical characteristics from the original overburden or coal strata. If spoil placed below the water table is less permeable than the original aquifer, the groundwater flow rate after completion of mining and reclamation may be lower than that before mining. This possibility is of greatest concern where the coal seam (or seams) being mined are themselves aquifers, which may occur when the coal is thick. This is of concern because the coal seam is replaced stratigraphically by spoil whose characteristics may not be well known.

Oxidation of acid-producing materials is not likely because of the relative lack of air flow in the spoil profile. However, groundwater flow through spoil materials may result in chemical changes in that groundwater due to dissolution of soluble minerals or trace elements. A way to prevent this kind of occurrence is to place the "best" spoil materials below the water table during the spoil placement phase of mining.

It should be noted that the effects of mining on groundwater, as well as on land use and other factors, may be beneficial rather than adverse. For example, groundwater recharge rate, flow rate, or quality may be improved by mining and reclamation. The purpose here, however, is not to present a carefully balanced cost-benefit analysis of surface coal mining, but rather to identify the known problems and the issues.

In addition to potentially affecting air and water, surface mining also affects land or, more specifically, land-use potential. After mining and reclamation, the land-use potential is usually dependent upon topography, drainage, vegetation, surface texture, and appearance.

As described above, in sidecast-area mining situations, topography suitable for the planned post-mining land use is restored by dozer grading of sidecast spoils. During the grading process, drainage ways can be reconstructed so that there will be positive drainage from reclaimed areas. If adequate drainage patterns are not restored, the use potential of the reclaimed land may be diminished because of the formation of bogs (among other adverse effects).

Revegetation of graded areas is a further means of improving land productivity. Burial of undesirable materials during spoil placement, reduction of slopes through grading, replacement of topsoil, fertilization or liming of spoil, seeding, mulching, and irrigation are all practices which may be used to aid in revegetating mined lands. The success of revegetation efforts depends on precipitation patterns, physical and chemical characteristics of surface and near-surface spoil materials, reclaimed topography, and vegetative species mixes.

In arid or semi-arid areas, supplemental irrigation may be required to establish vegetation. Acid-producing spoil materials, if present, must be covered to prevent oxidation and acid formation. Alternatively, or in addition, lime may be used to neutralize acid. If surface spoil materials are clayey in texture and therefore relatively impermeable, permeability may need to be improved, either by chemical treatment or covering of clayey spoils with suitable topsoiling material. Soil fertility may be improved by adding fertilizer. Mulch may be used to control erosion and retain soil moisture.

The texture of surface spoil or soil materials also affects use potential. In the case of timberland, rangeland, and possibly pasture, occasional rocks on land surfaces do not necessarily diminish the use potential. However, in agricultural and some pasture areas, rocks on the surface may degrade the use potential. If overburden, and thus spoil, is unconsolidated, or consolidated spoil materials weather rapidly, special rock-prevention techniques may not be needed. In other cases, rocky spoil materials must be buried during or after the spoil placement process.

The appearance of reclaimed lands affects use potential in a real but intangible way. Standard practice in area mining situations is to grade spoil piles so that reclaimed terrain blends with surrounding natural terrain. In addition, the appearance of the land is affected by types of vegetation and, to a lesser degree, by surface texture.

The general physical factors causally related to the environmental effects of surface coal mining are climate, topography, overburden, coal, hydrology, and land use. Each factor is described by a series of parameters, and each parameter may have several possible values. These parameters and possible values are shown in Table 1.2. The rationale for the choice of these parameters and possible values is described above.

Five "technological" factors or parameters are felt to have major effects on mining costs and reclamation performance. These factors are topography, mining method, type of stripping equipment, number of coal seams mined per pit, and type of stripping subsystem. The definitions of these terms are summarized in Table 1.3 and are discussed further below.

Topography, although physical rather than technological, has been included because of its effect on the applicability and environmental characteristics of alternative mining methods.

Five mining methods have been considered. These are area mining, modified open-pit mining, mountaintop removal, haulback mining, and conventional contour mining. It should be noted, however, that there is no sharp line dividing area mining methods from steep-slope mining methods where the topography is hilly or sharply rolling. The distinction has been made here on the basis of ground slope angles, number of stripping cuts, and type of overburden removal and placement (stripping) equipment.

Three types of stripping equipment have been considered. Draglines and stripping shovels are of the sidecast type. Open-pit equipment consists of loading shovels for overburden removal and off-highway trucks for spoil haulage and placement. Construction equipment includes dozers, end loaders, scrapers, and — if loaded by end loaders — off-highway haul trucks.

MINING AND THE ENVIRONMENT 13

Table 1.2. Summary of Possible Values of Important Physical Parameters

Environmental Category	Physical Parameter	Possible Parameter Value	Remarks
Climate	Climate	Arid or semi-arid Humid	
	Wind speed	High Low	
Topography	Topography	Flat to hilly Steep or mountainous	Slopes of up to 17° from the horizontal Slopes in excess of 17°
Overburden	Soil	High productivity Low or medium productivity	Encompasses fertility, erodability, texture drainage, depth
	Overburden	Acid producing Non-acid producing Unconsolidated	Alluvium, loess, glacial drift (sand, silt, clay) Hard clays, soft shales Limestone, sandstone, shale, slate
		Semi-consolidated Consolidated	
	Stratigraphy	Undesirable strata immediately above coal Undesirable strata not immediately above coal	
Coal	Average sulfur content(% by weight after washing)	High Medium Low	More than 2% 1 to 2% Less than 1%
	Average individual seam thickness (ft)	Very thick Thick Thin Very thin	More than 40 ft 8 to 40 ft 3 to 8 ft Less than 3 ft
	Dip	Flat or slightly dipping Pitching	Up to 5° More than 5°
Hydrology	Surface drainage (on mine site)	Perennial (continuous) Ephemeral (intermittent)	
	Height of water table (in mining area)	Shallow Deep	Up to 150 ft deep More than 150 ft deep

Table 1.2. (Continued). Summary of Possible Values of Important Physical Parameters

Environmental Category	Physical Parameter	Possible Parameter Value	Remarks
	Groundwater use (in mining area)	Stock or domestic Drinking	
Land Use	Land use	Row crops	Irrigation Corn, soybeans, sugar beets, other crops grown in rows
		Closely spaced crops	Small grains and other close-seeded crops not usually grown in rows and tilled
		Hay or pasture	
		Range	Cattle and sheep grazing
		Forest	Productive or nonproductive forest land
		Recreation	
		Unused	

From Cook, F. Evaluation of Current Surface Coal Mining Overburden Handling Techniques and Reclamation Practices, U.S. Department of Commerce, National Technical Information Service, PB-264-111 (1976), pp. 60–123.

MINING AND THE ENVIRONMENT 15

Table 1.3. Definition of Technological Parameters

Parameter	Parameter Value	Definition	Remarks
Topography	Rolling terrain	Average natural ground slope angles up to and including 17° from the horizontal.	Includes flat, gradually rolling, and hilly terrain.
	Steep slopes	Average natural ground slope angles greater than 17° from the horizontal.	Includes sharply rolling and mountainous terrain.
Mining method	Area mining	Method used in rolling terrain. Three or more parallel cuts are made.	
	Modified open-pit mining	Area-type method based on the use of loading shovels for overburden removal and trucks for spoil haulage and placement. Terrace type of pit or benched highwall is used.	Used only where the coal is very thick.
	Mountaintop removal	Method used in steep slope areas to mine high-lying coal seams by complete removal of mountaintop. Usually, there are no final highwalls after mining has been completed. Reclaimed topography is usually flat or gently rolling.	
	Haulback mining	Method generally used in steep slope areas. One or two cuts are made along the contour. Some spoil is hauled back to a previously mined area and placed on the solid bench.	
	Conventional contour mining	Steep slope method in which a single cut is generally made along the contour. Spoil is placed by pushing it to rest on natural ground immediately below the elevation of the coal seam being mined.	
Stripping equipment[a]	Sidecast stripping equipment	Dragline or stripping shovel as the predominant stripping machine.	In classifying mining situations, the category Sidecast Stripping Equipment includes the following situations: • single dragline or stripping shovel • draglines and stripping shovels in tandem with one another • draglines or stripping shovels in tandem with bucket wheel excavators

Table 1.3. (Continued). Definition of Technological Parameters

Parameter	Parameter Value	Definition	Remarks
			• draglines or stripping shovels in tandem with construction equipment In each situation, the sidecast machine is considered to be predominant if it accounts for at least two thirds of the total overburden removal yardage. Front-end loader/truck combinations are considered to be construction equipment, rather than open-pit equipment.
	Open-pit equipment	Loading shovel and rock haul truck combinations.	
	Construction equipment	Dozers, end loaders, scrapers, rock haul trucks.	

[a]Bucket wheel excavators (BWE) were observed in use only in tandem with sidecast stripping machines. In each case, the sidecast machines were the prime movers. For this reason, bucket wheel excavators were not included here as an equipment type.

From Cook, F. Evaluation of Current Surface Coal Mining Overburden Handling Techniques and Reclamation Practices, U.S. Department of Commerce, National Technical Information Service, PB-264-111 (1976), pp. 60–123.

The number of coal seams mined per pit may be single or multiple. Multiple seam mining is the mining of two or more coal seams in one pit. Practically speaking, however, multiple seam mining usually connotes two-seam mining. The non-coal strata separating two coal seams or splits in a given seam are termed partings if the total thickness of all strata is 5 ft (1.5 m) or less. In such cases, a mining situation is classified as single seam because removal of the parting usually presents neither mining nor reclamation problems. Where the total thickness of all separating strata exceeds 5 ft, the strata are collectively termed interburden, and the mining situation is classified as multiple seam.

The fifth and last technological parameter is the stripping subsystem, which may be single, tandem, or dual. A single-machine stripping subsystem is one in which one machine is used to remove and place all overburden materials. Prime examples are single dragline and single stripping shovel subsystems. In such cases, construction equipment used for clearing and grubbing, topsoil removal, or construction of shallow dragline benches is considered to be auxiliary equipment only. Infrequently used single machine subsystems include those in which a scraper or end loader is used for all overburden removal and spoil placement.

A tandem machine subsystem is one in which two or more machines are used for overburden removal, and one machine follows or works behind another. Tandem machine stripping is by definition always a multi-lift overburden removal operation, with the first machine removing the top overburden lift, and so on. Tandem machine stripping may be used in both single and multiple seam situations. In a situation in which two draglines are used in tandem to strip two coal seams in one pit, the first dragline removes all overburden. After removal of the upper coal seam, the second dragline follows and removes the interburden.

WATER POLLUTION

During underground operations, exposed material within the mine oxidizes and gives rise to acid production, which contaminates the water. The most common case is the exposure of a sulfide to air and water, which promotes a chemical reaction resulting in sulfuric acid. Problems of AMD are predominantly chemical and extremely complex and originate with the extraction of sulfide minerals. Many reactions are possible, depending upon the environmental conditions existing in the mine and the nature of other minerals that may be present. Not all mines have AMD problems. It is normally associated with sulfide-bearing metallic ores (pyrite, pyrrhotite, chalcopyrite, sphalerite, marcasite, arsenopyrite, etc.).

Acid contamination is increased by water entering the mine from external sources. In addition to seepage from natural watercourses, which constitutes the larger part of the external supply, water (usually one third by weight) may be used to transport mill tailings underground for the purposes of backfilling excavations and for processing and servicing needs. The average composition of underground drainage waters has been estimated at 51% from natural watercourses, 14% from mine backfill (where employed), 34% from service and process sources, and the remaining 1% from other unspecified sources.

As the total average water flow into an underground mine is estimated at 1000 L/min, pumping must be maintained to ensure continuous operation. The estimated flow is an average for the industry and is subject to wide variations between mines and over time in any particular mine. Following pumping, the drainage water, which frequently contains significant quantities of highly toxic dissolved minerals, is impounded at the surface prior to discharge and recycling. The impoundment areas may range from small ponds accepting a few tonnes per day to large tailings dams enclosing several square kilometers and receiving slurry wastes at the rate of thousands of tonnes per day. A sequential approach to water clarification is normally employed and involves the use of a single pond, or a series, to allow the settling out of the contained solids. The effluent is then subject to additional treatment to neutralize acids and remove heavy metals and radioactive wastes.

During impoundment, the effluent is partially depleted through seepage, percolation, runoff, and evaporation. The effect on the environment of water losses is dependent, to a large degree, upon the location of the mine within the localized drainage basin and its interconnecting watercourses. The nature of the materials discharged can also affect the environment profoundly. The heavier particulate matter in suspension will fall out under gravity relatively close to the point of discharge. Dissolved metals, on the other hand, are capable of being transported over much greater distances and can affect water quality at the regional scale.

Table 1.4 illustrates the effects of mining pollution on aquatic ecosystems in Canada. Together with Table 1.5, the values listed demonstrate the adverse effects of AMD, lowered pH values that eliminate aquatic organisms, and increased levels of acidity, sulfate iron, and total solids.

As in underground mines, surface operations (both open pit and strip) are also prone to the effects of AMD. While the mean flow of water into an underground mine is approximately 1000 L/min, for open-pit mining the mean value is 13,800 L/min, 48% derived from natural watercourse, 9% from service and process water, and 13% from other sources. The substantial quantities of water involved in surface mining operations enhance the potential for effluents to enter the natural environment. The problem is compounded by the runoff, leaching, and percolation processes acting upon the residuals contained in the solid wastes, which are considerably in excess to those produced in underground operations.

LAND USE

The nature and characteristics of the mining industry illustrate the wide range and intensity of land-use activities that can take place during the various stages of mining. Millions of hectares may be subject to wide-ranging exploration techniques,

Table 1.4. Some Details on the Effects of Mining Pollution on Aquatic Ecosystems in Canada

Locality	Source of Pollution	Chemical Effects	Biological Effects
New Brunswick			
S. Tomogonops R. and N.W. Miramichi R.	Heath Steele	Heavy metals, increased hardness acid generation	Decline in *Salmo salar* since 1965; reduced diversity and abundance of benthos
Nepisiguit R.	Brunswick 6	Heavy metals	*Salmo salar* fishery approaching zero
Ontario			
Crowe R. Basin	Bancroft area mines	TDS, SO_4^{2-} and hardness increased 8–10 × 4 in Bow L.	Aquatic biota affected only in immediate vicinity of tailings decant
Serpent R. Basin	Elliot L. area mines	Acid, high TDS, SO_4^{2-}, NO_3^- and Ca^{2+}. Most affected were Quirke L. and Pecors L. Biological effects mainly because of acidity.	Reduced productivity, altered phytoplankton and zooplankton communities. Elimination of *Stizostedion vitreum* and reduction of *Salvelinus namaycush*. Reduced benthos diversity.
LaCloche Mtn. lakes	Sudbury smelters	Increased acidity from aerial fallout	Extinction of fish populations
Manitouwadge L. area	Noranda and Willroy mines	Acidity, toxic concentration of NH_3, zinc, copper, iron; nutrient enrichment; high TDS, especially SO_4^{2-}. Mose, L. has become meromictic — O_2 deoxygenated below 20 ft.	Mose, L. total elimination of macroinvertebrates; reduced benthos in all lakes.
Manitoba			
Bernic L.	Tantalum Mining Co.	Slight increase in turbidity, TDS, and suspended solids	Decrease in abundance and diversity of benthos
Borden L., Clarke L. and Lily L.	Mannibridge	No effect	No effect
Grass R.	INCO, Thompson	Increased turbidity, hardness, TDS, copper and chloride	Reduced benthos, and absence of Ephemeroptera in a small area
Ospwagan Lakes	INCO, Thompson	Overburden into Upper Ospwagan L. Increase in most parameters in Lower Ospwagan L.	Major reduction in Amphipoda and Ephemeroptera in 1968, but recovering in 1969
Schist L.	Flin Flon Mines	Increased turbidity, siltation, CO_2, and metals. Pollution spreading into L. Athapapuskow.	Sparse benthos, entirely of chironomids. No *Stizostedion vitreum* or *Salvelinus namaycush* in affected area. Decreased angling success.

Table 1.4 (Continued). Some Details on the Effects of Mining Pollution on Aquatic Ecosystems in Canada

Locality	Source of Pollution	Chemical Effects	Biological Effects
Eldon R. and Cockeram L.	Lynn Lake	Siltation; toxic amounts of heavy metals, especially cadmium	Decrease in fish and benthos populations. Fish and benthos absent from area receiving drainage.
British Columbia Pend d'Oreille R.	Reeves MacDonald	Increased turbidity and siltation	Benthos severely reduced; fewer aquatic plants, indicating reduced 1° productivity
Benson L.	Coast Copper	Increased turbidity; has become meromictic	No benthos; fish population probably reduced

Source: Clarke.

Table 1.5. Typical Assays of Acid Waters from Mines (All Concentrations Are mg/L Except pH)

Type of Operation	Cu-Pb-Zn (Mine and Surface Drainage)	Cu-Pb-Zn (Mine Water)	Uranium (Seepage)	Cu-Zn (Active Mine)	Base Metal (Abandoned)	Uranium (Abandoned Mine)
pH	4.0	2.0	2.0	3.0	2.6	2.0–2.8
Suspended solids	8.8	690	Nil	—	—	25
Total less solids	79	24,000	—	—	9200	13,440
Hardness	293	2960	—	—	1390	—
Ca	—	—	416	—	454	—
Mg	—	—	106	—	178	—
Cu	17	11	3.6	0.0	2.5	2.2
Zn	118	1090	11.4	0.4	34	9.4
Pb	0.4	58	0.7	0.11	0.5	—
Fe (total)	79	1830	3200	11.7	11,300	300
Mn	21	0	5.6	0.4	8.2	3.6
SO_4	36	16,560	7440	885	4050	6900
COD	—	245	270	—	110	—

From Cook, F. Evaluation of Current Surface Coal Mining Overburden Handling Techniques and Reclamation Practices, U.S. Department of Commerce, National Technical Information Service, PB-264-111 (1976), pp. 60–123.

but only a fraction is directly affected by the development and production stages. In most cases, the factors determining the area of land affected are (1) the characteristics of the mineral being produced — depth to ore, density and type of material, (2) its ore grade and reserve, (3) its percentage recovery rate, (4) the method of mining, and (5) whether or not beneficiation and further processing take place at the mine site.

The widest geographic distribution of mining activity occurs at the exploration stage. Fluctuations in the supply of, and demand for, mineral products are such that very little activity may occur in a given region or indeed the whole country for extended periods, and then a sudden demand for certain minerals will cause a great expansion in exploration.

Depending on what mineral is being sought when, many areas were explored and re-explored, making it difficult to quantify all aspects of the exploration stage where they affect the land surface directly — through seismic lines, trenches, adits, or drill sites. For the purposes of this study, the major effort to quantify the effects of mining on land use, particularly land disturbances, will be in the development and production stages.

Once exploration has successfully identified a favorable zone worthy of further assessment, regulations require the establishment of a legal basis for development, either through claim staking, leases, grants, or licenses. The records of these legal requirements provide a good indication of the level of activity at this stage. The land area actually used and disturbed in the exploration, development, and production stages is much smaller.

The direct effects of most mining operations within the "mine site," while not negligible, are localized and relatively small compared to other forms of human economic activity such as agriculture, forestry, and urban settlement. That mining will have a considerable influence on the land surrounding its operations is inevitable, and this influence, referred to as the "shadow effect," is far more extensive than previously estimated. In addition, the indirect effects of mining can be quite large, especially the infrastructure developed for mining operations (rail, roads, housing power plants, water storage, and other facilities). It can significantly extend the land area directly influenced by all mine-related activities and often permits a large number of other activities that would be difficult or impossible to undertake without it.

Indeed, the interplay of the direct uses at the "mine site" and other land uses in the "shadow" zone illustrates the areas of increased responsibility that have been assigned to mining operations. In the past decade, the responsibility of a mining company for environmental protection has been dramatically projected beyond the visible perimeters of their working operations to substantially larger neighboring lands. In effect, changing environmental regulations have assigned an increasing degree of responsibility to the mining company to minimize the environmental impact of its operations on the use of neighboring lands. It has resulted in a continued and increasing interaction, desired or not, with landowners and resource managers

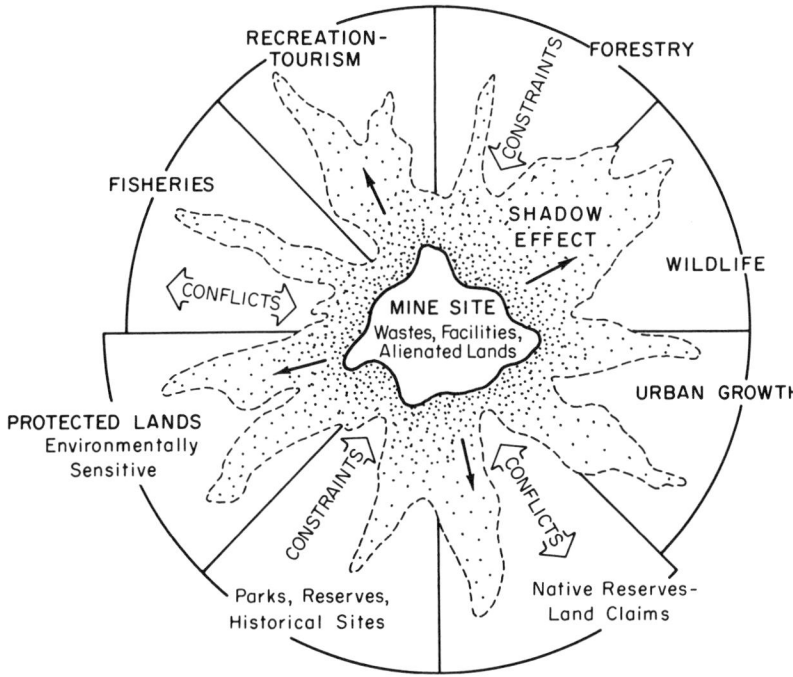

Figure 1.1. Mining and land use.

responsible for the land-use activities bordering mine sites (Figure 1.1). The range of interaction between mining and other land uses has become increasingly complex. No longer is it a straight economic or technological choice between competitive uses. Demands for resource conservation, environmental protection, and restricted use, coupled with the wide range of traditional uses — agriculture, forestry, settlement, and recreation — have all increased the potential constraints on development and the actual number of conflicts.

Over the past few decades, there has been a growing indication that mining developments are being given lower social approval than competing land uses. Much of this, no doubt, has been due to a past history of abandoned operations and their adverse effects on the local environment. Public attention to the allocation of land to certain uses has also been heightened by increased publicity about the scarcity of certain land resources and threats to them from an increasing array of degradation processes, both real and perceived.

Although the public has been aware for some time of the effects of mining operations, the deterioration of the land resource base attributable to other land-use activities, particularly those centered around the agriculture and forest industries, is now believed to be equally extensive, if not greater. This has only served to heighten the pressure on new industrial developments (particularly those associated with minerals and energy resources) and clouds their acceptance as a legitimate option in the highly emotional process of land allocation.

Mining options are limited to the presence of commercially developable mineral deposits. This is similar to other resource-dependent activities: National parks are located where the natural features or phenomena occur; and hydroelectric developments are limited to those waterways where sufficient head, flow, and drop exist. Approached from the same perspective, agriculture and forestry seem far more flexible in location, yet they, too, have specific requirements such as soil and temperature. Their high profile in our consumptive pattern becomes the focus of attention when competition for land occurs. It is imperative then that society be aware of the interrelationships, needs, and problems arising from the various economic sectors of our society, including mining, in order that decisions on land allocation reflect the needs and concerns of all.

Many of the processes in the mining industry that lead to a deterioration in land quality — physical alteration, chemical degradation, and biological, aesthetic, and cultural disruption — are evident to varying degrees in all areas of human activity. Those processes can be separated in two broad groups — local and regional. Land degradation, in its broadest sense, includes not only man-induced but also natural processes. Many of the natural processes involved in land degradation — for example, flooding, wind, and water erosion — are accelerated by exploitation and poor management, or a lack of proper land-use planning.

Although mining developments may not come into contact with a park, wilderness zone, Indian reserve, or native settlement, the suggestion of potential harm to the environment may be enough to arouse questions that may delay, or even exclude, exploration and development. We are only now beginning to see the magnitude of long-term effects of certain management practices. Our limited knowledge of the long-term capacity of the earth to absorb disturbances, certain wastes, or airborne contaminants in any medium has raised doubts and fears about the location of new mines, smelters, or refineries. Therefore, the mere presence of one of the above-mentioned protected areas in a drainage basin or downwind from a proposed mine development is sufficient to arouse considerable opposition, demand for a public inquiry, or outright exclusion. Thus, protected or limited-use lands can play a role disproportionate to their actual size, due to their multiplicity and wide distribution. They can, and have, become important factors influencing mine developments. In the case of native groups, a new development can become the focus of attention of a much wider land-claims issue.

The mining industry has been concerned with what appears to be a trend towards increased land withdrawals from any use whatsoever or the banning of exploration and development activities from certain lands.

Proponents of unrestricted access to land by mining believe that most exploration techniques have zero environmental impact; drilling can be of slight temporary impact, less for intrinsic reasons than because of the need to construct roads or helicopter pads in remote localities. Trenching, pitting, and shaft exploration can be intrusive but, even when carried out intensively, can easily be remedied by appropriate restoration; the underlying motive seems to be that mining might follow. It is better not to know what lies in the ground.

The argument supporting unrestricted access is that effective land-use planning and management can only be made after all the resource potential has been confirmed. The industry questions the arbitrary assigning of "values" to wilderness areas, parks, etc., without any consideration of the potential mineral value. It would appear to most proponents of mining developments that banning of mineral exploration is nothing less than an admission that if mining proposals do follow, the legal and administrative machinery is inadequate to enable the full and mature assessment of the merits of mining in relation to the merits of wilderness areas.

The dilemma facing the various levels of government is how to balance the overall needs for resource development with the demand for conservation and protection. Some workable mechanism for trade-offs among the various interested parties is still necessary. This requires the very difficult task of reconciling the multiple objectives of a wide range of agencies and departments among the varying levels of government. The task is going to become increasingly difficult as pressure for more parks, wilderness areas, increased environmental protection, and native claims, coupled with the demand for mineral and energy resources continues to accelerate in the next two decades. Jurisdiction and land allocation will continue to be the focus of attention for all parties concerned, and the determining factors in any attempt to reconcile competing demands on land resources.

In this regard, a major problem facing the mining industry is its unfavorable public image. Past perceptions of mining and its environmental impacts seem to be permanently etched in the minds of the general public. It appears that the attempts of the mining industry to improve its image and the understanding of the public for its activities have progressed only marginally. It is this general perception of mining as a singular source of land degradation and pollution which has caused so much attention to be focused on it in the form of government regulation or restricted zoning. The industry continues to be confronted with its image every time it enters a public debate or inquiry. Now that it is almost mandatory that environmental impact assessment and public hearings occur prior to any new mine development, public perceptions are playing an increasingly important role in the decisionmaking process.

The standards for all new developments have continued to become more stringent, since the quality of life as well as the environment have become increasingly important social and economic issues. The situation today has become considerably more complex as socioeconomic and political issues merge with the more traditional conservation and environmental issues. The links between resource developments, the environment, lifestyles, and institutional values can no longer be ignored.

At some public hearings, some of the major issues raised were identified as follows:

- waste management and its long-term storage and disposal
- health; effects of radiation
- land-use conflicts and their adverse economic impact
- lack of faith in government and regulating standards
- lack of public access to information
- commercial interests versus human concerns
- poor past record, in both mining and government
- enforcement and regulation problems

It is evident that the scientific community, regulating authorities, and elected officials also suffer from a lack of trust and a greater demand for public involvement in decision making.

The views of all the various pressure or lobby groups have become highly relevant within their own frame of reference, since there is no absolute way of measuring the relative importance of different environmental problems. Of overriding importance is the way in which individual decision makers and pressure groups perceive the situation in which they are confronted. The extent to which they diverge on issues may determine whether or not any decision is achieved. Differences in perception can occur on a personal basis or on up through various levels of authority and society on a local, regional, provincial, or national basis, and associations or pressure groups, although sincere, are often partisan by nature.

Mining land-use problems and conflicts often fall into one or more of the following generalized categories:

- Conflicts that arise when two or more proponents compete for the same parcel of land; for example, coal strip mining and agriculture in the western prairies of Saskatchewan or Alberta.
- Conflicts that arise when a particular land use on one parcel of land adversely affects the use of land on adjacent properties. This usually raises the question of compatibility or incompatibility of uses; for example, the impact of metallic mine wastes on the use of neighboring water resources, fishing in particular.
- Conflicts that arise between those who wish to develop a particular mineral or energy resource and those who desire the maintenance of a pristine environment through conservation and protection; for example, the establishment of parks or wilderness preserves.

SUBSIDENCE OF MINED LAND

Subsidence has been defined as "vertical and horizontal movements of surface, sub-surface and underground points as a result of various natural and manmade activities." Because of the inherent impacts on the surface structures/features, sub-

Table 1.6. Causes of Land Subsidence

Endogenic	Exogenic	
Volcanism	Removal of support	
Folding	Weakening of support	Increase in effective and actual loading
Faulting		
Continental drift	Increase in loading	

From Cook, F. Evaluation of Current Surface Coal Mining Overburden Handling Techniques and Reclamation Practices, U.S. Department of Commerce, National Technical Information Service, PB-264-111 (1976), pp. 60–123.

Table 1.7. Activities Causing Subsidence

Natural	Man-made
Solution of rocks, minerals, etc.	Pumping of water and petroleum from below ground
Drainage of subsoil	
Drifting of subsoil and sliding of rocks	Underground extractions for mining and other purposes
Rodents	
Frosting and defrosting	Settlement of foundations as a result of seepage of water and otherwise
Tectonic movement	

From Cook, F. Evaluation of Current Surface Coal Mining Overburden Handling Techniques and Reclamation Practices, U.S. Department of Commerce, National Technical Information Service, PB-264-111 (1976), pp. 60–123.

sidence has been a subject of deliberations on an international level since 1965 when UNESCO announced its International Hydrological Decade. The subsidence of the land surface may be due to endogenic causes (i.e., those within the earth crust itself) and exogenic causes (i.e., factors that originate from outside the planet).

The different causes that contribute to subsidence have been classified. In the case of endogenic causes, these have been classified in accordance with whether they have been caused by natural phenomena or by the activities of man (Tables 1.6 and 1.7).

While landslides have been classified as cases of land subsidence, here we are concerned with subsidence of the land surface resulting from the activities of extraction of the mineral. Though vertical subsidence is a common feature in mining areas, cases are on record where there has been lateral drift. These are the areas where the strata are highly water bearing with very high specific yield. Besides these physical ramifications, these have, as do other effects of subsidence in general, considerations of safety attached to them when the highly water-bearing strata, with high specific yield, discharge the water to the workings below.

The subsidence can vary from a couple of millimeters, in the case of water withdrawal from the aquifers below, to above 6 to 7 m in the case of extraction of coal from thick seams or due to mine fires in these seams. Subsidence, which was caused by pumping of water and petroleum, of as much as 15 m (in one case) has been reported.

Subsidence, due to shallow mining, causes direct air circulation through the goaved-out areas, which if not checked, coupled with the coal available, causes spontaneous heating and fires within the goaf areas. These engulf large areas and sometimes reach the surface with devastating effects on the land surface and other features. Fires starting in one seam may travel to upper seams and across the barriers to the neighboring mines. The subsidence in regular workings has been found to follow a regular pattern that becomes severe when other factors come into play, for example:

- presence of faults
- presence of dykes/sills having a breaking effect on the subsidence pattern against extension resulting from faults
- presence of old workings in the vicinity making things very severe and unpredictable, particularly where the interaction is with workings that were carried out before the current statute applied

The physical impacts of subsidence may vary from simple negligible lowering of the ground surface to severe damage to buildings or other surface features by wide and deep cracks. When subsidence is caused by fires, it progresses very fast within short periods of the fires starting. Other commonly observed impacts are as detailed:

- gross changes in surface topography resulting in undesired depressions
- disruption/disturbance in aquifers and thereby reduction in availability and contamination of water
- retardation in growth of vegetation due to poor availability of water
- waterlogging in the central portion of subsided areas
- underground fires in coal left below ground at shallow depths
- contamination of surface atmosphere by gases produced as a result of underground fires
- development of cracks leading to joining with surface and/or underground waterbodies, resulting in increased make of water in underground workings
- adverse flow condition and waterlogging in rivers, small waterlogging in rivers, small watercourses
- abrupt changes in road gradients
- damage to underground pipelines, cables, and drainage systems
- tilting and damage to buildings, structures, plants, pylons, etc.
- depression in ground level below HFL of watercourses nearby

In overlying seams, the impact of subsidence movements due to extraction below them could be as listed below:

- collapse of old workings standing on stocks/smaller pillars, and thereby increased subsidence on surface
- damage to overlying coal seams, which may sometimes be rendered unworkable
- floor lifting, spalling, and roof falls in workings

ENVIRONMENTAL AUDITS

As with most industries, mining must now include in its plan of operation mechanisms to address a project environmental impact. New regulations have focused attention on the potential environmental impacts of industry, particularly the handling of wastes.

One of the most effective tools in managing the regulatory requirements is an environmental audit. An environmental audit can be a literature review to determine regulatory requirements. The audit can also take on the form of a phase-one site assessment. This is a limited investigation to determine whether there are specific contamination problems at a site.

Phase-two or -three assessments are more detailed. They evaluate the feasibility of remedial action alternatives. Phase-three audits are generally performed as the result of regulatory requirements. It is a remedial investigation and feasibility study (RI-FS) if performed in the context of Superfund. Or, a phase-three audit can be a Resource Conservation and Recovery Act (RCRA) facility investigation or assessment (RCI-RCA) if performed under RCRA.

New regulations that address the storage, treatment, transport, and disposal of hazardous materials have made site assessments a major risk-reduction tool for those involved in real-estate transfers or in a business that uses hazardous materials. Most metals and many products used at mine facilities are defined as hazardous.

The concerns of purchasers arise primarily from the scope of liability under the Comprehensive Environmental Response, Compensation and Liability Act (CERCLA) or Superfund. CERCLA extends liability for cleanup costs and damages to current and former owners and operators, regardless of who caused the contamination.

Environmental risks and associated liabilities can occur due to soil, groundwater, surface water, or air quality contamination resulting from present or past land-use practices. Contamination may also occur due to past or future migration of chemicals onto a prospective site from adjacent properties. Most comprehensive insurance today will not cover any of these potential liabilities.

Lenders can also be exposed to liability even if they never actually take possession of a site. They may be considered operators if they get too involved in day-to-day decision making. They may be indirectly affected by lending money to a client whose collateral property holdings are found to be contaminated or are included in a statutory superlien to cover other cleanup costs. Lenders may be indirectly affected if the borrower's ability to repay is impaired by the cost of cleanup at another site.

Those contemplating the purchase of a site, acting as lender for a buyer, managing a real-estate based trust, or repossessing a site through foreclosure should ensure that they:

- exercise due diligence under CERCLA-RCRA regulations (They should make all appropriate inquiries regarding potential contamination before acquiring the property.)
- know the facility or site and what regulations apply
- know about any potential for, or presence of contamination that might require future expenditures

A site assessment can provide this information. It gives the present and future property owners and the lender a way to reduce their financial exposure.

Mine operators must understand the requirements of all environmental laws and regulations applicable to their facility. This is necessary to ensure compliance or at least to recognize noncompliance in day-to-day operations. A complete understanding of the regulations will help to protect mine operators' interests should they be subject to a regulatory agency review and inspection.

Regulatory requirements continue to grow with passage of state and local laws that must be reviewed on a site-by-site basis. Federal laws, however, are applicable to all mining facilities.

CERCLA establishes a mechanism of response for the immediate cleanup of hazardous waste contamination, accidental spills, or chronic contamination (abandoned, hazardous waste disposal sites). The Environmental Protection Agency (EPA) has promulgated regulations that establish the quantity of any hazardous substance that, if released, should be reported to the National Response Center.

The Superfund Amendments and Reauthorization Act (SARA) — Title III has two main components. Subtitle A establishes the framework for emergency planning by state and local governments. It creates a state emergency response commission, as well as local emergency planning committees. This section requires these local panels to work with representatives of facilities covered by the law on emergency response plans.

Subtitle B requires certain facilities to provide information to appropriate state, local, and federal officials on the type, amount, location, use, disposal, and release of chemicals. There are several reporting provisions contained in Subtitle B:

- Section 311 applies to facilities subject to the Occupational Safety and Health Act. These facilities must submit material safety data sheets or a list of the chemicals for which the facility is required to have material safety data sheets to local emergency planning committees, state emergency response commissions, and local fire departments.
- Section 312 establishes an inventory of toxic chemical emissions from facilities meeting certain criteria. Facilities subject to this reporting requirement must complete a toxic chemical release form for specified chemicals.

The RCRA hazardous waste program regulates all aspects of the management of hazardous waste from generation to disposal. Major components of the program include regulations for the identification of a hazardous waste, notification of any hazardous waste activities, compliance with standards for generators, transporters, and treatment-storage-disposal (TSD) facilities. Owners and operators of TSD facilities are also required to obtain a permit.

Mining waste regulations under RCRA are currently being developed.

The Clean Air Act provides the basic framework for modern air pollution control. Key elements of the act include establishment of human health-based ambient air quality standards and technology-limited uniform national emission standards. It also provides for prevention of significant air quality deterioration. Each state has specific air regulations to implement these programs.

The Clean Water Act addresses point and nonpoint sources of pollution. Point sources include industrial discharges. Nonpoint sources include mining and other construction activities that cause runoff into streams. Point sources are subject to five different effluent limitations administered through the National Pollutant Discharge Elimination System (NPDES). Control of nonpoint sources of pollution is left to the states. They are required to formulate plans that contain land-use regulations to control nonpoint sources. Hazardous waste and oil spills are also addressed in this act. Facility operators are liable, without fault, for the costs of cleaning up spills and for civil and criminal penalties.

The Safe Drinking Water Act (SDWA) was enacted to ensure safe drinking water supplies, protect especially valuable aquifers, and protect drinking water from contamination by underground injection of waste. Under this act, the EPA established a series of drinking water standards to protect the public health. These standards apply to public water that regularly supplies water to 15 or more connections or 25 or more individuals at least 60 days a year. This definition applies to most industrial establishments that supply water to employees.

The SDWAs most direct effect on industry is through the regulation of underground injection to protect usable aquifers from contamination. The regulations address hazardous waste disposal, the reinjection of brine from oil and gas production, and certain mining processes. The underground injection-control program is administered through a permit process with substantive requirements depending on the type of injection taking place. The most stringent conditions are for well injection waste classified as hazardous under RCRA.

The Toxic Substances Control Act (TSCA) provides the EPA with authority to require testing of chemical substances, new and old, entering the environment and to regulate them where necessary. This authority supplements sections of existing toxic substance laws, such as Sections 112 and 307 of the Clean Water Act and Section 6 of the Occupational Safety and Health Act. These already provide regulatory control over toxic substances.

Although the heart of the TSCA is the requirement for premanufacture notification, the area of interest in an environmental audit relates to the regulation by TSCA of polychlorinated biphenyls (PCB). Section 6(a) of the TCSA directed the EPA to phase out PCB manufacture and prevent the process, distribution, or use of PCBs except in a totally enclosed manner. The EPA regulations require transformers containing PCBs to be appropriately labeled. The agency sets standards for the transportation of PCBs and recommends certain disposal techniques.

Types of Environmental Audits

There are several types of environmental audits that can be tailored to suit specific needs.

Permit performance audits (compliance and monitoring). This is a review of environmental quality assurance plans, environmental permits, and agency-required operating restrictions-procedures. It assesses possible or actual nonconformance (especially regarding air and water emissions and hazardous materials management). This type audit also interprets regulatory agency permit conditions and suggests measures for ongoing permit conformance. It may also involve long-term monitoring of environmental activities.

Regulatory requirement audits. This provides a detailed evaluation of facility operations that are or may be governed by local, state, or federal environmental regulations. It identifies applicable regulations to pinpoint potential noncompliance or conflicts with such regulations. Procedures are also recommended for coming into compliance.

Environmental management practice audits. This type of audit examines existing management structure, procedures, and policies used by the client to implement environmental compliance and to communicate environmental-regulatory awareness (including health and safety) to work-force personnel. Recommendations are also provided for remediation of deficient practices.

Technical processes-practices audits. Production practices and facility conditions are reviewed to determine whether design or process modifications should be made to accomplish specific environmental goals (minimizing hazardous waste, waste stream treatment, or technology transfer).

Risk management audits. Practices, procedures, and policies are surveyed to identify sources of risk. It suggests how risks of environmental (or health and safety) incidents, accidents, and liability exposure can be reduced or eliminated. A risk management audit may also include a formal risk assessment study or contingency planning component.

Special purpose audit. This is a one-time audit conducted in response to unusual circumstances or requirements, such as an SEC 10k report, an EPA consent decree, insurance-liability impairment determination, or emergency response plan.

Site assessment audits. This consists of a thorough examination of previous and current environmental hazards and physical conditions on or surrounding a facility-site. Its purpose is to assess potential on-site problems or sources of external encroachment, contamination, or threat. This audit includes measures to remediate or reduce such problems before they affect operations. A site assessment audit is particularly useful as a planning and predevelopment decision-making tool for suspected problem sites. It is necessary before property transfer or asset sale acquisition.

REFERENCES

1. Cook, F. Evaluation of Current Surface Coal Mining Overburden Handling Techniques and Reclamation Practices, U.S. Department of Commerce, National Technical Information Service, PB-264-111 (1976), pp. 60–123.
2. Philbrook, J. N. Environmental Audits: Determining the Need at Mining Facilities, *Min. Eng.* 43(2):207–209 (1991).

CHAPTER 2

Surface Coal Mining with Reclamation

DRAGLINE OPERATION

A simplified dragline operation is illustrated in Figure 2.1. The overburden is excavated, and the coal seam is uncovered along a mining panel. The spoil material is cast into the adjacent mined-out pit area. Usually, the dragline cuts a trench referred to as a keycut, adjacent to the newly formed highwall. The keycut is made from Position 1. The distance between the previous keycut position and the present keycut position is known as the digout length. Sometimes, this is also referred to as "length of block." The keycut material is deposited in the bottom of the keycut, mined-out pit. Sometimes, the keycut material can be used to term and extended bench. The extended bench method is described in subsequent sections.[1]

When the keycut has been completed, the dragline moves to Position 2. From this position, the dragline excavates the digout and casts the spoil into the adjacent area of the pit. Position 2 is known as the production cut. When the excavation is completed, the dragline moves to Position 3 to start a new keycut.

The operating cycle of a dragline is separated into five distinct steps:

1. The empty bucket is placed in a position ready to be filled in the cut. This position may vary from the maximum position, which is the interface between the coal seam and the overburden, to any position in the overburden above the coal.
2. The bucket is dragged towards the dragline to fill it.
3. The filled bucket is hoisted up, and the boom is swung out towards the spoil pile. When swing has to be slowed to permit hoisting, the operation is said to be hoist critical. When hoisting to the dump point is completed before the boom reaches the dump position, the operation is considered as swing critical. While swinging out to the spoil, the drag

36   ENVIRONMENTAL IMPACTS OF MINING

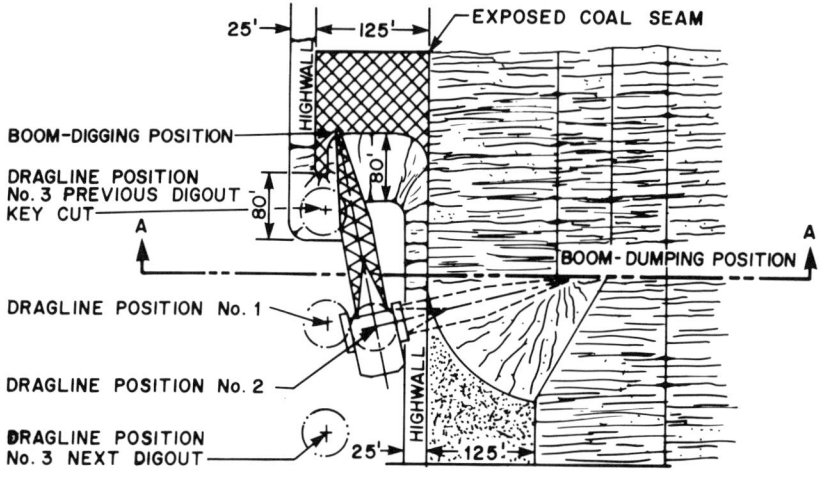

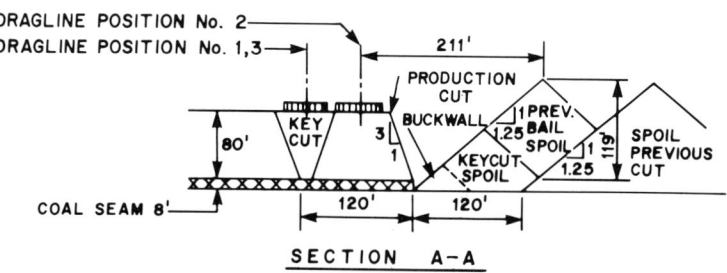

Figure 2.1. A typical dragline operation.

cables are released so that the filled bucket can be lowered down to the dumping position under the boom point. If the swing is slowed to permit payout of the drag cables, the operation is said to be payout critical.

4. The bucket dumps the spoil material it is carrying.
5. The bucket is lowered, and the boom swings back to the cut. In the return part of the cycle, the empty bucket is lowered from the dumping position to the digging position by reeving in the drag cables. If the return swing is slowed to permit reeving in the drag cables, the operation is said to be retrieve critical. If the return swing is retarded to allow lowering of the bucket, the operation is said to be lower critical.

Optimization of the dragline operating cycle requires minimization of the time required to position, drag, and dump; and synchronization of hoist/swing-out and retrieve/swing-in to the minimum critical state.

SURFACE COAL MINING WITH RECLAMATION

Keycutting and Layered Cutting

When a new digout is initiated, the dragline is positioned directly over the toe of the future highwall (Position 1, Figure 2.1). In this position, a clear and safe highwall can be established in stable ground.

A keycut is made from this position over the new highwall. The keycut separates the new highwall from the remaining overburden to be excavated. The keycut is about one to two buckets wide at the bottom. The objective of making a keycut is to establish a steep and safe highwall. A keycut may not be needed when the overburden falls truly from the highwall and a relatively clean highwall slope can be formed by the digging actions of the dragline.

The panel width and dragline size generally dictate a two or more position digout. Only the largest dragline operating within ideal geometric parameters allows digout from one position while providing a panel wide enough for efficient coal production. In the first position, the keycut material is cast short, adjacent to the rib of the coal being stripped, so as to minimize the cycle time. This short-casted material forms a "bucket-wall" slope. The dragline then moves out towards the old highwall (Position 2, Figure 2.1) and completes the digout. The material is cast behind the apex of the keycut spoil pile.

In some situations where the dragline dumping radius and dumping height are compatible with overburden depth, the dragline may be able to excavate the complete digout from the position over the toe of the new highwall. The width of the digout is excavated in layers called "layered cutting." The keycut is formed as the layers are removed. Figure 2.2 illustrates the cutting sequence for layered cutting. Each layer is usually one bucket deep. The numbers in the figure refer to the digging sequence. The highwall is "dressed" prior to completion of each layer.

The layered cutting can be performed in another manner in deeper overburden by positioning the dragline in the production boiling position. A tractor-dozer is needed to dress the highwall slope while the dragline is digging.

In some mines, the keycut is dug first, and then the production cut is excavated by layers. A small swing-critical dragline is used to excavate wide panels in deep overburden. An extension of the bench is necessary to provide spoiling reach. Between each layer of the production cut the dragline would take a few steps toward the spoil pile to which it was casting (termed layered walk). This walking action between layers helps to minimize swing angle while digging down through the layers.

Panel (Pit) Width

Panel spill width is primarily a function of the dragline size. Other factors influencing optimum panel width include dragline operating rate, overburden depth, swell of spoil, and coal production requirements.

Small draglines operating on wide panels become swing critical, causing a large increase in cycle time. Medium and large draglines usually have fast swing rates but comparatively slower hoist, payout, and drag receiving rates. Therefore, they are

38 ENVIRONMENTAL IMPACTS OF MINING

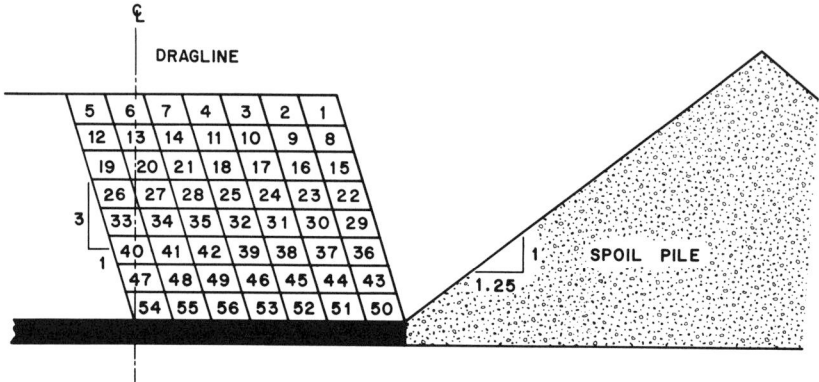

Figure 2.2. Layered cutting sequence. (From Fluor Engineers, Inc. Optimal Dragline Operating Techniques, U.S. National Technical Information Service, DE83-006980 (1982), pp. 10–52.)

usually hoist, payout, and return critical. But, this does not affect cycle time as much as its being swing critical. The cycle time of medium and large draglines does not increase significantly with an increase in panel width. Due to its short reach, a small dragline cannot strip a wide panel. For working on a wide panel, a small dragline may be required to rehandle spoil (e.g., extended bench). A large dragline has only to increase its swing angle to operate efficiently on a wide panel.

The optimum panel width is determined from consideration of the following factors:[2]

- Coal loadout: The practical minimum width for a panel is 28 m. Panel width less than 28 m. greatly hampers coal haul truck maneuverability.
- Slope stability: Wide pits are considered safe for men and equipment as they allow more room for movement. Slope failures in wide pits allow greater safety than in narrow pits. However, wide panels tend to create higher spoil piles, which can cause spoil instability.
- Cycle time: With small, swing-critical draglines, narrow panels are preferred. For medium and large draglines which are not swing critical, wide panels give better productivity.
- Spoil regrading: In wide panels, the spoil peaks are farther apart. Also, the vertical height between the peaks and villages is greater than in narrow panels. The amount of dozing to level the spoil piles is greatly increased with wide panels.
- Walking: Wider panels imply less walking of the dragline in a given mining area. It may not produce a significant difference in time spent in walking unless there is a great difference in panel widths.
- Spoiling at entryways: Narrow panels allow greater flexibility in building haul roads through spoil, inside curves, etc. Narrow panels require a shorter spoiling radius.

SURFACE COAL MINING WITH RECLAMATION 39

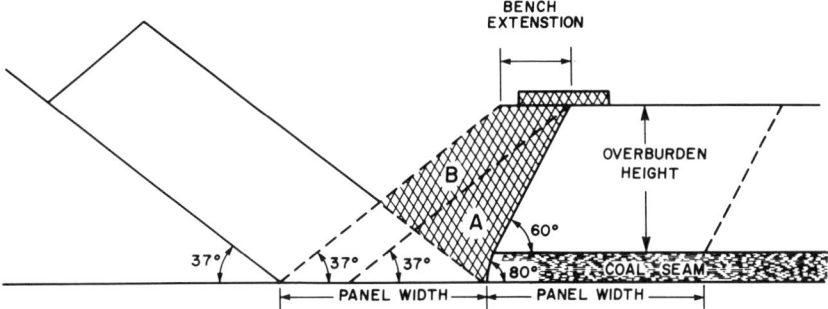

Figure 2.3. Sectional view of an extended bench. (From Fluor Engineers, Inc. Optimal Dragline Operating Techniques, U.S. National Technical Information Service, DE83-006980 (1982), pp. 10–52.)

The cycle time for a small dragline can increase up to 20% when panel width is increased from 22 to 53.3 m. When a large dragline is used on the same depth of overburden, cycle time is increased by only 1.6% when panel width is increased from 75 ft (22.8 m) to 175 ft (53.3 m).

Extended Bench

This method is usually used to increase dragline reach. The bench width is extended with spoil, so the dragline is capable of dumping spoil off the coal. Figure 2.3 illustrates a typical extended bench with the dragline positioned on the spoil extension. The material shown in crosshatching will require rehandling by the dragline.

Sometimes an extended bench becomes indispensable for a particular stripping situation. Reasons may be as follows:

- overburden depth
- highwall stability
- necessity of a wide pit for safety during coal removal
- high swell of spoil material
- the limited reach of the dragline

The main disadvantage of the extended bench is the rehandling of the spoil, which increases the total volume of material that must be moved to uncover coal.

DRAGLINE STRIPPING PROCEDURES

Dragline stripping procedures are divided into different systems depending upon the physical and technological parameters of the coal property. The classification of dragline systems is presented in Figure 2.4. As shown in the figure, operating procedures are dependent upon the number of coal seams mined per pit, the spoil placement method, the number of overburden and interburden lifts, and the number of dragline passes per pit. The dumping radius (range) of the dragline is a basic parameter, because the classification of the overburden as shallow or deep is relative to the range of the machine. Identified in Figure 2.4 are 17 different classifications.

Single Seam Stripping with Nonselective Spoil Placement

When a single coal seam is mined and the overburden is placed nonselectively, the practice can be described as the least complex of the dragline operating systems.

Single Machine Subsystems

There are five basic procedures for mining a single coal seam using a single dragline and nonselective overburden placement. These procedures depend upon overburden depth, the number of overburden/interburden lifts, and the number of dragline passes per pit.

Shallow Overburden, One Lift, One Pass

The dragline operating in a shallow overburden occurs early in the life of a mine when mining takes place in a shallow overburden. Basic operating procedures are shown in Figure 2.5. The dragline operates from a position on the natural ground surface, either after a thin layer of top soil has been removed by scrapers or from a position on a bench.

The dragline is positioned directly over the desired location of the new highwall. A keycut is made. The keycut is a narrow trench, about one or two buckets wide at the bottom. It is excavated to establish a safe, new highwall. The dragline must be placed directly over the keycut in order to establish a clean, safe highwall. The spoil material from the keycut can be placed close to the old highwall in the adjacent open cut, or the dragline can swing and dump over the apex of the final spoil pile. The former procedure, called logging the spoil, minimizes dragline swing angle and cycle time.

In shallow overburden, the dragline has sufficient range to excavate the overburden remaining in the keycut. Two practices are commonly observed. In the first, the spoil is dumped on a fixed apex point, resulting in the deposition of the spoil in conical piles. In the second method, the boom is swung only the required minimum distance on each cycle, and the spoil is deposited along a curvilinear ridge line. The second method is considered superior to the former as swing angles are minimized, resulting in minimum cycle times and maximum productivity, as well as more efficient use of spoil room.

After having excavated an entire block of overburden which may be from 13 to 60 m deep (depending on the size of the dragline), the machine is moved back to a

SURFACE COAL MINING WITH RECLAMATION   41

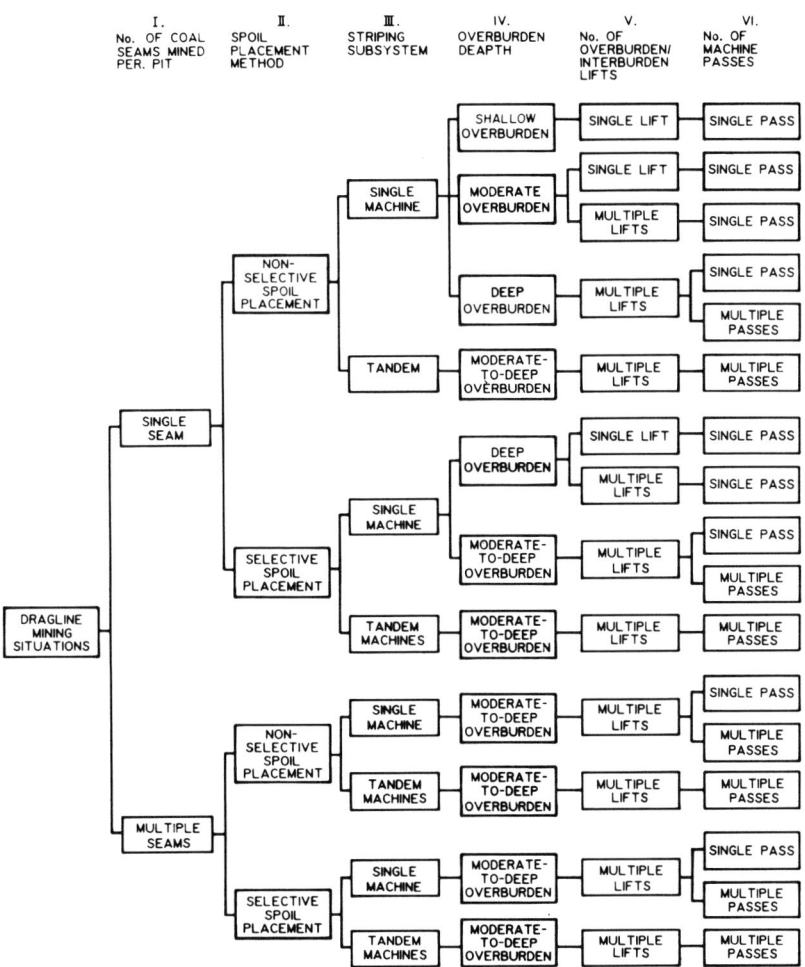

Figure 2.4. Classification of a dragline mining system. (From Math Tech, Inc. Evaluation of Current Surface Coal Mining Overburden Handling Techniques, U.S. National Technical Information Service, PB-264 111 (1976).)

new place as indicated by Position 2 in the plan view of Figure 2.5. The distance the machine is moved is called the length of the digout, the digout being the block of overburden excavated from a given machine position. The length of the digout affects both dragline efficiency and spoil grading costs. Many operators tend to make mistakes. A spoil ridge line becomes undulating because spoil is deposited in widely separated conical piles. The alternative is to shorten the digout and to develop a sharp spoil ridge line. The widely spaced spoil piles may have two major disadvantages.

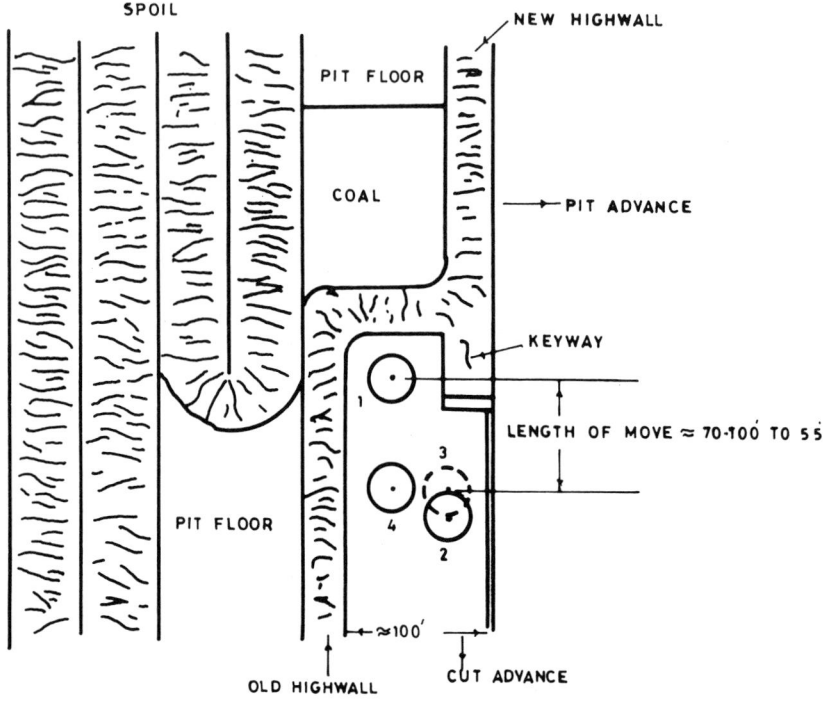

Figure 2.5. Plan view of dragline operating in shallow overburden. (From Math Tech, Inc. Evaluation of Current Surface Coal Mining Overburden Handling Techniques, U.S. National Technical Information Service, PB-264 111 (1976).)

Some part of the spoil storage room is lost in the gaps between spoil piles as shown in Figure 2.6. During grading of the piles, dozers must be used to flatten the ridge line before the main grading activities can be performed. When the spoil ridge line is even, this procedure is fairly straightforward.

Any increased length of the digout increases the dragline cycle time. The longer the digout, the farther the bucket will be dragged towards the machine before hoisting. The solution lies in determining an optimum digout length. The dragline may move to Position 3 from Position 2 to complete the keycut, so as to prevent the drag ropes from dragging on the highwall with increasing depth of keycut. Then the dragline moves to Position 4 from Position 3 and completes stripping setup.

Figure 2.7 shows the final cut removal in the simple sidecasting method.

When the dragline reaches the end of the pit, two alternative procedures can be followed. Where the coal is not too thick and there is no need to maintain a stock of exposed coal seams in the pit, coal extraction closely follows the dragline. At the end

SURFACE COAL MINING WITH RECLAMATION   43

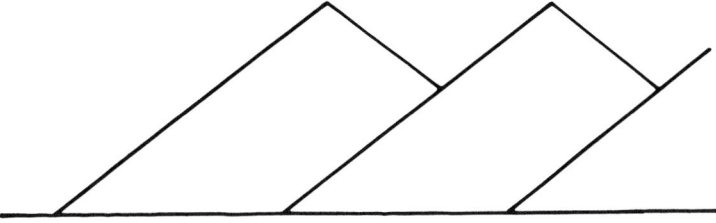

Crest-to-Crest Spacing: 90 feet.

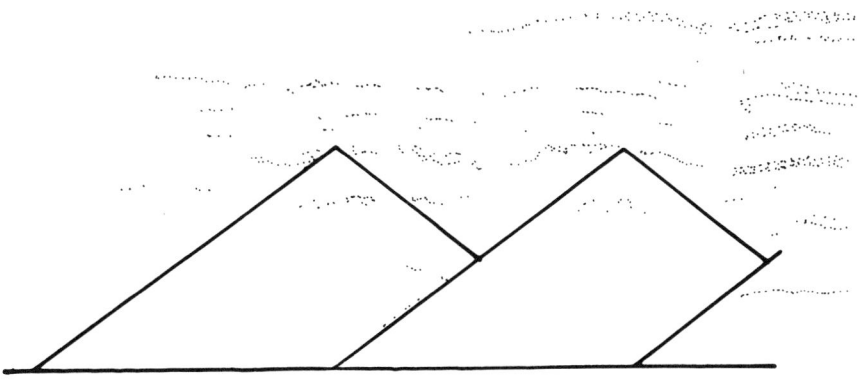

Crest-to-Crest Spacing: 120 feet

Figure 2.6. Comparison of spoil piles with 90- and 120-ft crest-to-crest spacing. (From Math Tech, Inc. Evaluation of Current Surface Coal Mining Overburden Handling Techniques, U.S. National Technical Information Service, PB-264 111 (1976).)

of this pit, the dragline is turned around, and stripping commences back in the opposite direction. Where the coal seam is thick and the dragline stripping operation is far ahead of the coal loading, or where it is desired to maintain an inventory of exposed coal seams in the ground, the dragline is deadheaded back to the other end of the pit, and stripping begins again.

The width of the pit affects dragline productivity and spoil grading costs. The effect of pit width is less critical in shallow overburden than in deep or moderately deep overburden. Narrowing of the pit results in a decrease in dragline cycle times as swing angles are reduced. If the pit is narrowed too far, the swing angle becomes so small that the loaded bucket cannot be fully hoisted during the swing. So narrowing the pit beyond certain limits may reduce dragline productivity.

44 ENVIRONMENTAL IMPACTS OF MINING

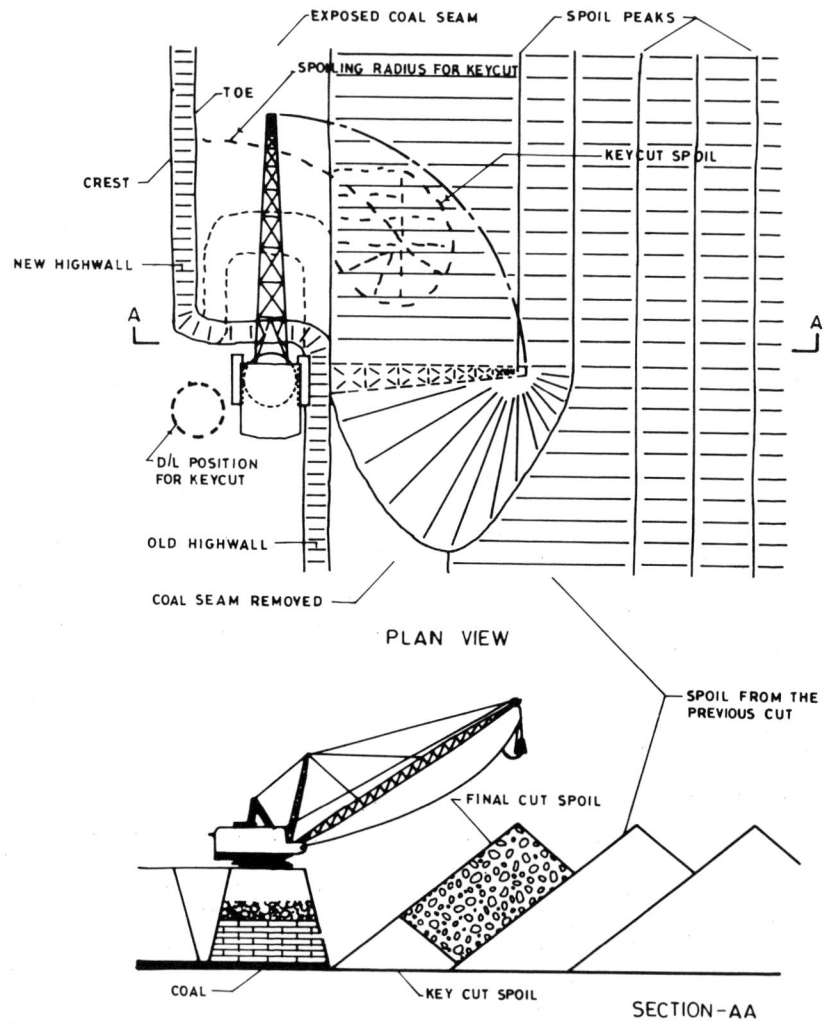

NOTE: DRAGLINE POSITION, LAST SPOILING RADIUS, & KEY CUT SPOIL LOCATION SHOWN DOTTED IN PLAN

Figure 2.7. Sidecasting — final cut removal. (From Math Tech, Inc. Evaluation of Current Surface Coal Mining Overburden Handling Techniques, U.S. National Technical Information Service, PB-264 111 (1976).)

Narrowing of the pit in a specific situation reduces spoil grading volumes and costs. As the pit is narrowed, the crest-to-crest spacing of spoil piles and the depth troughs between piles are reduced. The grading cost per hectare of spoil piles with 40-m crest-to-crest spacing is roughly double that for spoil piles with 30-m spacing.

The minimum pit width is determined primarily by the room needed for coal leading and haulage equipment. At large mines, 30 to 35 m is the practical minimum. In very deep mines, wider pits are used for safety reasons. A rule of thumb often used is that the pit should be at least as wide as it is deep. The maximum pit width is determined by many factors, such as the dumping radius of the dragline, the overburden depth, and the volume of rehauled spoil.

As the height of the spoil pile increases, the bucket hoist time increases. Usually, the short-swing-angle sections of the overburden are dug early in the digout when the height of the spoil pile is small.

The choice of pit widths is influenced also by topography. If the overburden depths quickly increase in successive cuts, the early cuts in shallow overburden may be made fairly wide, and the widths of successive cuts can be progressively narrowed.

Summarizing, with shallow overburden, operating decisions must be made according to the following factors:

- width of the keycut
- spoil casting pattern — conical piles or curvilinear ridge lines
- bucket loading method — layer loading or non-layer digout
- length of the digout
- pit width
- digging and casting pattern

Box Pits

In some cases, the first cut to begin a new mine is commonly a box cut. This produces an initial highwall. This pit gets its name from the long box-like shape of the completed cut. There are two distinct methods of making the box pit without rehandling material:

- end cut
- side cut

For very thick overburdens relative to dragline size, methods which entail certain quantities of rehandle are necessary:

- spoiling on both sides of the box cut
- first producing a borrow pit on the spoil side of the box cuts

46 ENVIRONMENTAL IMPACTS OF MINING

Each of these four methods are discussed below.

End Cut

The end-cut box pit is produced with the dragline positioned at the end of the pit, as shown in Figure 2.8. This method is used in relatively thin overburden compared with the dragline dimensions. Thus, it is possible for the machine to spoil all the material to the side away from the coal deposit or what becomes the opposite side of the cut from the new highwall. All the material can be spoiled in this case using a maximum dragline swing angle of 60°. The dragline operating cost is greatly dependent on the angle of swing, as well as the percentage rehandle. The end-cut method has a very favorable dragline swing angle (maximum 90°), especially when compared with the side-cut method. However, the end-cut method is restricted to relatively shallow overburden thicknesses because of the limit on the dragline dumping radius. As the cut increases in depth, the waste pile increases in volume, until it can no longer be spoiled completely clear of the box pit.

Rehandle in these box-pit cuts is either 100% or nil, depending on whether the coal seam does or does not continue under the spoil pile. If the coal extends under the spoil pile because the box pit was started at a mining lease boundary, then it may be economical to recover the covered coal later. One method frequently used is the auguring technique. However, on occasion this coal is lost.

Side Cut

The side-cut box pit is produced with the dragline positioned at the side of the pit, as shown in Figure 2.9. This method is used when the overburden is of such a depth that it is impossible for the dragline to use the end-cut position. The side position allows the spoil to be dumped at a greater distance from the pit, making it possible to widen the spoil pile at the maximum dumping height, as required.

Figure 2.9 shows that the swing angle can be increased to the theoretical maximum of 180° for certain parts of the cut. In practice, the highwall side of the pit would be allowed to lag behind the near side, reducing the maximum swing angle to approximately 130°. This is still considerably larger than for the end-cut method, but for thicker overburdens and this given machine size, it is the only way the box pit can be produced without rehandle.

Rehandle (End Cut)

An extension of the regular end-cut method incorporates spoiling from the end position to both sides of the cut. This method can be used to allow the end-cut position to be maintained as the overburden thickness increases. However, the attraction of reduced swing angles, as compared with the side-cut method, does not usually outweigh the additional material rehandle. It can also be used producing a regular end cut which was temporarily being produced in material of lower angle of repose. Thus, the end cut could be continued, with the excess spoil being placed on the highwall side. Figure 2.9 shows a plan and cross-sectional view of a box pit produced in this manner. Again, as with the regular end-cut method, the angle of swing is seen to be a maximum of 90°.

Having produced this type of box cut, the material above the coal on the highwall side, which would be up to a maximum of 50%, must be rehandled in some manner. There are many methods available, including using a shovel/truck or loader/truck operation. However, the most common method is the use of dozers or dozer/scraper combinations. In this case, material is then spread out on the highwall side to enable the dragline to rehandle it during later cuts. This offers the possibility of an advantage for this method. If the ground surface is pitted and uneven, the use of this extra spoil as fill can enable the production of a good dragline floor for use in later cuts. This, then, could even reduce the later work of dragline pad production. Similarly, if the ground surface is unsuitable for the dragline bearing pressure in some areas, then selected material from the box cut can be spoiled on the highwall side.

Rehandle (Borrow Pit)

This rehandle method does not entail the use of auxiliary equipment to aid in the rehandle operation. The aim is to spoil all the overburden on the waste side of the box cut by first producing a borrow pit.

The initial cut for this method produces a borrow pit parallel to, and on the waste side of, the proposed box pit, as illustrated in Figure 2.10.

The borrow pit encompasses the sum total of the rehandle. The volume of rehandle depends solely on the amount of extra volume required to spoil the final box cut into a single pile. The final box cut itself can be an end or side cut, depending on the overburden cover. The side cut again provides the means of digging the maximum overburden depths.

Figure 2.10 shows the box cut being spoiled into the borrow pit. This figure suggests the dragline has deadheaded to the beginning of the borrow pit and is now returning in the same direction of advance. This is certainly not essential. For most operations, the dragline would simply turn round and dig the box cut into the borrow pit with no deadheading. Suitable access would be designed into the cut from the appropriate end or, preferably, from both ends.[3]

In general, the larger material is kept near the base of the pile, whereas the clay would be dumped on top. It is frequently advantageous, and sometimes essential for reclaim purposes, to ensure that certain materials remain well buried after spoiling. Strongly acidic rocks are common just above the coal seams. Assuming these initial spoil piles are to be leveled at a later date during land reclamation, then only the borrow pit method offers the opportunity to get this material below final grade immediately. The other methods can require considerable, additional work in this area.

48 ENVIRONMENTAL IMPACTS OF MINING

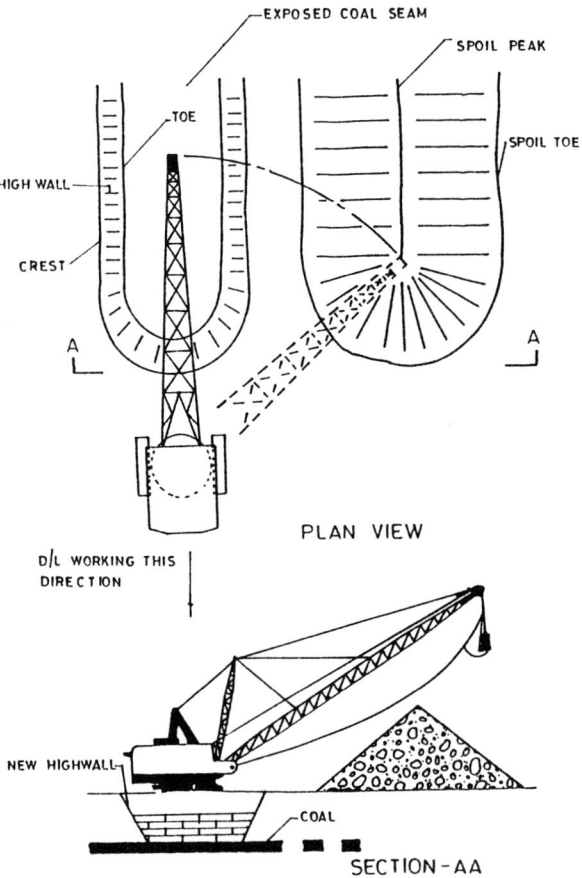

Figure 2.8. Box pit — end cut methods. (From Bucyrus-Erie Co., Surface Mining Supervisory Training Program, 1977. With permission.)

SURFACE COAL MINING WITH RECLAMATION   49

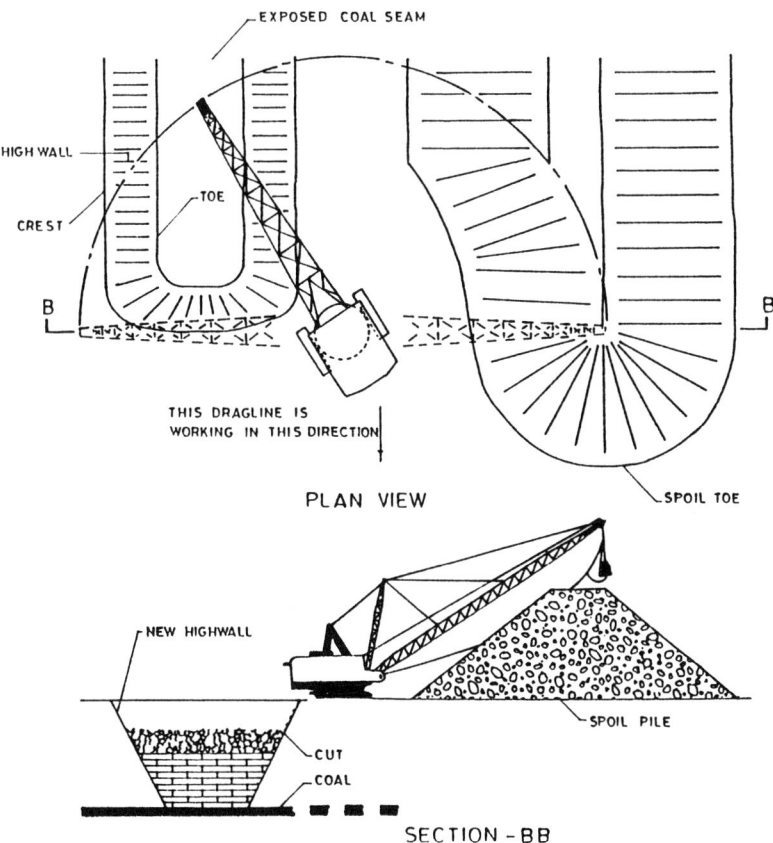

Figure 2.9. Box pit — side-cut methods. (From Bucyrus-Erie Co., Surface Mining Supervisory Training Program, 1977. With permission.)

50   ENVIRONMENTAL IMPACTS OF MINING

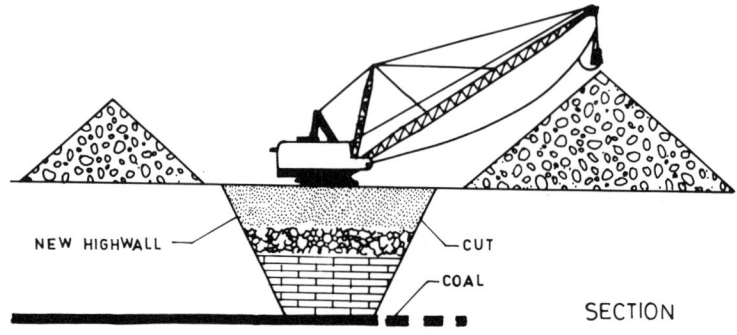

BOX PIT – BORROW PIT METHOD

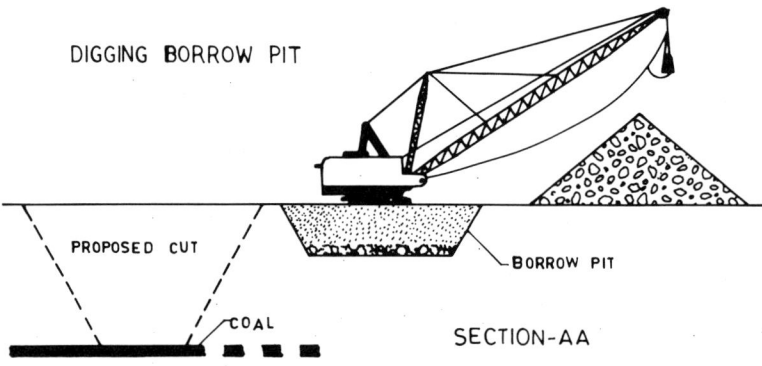

DIGGING BOX PIT – END CUT METHOD

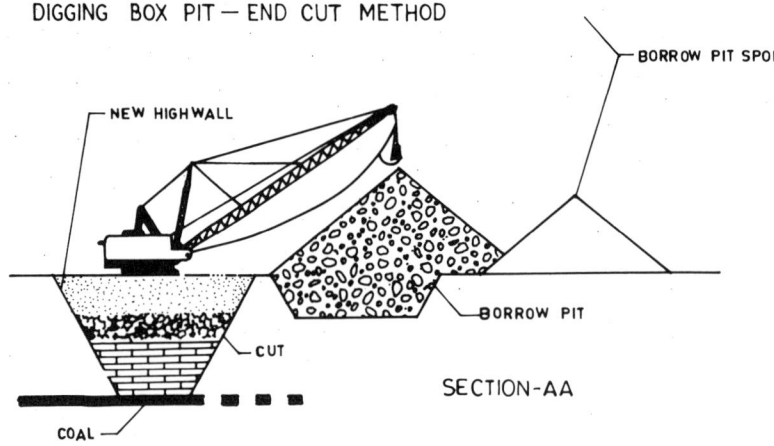

Figure 2.10. Box pit — end cut with rehandle, borrow pit, and digging box pit — end-cut method. (From Bucyrus-Erie Co., Surface Mining Supervisory Training Program, 1977. With permission.)

The box pit is used for making the initial cut when opening a new deposit. For a given property, only one such cut should be necessary. These first stripping methods, therefore, represent only a very small part of the total.

Moderate Overburden Depth, Single Lift, Single Pass

A one-lift, one-pass dragline procedure in a moderate overburden depth is illustrated in Figure 2.11. The moderate overburden depth refers to the depth in which the machine is designed to operate with little or no spoil rehandle. The figure depicts a one-lift operation where the dragline operates from the natural ground surface.

Procedures are similar to those used in shallow overburdens except for two major differences. A keycut is made in the same manner as in the shallow overburden. However, it is not possible to excavate and cast all the remaining overburden material from the keycut position of the dragline, because the top of the spoil pile would ride up the highwall, needing subsequent rehandling of the spoil. To avoid this, after completion of the keycut, the dragline is moved sideways from Position 1 to Position 2 as in Figure 2.11. The move extends the spoil-developing range of the machine. The remaining overburden is then excavated from this new position. Thereafter, the machine is moved diagonally to keycut Position 3 for the next digout, and the cycle is repeated.

In some situations, there may be more than two dragline positions for a given digout. The beginning position is always over the keycut, and the finishing position is at the edge of the old highwall.

When the machine is crawler-mounted, the sideways and diagonal movements take little time. Where a walking dragline is used, the sideways and diagonal walkings may take 5 to 15 min on each digout. The total nonproductive time due to sideways and diagonal walking can be optimized by controlling the number of digouts per time period at a given mine. The number of digouts which can be completed in a given time period depends on the length of the digout and the width of the pit.

In extreme cases, this type of reasoning can result in loss of dragline productivity and increase in spoil grading costs. For example, in one mine the length of the digout was decreased in an effort to reduce spoil grading costs. It was expected that dragline productivity would decrease because of the increased nonproductive dragline-walking time associated with shorter digouts. As it happened, grading costs were reduced, and dragline productivity was increased by 2%. The increased productivity resulted from the reduction of the distance over which the loaded bucket was dragged being hoisted.

Widening the pit can be another means to reduce the number of digouts in a given time period. There are limits on how far the pit can be widened for this purpose because widening of the pit beyond a certain extent will require spoil rehandle. In addition, the dragline swing angle increases as the pit is widened. For example, consider an area 600-m square which is to be mined using pits 600-m long with a digout length of 30 m. If the pit width is 30 m, the number of digouts to mine the

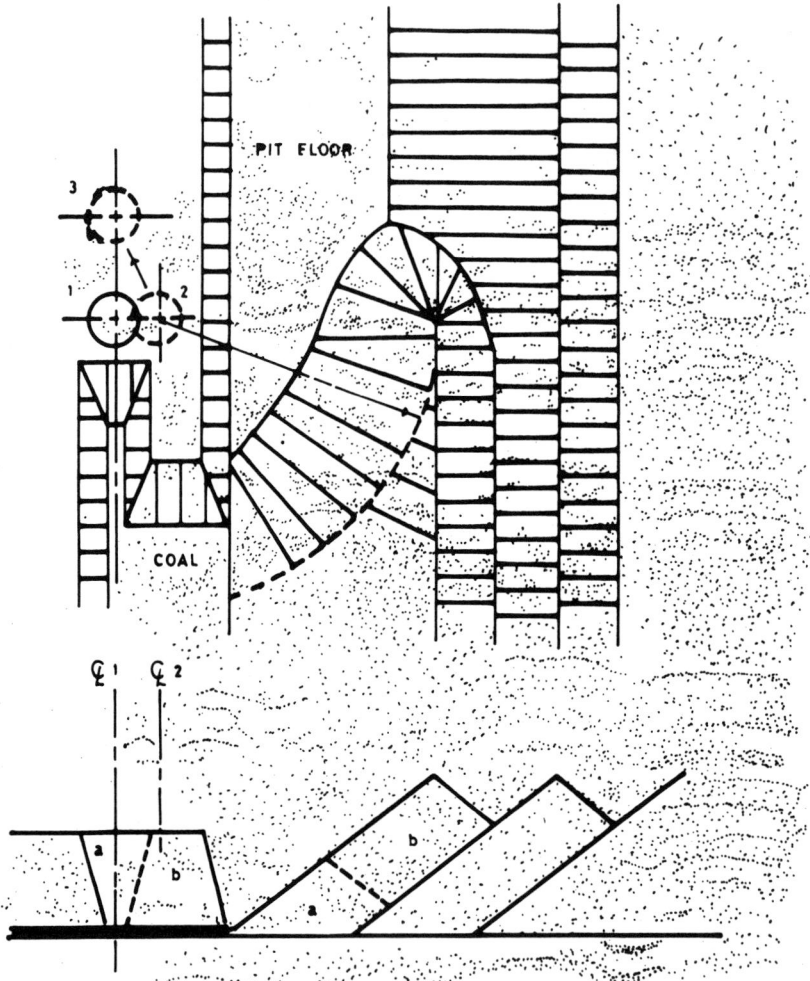

Figure 2.11. Dragline operating procedures in moderate depth overburden. (From Math Tech, Inc. Evaluation of Current Surface Coal Mining Overburden Handling Techniques, U.S. National Technical Information Service, PB-264 111 (1976).)

entire area will be approximately 2640. When the pit width is increased to 50 m, if this is feasible without resulting in unnecessary spoil rehandle, the number of digouts is reduced to 1760. This is a reduction of 33%. If the dragline is deadheaded at the end of each pit, the wider pit would result in less deadheading time over the life of the mine as well.

With moderate depth overburden, the decision variables are the same as with shallow overburden, but the sideways movement of dragline on each digout complicates the decisionmaking process.

SURFACE COAL MINING WITH RECLAMATION

If the advantages and disadvantages of long and short digouts and wide and narrow pits are known, it may be difficult to arrive at an optimum decision. However, mine operators do take these decisions with varied results.

In a moderate overburden situation, the effect of pit geometry on productivity is more pronounced than in a shallow overburden situation. As the overburden becomes deeper, it becomes more difficult to keep the coal haulage roads open.

Moderately Deep Overburden, Two Lifts, One Pass

Very often, the dragline works from a machine-supporting bench, rather than from the natural ground surface. The natural ground surface can consist of unconsolidated material which may be fairly deep and might not support the dragline when it is positioned close to the highwall edge. The unconsolidated surface material becomes muddy in wet weather, making it difficult to walk the dragline from one position to another.

The dragline works from a bench — known as the established bench — which typically is 3 to 6 m below the natural ground surface. The bench is established over consolidated material. Sometimes competent shale from the bank is used to form a road or a pad over the bench surface in order to eliminate muddy conditions which complicate dragline walking.

On each pass of the pit, overburden below the bench height is excavated and spoiled as discussed in previous sections. Additionally, however, a side bench, that is the bench to be used on the subsequent pass, is excavated by dragline. Figure 2.12 illustrates the side benching procedure.

The dragline works from the established bench and is turned to the highwall side, with the boom at right angles to the highwall. Then the unconsolidated side bench overburden is excavated. The boom is swung through a 180° angle, and side bench material is dumped onto the top of the existing spoil pile. This can be sometimes very desirable from a reclamation standpoint and is an example of one way in which spoil is selectively placed by the dragline. Of course, the procedure is adopted for production reasons, but the reclamation advantages are notable. In fact, the side benching procedure is also used for reclamation purposes.

Side benching has several disadvantages. A relatively large swing angle is required in cutting the side bench and spoiling the bench material. This reduces dragline productivity. An overhand chopping motion is required to excavate the side bench. A chopping motion is one in which the digging is done above rather than below the elevation of the machine-supporting bench. This does not present significant problems when the bench is high but may cause problems when benching is deeper.

When the side bench spoil is shallow, the reclamation advantage may be apparent. When the side bench spoil is fairly shallow on the surface of the spoil pile, the grading dozer will cut through the material and expose the underlying spoil. As shown in Figure 2.13, the side bench material on the spoil pile should be sufficient

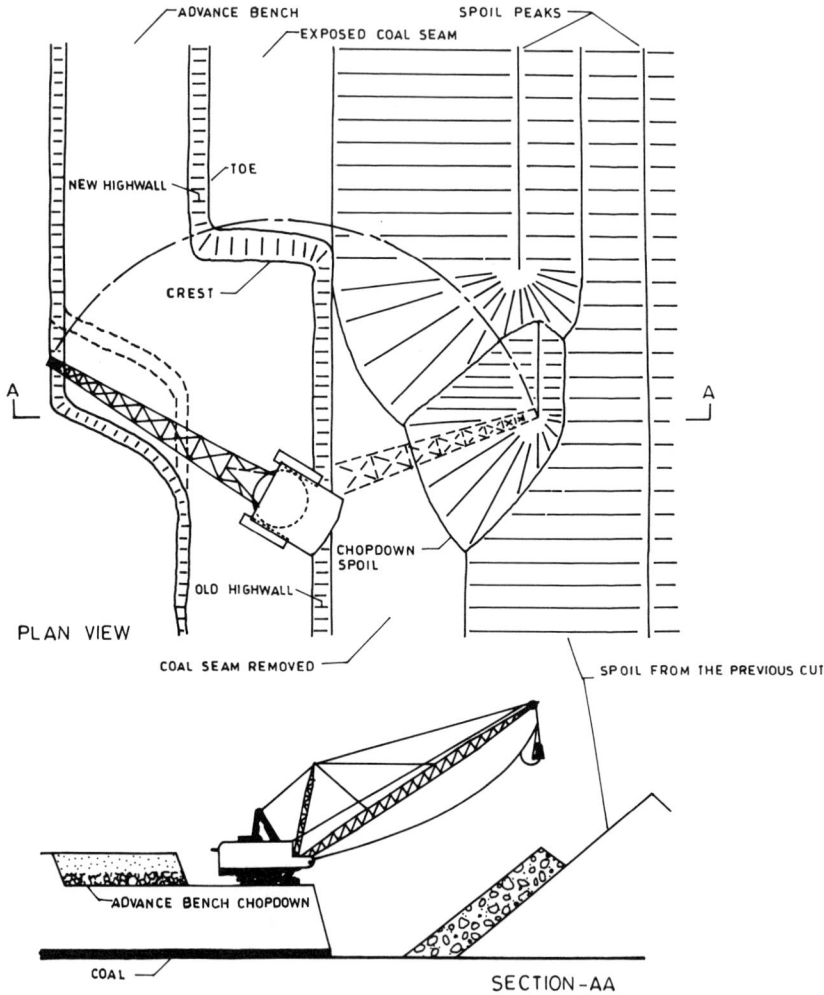

Figure 2.12. Dragline making chopdown cut using advance bench. (From Math Tech, Inc. Evaluation of Current Surface Coal Mining Overburden Handling Techniques, U.S. National Technical Information Service, PB-264 111 (1976).)

to prevent breakthrough by dozers during grading. If the side bench material is not desirable from a reclamation standpoint, then the side benching procedure may not solve the reclamation problem.

Deep Overburden, Split Bench Mining

A section view of this method is shown in Figure 2.14. On the first lift, which consists of about half the overburden depth, the operating procedure is similar to the

SURFACE COAL MINING WITH RECLAMATION   55

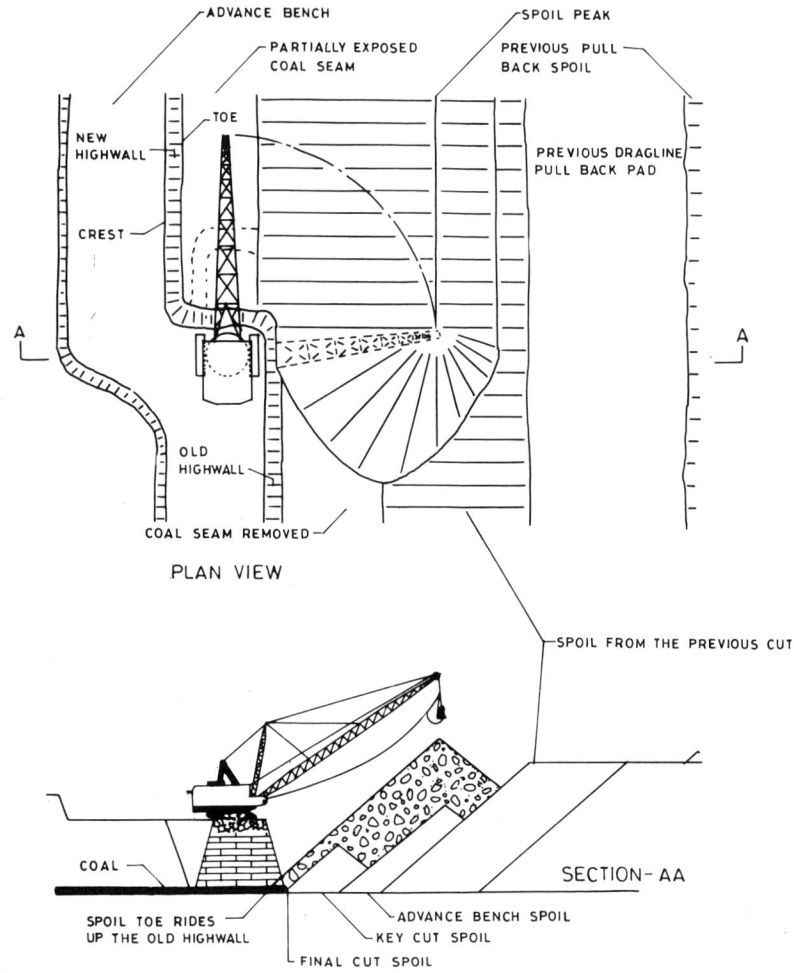

Figure 2.13. Dragline performing the last cut of an advance bench as part of the pullbackstripping method. (From Math Tech, Inc. Evaluation of Current Surface Coal Mining Overburden Handling Techniques, U.S. National Technical Information Service, PB-264 111 (1976).)

previously described conventional practice. The top lift spoil is cast close to the highwall to minimize dragline swing angles. When the top lift stripping has been completed and the dragline has reached the end of the pit, the dragline is walked down a ramp to the lower lift, and the stripping of the second lift proceeds in a direction opposite to that of the first lift.

There are several alternative procedures to make the keycut in the lower lift overburden. The keycut cannot be made immediately adjacent to the upper highwall

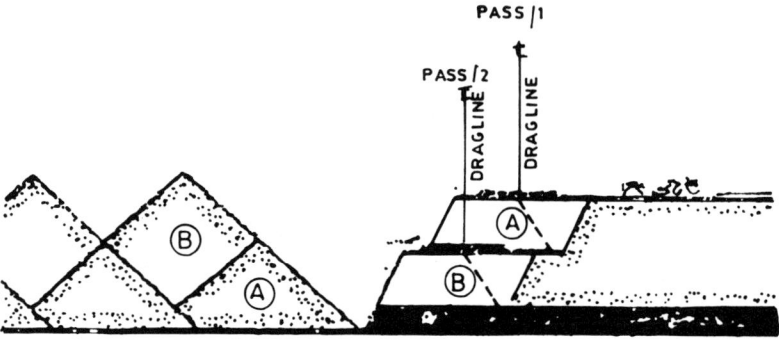

Figure 2.14. Section view showing the split bench method with two dragline passes per pit. (From Math Tech, Inc. Evaluation of Current Surface Coal Mining Overburden Handling Techniques, U.S. National Technical Information Service, PB-264 111 (1976).)

because the dragline course would hit the upper highwall during rotations of the boom to cast the keycut spoil. The lower lift keycut has to be made at some distance from the upper highwall. To accomplish this, the top lift bench must be made wider than the pit.

An alternative procedure is to make the lower lift keycut using other equipment: typically dozers and front-end loaders. Dozers make the keycut by pushing the overburden down into the pit, where it is loaded and placed using a front-end loader. When this procedure is followed, the lower lift keycut can be cut adjacent to the upper highwall. When the overburden is deep, this procedure is not suitable for safety reasons.

The highwall is cut on a slope of 65 to 75° from the horizontal. The dragline is closer to the second lift than to first lift. This can be seen in Figure 2.14 by comparing the dragline centerlines for Passes 1 and 2. This situation effectively increases the dumping radius of the dragline. For example, in a 25-m overburden depth with a 13-m bench depth and 65° highwall, the dragline would be 9 m closer to the spoil side on the second lift than on the first. This increases the effective dumping radius of the dragline, enabling no-rehandle stripping of overburden deeper than the design limit of the machine. The design limit of the dragline is defined as the maximum depth of overburden that can be excavated and spoiled without spoil rehandle. This limit is dependent upon the dumping radius of the machine, the distance of the machine centerline from the highwall during stripping, the angle of repose of the soil, the pit width, the spoil swell factor, and the highwall angle. For example, a maximum overburden depth of about 25 m can be handled without spoil rehandle by a dragline with a dumping radius of 75 m. A range extension of 9 m would increase this overburden depth to about 28 m. This is the main advantage of the split bench method.

SURFACE COAL MINING WITH RECLAMATION

The split bench method suffers from several disadvantages. The deadheading time for the dragline is approximately double that of the single pass case because the dragline is deadheaded twice in each cut. Ramps must be constructed in each pit to enable movement of the dragline between the upper and lower benches. During the second lift stripping, the dragline is 13 m below the ground surface. So the bucket hoist heights during spoiling of second lift material are higher than usual. This may increase cycle time during second lift stripping operations.

The decision variables in split bench mining are similar to those in side-benching. But determination of an optimum pit width, bench height and keycutting methods may be somewhat more difficult than in side bench mining.

In deep overburden situations, most mines prefer a one-pass extended bench method to the split bench method.

After placing the keycut spoil against the existing highwall, the spoil is leveled by a dozer to form a level bench extension. After completing the keycut, the dragline moves to a position on this extended bench to complete the digout. The digout includes a portion of both the bench extension from the previous digout and the side bench. The width of the extended bench determines to what extent the sidebench can be cut as the last activity in the digout. If the side bench cannot be cut, then the rehandle material is placed on the pits. Figure 2.14 shows section views of an extended dragline bench.

Usually there are two different approaches to the extended bench method — the high bench and deep bench. The high bench situation is illustrated in Figure 2.15. The alternative is to work the dragline from a bench whose depth is one- to two-thirds of the overburden depth. In 30-m overburden, the dragline bench would be approximately 10 to 20 m below the natural ground surface. The main advantage of this deep bench situation is that the spoil rehandle percentage decreases as the bench depth increases. For example, with a overburden depth of 30 m, a 15-m wide extended bench and a bench depth of 6 m, the rehandle is 30%. But with a bench depth of 15 m, the rehandle is 14%. These situations are illustrated in Figure 2.16.

During the rehandling operation, the spoil must be cut back to its natural angle of repose, which is usually between 35 to 40°. In some instances, the compaction of the extended bench spoil may allow oversteepening the spoil during the rehandling operation, thereby reducing rehandle percentages. The degree of oversteepening depends primarily on the characteristics of the spoil, the method of spoil placement, and the height of the extended bench, as shown in Figure 2.17.

The deep bench method also suffers from several disadvantages in comparison with the high bench method. The average bucket hoist height is greater for the deep bench than for the shallow bench. Where the swing angles are sufficiently small, the overall swing and hoist time for the deep bench is determined by the hoist time. As a result, the total cycle time for the deep bench may be greater than that for the high bench.

A second disadvantage of the deep bench method is that as the bench depth is increased, the percentage of total overburden volume that must be excavated by an

58 ENVIRONMENTAL IMPACTS OF MINING

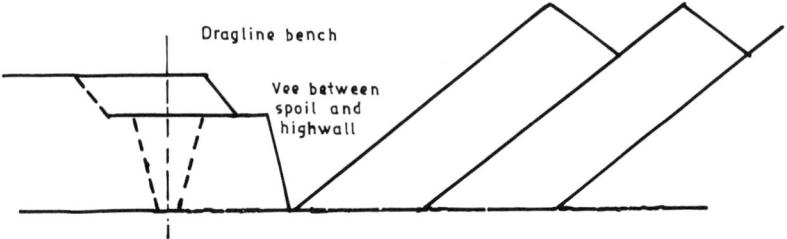

Section View: Vee Between Highwall and Spoil

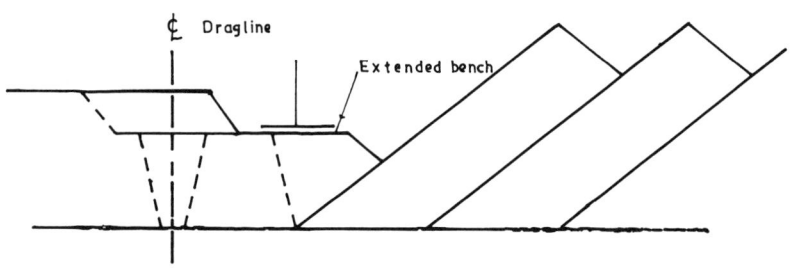

Section View: Vee Filled with Spoil to Extend the Dragline Bench

Figure 2.15. Section views showing an extended dragline bench. (From Math Tech, Inc. Evaluation of Current Surface Coal Mining Overburden Handling Techniques, U.S. National Technical Information Service, PB-264 111 (1976).)

overhand chopping motion also increases. Many operators do not like this, as the dragline productivity in overhand chopping is less than in the conventional mode of operation.

Deep Overburden, Multiple Lifts, One Pass (The Extended Bench Method)

The primary objective of the extended bench method is to increase the effective range of the dragline by extending the dragline bench out into open pit and filling the trough between the highwall and the adjacent angle of repose of the spoil pile. The dragline can then be moved out onto this extended bench beyond the edge of the existing highwall, thereby increasing the effective spoil-dumping radius of the machine. Ultimately, some or all of the extended bench material will have to be rehandled by the dragline to clear the spoil away from the highwall. The spoil

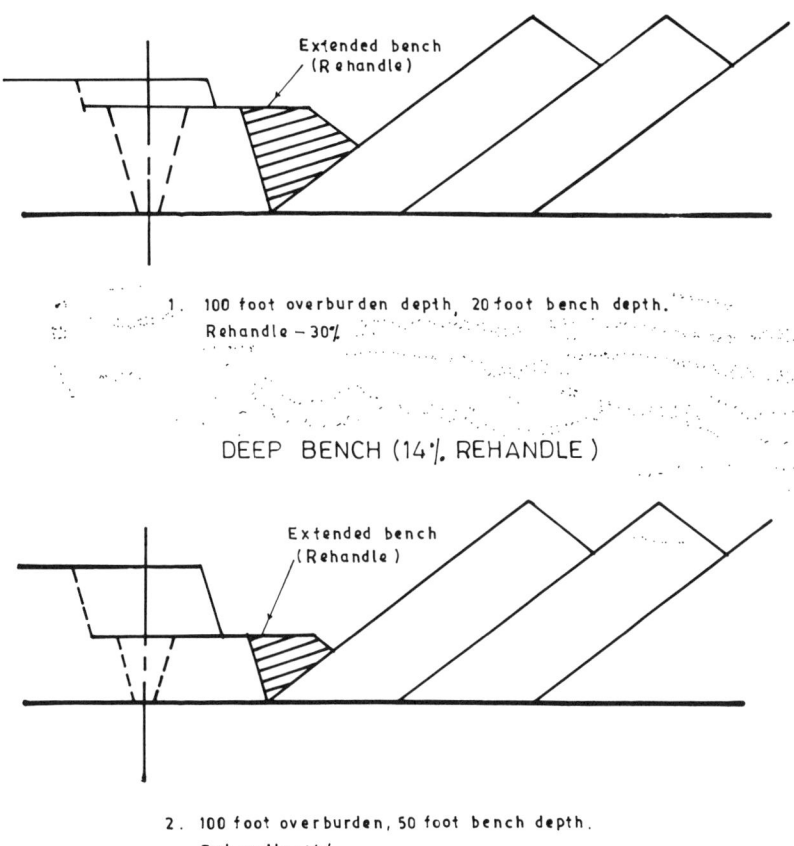

Figure 2.16. Comparison of rehandle methods for high and deep benches. (From Math Tech, Inc. Evaluation of Current Surface Coal Mining Overburden Handling Techniques, U.S. National Technical Information Service, PB-264 111 (1976).)

rehandle requirement is a characteristic of the extended bench method. In most single-seam, single-dragline extended bench methods, the keycut material is used to extend the dragline bench. The main difference between the extended bench method and the side bench method is that instead of logging the keycut spoil, the boom is swung at approximately 120°, and keycut spoil is placed against the existing highwall in the direction of stripping. This procedure, known as leading the keycut spoil, is shown in Figure 2.18. Figure 2.19 illustrates a typical method of extending dragline bench. Figure 2.19 shows a dragline on an extended bench.

The choice of pit width is also an important decision in the extended bench method. The rehandle percentage is largely independent of the pit width. Mine

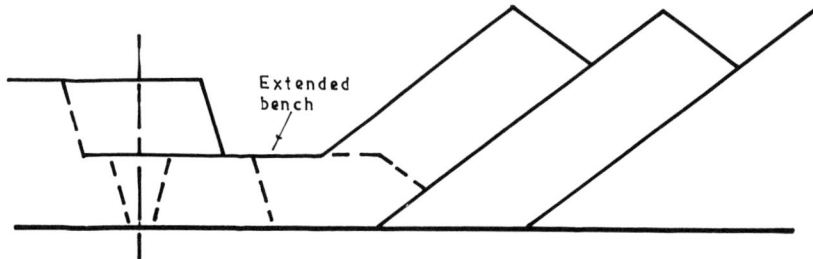

1. Extended bench: before rehandle

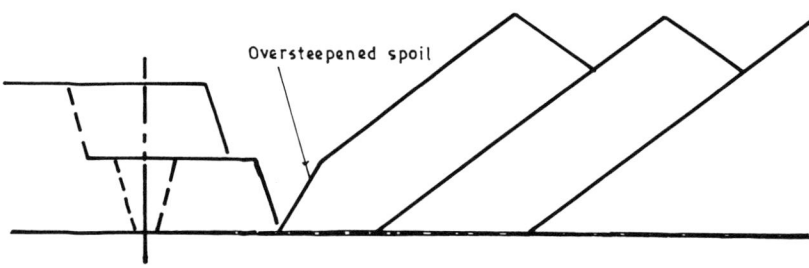

2. Oversteepened spoil: after rehandle

Figure 2.17. Section views showing extended bench and oversteepened spoil. (From Math Tech, Inc. Evaluation of Current Surface Coal Mining Overburden Handling Techniques, U.S. National Technical Information Service, PB-264 111 (1976).)

operations prefer wide pits because the wider the pits are, the fewer pits in a given acreage. Widening of the pits also reduces the rehandle volumes per acre. However, wide pits increase dragline cycle time and spoil-grading costs.

Extending the effective range of the dragline also yields other advantages compared to conventional operating methods. In conventional stripping of deep overburden using a long-boom machine, the spoil piles are very high, causing spoil slides into the pit. With the extended bench method, the spoil can be cast into the troughs between the spoil piles rather than on the tip of the pile closest to the pit. This results in better stability and better spoil grading.

The extended bench enables the use of a dragline with a steeper boom angle than is required if the bench is not extended. As the boom angle is increased in the dragline, the bucket size can be increased, and the machine radius can be decreased. Reduction in dump radius increases boom acceleration with radius boom swing time on each cycle. Additionally, because of the larger bucket, the volume of overburden

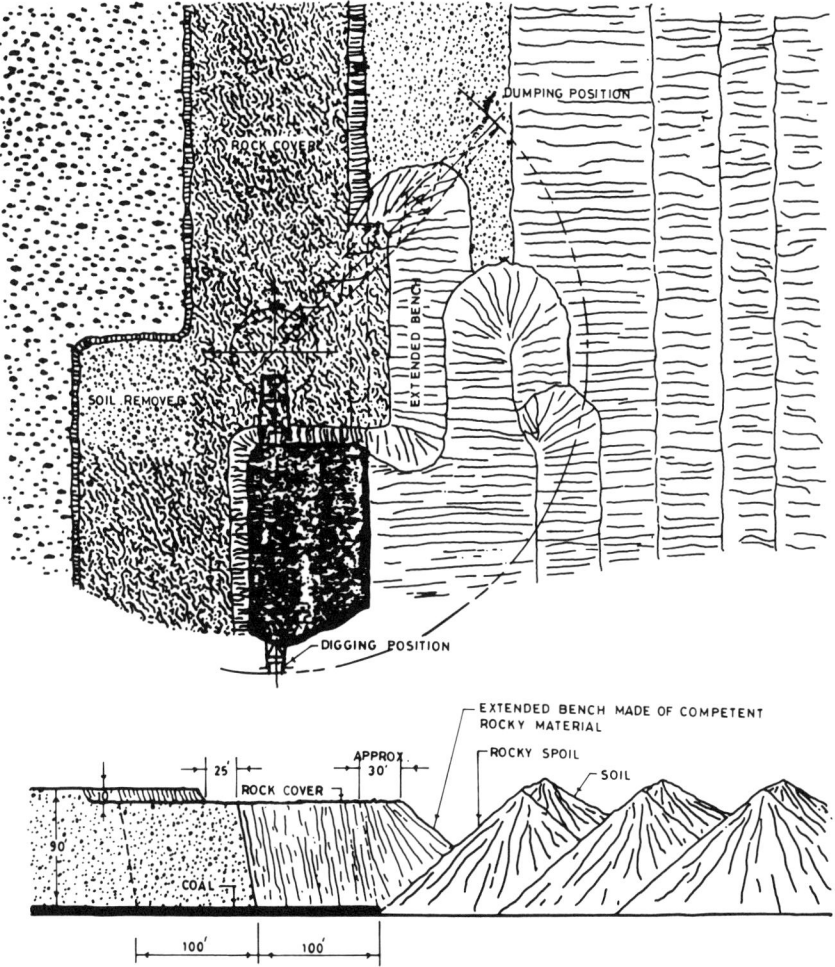

Figure 2.18. A typical method of extending the dragline bench. (From Math Tech, Inc. Evaluation of Current Surface Coal Mining Overburden Handling Techniques, U.S. National Technical Information Service, PB-264 111 (1976).)

material moved in each cycle is increased. This yardage advantage has to be compared with the volume of material to be rehandled in the extended bench method.

Most mines prefer to trade off bucket capacity for boom length. They prefer machines with long booms and moderately sized buckets. Some long-boom machines are capable of handling overburden depths up to 40 m without spoil rehandle. However, machine operators do not like to operate a large machine from the edge of a 40-m highwall, indicating that benching will be used in deep overburden regardless of the range of the dragline.

62 ENVIRONMENTAL IMPACTS OF MINING

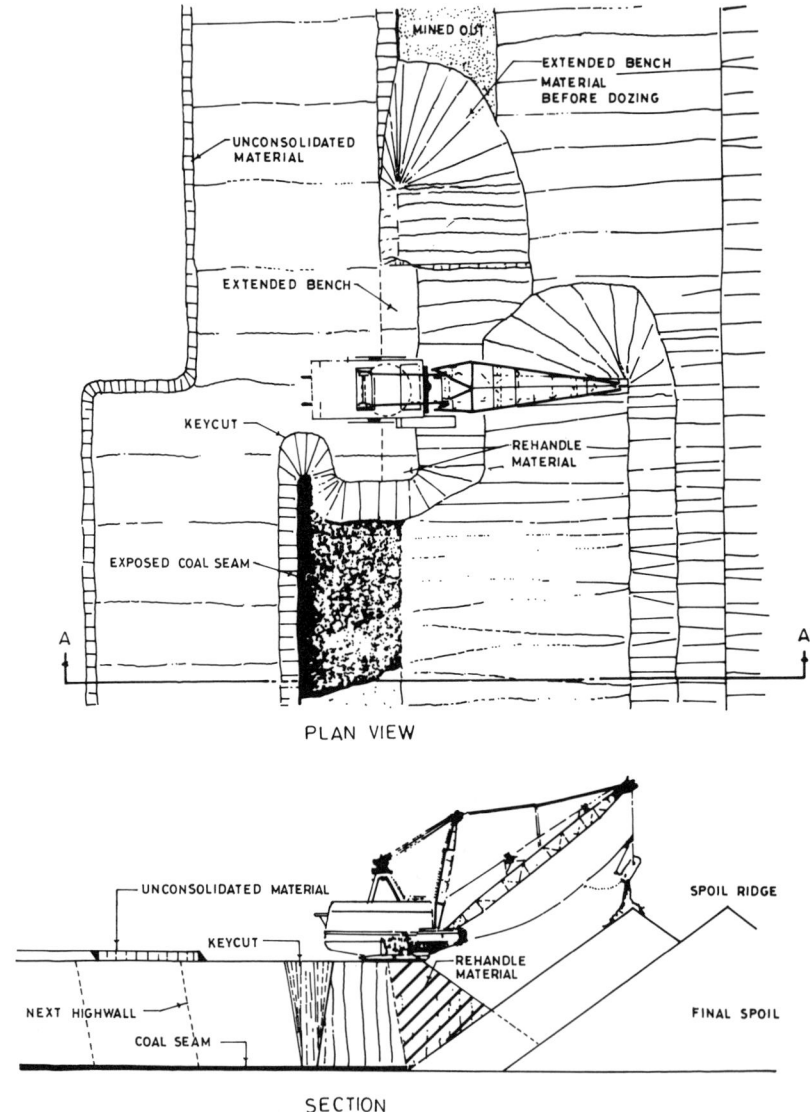

Figure 2.19. Dragline on extended bench. (From Math Tech, Inc. Evaluation of Current Surface Coal Mining Overburden Handling Techniques, U.S. National Technical Information Service, PB-264 111 (1976).)

Two-Pass Extended Bench Method

In extended bench mining, two passes can be made in each pit. This does not require the use of the overhand chopping procedure. The advantage of the two-pass method over the single-pass method is the reduction in rehandle percentages without loss of any productivity, as in overhand chopping. The disadvantages include increased dragline deadheading time, ramp construction, and ramping of the dragline between the two bench levels.

The first pass is made on a very high bench or on the natural ground surface. Top lift key way material is placed against the lower highwall as the lead spoil to form the extended bench for the second pass. When the pit end is reached, the dragline is ramped down to the lower bench, and stripping continues in the direction opposite to the direction of the top lift stripping. The method is illustrated in Figure 2.20.

Pullback Method

The pullback method is an alternative approach to extended benching under similar pit conditions. It has additional use in multiple-seam applications. A typical pullback operation is illustrated in Figure 2.21. The pullback method can utilize a single dragline or two draglines. In the latter case, both machines are assigned a specific part of the operation.

The method involves first either side casting from the original ground surface or from an advance bench. Figure 2.21 shows the last cut being spoiled for the advance bench situation in the pullback method. In the pullback method, the dragline is unable to spoil all the overburden clear of the coal. As a result, this material rills over the coal seam at the toe of the spoil pile. This material has to be rehandled to completely uncover the coal.

The rehandle is performed with the dragline sitting on a prepared pad on the spoil pile itself. This pad is prepared using dozers or by the dragline removing the spoil peak using chop down. Final surface leveling is performed by a dozer. Figure 2.21 illustrates the pullback operation. The spoil pile is dug back away from the highwall and spoiled behind the dragline on the top of previous spoil.

The spoil pile can rill over and cover the coal to a greater or lesser extent depending on the relative dragline and pit geometry. The rehandle can be theoretically calculated as the area of the spoil pile above a line intersecting the coal toe at the spoil pile, as shown in Figure 2.21. The mutual area, although it is removed in the pullback operation, is not rehandled, as it need not be removed during the initial spoiling operation.

The theoretical rehandle is not necessarily the most desirable from an economic standpoint. When the spoil pile slope has been pulled back to the top edge of coal, further rehandle exposes proportionally less coal. There is a break-even point or an economic limit to the extent of uncovering of coal.

The main advantage of this method is to enable a dragline which has a limited operating radius to handle overburden covers of greater depth than would normally be feasible.

The pullback method can be performed with either one or two draglines. With a single dragline, the machine has to move periodically across to the spoil pile, either around the pit or on a section of extended bench.

64 ENVIRONMENTAL IMPACTS OF MINING

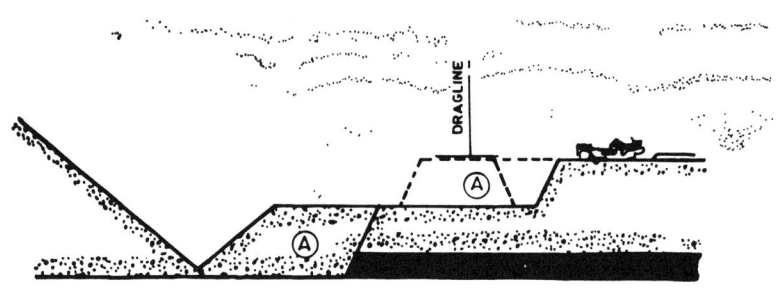

PASS NO. 1

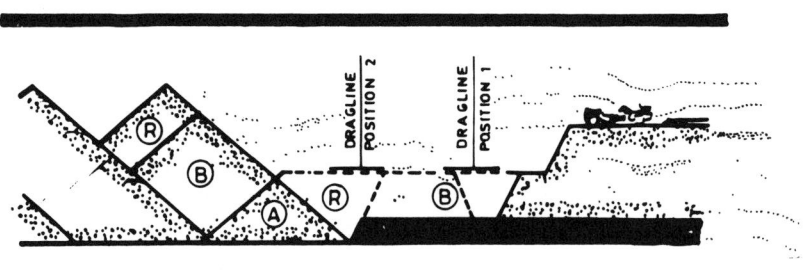

PASS NO. 2

Figure 2.20. Section views showing a two-pass extended bench method. (From Math Tech, Inc. Evaluation of Current Surface Coal Mining Overburden Handling Techniques, U.S. National Technical Information Service, PB-264 111 (1976).)

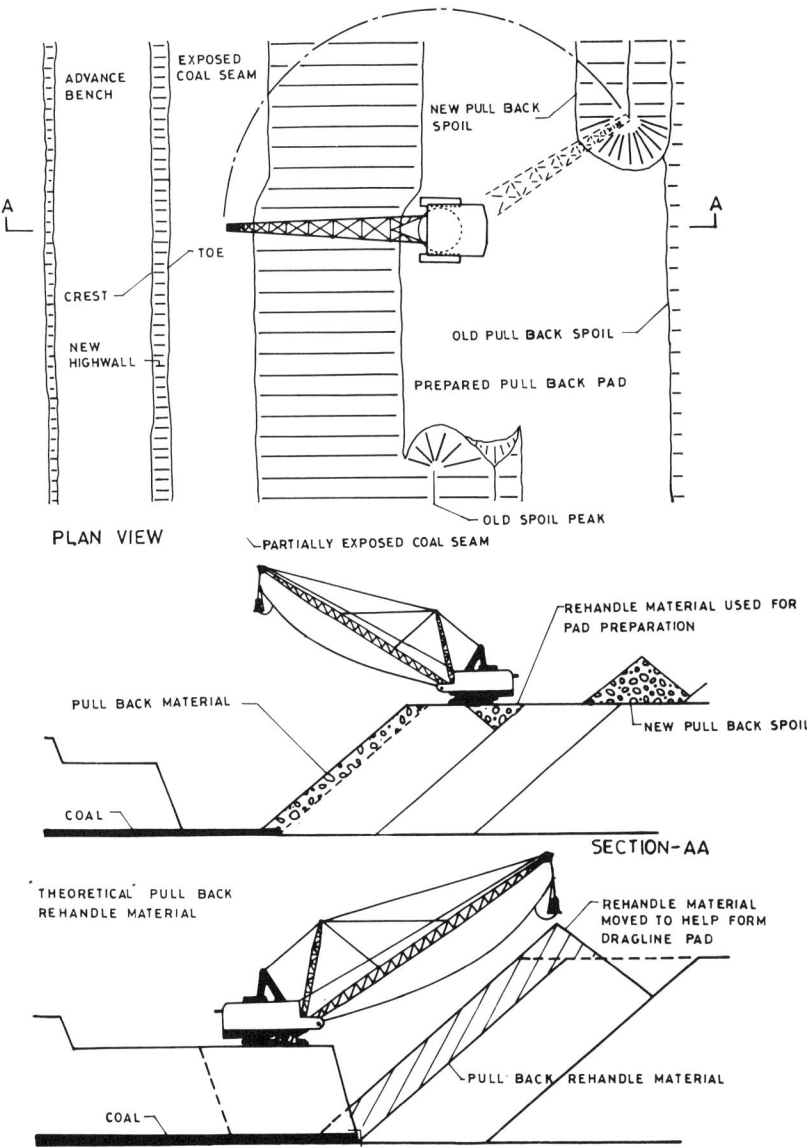

Figure 2.21. Dragline using the pullback method. (From Math Tech, Inc. Evaluation of Current Surface Coal Mining Overburden Handling Techniques, U.S. National Technical Information Service, PB-264 111 (1976).)

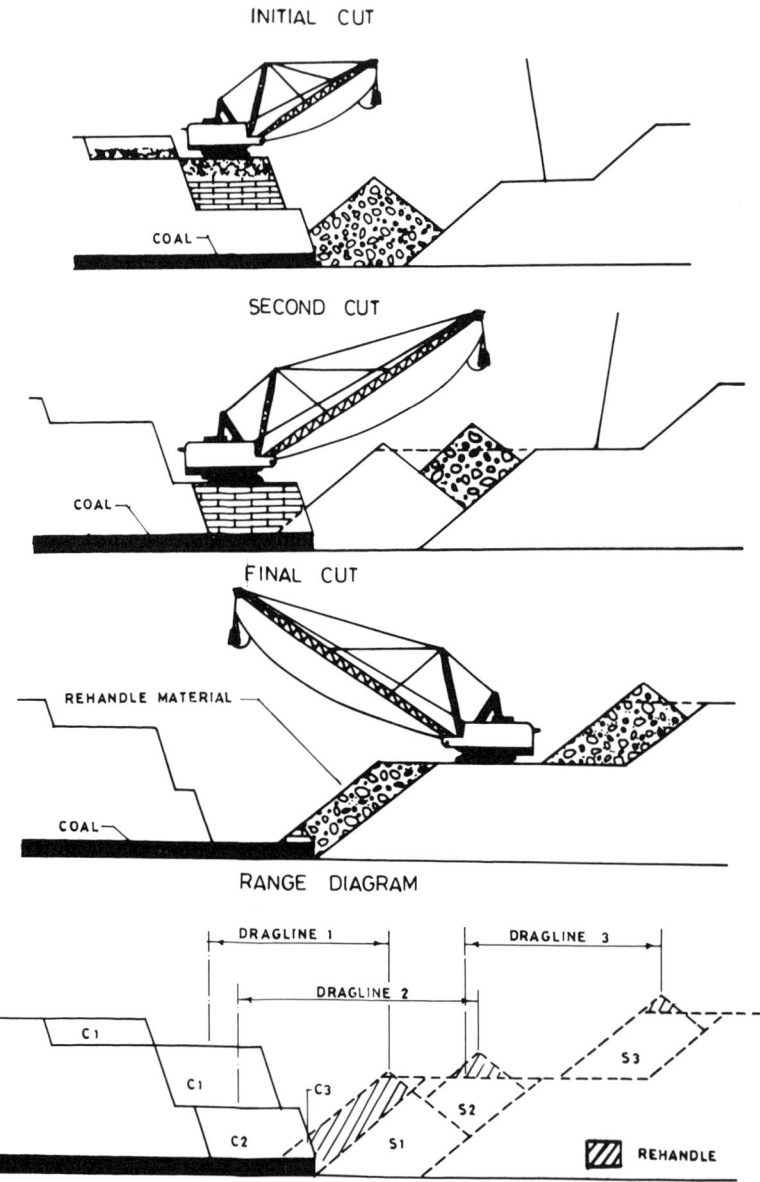

Figure 2.22. Terrace mining. (From Math Tech, Inc. Evaluation of Current Surface Coal Mining Overburden Handling Techniques, U.S. National Technical Information Service, PB-264 111 (1976).)

SURFACE COAL MINING WITH RECLAMATION

The alternative is to use two machines: one permanently situated on the highwall, the other permanently on the spoil pile. The productivity of the two machines must be carefully matched so that they operate as a team.

In both these alternatives, the spoil pile stability is an important problem. Where spoil piles have inherently poor stability, it is inadvisable to place a dragline on the spoil pile. It would be better to use the extended bench method.

TERRACE MINING

This method enables deep overburden depths to be stripped. The operation usually incorporates three or more machines. Because of their height, both highwalls and spoil piles are stepped to maintain slope stability.

Figures 2.22A to C show cross-sections and plans of a typical mining sequence, one machine at a time. Figure 2.22A shows the first dragline starting the stripping sequence. This machine has a shorter boom and, consequently, a larger bucket than the two subsequent machines. Figures 2.22B and 2.22C show the other two machines and their relatively lighter workloads.

This method can be applied to strip greater and greater depths using any number of machines. However, as the overburden depth and number of machines increase, so does the rehandle which can easily become more than 100%.

TANDEM MACHINE SYSTEMS

Tandem machine systems include the use of the dragline in conjunction with dozers, scrapers, front-end loaders, stripping shovels, bucket wheel excavators, and other draglines. In a single-seam mining situation, the tandem system most frequently used is a dozer/dragline combination. Other tandem systems are used in multiple-seam situations and are described in subsequent chapters.

Tandem Dozer/Dragline Stripping

In many deep overburden situations, stripping is carried out using large dozers in tandem with relatively small, crawler-mounted draglines. These draglines typically have lengths of 35 to 45 m and bucket capacities of 7 or 8 m^3. Under normal operating conditions, a machine with a 45-m boom can handle overburden up to about 15 m in depth, using a conventional no-rehandle procedure. But when operating in tandem with a dozer, these draglines can strip overburden 25 to 33 m deep, using the extended bench method.

In the extended bench method, maximum use of the force of gravity is made to move overburden. In the conventional situation using a single dragline, the top lift overburden is dragged down toward the pit, so that the overburden falls into the cut

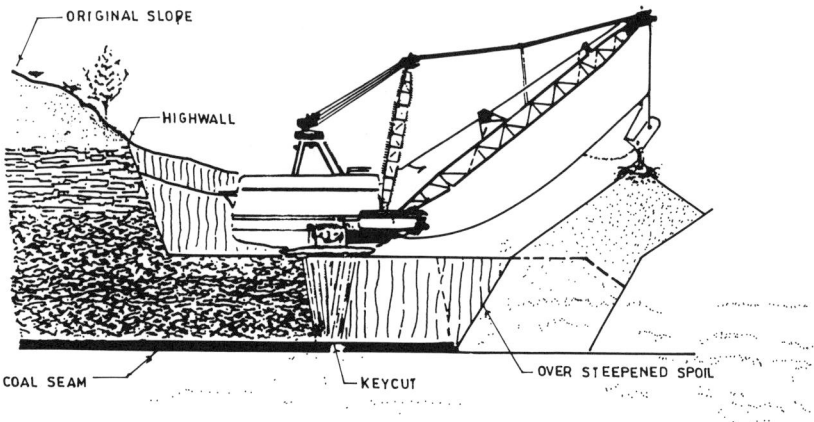

Figure 2.23. Dragline making keycut in second lift overburden. (From Math Tech, Inc. Evaluation of Current Surface Coal Mining Overburden Handling Techniques, U.S. National Technical Information Service, PB-264 111 (1976).)

rather than interfering with the dragline bucket. This gravity principle can also be applied by using dozers to push the top lift overburden into the pit. So the dozers make the bench for the dragline, which then strips the second lift.

The stripping procedure for the top lift is shown in Figure 2.23. After the overburden is blasted, the material is pushed over the edge of the existing highwall into the open cut. As the depth of the overburden excavated by the dozer increases, the extended bench is automatically prepared by the dozer. The extended bench is essential as the range of the dragline is limited. The dozers compact the bench material fairly tightly. This allows for extreme oversteepening of the spoil during the dragline rehandle phase in order to minimize spoil rehandle percentage.

The dragline strips the second lift. The machine is positioned directly over the keycut outerline as shown in Figure 2.23. The keycut is made at some distance from the highwall. To complete the digout, the dragline is moved out to the extended bench. During the rehandle phase, the spoil can be cut at an angle as high as 80°. The spoil can stand this angle, as it has been compacted. Because of this steep spoil angle, the rehandling is reduced from 15% to 5%. Figure 2.24 shows views of dragline on extended bench stripping second lift overburden.

The use of dozers in the top lift offers economic advantages. When the dozers operate level or downslope and push over fairly short distances, they are less costly than the small dragline. This may not be true for large draglines.

The important decision variables are the bench depth and the pit width. As the bench is deepened, the dozer push-distance increases. At some point, the dozer has to push uphill over long distances, a procedure which can be costly. The decision involved is to determine at which point the dozer cost exceeds the dragline cost. This is explained in Figure 2.24. For shallow benches, the dozer costs are lower than those

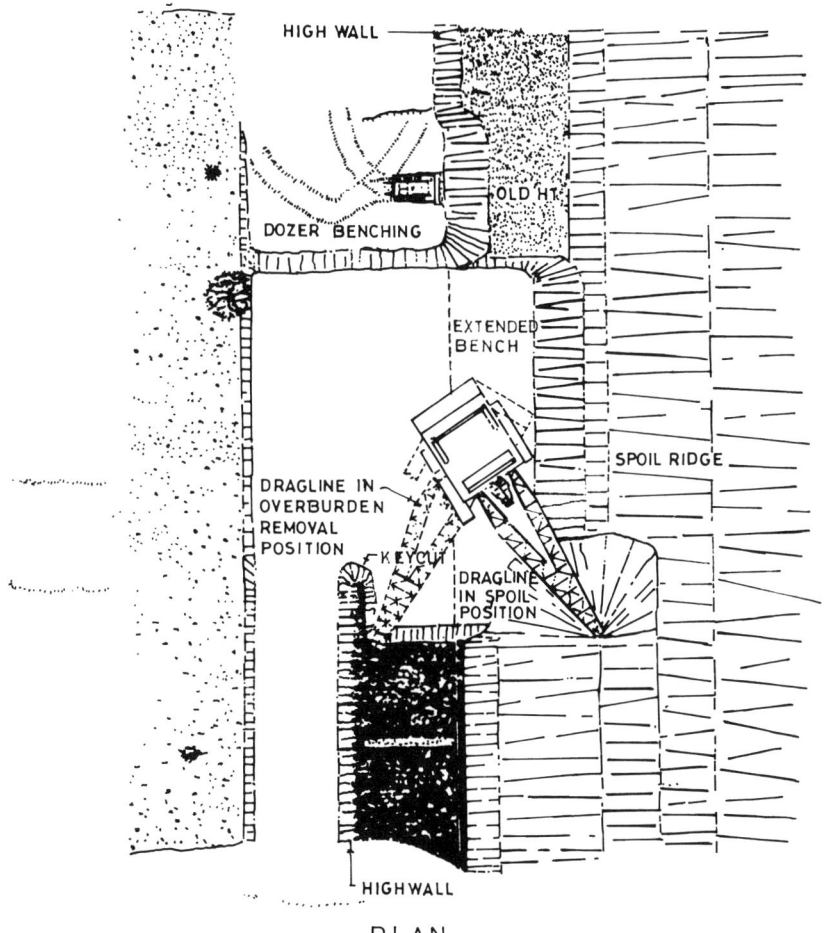

Figure 2.24. Dragline on extended bench stripping second lift overburden. (From Math Tech, Inc. Evaluation of Current Surface Coal Mining Overburden Handling Techniques, U.S. National Technical Information Service, PB-264 111 (1976).)

for the dragline as push distances are short and considerable amounts of overburden are moved downslope. At certain bench depths, dozer costs equal dragline costs, and this depth is the optimal depth for the top bench.

Similar reasoning can be used to determine optimal pit width. The problem is complicated as pit width, stripping costs, spoil-grading costs, and rehandle percentages are interrelated.

There are alternative methods for making the keycut in the lower bench. Economic and safety factors determine the choice. When the dragline is used to make the keycut, the bench must be wider than the pit. Then the average dozer push distances are longer than when the keycut is made adjacent to the upper highwall. This results in a bench width equal to the pit width. But if the keycut is made adjacent to the upper highwall, and dozers and front-end loaders make the keycut, the costs of this procedure may be higher than for having the dragline make the keycut. If the keycut is made adjacent to the upper highwall, then the highwall will not be benched. In a deep overburden situation, this may be undesirable.

Single-Seam Stripping with Selective Spoil Placement

Selective spoil placement usually has one or more of the following four objectives:

- burial of acid, alkaline, or rocky spoil material
- placement of chemically suitable, unconsolidated material on spoil surfaces
- placement of materials high in soluble minerals above the water table
- placement of competent spoil materials on the pit floor

A dragline can place the spoil selectively in one or more of the following ways:

- use of lead and lag principles in spoil placement
- use of the side bench method to place side bench material on the top of the spoil piles
- in extended bench situations, advantageous use of rehandle material

Moderate Overburden Depth, One Lift, Single Pass

Two kinds of spoil materials may be undesirable. Alkaline shales and clays would become impermeable if placed as spoil surfaces, thereby impeding revegetation of graded spoils. These types of materials should be buried in spoils if possible. Materials which are high in soluble minerals or trace elements should not be placed below the water table. Where the coal or overburden are below the water table, such materials should not be placed on or near the pit floor. If a given overburden stratum is clayey in texture, high in soluble minerals and exchangeable sodium, and the coal seams are below the water table, then the material should be placed neither too high nor too low in the spoil profile.

If soft shales or clays occur in the overburden strata immediately overlying the coal, such material should be buried in the spoil profile. If placement of the material on the pit floor is permissible, burial may not be difficult. When the overburden strata are excavated using the layer loading technique, the overburden materials immediately overlying the coal can be buried by leading the spoil in the direction of the stripping advance and placing it on or near the pit floor, ahead of the main spoil pile.

In some cases, most of the overburden below the top 3 m or so is clayey in texture and high in sodium. The topsoil is salvaged and replaced in order to bury the undesirable material. The side bench method is desirable under these conditions.

Moderate Overburden Depth, Two Lifts, One Pass

The backwall is a pile of competent spoil materials which are placed on the pit floor to buttress other, less competent materials subsequently cast onto the top of the backwall. In side bench situations, construction of the backwall is simple. The dragline bench is usually cut deep enough so that all overburden strata below the elevation of the bench are competent. The bench surface is laid out at the interface of unconsolidated and consolidated materials. The keycut is made following the conventional practice, and this is the first activity on a given digout. Since all keycut materials are competent, they are used to build the backwall. This is done by leading the spoil at a boom swing angle of about 120°, so that the keycut spoils are placed on the pit floor, near the highwall, ahead of the remaining spoil materials which are cast behind and on top of the backwall. Sometimes, the backwall is called the lead spoil.

Acid material can also be buried using the lead spoil method. Acid materials can be rendered harmless by burying them, as both air and water are needed for acid production. Usually, the acid materials consist of slates and dark shales immediately overlying the coal. Where these do not comprise a large proportion of the overburden, they can be buried using the lead spoil principle. Placement of side bench spoil on tops of spoil piles gives additional insurance against acid production.

Deep Overburden, Two Lifts, One Pass
(The Extended Bench Procedure)

In the extended bench method, successful spoil segregation and burial can be achieved if desirable spoil material with which to bury undesirable material is available. The lead spoil and side bench procedures can also be used to cover undesirable spoil materials with the rehandle material. The rehandle material should have desirable characteristics from a reclamation standpoint. Since the keycut material is used to prepare the extended bench, it may not always have desirable reclamation characteristics. This means that consolidated materials high in the bank are used to extend the bench. During the rehandle phase, these materials are placed on the surface of the spoil piles. A practical problem may arise in that during grading a dozer may cut through the surface material. A second problem may arise because the use of top, unconsolidated material in the extended bench can cause stability problems.

Multiple Seam Stripping with Draglines

Different approaches can be adopted for mining multiple seams with draglines. With a given set of equipment and physical conditions, the choice of the mining method is dependent on the following factors:

- relative depths of the overburden and the interburden
- dumping radius and height of the dragline
- thickness of the coal seams being mined

One major difference between the single-seam mining system and the multiple-seam system is that the dumping height of the dragline is often important for the latter system and not for the former. An additional difference is that segregation and burial of undesirable materials is more frequently necessary in multiple-seam mining than in single-seam mining.

Multiple-seam mining requires multiple lifts. While stripping two coal seams, removal of the overburden requires at least one lift, and the removal of the interburden requires another lift.

Multiple-seam mining systems may be single or tandem machine systems. In subsequent sections, the use of tandem machines in multiple-seam situations will be viewed as advantageous than in single-seam situations.

Single Dragline, Two-Seams, Nonselective Spoil Placement

A single dragline system stripping two coal seams is frequently used.

Moderate Overburden and Interburden Depths

This method involves two lifts — one for overburden and the other for interburden and two passes. No rehandling of spoil is needed. The method is shown in Figure 2.25. On the first pass, the dragline is placed on the natural ground surface or a very shallow bench. The overburden is stripped and placed as in the conventional method. The top coal seam is mined closely, following the stripping operations. When the end of the pit is reached, the dragline is deadheaded back to the other end of the pit and ramped down to a position on top of the interburden. However, if the top coal seam can be mined closely in pace with the stripping of the overburden, then deadheading may not be needed. The machine can be ramped down after reaching the end of the pit and placed on top of the interburden. The bottom coal seam is then stripped in the direction opposite to that of the first pass.

The effective range of the dragline is increased while stripping the interburden because of the highwall slope angle. The interburden is stripped, and spoil is placed as in the conventional procedure. At the end of the pit, the dragline is ramped up to the top of the overburden, and a new top pass begins.

The same decision variables which are applicable to the single-seam situation are applicable in the multiple-seam situation. The only exception is that in multiple-seam

SURFACE COAL MINING WITH RECLAMATION

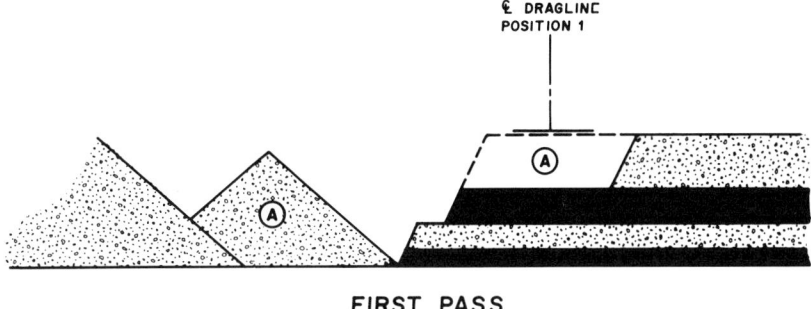

FIRST PASS

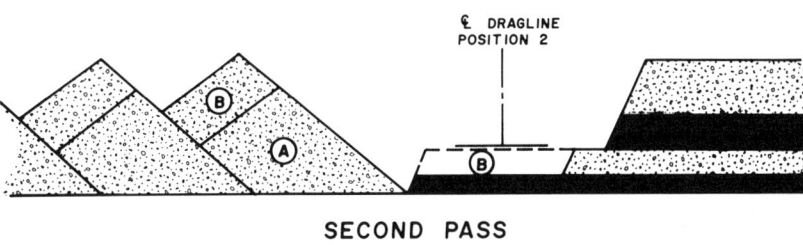

SECOND PASS

Figure 2.25. No-rehandle method of stripping two coal seams. (From Math Tech, Inc. Evaluation of Current Surface Coal Mining Overburden Handling Techniques, U.S. National Technical Information Service, PB-264 111 (1976).)

mining it is difficult to control the depth of the bench on the second pass. The depth of the second pass is equal to the combined depth of the overburden and the top coal seam. This has an important implication, in that while stripping the interburden, the dragline should have sufficient dumping height so that all interburden material can be spoiled from a dragline position on the interburden. Additionally, the dragline should have sufficient range so that an extended bench is not needed in the second pass.

One of the disadvantages of this method is a high hoist height in the second pass. A second disadvantage is that in some cases final spoil piles may be very high, which may cause spoil slides. The interburden material is placed on top of the spoil piles. This may not be always desirable if the interburden material is not environmentally desirable. It may be costly or difficult to bury this material.

The Extended Bench Method for Two Seams

In a two-seam situation, when the dumping radius of the machine is short or where the overburden is thick, the use of the extended bench method can be efficacious.

A typical operating procedure is shown in Figure 2.26. The first pass on overburden is conventional. Some spoil material is cast close to the lower highwall to extend the second pass bench, and the swing angles are usually greater than 90° in these spoil casting activities, thereby reducing dragline productivity.

After completion of overburden stripping and mining of the upper coal seam, the dragline is ramped down to a position on the interburden, and the interburden is stripped with conventional extended bench procedures. The rehandle material is excavated as the last activity of a given digout.

The extended bench method has both advantages and disadvantages when compared with the conventional two-seam no-rehandle stripping method. One advantage is the extension of the effective dragline range so that deeper overburden or interburden can be stripped. In the extended bench method, the interburden spoil material can be buried by the rehandle material. This is an important advantage from the environmental control standpoint. The extended bench method also improves distribution of the spoil.

The disadvantages include requirements of spoil rehandle and high bucket hoist heights on the second pass.

The One-Pass Extended Bench Method for Two-Seam Stripping

This method is similar to the single-pass extended bench method in single-seam mining. With multiple-seam situations, the dragline is always placed over the interburden and never over the overburden. All overburden is stripped by overhand chopping.

The method is illustrated in Figure 2.27. At first, a bench is established on the interburden. The keycut is made in the interburden by a dragline, but keycut spoil is led from the backwall. The interburden material has to be competent. Next, the surface overburden material (#2 in Figure 2.27) is removed by overhand chopping and is used to extend the dragline bench. The reason for using this surface material in the extended bench is that it will be utilized to bury the interburden material. Subsequently, the remaining overburden, the interburden, and the bench extension are removed by the dragline. The material of the extended bench is then placed on top of the spoil pile.

The main advantage of this one-pass procedure is that dragline walking times, which are unproductive, are reduced. Since a backwall is constructed with competent material, the extended bench can be built with environmentally desirable, unconsolidated material.

The disadvantage of the method includes the use of overhand chopping. The one-pass method becomes more attractive as the overburden depth decreases relative to the interburden depth. For example, the method may work efficiently when the overburden depth is 14 m and interburden height is 25 m. However, if the overburden depth is 25 m and the interburden depth is 14 m, the method may become undesirable for the following reasons:

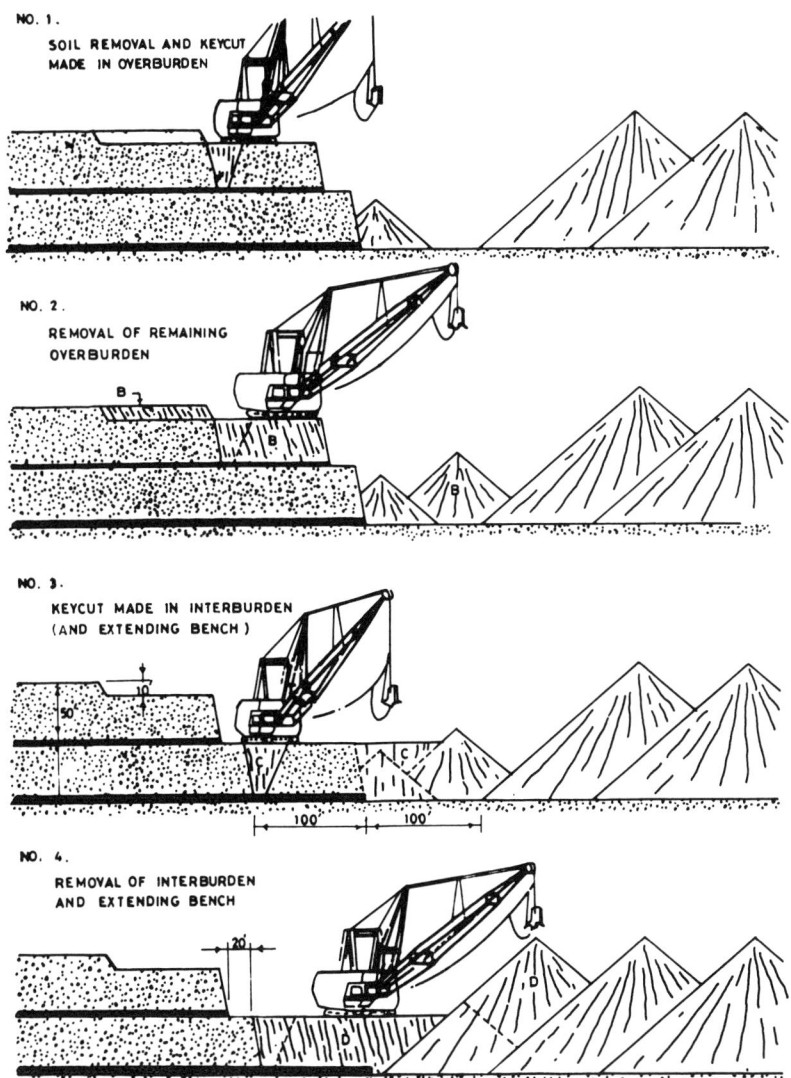

Figure 2.26. Stripping of two coal seams by a single dragline using an extended bench method. (From Math Tech, Inc. Evaluation of Current Surface Coal Mining Overburden Handling Techniques, U.S. National Technical Information Service, PB-264 111 (1976).)

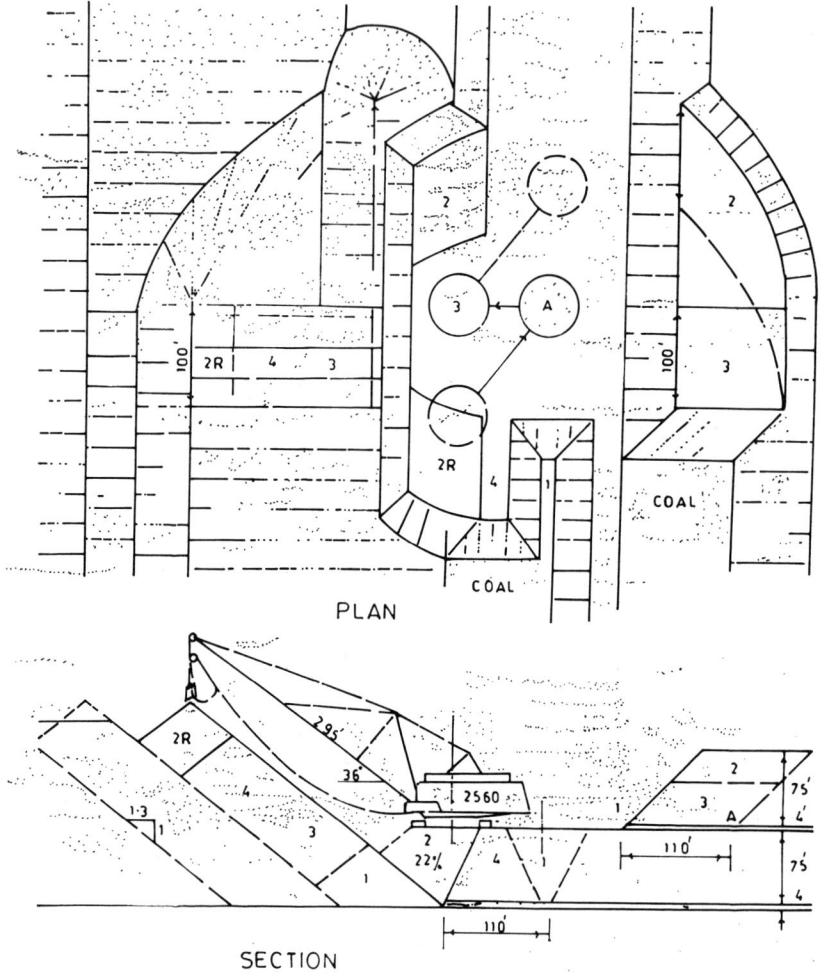

Figure 2.27. A one-pass extended bench method of stripping two coal seams. (From Math Tech, Inc. Evaluation of Current Surface Coal Mining Overburden Handling Techniques, U.S. National Technical Information Service, PB-264 111 (1976).)

1. The dragline may not have sufficient dumping height to spoil the interburden when it works from a bench which is 25 m below the ground surface.
2. Even if the machine has sufficient height, it may not be desirable to remove overburden by chopping from a bench which is 25 m below the ground surface.

Under favorable conditions, the method works excellently, from both production and reclamation standpoints. The method is generally applicable to situations where the average overburden depth is less than the average interburden depth.

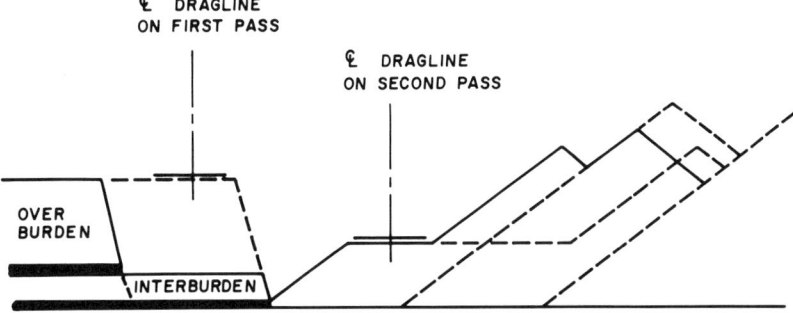

Figure 2.28. Bench is spoil pile from which the lower of two seams is stripped. (From Math Tech, Inc. Evaluation of Current Surface Coal Mining Overburden Handling Techniques, U.S. National Technical Information Service, PB-264 111 (1976).)

The Two-Pass Spoil Side Method

The multiple-seam mining methods which have been discussed previously are feasible only when the dumping height of the dragline is large enough so that all interburden material can be placed on the top of the spoiled overburden material from a dragline operating on top of the interburden. In some situations, the dragline dumping height may not be sufficient to achieve the placement of interburden material over the overburden material.

The solution to this problem is to increase the effective dumping height of the existing dragline in the second pass by raising the height of the second pass bench. There are two possible ways to achieve this.

In the first method, a high bench is constructed over the spoil pile of the stripped overburden material. This is illustrated in Figure 2.28. The spoil pile can sometimes be oversteepened. Additionally, the bench can be extended toward the highwall side if the range of the dragline is not sufficiently long.

In the first pass, the dragline is positioned on the overburden and the overburden is stripped using conventional procedures. When the end of the pit is reached, a ramp is constructed and the dragline is moved across the spoil side of the pit and used to cut the initial bench on the spoil pile. Alternatively, dozers can be used to prepare the bench. After this upper coal seam is mined out, the machine is moved onto the bench on the spoil pile to strip the interburden.

When the dragline is stripping the interburden, its boom is typically at an angle of 60 to 90° to the highwall in contrast to the angle in the conventional practice, where the boom is roughly parallel to the highwall during the digging portion of the cycle. The interburden is excavated by chopping downward and dragging the bucket toward the dragline tub. The boom is then swung through an angle of 120 to 150°, to spoil the interburden material. Some of the interburden spoil can be used to build the bench for the next digout.

The primary advantage lies in the fact that the method increases the effective dumping height of the dragline so as to enable stripping of two coal seams rather than one. Additionally, in contrast to previously described methods, bucket hoist heights in spoiling of interburden materials are reduced because the bench is raised above the top of the interburden.

The disadvantages include the need for chopping, the relatively large swing angle required for placement of interburden spoil, and the need to construct a bench in the spoil.

There are procedures to reduce or eliminate the need for chopping. One common procedure used to improve productivity of the dragline in chopping is to fragment the interburden by blasting it very hard. A second method is to delay the shots from the spoil side of the interburden inward, thereby using the explosive to move the interburden material away from the lower highwall. This creates a pocket in the interburden. The dragline bucket is lowered into this pocket to dig out the interburden. If rehandle is not otherwise needed, this blasting makes it necessary, as the interburden material will be cast by the blast into the trough between the lower highwall and the spoil pile. In most instances, rehandle is needed and the blasting technique does not create any additional rehandle.

A second procedure used to reduce the need for chopping involves keycutting the interburden prior to the main excavation activity. The dragline bucket enters the keycut to begin the digout. If the keycut is wide enough, it virtually eliminates the chopping method. However, as the keycut is narrower at the bottom than at the top, chopping may be necessary in the lower corner of the highwall. The keycut in the interburden can be made using dozers or loaders or the dragline.

Procedures for preparing the keycut using a dragline are as follows. In the second pass, the dragline is positioned on the interburden and the keycut is made for several digouts. A ramp is then constructed across the pit, and the dragline is moved over to the bench in the spoil pile and back along the pit to the starting point of the keycut. The interburden is then removed as described earlier. The productivity gains resulting from reduction or elimination of chopping must be weighed against the dragline walking time.

Elevated Bench Method

The elevated bench method offers an alternative for eliminating underbench chopping and large boom swing angles needed to dispose of interburden materials from a position on a bench spoil. The method is illustrated in Figure 2.29. It shows a one-pass extended bench method where the dragline is always placed on the interburden. Some overburden is spoiled on top of the interburden to raise the dragline bench.

The major advantage of the method is the elimination of interburden chopping and reduction of boom swing angles during the spoiling of the interburden. The main disadvantages include a spoil rehandling requirement and the need for overhand

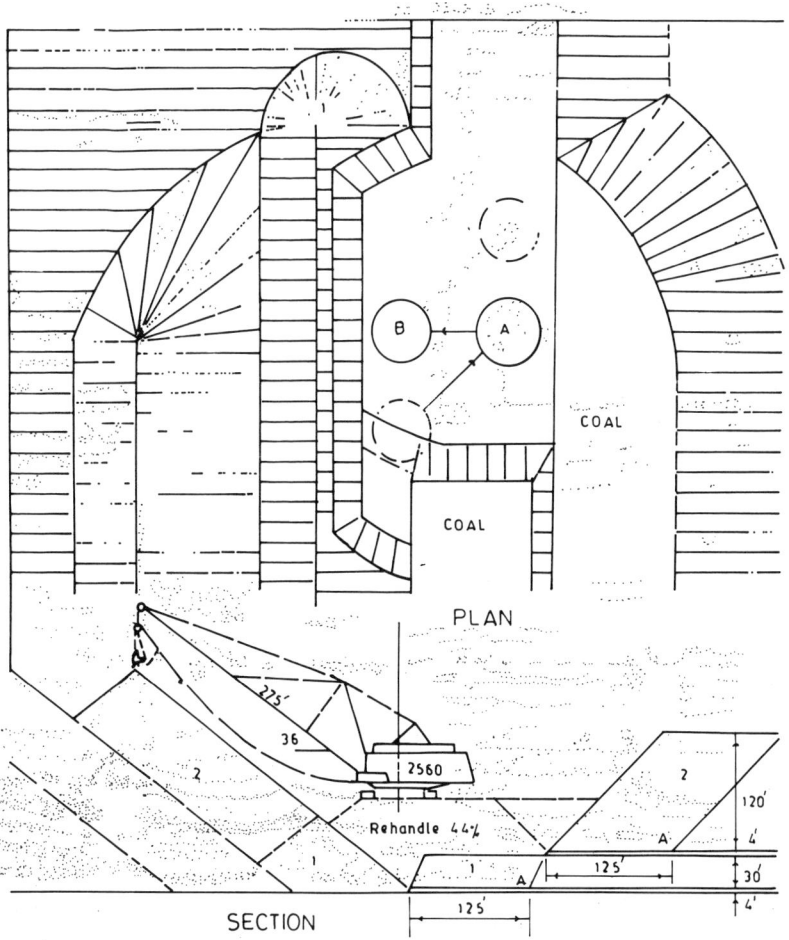

Figure 2.29. A method for increasing the effective dumping height of a dragline. (From Math Tech, Inc. Evaluation of Current Surface Coal Mining Overburden Handling Techniques, U.S. National Technical Information Service, PB-264 111 (1976).)

chopping of the overburden. The actual rehandle percentage is smaller since some of the elevated bench material will have been moved by gravity. The spoil rehandle percentage decreases as the height of the elevated bench decreases.

In comparing the spoil bench and elevated bench methods, the tradeoffs can be obvious. In the elevated bench method, chopping of the interburden and large swing angles are involved. In the bench on the spoil method, rehandling of spoil and overhand chopping of the overburden are major disadvantages.

80 ENVIRONMENTAL IMPACTS OF MINING

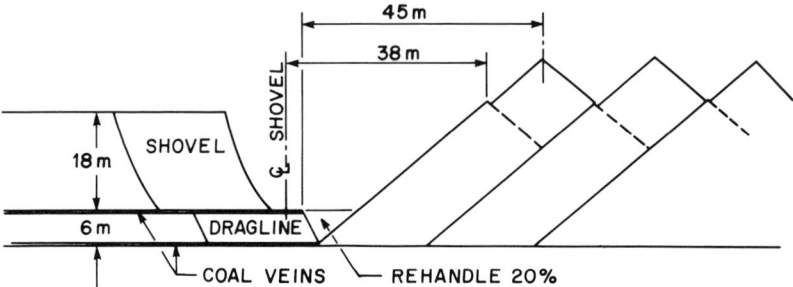

Figure 2.30. Two-seam tandem dragline/shovel method. (From Math Tech, Inc. Evaluation of Current Surface Coal Mining Overburden Handling Techniques, U.S. National Technical Information Service, PB-264 111 (1976).)

Tandem Machine, Two-Seam Condition, Nonselective Spoil Placement

In two-seam situations when tandem machines are used, one machine strips the overburden and the second machine strips the interburden. The tandem machine combinations are dragline/shovel, dragline/dragline, and shovel/shovel. A major advantage of the tandem machine subsystem is that each of the stripping machines can be tailored for a specific task. In most multiple-seam situations, the overburden depth is greater than the interburden thickness. The stripping capacity required for overburden exceeds that for the interburden. When stripping of the interburden is done by a machine needs more range and dump height than the overburden machine. Tandem machine system offers highest efficiency for making these requirements.

The Tandem Shovel-Dragline System

The system is shown in Figure 2.30. The overburden depth is 20 m with interburden 6-m thick. Overburden is removed by a stripping shovel which has a 45-m boom and a 45-m3 bucket. The interburden is stripped by a dragline with a 55-m-long boom and a 10-m^3 bucket. The range of the dragline exceeds that of the shovel so that the interburden can be spoiled beyond the ridge of the shovel spoils. The ratio of the shovel and dragline bucket capacities is 3.3:1, roughly the same ratio as the ratio of overburden depth to interburden depth. The rate of stripping advance for the two machines will therefore be approximately the same.

As explained earlier the main advantage of this tandem system over a single-machine system is economic. A single shovel or a dragline suitable to strip both the seams would require a 55-m-long boom and 45-m^3 bucket. The capital cost of such a machine would exceed the combined capital costs of the smaller shovel and dragline. An additional advantage of the tandem-machine subsystem is related to the rates at which the two coal seams are exposed. In this example, the stripping ratio for the upper seam is 28:1, while the ratio on the lower seam is 12.5:1. If a single

SURFACE COAL MINING WITH RECLAMATION 81

machine is used to strip both seams, the lower seam would be exposed roughly twice as fast as the upper one, on a tonnage basis. This could complicate production scheduling in situations where the coal from both seams must be blended. A tandem machine can achieve a balanced production.

Ramping of the stripping machine between the overburden and interburden is not required when tandem machines are used since each machine works on one level only.

Disadvantages of the tandem system include higher inventory requirements for spare parts and lower availability compared to the single-machine case. Failure of either stripping machine may mean failure of the whole system.

Dozer-Dragline System

When the interburden is relatively deep, all overburden is removed by D-dozers as shown in Figure 2.31. The second pass bench is extended and used by the dragline. After the upper seam has been loaded, the dozers are used to remove some of the interburden to make a level bench for the dragline.

Subsequently, the interburden is stripped by dragline. This can be done either from an extended bench or a bench in spoil. Rehandle will be required.

The method has several advantages. Gravity is utilized to move overburden. The dragline always sits on the same bench level, thereby eliminating the need for ramping. The topsoil, which is the rehandle material, is placed on the top of the spoil pile, burying the interburden spoil.

When the overburden is thicker than the interburden, the dozer and dragline are used in tandem to strip the overburden, and then the dragline is ramped down to the interburden level. All interburden is then stripped by dragline.

Tandem Machine-Multiple Dragline

Figure 2.32 illustrates that two draglines can be simultaneously used to strip two coal seams. One dragline strips the overburden and the other dragline strips the interburden. The capacity of the two machines has to be matched to their stripping requirements so as to maintain their required rate of progress. The leading machine over the overburden has to maintain its lead ahead of the second machine.

The spoil of the leading machine does not abut the second seam. This allows the second machine to spoil the interburden on the pit floor without covering the seam.

Selective Spoil Placement in Two-Seam Condition

In most two-seam dragline situations, spoil rehandle material is the last material placed on the spoil pile during a digout. The rehandle material, if it is topsoil, is suitable for reclamation. Therefore, where rehandle is required for production requirements, segregation and burial of the spoiled interburden may have little effect on operating procedure.

82 ENVIRONMENTAL IMPACTS OF MINING

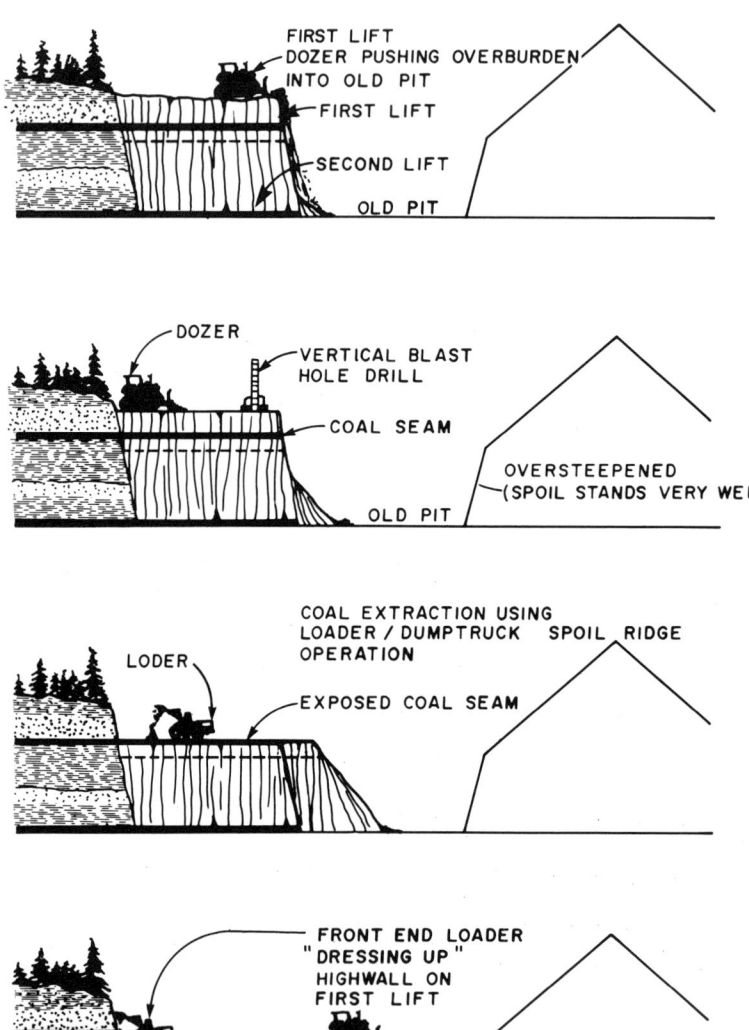

Figure 2.31. Removal of overburden by dozer in tandem dozer/dragline stripping of two seams. (From Math Tech, Inc. Evaluation of Current Surface Coal Mining Overburden Handling Techniques, U.S. National Technical Information Service, PB-264 111 (1976).)

If spoil rehandle is not involved for production reasons, then a requirement for burial of the interburden material may make it necessary. Figure 2.33 illustrates a situation where an extended bench was not required for the range of the dragline.

SURFACE COAL MINING WITH RECLAMATION

MULTIPLE SEAM — TANDEM MACHINES

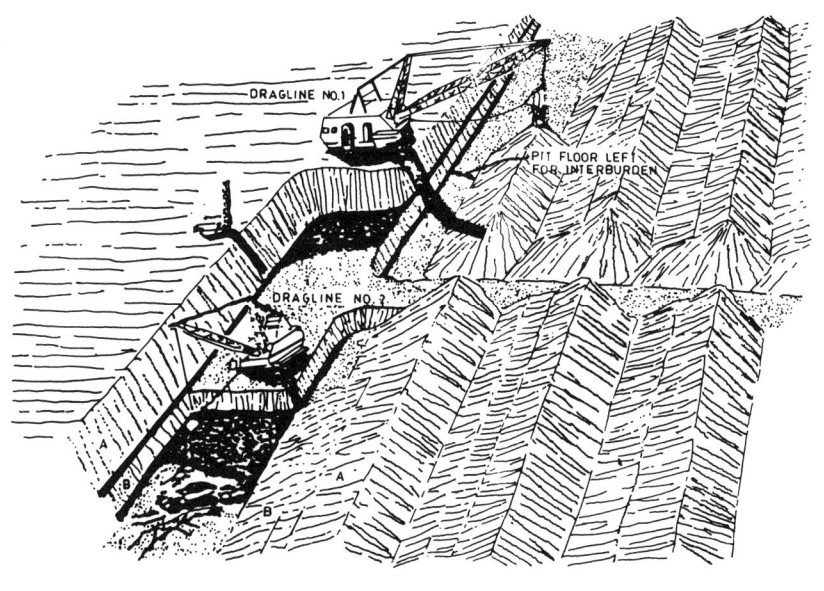

NOTES:

IN THIS CASE, SPOIL OF LEADING MACHINE DOES NOT ABUT 2ND SEAM, ALLOWING 2ND MACHINE TO SPOIL INTERBURDEN ON PIT FLOOR WITHOUT COVERING SEAM.

WHEN LEVELING, OVERBURDEN (A) WILL BE DOZED OVER INTERBURDEN (B) IN SPOIL.

Figure 2.32. Multiple seam — tandem machines. (From Math Tech, Inc. Evaluation of Current Surface Coal Mining Overburden Handling Techniques, U.S. National Technical Information Service, PB-264 111 (1976).)

Nonetheless, in order to satisfy the requirement that the interburden spoil be buried, the rehandle material was used.

Stripping of Unstable Overburden Material

Some surface coal mines are located in swampy lowland areas. Draglines are used for overburden removal. Unstable overburden consists of a thick layer of mud. Conventional dragline procedures cannot be used in such situations in order to avoid spoil slides into the pit.

84   ENVIRONMENTAL IMPACTS OF MINING

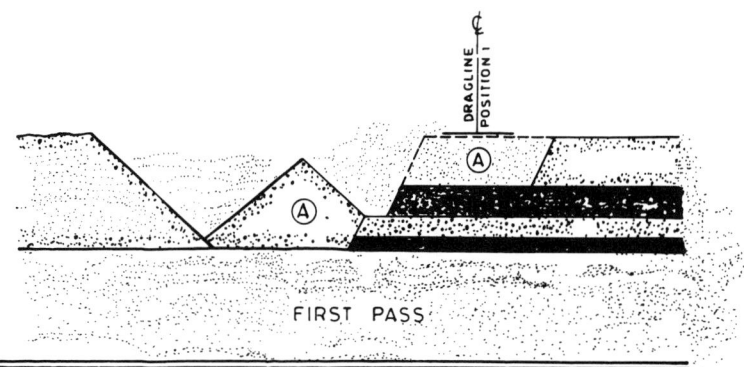

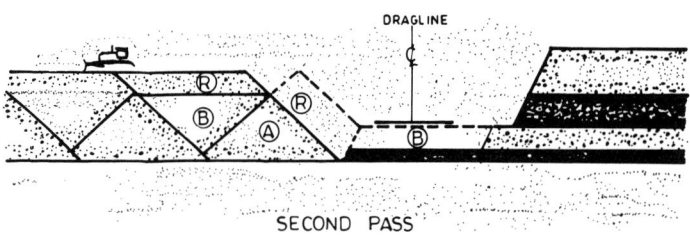

Figure 2.33. Use of rehandle material to bury spoiled interburden material. (From Math Tech, Inc. Evaluation of Current Surface Coal Mining Overburden Handling Techniques, U.S. National Technical Information Service, PB-264 111 (1976).)

The first step in the special mud handling procedure is shown in Figure 2.34. The dragline works on an established bench (not the extended bench) and the boom is turned to the highwall side. The mud in the overburden is scooped out by the dragline. The boom is then swung through a 180° angle and the mud is placed in the vee between first and the second spoil piles that are adjacent to the pit.

After the excavation of the mud hole has been completed, the hole must be filled with competent spoil material to force the machine-supporting bench for the next pass. Competent material is excavated from the overburden bank and the boom is turned 90° to the highwall side to deposit the material in the mudhole.

Thereafter, the dragline is moved out to the edge of the extended bench and the bucket is lowered into the vee between the first and second spoil piles, and the mud which has been placed recently is excavated and cast further away from the pit to rest in the vee between the second and third spoil piles. The movement prevents spoil from sliding into the pit.

SURFACE COAL MINING WITH RECLAMATION

STEP 1: DIGGING MUD HOLE & CASTING TO SPOIL

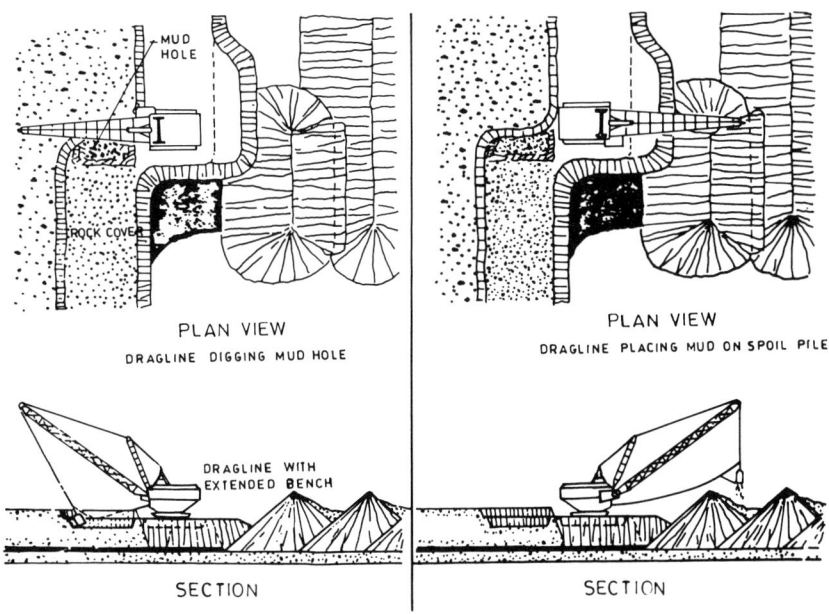

STEP 2: REHANDLING MUD (OPT.)

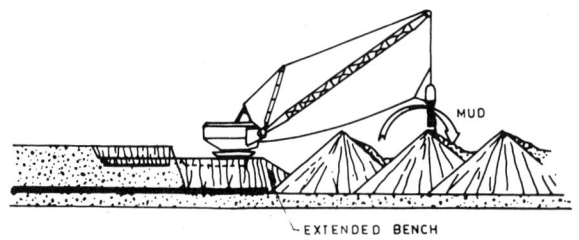

Figure 2.34. Dragline mud-handling procedure. (From Math Tech, Inc. Evaluation of Current Surface Coal Mining Overburden Handling Techniques, U.S. National Technical Information Service, PB-264 111 (1976).)

MULTIPLE SEAM SYSTEMS

Up to this point the discussion has centered primarily on two-seam systems. It remains now to describe a multiple-seam system mining, say five seams at the one operation.

86 ENVIRONMENTAL IMPACTS OF MINING

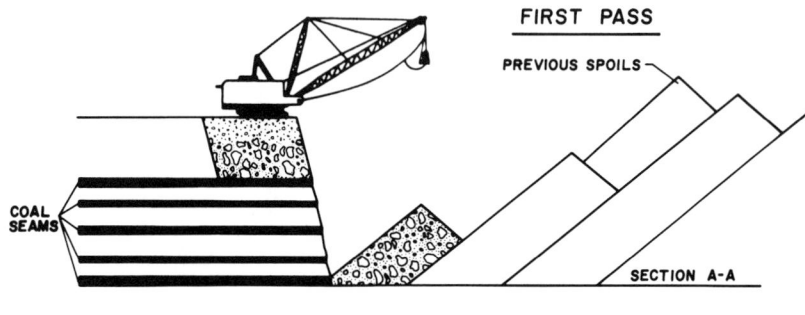

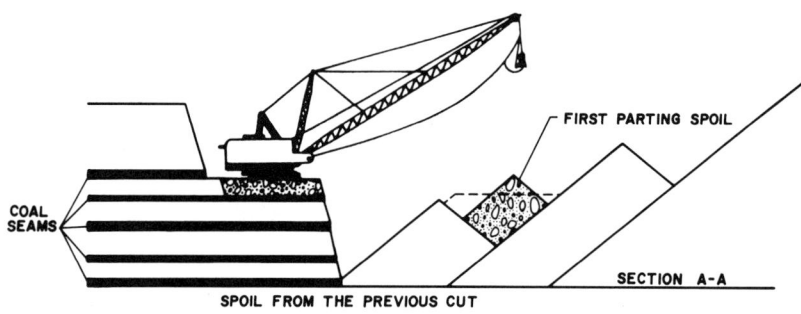

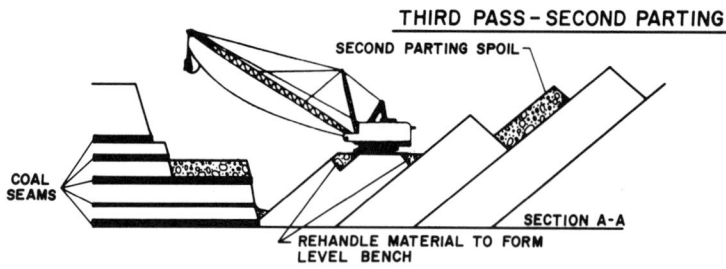

Figure 2.35a–c. Multiple seam mining — five seam operations. (From Bucyrus-Erie, Co., Surface Mining Supervisory Training Program, 1977. With permission.)

Figure 2.35A shows a dragline taking the top cut of a five-seam operation using the simple side casting method. As in all these diagrams the dimensions are representative of actual practice, and specific values are not usually significant. In this case, however, the operation is more unusual. These particular drawings illustrate a 40-m pit width being taken, the top cover being 35 m, the coal seams being in the range 2- to 4-m thick, and the parting thicknesses in the range of 6 to 9 m. The overall operation has a height in excess of 60 m.

SURFACE COAL MINING WITH RECLAMATION

FOURTH PASS – THIRD PARTING

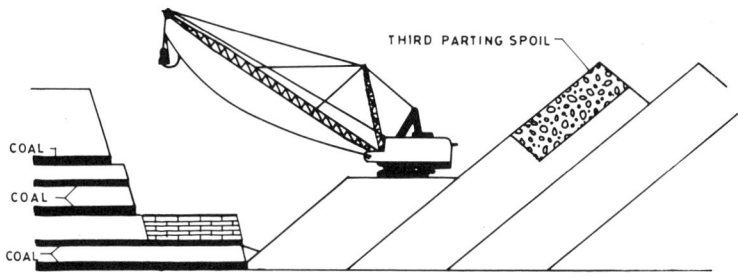

FIFTH PASS — FOURTH PARTING

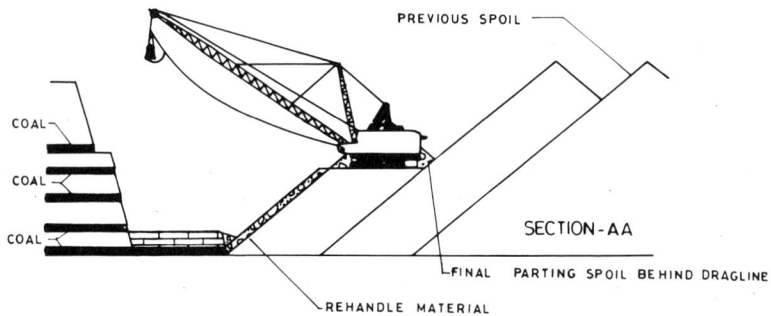

Figure 2.35d, e. Multiple seam mining — five seam operations. (From Bucyrus-Erie, Co., Surface Mining Supervisory Training Program, 1977. With permission.)

The various partings, other than the uppermost, are taken using a second machine situated on the spoil pile produced with the upper overburden. Figure 2.35B is a cross-section illustrating the first parting being taken. Figure 2.35C illustrates the second parting being taken by another machine. In preparation for this operation the spoil pile top had to be leveled. Figure 2.35D shows a pit cross-section illustrating the third parting being dug, while Figure 2.35E shows the final parting and some rehandle being taken.

Spoiling dragline swing angles are approximately 90° for the top cut and 180° for the partings, other than the last parting. Here, the final material, including some rehandle, is spoiled at 90° directly on the working pad beside the dragline.

Rehandle varies with pit geometry and the reach capabilities of the first dragline. The rehandle will be further increased by material sloughing and falling from the

88   ENVIRONMENTAL IMPACTS OF MINING

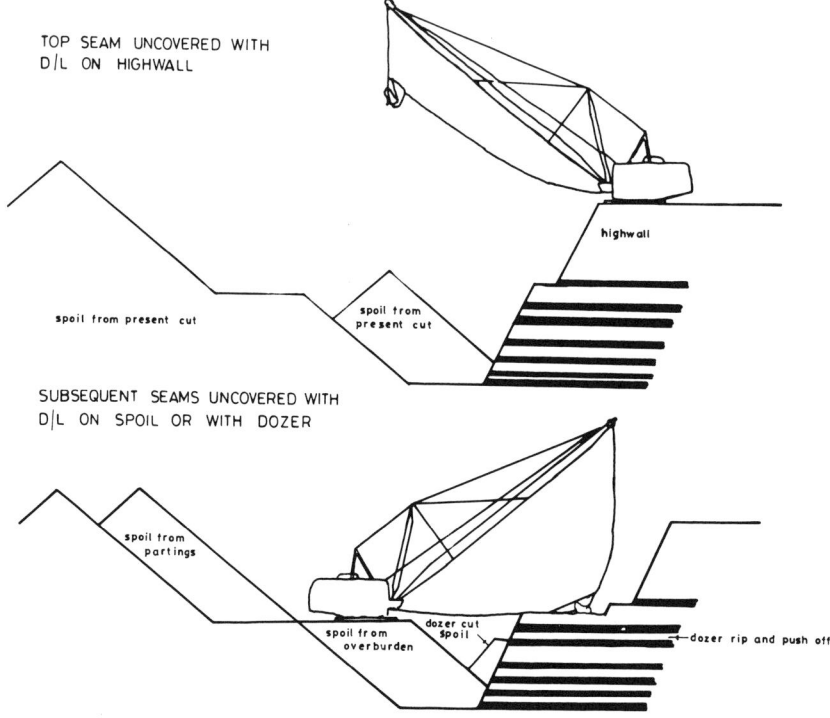

Figure 2.36. Dragline/dozer operation, seven seams. (From Bucyrus-Erie, Co., Surface Mining Supervisory Training Program, 1977. With permission.)

bucket of the draglines while they take the upper partings. The majority of the rehandle is necessary in order to take the lowest seam. The economics then balance on whether the fifth seam is worth mining. However, in certain locations government agencies may require that this seam be taken, even if it is not economical.

For properties having these types of coal resources this method offers a means of strip mining that can handle large bank heights at relatively low overall strip ratios.

The coal seams are successively mined after each stripping operation. Inclines are provided at suitable intervals. For the upper seams these access roads take the form of temporary bridges off the spoil pile across to the coal. These bridges are formed where required by the dragline dumping extra spoil, followed by suitable dozer work. The frequency of access roads should be such that two accesses are always between each dragline stripping operation.

A series of multiple seams with thin interburdens can be stripped by dragline-dozer combination, as shown in Figure 2.36.

The top overburden is extracted by dragline placed on the highwall. The subsequent interburdens are extracted by dragline placed on the spoil or with a dozer.

SURFACE COAL MINING WITH RECLAMATION

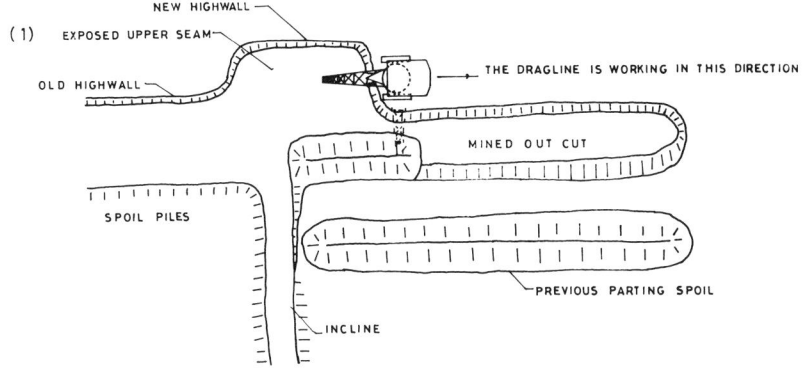

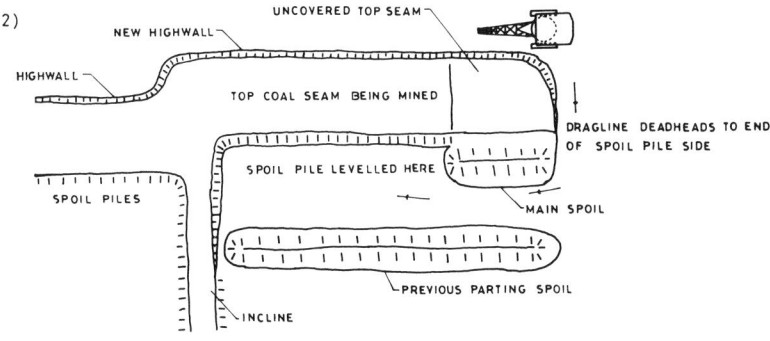

Figure 2.37. Horseshoe mining sequence. (From Bucyrus-Erie, Co., Surface Mining Supervisory Training Program, 1977. With permission.)

HORSESHOE MINING SEQUENCE

The "horseshoe" mining sequence is a method of dragline operation that can be used to mine one end of a two-seam cut with a single machine. The stripping method used would be double pass with either rehandle or no rehandle.

This mining sequence is illustrated in Figures 2.37A through 2.37D. In this case the end of the pit is assumed to have no access, but the method can be equally well used when end access is provided. The dragline move to the spoil pile may require

90 ENVIRONMENTAL IMPACTS OF MINING

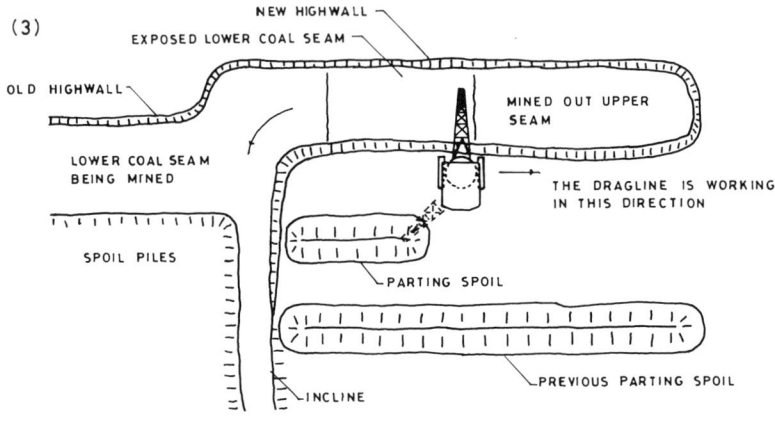

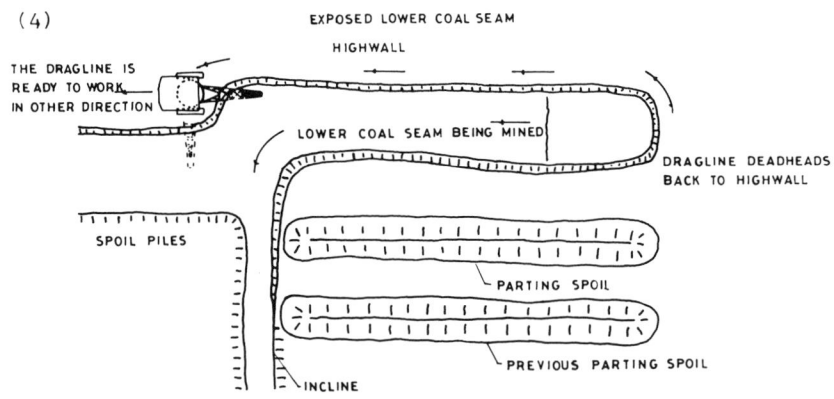

Figure 2.37, continued. (From Bucyrus-Erie, Co., Surface Mining Supervisory Training Program, 1977. With permission.)

ramps, but parting stripping need not be held up until all the upper coal seam is mined because of the alternate haulage route via the end access.[1]

Figure 2.37A shows the first step — stripping of the upper seam. As this operation proceeds the upper coal seam can be mined out. Upon completing this initial cut the dragline is deadheaded around the end of the cut and along the top of the prepared spoil pile to commence the parting stripping (Figure 2.37B).

When the upper seam has been mined out, the parting can then be stripped. Any waiting period here can usually be used for some scheduled maintenance. Figure 2.37C shows the parting spoil being taken. Mining of the lower coal seam follows after the dragline.

Figure 2.37D shows the parting spoiling complete and the dragline deadheaded back to the opposite area of the pit. This section may well be taken by an alternative machine, especially if this is just one end of a larger pit with only one incline. In this case the dragline would not be in a position to repeat the sequence on the other arm.

STEEP SLOPE MINING SYSTEMS

Three mining methods are commonly used in steep slope areas. These are the conventional contour mining method, the haulback method, and the mountaintop removal method. The environmental damage caused by past use of the conventional contour mining method, primarily erosion, sedimentation, landslides, and aesthetic degradation, was severe and extensive. As a result, use of the method is now prohibited by law in all states.

There have been many dramatic changes in steep slope mining practices over the past several years, and there will be more changes in the future. Spoil haulage trucks, until several years ago used only at the largest mines, are now a virtual necessity for mining in compliance with the reclamation laws in all states. Although construction equipment is used for overburden removal and spoil placement at an estimated 97% of the mines, it is likely that draglines will be used to mine large mountaintop areas.

Although only three basic mining methods are in use in steep slope areas, there are many variations of each method. Additionally, state reclamation laws cause still further variations of the basic mining methods.

A description of the sequence of procedures used in mine development, mine operation, and reclamation is presented next. Following that are discussions of conventional contour, haulback, and mountaintop removal mining methods.

General Sequence of Mining and Reclamation Operations

Many mine planning and development activities are common to all mining methods in three of the four central Appalachian states. Prior to mining, sediment basins are constructed in selected drainage ways. The capacities and methods of construction for these basins generally are regulated by the states. If hollow fills are to be used, the hollows to be filled are selected and the storage capacities of the hollows are computed. If spoil is to be placed downslope of the coal seam elevation in other than hollow fill areas (this is prohibited only on very steep slopes), the legal limit for the position of the spoil toe is marked on trees downslope of the coal seam elevation in Kentucky and Tennessee. Areas upon which spoil is to be placed are generally scalped of trees, brush, and topsoil prior to placement of spoil on the area.

The vegetation and soil are windrowed at the projected position of the toe of spoil.

Where hollows are to be filled, more extensive preparation measures are used. In addition to scalping the area, a drainage way is cut in the hollow and is then filled with rocks. After the fill has been placed, this rock-lined trench will serve as the drainage channel for the fill area.

Coal haulage roads are built to the coal cropline. In the past, these roads were typically narrow, winding, poorly surfaced, and were purportedly a major source of sedimentation. Today, in most states, the roads are better constructed. Drainage from the roads is controlled by ditches and culverts, as specified by reclamation regulations. Nonetheless, even today, roads at some of the smaller mines are poorly constructed and maintained. In some cases, bad road conditions caused by winter freeze-thaw cycles result in the closing of mines during winter months. Roads conditions are especially bad at some of the small mines. Because the acreage at many of these mines is small, there is not enough room to construct a road that has a gradual grade. As a result, roads at some small mines are very steep and winding. A little rain makes such roads impassable.

After site preparation and haul road construction have been completed, mining activities begin. Drill benches are constructed using dozers, overburden is vertically drilled and blasted, overburden is removed, coal is loaded out and hauled to the tipple by contract truckers, and the area is reclaimed. Overburden removal and placement methods and reclamation practices vary widely and are described in subsequent sections of this chapter.

One-Cut, Single-Seam Conventional Contour Mining

Conventional contour mining, called the shoot-and-shove method by some mine operators, is based on the use of dozers and end loaders for overburden removal and placement. At one time, this was the only surface coal mining method used in steep slope areas, but now it accounts for only a relatively small percentage of total production. Mining usually commences as soon as the coal haulage road has been extended to the coal seam cropline. Mining advances in such a way that pits become longer as mining progresses.

Typically, only one cut, averaging 60 to 80 ft in width, is made into the hillside. Mine operators use long-established rules of thumb to decide upon pit widths, or, more correctly, to decide upon the height of the final highwall. A commonly used rule is that the final highwall height should be 1 ft for every inch of coal thickness. Thus, if the coal seam is 50 in. (127 cm) thick, the final highwall should be 50 ft (15.2 m) high. If the ground slope angle is 2:1, the resultant pit width would be 100 ft (30 m). Generally, ground slopes are steep in hollows (inside curves) and more gradual on points (outside curves). For this reason, when using the aforementioned rule of thumb, the pits are generally narrowed in hollows and widened on the points.

For a given ground slope angle, the average stripping rates can be reduced by narrowing the pit. In the authors' opinion, this procedure, known in the industry as creaming, is used by some conventional contour miners as a means of maximizing profit per ton of coal mined. The obvious disadvantage of use of this procedure is that coal which could profitably be mined is left in place.

Overburden is excavated in blocks that are usually 250 to 1500 ft (76 to 457 m) long. The length of the block is determined in part by the number of dozers that are used for overburden removal. The first step in the overburden removal process is construction of the drill bench. Dozers are used to cut a solid bench and make a fill bench; i.e., an extension of the solid bench made from excavated material. The drill bench can be cut only as deep as the unconsolidated or semiconsolidated material near the ground surface. If this material is thin, a wide drill bench cannot be made, and the overburden must be drilled in two or three lifts. Sometimes, however, the overburden is drilled in one lift with the result that the lower outside corner of the overburden is not well fractured. The result is the huge sandstone boulders that are commonplace at conventional contour mines.

At some mines, federal safety inspectors require the construction of a safety berm above the highwall. The purpose of the berm is to catch falling rocks. The berm is cut on a 1:1 slope using dozers. Where the unconsolidated surface material is thin, construction of the berm causes problems for mine operators.

Overburden blast holes are drilled on 10- to 15-ft (3- to 4.5-m) centers and are loaded with bulk or bagged ANFO. Blasting is delayed by row from the outcrop in toward the highwall; i.e., the outermost row of holes is shot first. This is done so that the explosive charge will move some of the overburden downslope, to rest on the ground surface below the elevation of the coal seam. Large boulders, termed flyrock, are sometimes cast far downslope when this blasting technique is used.

Subsequently, dozers are used to push the shot overburden down the hill. If the area has not been scalped prior to mining, trees are knocked down and buried in the spoil. Eventually, these trees will rot, leaving voids in the soil. These voids may eventually contribute to the instability of the fill.

Removal of overburden in this manner is relatively inexpensive since the dozers push downhill over short distances. According to mine operators, methods involving haulage of spoil are far more expensive than the dozer push method. They are understandably reluctant to use haulback methods.

It is difficult for the dozers to work right next to the highwall, so front-end loaders are generally used to square up the highwall. Spoil excavated by the loader is carried to the edge and dumped over, or is stacked for eventual use as backfill material.

If the overburden is drilled in two lifts, a second drill bench is constructed. The lower lift is drilled and blasted, and more spoil is pushed downslope by the dozers. At some mines, the overburden in the outer half of the pit is cleared first and half of the coal is loaded out. The front-end loader can then strip the remaining overburden from the front, turning and stacking the spoil on the fill bench for eventual use as backfill material.

When all overburden has been removed from a block, a fill bench and long slope outslope are left permanently in place. The uncompacted spoil rests on the steep slopes at its natural angle of repose. Landslides sometimes are the result of these factors in combination with the heavy rainfall typical of the area. Tension cracks in the fill bench are early indicators of spoil movement.

After all the coal has been removed from a block, the block is backfilled, as shown in Figure 2.38. The backfill is sloped back toward the highwall at an angle of 2 to 4° as a means of reducing runoff and erosion on the spoil outslope. In theory, ditches parallel to the highwall are constructed so that water which collects near the highwall will be carried to controlled discharge areas. In practice, water collects near the highwall, seeps into the spoil, and possibly contributes to slope instability.

After backfilling, the bench area is graded. The bench and spoil outslope are then fertilized and seeded using hydroseeders. With the exception of the lower third of the outslope, which is difficult to reach with the hydroseeder, revegetation efforts are usually successful. Erosion on the spoil outslope is nonetheless a problem (Figure 2.38).

Although the reclaimed bench may have greater use potential after mining than before, the use potential of adjacent land may be reduced due to the danger of slides, increased erosion and sedimentation, and degraded aesthetics.

The major environmental impacts of conventional contour mining are landslides, erosion, sedimentation, and aesthetic degradation, which results from exposed highwalls and eroded outslopes. Reduction of the environmental effects of conventional contour mining is needed. Modified methods, such as the slope reduction and parallel slope fill methods, were proposed years ago as a means of reducing erosion and slope instability, but they were used only on occasional experimental bases and reportedly did not work well. These kinds of modifications, although widely reported in the literature, were never used on a production basin in the past, nor are they used today.

In devising research programs whose goal is reduction of the environmental impact of conventional contour mining, economics are of paramount importance. In conventional mining, dozers push downhill or on level ground for one-way distances averaging less than 100 ft (30 m). The capital and operating costs of this method are relatively low. No spoil haulage trucks are required. Mine planning requirements are minimal. If a modified method could be devised wherein the amount of spoil placed downslope was significantly reduced, but the costs of mining and reclamation were similar to those of conventional contour mining, then research will have served a useful purpose.

The solutions devised are the full and partial haulback methods, but according to mine operators, haulback is as much as 60% more costly than conventional contour mining. Both capital and operating costs are affected. Since haulback mining is believed to be a means of reducing the environmental impacts of steep slope mining to acceptable levels, research should be conducted to devise a haulback method that has the same cost characteristics as the conventional contour method.

SURFACE COAL MINING WITH RECLAMATION 95

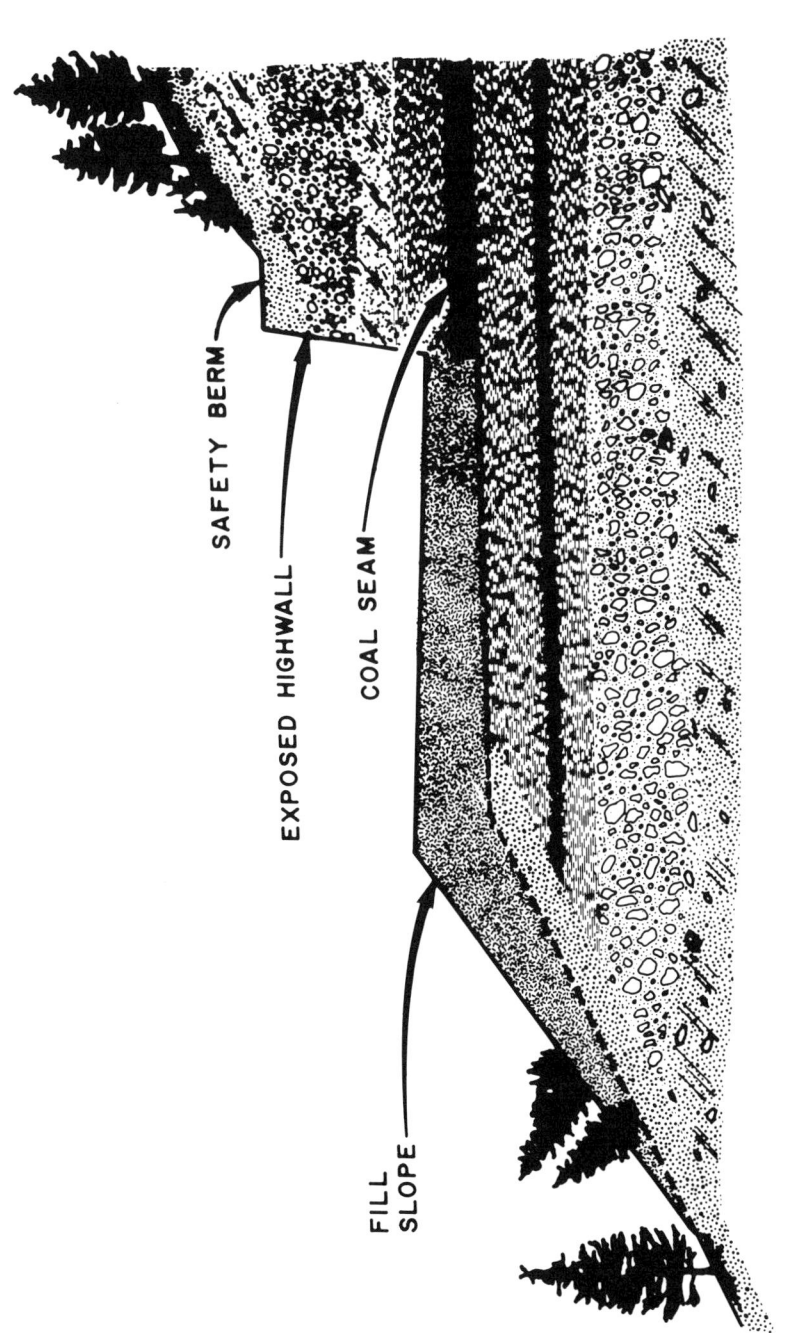

Figure 2.38. Conventional contour mining final grading. (From Math Tech, Inc. Evaluation of Current Surface Coal Mining Overburden Handling Techniques, U.S. National Technical Information Service, PB-264 111 (1976).)

Other Conventional Contour Mining Situations

In some situations, multiple seams are mined by the conventional contour method, or two cuts are taken on a single seam. There are three kinds of multiple-seam mining. The first of these, which apparently occurs infrequently, is the case where two seams appear in the first cut highwall.

Where slopes are gradual or coal is of a very high quality, two cuts are sometimes made in conventional contour mining. After the coal has been loaded out the first cut, second-cut overburden is blasted, pushed, and carried into the open first cut. It should be possible through modification of operations procedures to reduce the amount of spoil pushed downslope in this situation.

Haulback Mining Methods

The haulback method, as the name implies, involves haulage of spoil laterally back along the bench, where it is placed on the pit floor. The method is now widely used to comply with regulations that prohibit or limit downslope placement of spoil and require that the final highwall be completely or partially buried.

Unlike the conventional contour method, which is essentially the same wherever practiced, haulback methods are state- and site-specific. Four parameters are important in defining haulback situations; these are the percentage of spoil that can be placed downslope, the permissible height of the reduced final highwall, the number of cuts made into the hillside, and the number of coal seams mined per pit. The values of the first two of these parameters are controlled by state laws, the third by mining economics, and the fourth by overburden and coal stratigraphy.

In steep slope areas, no spoil can be placed downslope, except in hollow fill areas. Maximum permissible height of the reduced highwall is 30 ft (9.1 m). Some spoil can be pushed downslope, the amount depending on ground slope. In very steep areas, 28° and above, no spoil can be placed downslope. Reduction of the final highwall may be required. A determining factor is the competence of the highwall. In areas less than 28°, a limited amount of spoil can be pushed downslope — the straight-line distance from the outer edge of the solid bench to the toe of spoil cannot exceed 50 ft (15.2 m) — and complete highwall burial is required. An exception to the latter rule is made where a previously stripped area is mined. Then the final reduced highwall can be as high as that left after the original mining operation.

Single-Cut Haulback Mining of Single Seams

Figure 2.39 shows an artist's conception of a single-seam, single-cut haulback method based on the use of dozers and loaders for overburden removal, and rock haul trucks for spoil haulage and placement. Many features shown in the figure are common to all haulback situations.

A drill bench is first cut by dozers and overburden is vertically drilled, using procedures identical to those used in conventional contour mining. Haulback blasting procedures are different than conventional procedures, however, because spoil

SURFACE COAL MINING WITH RECLAMATION   97

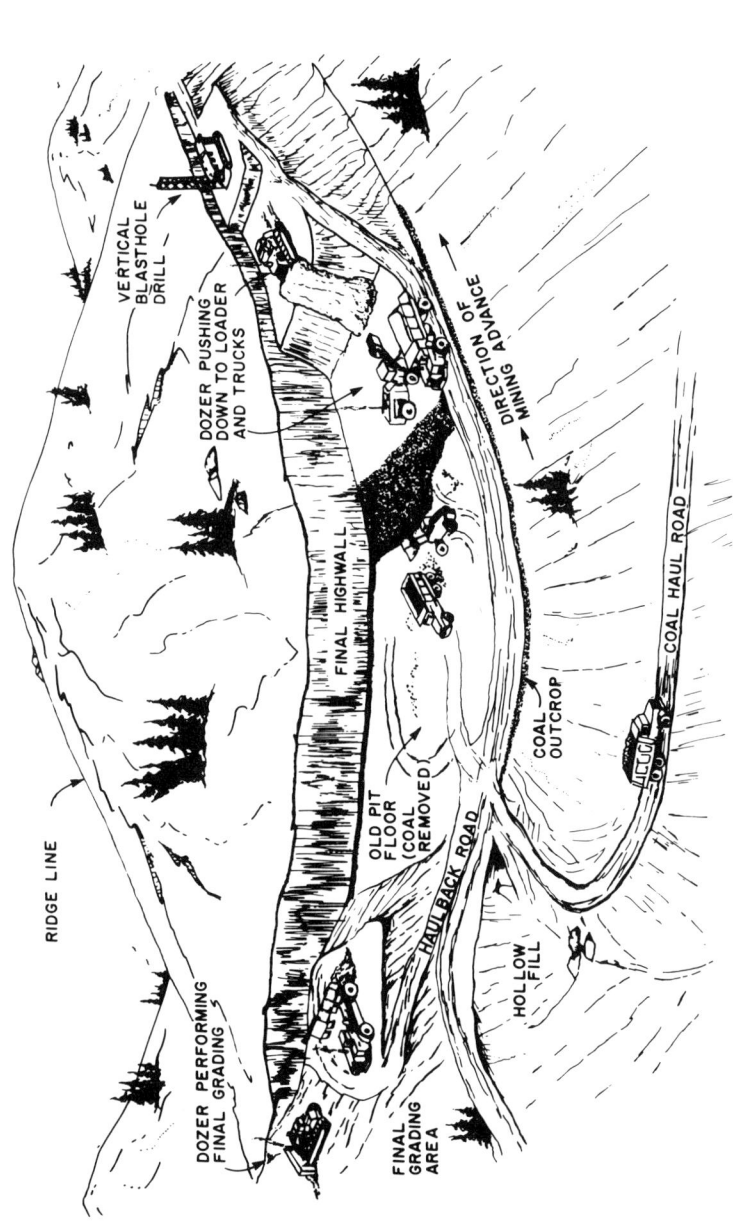

Figure 2.39. Haulback mining in single seam. (From Math Tech, Inc. Evaluation of Current Surface Coal Mining Overburden Handling Techniques, U.S. National Technical Information Service, PB-264 111 (1976).)

must not be cast downslope by the blast. This is true even where limited amounts of spoil can eventually be placed downslope. Operators of most haulback mines visited during the field survey delayed the shots so that the overburden would be lifted but not moved outward during blasting. A special haulback blasting procedure had been devised by one West Virginia company. This procedure consists of delaying the shots in curvilinear "rows" so that overburden is thrown laterally back into the open pit by the blast. According to a company official, several years were required to develop and perfect the method, which now works well.

The initial block cut is generally located adjacent to a hollow which is to be filled with spoil during mining. This method of opening the initial block depends on prevailing reclamation regulations. Dozers push spoil from the first block directly into the hollow in conventional fashion. When working room has been established, loaders and trucks may also be used to dump spoil into the hollow. As this is done, dozers may be used to work the outslope of the fill to about a 2:1 slope. The hollow fill must be constructed in lifts, from the bottom up. This necessitates truck haulage of first-block spoil down the hollow to the projected position of the toe of the fill.

After a sufficient length of pit has been opened, the haulback operation begins. In a full haulback situation, in which spoil cannot be placed downslope, a dozer works on the shot overburden, pushing it down to an end loader. Typical loader capacity is 12 cubic yards (9.17 m^3). According to one engineer, with the exception of only the largest loader built, dozers must push to the loader because the loader does not have breakout force sufficient to excavate the sandstone overburden without the dozer assist. This is but one of the factors that makes haulback mining more costly than conventional contour mining.

It is easy to segregate spoil materials when this method is used. First, the dozer pushes all soil down to the loader, which loads it into haul trucks. Typical truck capacity is 35 to 50 tons. Usually, two trucks work with one loader. The soil is hauled back a relatively long distance and is placed on top of graded soil. Next, the bulk of the remaining overburden is pushed down to the loader, loaded into the trucks, hauled back, and placed against the highwall. Finally, the overburden directly overlying the coal seam is loaded, hauled back, and placed on the pit floor near the highwall, where it subsequently will be covered with other spoil. Not all mine operators use this method, but it may not be expensive to do so.

Stripping procedures are somewhat different in a partial haulback situation, where some spoil can be pushed downslope by dozers. In this case, after blasting, dozers push some spoil downslope in conventional contour fashion. The amount of spoil placed in this manner is fairly small, however, and in the authors' opinion, poses no major landslide or erosion problems. It does, however, eliminate the possibility of using topsoil as dressing on reclaimed areas, since the soil is pushed downslope. After the permissible amount of spoil has been placed downslope, the previously described haulback procedures are used.

Occasionally, in situations where some spoil can be placed downslope, after the initial block has been opened, no further spoil will be placed in hollow fill areas.

More typically, however, because of the 30% spoil swell factor, as mining progresses, it will be necessary to store additional spoil in a hollow. This is always the case where spoil cannot be placed downslope. Operating procedures used to deal with this problem vary. One mine operator opens a long initial block, storing all the first-block spoil in a large hollow fill area. Then, for a while as mining advances, all spoil is hauled back and placed on the solid bench; none is placed in hollow fill areas. Because of the spoil swell, the spoil eventually "catches up" (this is the mine operator's phrase) with the stripping and loading operation. This means that the length of the open cut gets progressively shorter as mining advances. When the open cut reaches minimum acceptable length, another hollow fill is started. With some planning, this point will be reached directly adjacent to the hollow which is to be filled.

Another design includes a hollow fill area at least every 3000 ft (914 m) along the contour. Although mining proceeds in only one direction, spoil is hauled to hollow fill areas in two directions, ahead of the block being stripped and behind it. Suppose, for example, that mining is proceeding in an easterly direction. One hollow fill area would be located west of the active pit. Spoil from the active pit would be hauled to this area in a westerly direction back along the contour on the road used by coal haulage trucks. The second hollow fill area would be located east of the active pit, in an area which had not been stripped. In order to enable haulage of spoil to this fill area, an advance road would be constructed on the contour ahead of the active pit area. In practice, spoil is hauled the shortest possible distance. Sometimes this is to a position on the solid bench, sometimes to the advance hollow, and sometimes to the "retreat" hollow. The hollow fill areas are deliberately spaced no more than 3000 ft (914 m) apart so that the haul distance to a hollow will never exceed 1500 ft (457 m).

Spoil haul road profiles are dependent primarily on the following factors:

- number of overburden lifts
- haulage distance
- height to which spoil is stacked on the solid bench

If overburden is drilled and stripped in one lift, the end loader will work from a position a few feet above the top of the coal seam. Haul trucks approach the loader on relatively level ground. If, on the other hand, two or three overburden lifts are used, when the top lift is stripped, the loader will work on the upper bench and trucks will have to be ramped up to this loader.

At southern West Virginia full-haulback operations, the coal haulage road is constructed on the solid bench as shown in Figure 2.40. The road is not built on the fireclay pit floor, however, but is built up using competent spoil materials. The road, which is 15- to 20-ft (4.5- to 6-m) wide, is inclined downward toward the highwall so that runoff from the road will flow into the drainage ditch shown in the figure. Sumps are located in the ditch at intervals along the contour. Water that collects in

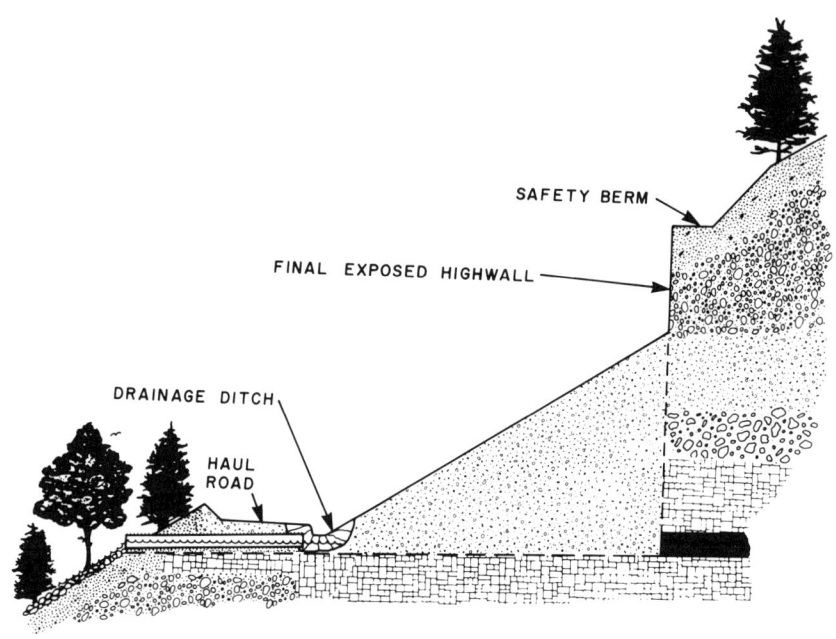

Figure 2.40a. Reclamation with no spoil downslope (section). (From Math Tech, Inc. Evaluation of Current Surface Coal Mining Overburden Handling Techniques, U.S. National Technical Information Service, PB-264 111 (1976).)

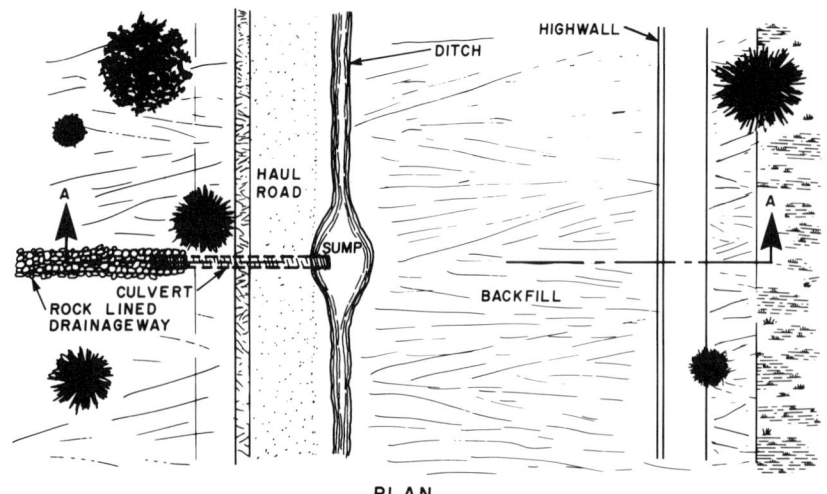

Figure 2.40b. Reclamation with no spoil downslope (plan). (From Math Tech, Inc. Evaluation of Current Surface Coal Mining Overburden Handling Techniques, U.S. National Technical Information Service, PB-264 111 (1976).)

the sumps is taken off the bench through culverts that have been installed under the haul road. The culvert discharges to a rock-lined drainageway. This is a good drainage-control system.

Some mine operators like to have two active pits at each mine. In haulback mining, this is easily accomplished by mining outward in two directions from the initial block. If sufficient stripping equipment is available, both pits will be worked simultaneously, with spoil being hauled back in the direction of the initial block. Except for the sharing of coal loading and haulage equipment, these are merely two separate haulback operations. In some cases, if there is not enough stripping equipment to work two pits simultaneously, the haulback can still be worked in two directions by shuttling equipment between the two pits. Coal is loaded from one pit while the other is being stripped, and vice versa. Some have called this latter procedure the block cut method; others have called it the modified block cut method, although no one seems to know what has been modified. Practically speaking, it is just single-cut haulback mining.

Mining proceeded in only one direction at each of the 11 field survey haulback operations, but several operators were considering the use of the two-directional method.

Backfilling and grading activities are an integral part of the haulback mining process. After spoil has been dumped on the solid bench by the spoil haulage trucks, dozers are used to grade the spoil. One method entails grading of spoil on the solid bench to about a 2:1 slope, the maximum slope that can be navigated by a large dozer. After grading, topsoil may be spread over the spoil surface. This is done at 3 of the 11 survey haulback operations. Fertilizer, seed, and sometimes mulch are applied by hydroseeder (Figures 2.40a and 2.40b).

REFERENCES

1. *Surface Mining Supervisory Training Program* (Bucyrus-Erie Co., 1977).
2. Fluor Engineers, Inc. Optimal Dragline Operating Techniques, U.S. National Technical Information Service, DE83-006980 (1982).
3. Math Tech, Inc. Evaluation of Current Surface Coal Mining Overburden Handling Techniques, U.S. National Technical Information Service, PB-264 111 (1976).

CHAPTER 3

Reclamation and Revegetation of Mined Land

INTRODUCTION

Federal and state reclamation laws and regulations sometimes require, or tend to imply, that mined land be restored to premining use. This has been understood to require restoration of the mined land to approximate premining topography, respreading of all or most of the original soil material, and reestablishment of the same or similar vegetative cover.

The goal of reclamation should be to establish a permanently stable landscape that is aesthetically and environmentally compatible with surrounding undisturbed lands. Postmining land use should be one that contributes most effectively to the productive capacity and stability of the greater ecosystem, of which the mined land is one component. The size of the ecosystem depends upon practical considerations at each site. In areas where land ownership is predominantly private, each privately owned block could comprise an ecosystem. Where the ownership is public, the ecosystem may be much greater in extent. We cannot consider each mined area as a separate entity but, rather, as one component of the ecosystem in which it occurs. Reshaping the mined land to a topography that contributes best to the stability and productive capacity of the entire area can be more important than reshaping to premining contours. Revegetation should use species that will contribute most to the stability and utilization of the entire system. The land should be reclaimed so that it is suited for as many alternative uses as may be practically feasible.

The postmining productive capacity depends upon the inherent properties of the soil and/or soil materials replaced in the root zone. The actual productive potential of a reclaimed soil depends upon its topographic location and the degree of restoration of specific root zone properties that were disrupted in the mining and reclamation processes. A guiding concept was that soils existing prior to mining could be replaced on the mined area. With minimum change in properties that influence land productivity, the soils of the mined land should approach the productivity of the original soils at the site. Even if the exact thickness of topsoil and subsoil were replaced, two soil properties that likely would not be the same as those premining are

- Size, shape, and volume of pores
- Number and kind of soil microorganisms

The latter can be avoided by direct, immediate replacement of soil without stockpiling. The number of microorganisms in the stockpile decline due to the lack of oxygen.

Some natural soil horizons are not particularly suitable for use in reclamation, so in some instances substitution of soils for both the sub- and the topsoil may be necessary. Some cases exist in which the bank-run mine spoil or soil mixes that could be created with specific mining operations are as adequate as the subsoil and could be provided at much less cost to the mine operator.

The potential productive capacity of a reclaimed soil is dependent upon the inherent properties of the soil and/or spoil materials in the root zone. Among the many inherent chemical properties that effect plant growth are

- pH
- Toxic elements
- Cation exchange capacity
- Nutrient availability

The most important inherent physical factor is the texture which determines water-holding capacity and the entrance and internal movement of air and water.

These soil properties have formed in soils during the long period of their development. The properties are partially or totally destroyed during mining and reclamation. They must be reestablished before the potential productivity of the soil can be reached.

Of the in-site properties, bulk density is used most frequently. It is often erroneously considered the only necessary measure of root zone physical properties. Bulk density is simply the weight of soil per unit volume, and as such can be used to estimate the total pore space in the soil. Pore size distribution is important because measurements of the space occupied by the smaller pores give an estimate of the available water-holding capacity of the soil, while measurements of the larger pores (macropores) give an estimate of the volume of pore space available for conducting water and air through the soil.

The root zone material must be regraded with proper compaction. The restoration of all surface mined land would require regrading, if only to level the spoil banks. The grading, however, should be performed in such a way that compaction is minimized.

Several mine operators have made progress in developing methods of handling the soil in order to minimize compaction. Use of a bucket wheel excavator and hauling by trucks with base-level dumping that avoids traffic on top of the soil being replaced has given encouraging results. The main difficulty in the compaction problem is the use of scrapers, especially in cases in which the lifts are thin and the equipment must make repeated passes over the area. Making the lifts 3 m or more thick is helpful.

Handling the soil when it is dry is very critical in reducing compaction. Soil that is excavated when wet is likely to be puddled during excavation and is not likely to dry while in stockpile. Puddling is then likely to recur when the soil is regraded. Wet soil is vulnerable to overcompaction, even with minimal compactive forces. Handling of top and subsoil materials should be planned for seasons when the soil has the lowest moisture content.

The bottom of the soil zone used by plants such as corn is in the range of 40 to 72 in. (1 to 2.5 m). Corn plants can extract significant water or nutrients below a depth of 48 in. (1.3 m) on a few soils with thick subsoils and well-developed soil structure. Root-restricting layers occur in some prime farmland soils and have presented a special problem — they may affect soil productivity. Such horizons should be considered a part of the root zone provided for in soil reconstruction plans.

Restoration of the mined land to its original contour can place some restrictions on final soil slope. Some surface land has poor natural soil drainage. Most regraded mine soils have slower permeability after mining than before regrading, and it is likely that if the regrading does not provide for better surface drainage than the original surface did, the reconstructed soil will be wetter than the original. Productivity might be much lower than the original soil because of higher wetness of the soil. Slopes of 2 to 3% are desirable.

RECLAMATION OF SURFACE MINED LAND IN AUSTRALIA

Reclamation planning is integrated with mine planning from the earliest conceptual stages so that reclamation constraints can be accommodated in the mining plans and vice versa. One important requirement is the development of a flexible reclamation plan which will accommodate changes in mining operations.[1] Specific attention is paid to the physical and chemical characteristics of the spoil, and to catchment size, drainage density, and channel capacity in landform design. Greater attention is given to quality control in the selection, stripping, and use of topsoil.

Drainage stability erosion control and pre- and postmining land use are primary considerations of the reclamation plan. The need to construct a suitable landform within the confines of an operable mining plan is important. To achieve this, interaction between environmental personnel and the mine planning group during the planning stages is required with respect to equipment scheduling, overburden volumes, swell factors, and locations for haulage ramps and roads.

The topography consists of a narrow ribbon of highly textile floodplain, no more than 2 to 3 mi wide. The floodplain is intensely irrigated and used for dairying and the production of oilseed crops. Rising away from either side of the floodplain are gently rolling footslopes with relatively infertile soils which are used for low-intensity beef cattle production on native pastures. Isolated areas of more fertile volcanic soil types have been used for grape growing and area crop production. Surface coal mines are located on this type of land.

The objective of reclamation is to return mined land to its previous capacity and productivity for beef cattle grazing. This requires constructing stable landforms and establishing maintenance for pastures. More vigorous and productive exotic grass and legume species are used in place of the native species because the seed of the native species is difficult to procure commercially, germination percentages are poor, and the plants are slow to establish. Native grass and legume species do not provide sufficient protection from soil erosion during the early stages. Native trees are planted in areas ranging from small woodlots serving as windbreaks and stock shelter to sizable areas of forest on steeper slopes.

The following is a discussion of typical reclamation planning procedures.

Material Characterization

Overburden material is analyzed from exploration borehole scrapes taken during the planning stage. This helps determine fertilizer requirements on reclaimed pastures and identifies any potentially toxic or otherwise intractable layers which may have to be selectively handled. Table 3.1 presents a typical physical and chemical analysis of the two principal overburden types: a poorly laminated argillaceous gray mudstone and a massive brown-gray lithic sandstone. The main physical difference between the two is that the mudstone weathers rapidly upon exposure to form a reasonably acceptable substitute soil or subsoil, while the sandstone is strongly resistant to weathering. Both materials are extremely deficient in nitrogen and phosphorous, indicating the need for heavy fertilizer application. Phosphorous is present in the rocks in the form of apatite or hydroxy apatite which becomes increasingly plant-available as weathering progresses. Analysis of spoils over 20 years old has shown levels of plant-available phosphorous sufficient to sustain plant growth, but not in toxic concentrations.

The principal deficiencies in the materials are high pH and high salinities. Salt levels are rarely sufficient to inhibit plant growth, but do cause dispersion of clays leading to surface crusting and sealing, which impedes moisture penetration. Sulfur levels are quite low, except in the northern end of the valley.

Table 3.1. Overburden Characteristics

Parameter	Sandstone	Mudstone
Particle size (%)		
Sand (2.0–0.2 mm)	14.2	12.8
Fine sand (0.2–0.02 mm)	48.8	33.6
Silt (0.02–0.002 mm)	9.4	19.6
Clay (0.002 mm)	27.6	34.0
Dispersal index	4.2	3.3
Moisture characteristics		
Field capacity (pF 2.0)	21.5	21.5
Wilting point (pF 4.2)	9.0	9.0
pH (1:5)	8.5	8.0
EC (mS)	0.16	0.12
Water–soluble cations (meq/100 g)		
Na	0.50	0.30
K	0.60	0.30
Ca	0.10	0.20
Mg	1.70	1.70
Exchangeable cations (meq/100 g)		
Na	1.0	0.5
K	1.8	0.6
Ca	4.8	5.1
Mg	12.0	13.7
Trace elements (ppm/0.5 mL HCl extract)		
Zn	9.0	2.1
Ca	1.1	0.9
Pb	n.d.	0.5
Mn	38.0	10.5
Mo	1.0	1.0
Cu	0.2	0.4

n.d. = not detected.

From Gordon, R. M. and J. G. Hannen. Reclamation of surface coal mines in the Hunter Valley, New South Wales, Australia, in *Innovative Approaches to Mined Land Reclamation* (Carbondale, IL: Southern Illinois University Press, 1986), pp. 525–559. With permission.

Landform Design

In the development of a final land surface, establishing a lateral drainage system is essential. The drainage system should be compatible with the off-lease drainage pattern, minimize flow into mine workings, and provide containment and treatment options to satisfy pollution control requirements. Interference between the drainage system and access to mine workings should be minimized.

The ultimate objective is to use major coal haulage roads and ramps as the principal drainage channel. The drainage density of the surrounding land, particularly those with similar slopes, vegetation, and land use, guide the determination of appropriate catchment sizes. One of the main objectives of the drainage design for the mine as a complete system is separating runoff coming from undisturbed areas from the drainage which may be contaminated by mine workings and spoil piles. All water flowing from reclaimed surfaces is contained and treated in sedimentation dams before release.

Attaining satisfactory slopes on reformed spoil piles is an important requirement. Influencing factors include the erosion potential of the reformed surface in relation to slope length, upslope catchment and maximum anticipated runoff, soil materials used for topdressing, and vegetation density.

Natural slope angles in the vicinity of several old mines were surveyed and the median angle was found to be 10°. This angle has been accepted as a standard. In practice, reformed slope angles range between 6° and 10°. A minimum slope of 0.5° is used to ensure free drainage.

In the development of the reclaimed landform, computer models are used to simulate mining and beef-filling operations over the operating life of the mine. In generating landforms, the computer operates within the defined constraints of slope angle, overburden swell factors, and elevation. The landform is designed to visually blend with the surrounding topography, to be stable in the long term, and to suit the proposed postmining land use of beef cattle grazing.

Use of Topsoil

Before any mining operation is conducted, a detailed soil survey is undertaken. These surveys are conducted in increments, covering the area to be mined during each 5- to 7-year open-cut approval.

Soil profiles are examined at exposed areas such as cuttings and gullies, and by means of coring. Soil types are classified according to horizon, color, texture, structure changes, and pH. In addition, representative samples are collected for laboratory analysis.

A topsoil stripping plan is prepared which indicates the required stripping depth or the thickness of soil to be removed within each portion of the mine site, including areas to be used for out-of-pit spoil emplacements and washery reject disposal. Strippings are indicated in increments of 15 cm, about the limit of accuracy at which the dozers and scrapers are able to operate.

Soils covering most of the rolling footslopes where the surface mines are located have developed more or less in-site from sedimentary rocks of the coal measures. They are relatively infertile duplex types in which a shallow sandy or loamy A horizon overlies a medium to heavy clay B horizon. The organic-rich A horizon is frequently <10 cm thick. On the lower parts of slopes, this is often underlain by a bleached, hard-setting A_2 horizon, which is quite infertile, highly dispersible, and of no value for reclamation purposes. The B horizon is also usually dispersible, often saline and of little value for reclamation. Table 3.2 presents physical and chemical analyses for a typical sodic soil profile.

The optimum thickness for spreading topsoil over reformed spoil piles is 10 cm. Over quite significant portions of the mine lease, none of the soil horizons are suitable for reclamation purposes because of high salinity, a high proportion of stone in upper horizons, or the absence of A_1 horizon due to erosion. To compensate for this problem, stripping depths are increased on better soil types in the lease.

Table 3.2. Analysis of Typical Solodic Soil Profile

Parameter	A_1 Horizon (0–10 cm)	A_2 Horizon (10–40 cm)	B Horizon (+40 cm)
Particle size (%)			
Sand	22	21	6
Fine Sand	57	62	19
Silt	8	9	2
Clay	11	8	72
pH (1:5)	6.2	6.3	8.3
EC (mS)	0.06	0.04	0.05
Water-soluble cations (meq/100 g)			
Na	0.1	0.1	0.2
K	0.1	0.1	0.1
Ca	0.1	n.d.	n.d.
Mg	1.8	0.3	0.1
Exchangeable cations (meq/100 g)			
Na	0.2	0.4	1.9
K	0.6	0.6	1.9
Ca	3.0	3.6	3.4
Mg	1.9	3.4	4.4

n.d. = not detected.

From Gordon, R. M. and J. G. Hannen. Reclamation of surface coal mines in the Hunter Valley, New South Wales, Australia, in *Innovative Approaches to Mined Land Reclamation* (Carbondale, IL: Southern Illinois University Press, 1986), pp. 525–559. With permission.

Suitable topsoil material may not always be available, however, to topdress all reformed spoil piles and other disturbed areas. The available material must be used in the most beneficial way so as to rapidly establish dense vegetation on areas susceptible to erosion, such as steeper slopes and the channels of newer watercourses. Preference is given to establishing trees directly into the spoil material. Some native trees establish and grow better in some spoil materials than on sites that have been topdressed with soil. The main use of topdressing with soil is to provide a better physical environment for the initial germination of pastures and as a source of native seed which helps increase the diversity of species in the final pasture. The relatively thin layer of applied topsoil means that vegetation depends upon the underlying spoil materials for long-term growth and survival.

Reclamation Procedures

Standing timber is cleared in strips ahead of mining operations before the topsoil is cleared. To avoid dust generation problems, the length of the strips is limited to a short distance ahead of the highwall. Self-loading scrapers are the most commonly used for removing topsoil. In some instances the topsoil is removed by bulldozers and transported by trucks.

Soil is stockpiled at strategic locations around the mine for subsequent recovery and for spreading over reformed spoil piles. However, after reclamation has commenced, the topsoil is directly transported to the areas being reclaimed.

In mines that utilize trucks and shovelers for waste handling, the desired landform can be roughly constructed as the material is dumped. Afterwards the bulldozers can build the landform by trimming the spoil piles. Dragline spoil piles require considerably more shaping.

The first step in the reshaping program is to erect slope profiles or survey pegs that indicate the required slopes in accordance with the landform plan. Slots are then bulldozed to achieve the working slope at selected points over the area to be reshaped.

At several mines, graded banks have been progressively introduced and have proven successful as an erosion control measure on longer slopes. The banks are designed with channel gradients of 0.5%, though it is believed that gradients could be increased to 1%. Differential settlement of the spoil leads to overtopping and bank failure. A limitation on the use of graded banks is the availability of suitable discharge points.

In most instances, scrapers used for topsoil spreading utilize material transferred from a point directly ahead of the cut or material reclaimed from stockpiles. Scrapers are usually worked on the contour to avoid creating tracks which may become sites for gully erosion after a heavy rain. Scraper movements over previously topdressed surfaces are minimized to avoid overcompaction of the soil. Topsoil is spread at a nominal thickness of 10 cm.

Revegetation Methods

Valleys experience a warm temperature climate with summer-dominant but winter-effective rainfall. The average annual rainfall is 0.6 m. Summer rainfall commonly occurs as short duration, high-intensity storms, often causing localized flooding and soil erosion damage.

The climate of the region presents particularly difficult conditions for pasture establishment as it favors warm-season species. However, the unreliability of spring rains and the hot, dry, early summer make their establishment in early spring a difficult process.

After some research, a technique was developed involving sowing in the early autumn to take advantage of the relatively reliable late summer and autumn rains, using a mixture of cool- and warm-season species sown simultaneously. The cool-season grasses and legumes germinate and establish quickly to provide early protection of the surface from soil erosion. The warm-season seeds remain dormant in the soil until the following spring or early summer when they germinate and establish through the mulch provided by the hayed-off, cool-season grasses. These warm-season species then develop into permanent, productive pasture.

All cultivation in preparation for sowing is carried out on the contour to retard surface runoff and promote moisture infiltration. The usual practice is to deep-rip the ground to a depth of 0.6 to 1 m using a dozer-mounted ripper with tines spaced about 1 m apart. This is followed by contour cultivation with a conventional chisel plow to a depth of about 22 cm. The purpose is to produce a very coarse, rough seedbed that retards the formation of a surface crust or seal on the soil with deep furrows for moisture retention.

Inoculation of legume seed by Rhizobium occurs via tractor or truck-mounted broadcasters immediately prior to sowing. Sowing is usually timed for the early autumn season (March to April) when soil moisture levels are high. The cool-season species (Barrel medic and ryegrass) germinate to provide a quick-growing cover crop, protecting the surface against erosion. The warm-season species germinate from the early spring onward to provide a Rhodes grass- and alfalfa-based long-term pasture.

Tree Planting

Sites for tree planting are chosen based on their physical suitability for trees, aesthetic appearance of the reclaimed surface, and their usefulness in providing stock shelter. Favored locations include ridgelines adjacent to major watercourses and dams, and in scattered woodlots.

The areas selected for tree planting are pegged out and topsoil is not placed on them, although normal cultivation work is performed. The sites are then deep-ripped on the contour a single time and are spaced at 4 to 5 m. The riplines are allowed to settle for 4 to 6 weeks prior to planting.

Planting methods use tubed seedlings 6 to 9 months old, which are raised in a commercial nursery. Some mines collected the seed of native tree species growing on the lease and gave them to a nursery to raise the tubed stock.

Seedlings are planted out May through June. Planting is carried out by hand, with seedlings being placed about 5 m apart; several slow-release fertilizer pellets are placed into the hole with some water.

Pasture Management

A management program is implemented to ensure the persistence of a productive pasture and tree lots on the reclaimed land. To maintain a balance of legumes and grasses in the pasture, livestock grazing is required as soon as possible after the pasture has become established.

Annual fertilizer topdressings are applied by aircraft or by ground broadcaster twice during the year.

Erosion damage is promptly repaired. Weeds can usually be controlled by appropriate pasture and grazing management, but spot spraying is necessary.

REVEGETATION OF A SURFACE MINED LAND IN MONTANA

Reclamation of coal-mined land in Montana is oriented toward rebuilding the land to its premining structure, function, and use. The state regulations require the establishment of suitable, permanent, diverse vegetative cover capable of funding livestock and wildlife, limiting soil erosion, and persisting under prevailing environmental conditions. Restoring productivity of the land to provide a renewable resource requires coherent functioning of soil and plant systems. These systems are interdependent, and their development during and following reclamation is of particular concern.[2]

Plant and soil successions on mined land contain elements of both primary and secondary succession. The substitute lacks normal microorganism populations, cycling, and biogeochemical balances, but it has water-holding capacity and can perform soil functions. Plant succession on abandoned spoil is not the same as on abandoned land. Some early successional stages associated with soil formation can be achieved in reclamation proves by topsoiling, fertilization mulching, and the use of seeding mixtures of climax and subclimax native and introduced plant species.

Concerns about reclamation success have included lack of diversity, declining productivity, and marked compositional characteristics. Difficulty in establishing adaptive native grass, forb, shrub, and tree species has posed a serious problem. Other problems affecting revegetation include alkalinity, salinity, excessive litter accumulation, insect and pathogen damage, and soil nutrient deficiencies.

Plant succession on seeded mine spoil is strongly influenced by the following variables:

- Species seeded
- Initial seeding success
- Cultural practices
- Weather

At the end of the first growing season on seeded mine spoil, annual grass and forbs provided about 90% of the vegetation cover. Seeded perennial grasses, forbs, and shrubs provided the other 10%. By the end of the second growing season volunteer and seeded annual and biennial species were dominant and provided most of the cover. Perennials, especially introduced grasses, increased in cover to about 40% of the total, depending on the success of stand establishment. By the third growing season perennials had assumed complete dominance. Perennial cover stabilized at about 40% in the fifth growing season. Litter cover and moss increased rapidly to over 95% and 5500 kg/ha, respectively, after three growing seasons.

On sites revegetated, introduced perennial grasses have commonly achieved dominance despite simultaneous seeding of native species. The most successful species have been crested wheatgrass, smooth bromegrass, and alfalfa. When seeded at moderate rates (<100 PLS/m^2), these species produced most of the cover and yield

RECLAMATION AND REVEGETATION OF MINED LAND 113

following establishment. Over time they have tended to exclude other species. Because of the competitive ability of introduced species, the advantages of seeding mixtures as well as topsoil containing native species were quickly lost.

Several native species have proven valuable in reclamation. Rosana western wheatgrass, Critane thickspike wheatgrass, and Wytene fourwing saltbush have occasionally performed well. These species have persisted on seeded raw spoils for 8 years when seeded without cover crops, crested wheatgrass, and other introduced grasses.

Repeated fertilization was probably an important factor in the failure of some native species to establish and persist. On several seeded stands in which introduced species showed vigorous stands, repeated fertilization and favorable growing conditions resulted in high accumulation. The thick mulch layer delayed pH development and decreased vigor and yield. On one site seeded, in 5 years the yield decreased from 2200 to 750 kg/ha. Wide C/N ratios on such sites indicated that the nutritional status of the spoil soil had been lowered.

Final preliminary observations indicate that successful reclamation in semiarid lands in Montana will not be easily or quickly achieved. Reclamation law requires establishing permanent diverse vegetative cover which must be composed of predominantly native species. Native species, particularly shrubs and forbs, have not been successful in reclamation seedings. Species diversity has remained or become low in recut mined lands in comparison to native rangeland and old, naturally revegetated spoil. Variations in yield, despite normal climatic conditions, suggest that adequate soil functioning was not attained. Management practices following stand establishment such as grazing, prescribed burning, and proper fertilization may improve successional development. Greater seeded stand diversity should occur through proper seeding formulation as well as species enrichment from topsoil and native plant communities.

FACTORS AFFECTING NATURAL REVEGETATION OF COAL MINE SPOIL BANKS IN OHIO

Strip mining spoil banks cover a large area of land surface in Ohio. These spoil banks erode severely and in many instances increase silting and the acidity of nearby streams. In addition to erosion, downslope displacement and slumping of surface soil can occur on unvegetated areas. Spoil banks are also characterized by extreme acidity.[3] A pH of 4.0 or less is common and is classified as toxic to most species in this area. Along with the high acidity are high levels of iron, aluminum, and manganese and a low supply of magnesium, calcium, potassium, and phosphorous. Toxic levels of aluminum, iron, manganese, copper, zinc, and nickel have been reported in the acidic soils of spoil banks. Calcium, potassium, nitrogen, and phosphorus ions may be deficient in soils of low pH. The soil and surface temperatures play important roles in the vegetation developments on strip mine banks where surface temperatures exceeding 60°C were reported. High surface temperatures cause high evaporation rates of soil water, reducing water availability and limiting plant establishment.

Past researchers have indicated that once the acid-tolerant species became established, the spoil banks became richer in organic and microorganisms and could support less tolerant species. On slopes exposed to the southwest, soil temperatures were too high for seedling establishment. Surface conditions, rather than pH or soil content, were more important in plant establishment on raw spoil. Factors that affected moisture in the environment, such as evaporation due to wind exposure and high surface temperatures, were important. In extremely acidic areas with a pH of 3.5 or less, no plants became established. Soil particle size (a fraction of the sample smaller than 2 mm diameter) <20%, soil pH <3.5, and a specific of 3 mm or higher were limiting factors for the invasion and growth of natural vegetation.

A special research project was conducted to determine whether soil conditions in both vegetated and unvegetated areas of coal mine spoil banks were correlated with the establishment of natural vegetation. The research site was a former coal strip mine, abandoned some 30 years ago. It was chosen because there was a distinct boundary that separated vegetated "islands." This zonation of natural vegetation can be attributed to soil acidity. In areas where pH is below 4.0, no vegetation existed.

A soil of 4.0 or less is usually considered a toxic level for plants in this region. Tolerance to soil acidity varies from species to species, so it is reasonable to assume that low pH values on this spoil bank excluded many intolerant species. Therefore, pH can be considered a major factor limiting the species that will become established.

Exchangeable aluminum ion concentrations of 1.0 meq/100 g or more are considered toxic to plants. In open sites significantly higher concentrations of aluminum were observed, whereas in the vegetated and open sites, toxic aluminum concentrations were found. Exchangeable manganese, magnesium, and calcium concentrations and water-soluble potassium concentrations for the soil extracts were near or below critical values given for deficiency levels. Other researchers reported that iron, copper, and zinc may also reach toxic levels on spoil banks. Nitrogen and phosphorus may be unavailable to plants in the spoil bank; however, nutrient levels in the soil may not be important to establishment on bare spoil for species that tolerate such conditions. Successful transplants were made on the bare spoil, indicating that mature plants can tolerate the higher concentrations of aluminum and lower concentrations of potassium.

Mature plants can survive the low pH, high aluminum concentrations and low nutrient levels, and seeds can germinate under these conditions. No vegetation was found on much of the spoil bank, because no viable seeds exist in the soil, conditions were not suitable for seed germination, or seedlings could not survive.

Field germination data revealed that on the soil surface, no seeds germinated in the open sites, whereas seeds did germinate in the vegetated sites. Buried seeds germinated in the open sites, whereas the surface seeds did not.

Soil moisture and soil temperature data indicated that water availability may be a problem at the soil surface. The effect of low soil moisture on plants varies with species tolerance to moisture stress and the textural properties of soil.

Where vegetation is absent, much of the water from rainfall may be lost in runoff. Evaporation of soil water may be greater in unvegetated areas. On warm days surface temperatures of the open spoil material can be much higher than soil temperatures measured below the surface. Shading the soil with plant cover and leaf litter decreases soil temperature, which in turn reduces evaporation and increases soil moisture.

Variation in soil temperature may explain field germination results. Buried seeds are not exposed to the extreme surface temperatures and low soil water levels and may be able to germinate. Once the buried seedlings in open sites break through to the surface, they may not be able to survive the high temperatures and the low moisture content. Such conditions did not prevail in vegetated sites, and seed germination and seedling survival do not appear to be inhibited.[4]

Robinia pseudo-acacia trees were planted on the bare spoil bank in an attempt to reclaim the site. Cores of these trees indicated that they were 12 to 15 years old. Leaf litter accumulated at the base of these trees and altered the edaphic conditions enough to allow seeds to germinate and for acid-tolerant species to become established. These plants in turn accumulated more leaf litter, and the spoil bank became richer in organic matter and could support less tolerant species. In this way the vegetated islands have grown to their present size.[5,6]

VEGETATION DEVELOPMENT ON OLD AND ABANDONED LEAD AND ZINC MINE WASTES

Establishment of a stable vegetative cover has been shown as the only feasible method of physically stabilizing metal-mine wastes and reducing pollution effects. Artificial vegetation has been attempted, but little is known about the processes that accompany natural revegetation of contaminated sites. Patterns of natural successional development can serve as models for artificial revegetation where the goal is to grow stable, well-established vegetation without extensive use of fertilizers and irrigation.

Ecosystem response to the deposition of metal-mine waste occurs at different levels. Individual organisms may exhibit specific sensitivity to metals which can be translated to population level response in the evolution process-tolerant vegetation. Compositional changes have been observed.

A research project was conducted on old and abandoned mill tailings of lead and zinc mines in Wisconsin. Some physical and chemical properties of the tailings are given in Table 3.3.

The mine flora was composed of primarily native species, many of which are characteristic of droughty waste places. No significant trends can be discerned in

Table 3.3. Physical Properties of the Tailings Material[5]

	(1900) Badger Mine	(1928) Middie Mine	(1945) Thompson Mine
Date of Abandonment			
% Gravel by weight	64	59	60
Textural class	gls[a]	gls	gsl[b]
pH	7.3	7.8	7.9
Total N (μg/g)	1400	1200	750
Available P (μg/g)	33	1.5	6.7
Available K (μg/g)	118	34	21
Total Zn (μg/g) $\times 10^{-4}$	1.6	1.7	2.1
Total Pb (μg/g)	450	540	760
% Soil moisture	—	6.6	7.1
% Organic matter	—	0.6	0.9

[a] Gravelly loamy sand.
[b] Gravelly sandy loam.

either species richness or variety. A slight decrease in dominance is observed with the increasing age of the site. Tree abundance was related to mine age. Aggregation of vegetation was extremely pronounced at each of the sites. Numerous bare areas existed interspersed with well-vegetated zones. The open areas supported sparse populations of annuals. The closed-turf areas were dominated by perennial grasses and sedges and perennial forbs.

Cover changes with time were observed. Little vegetation establishment occurs in the first decade after abandonment. The long lag in initial colonization can be attributed to the lack of species capable of tolerating or evolving a tolerance of extremely harsh sites. Evolution of tolerance is known to be a function of species sensitivity, metal concentrations, and time.

The initially slow rate of cover development was followed by rapid growth due to an increase in the frequency of turf-forming sedges and grasses. These species are among the earliest, but represent only a small contribution to total cover.

Subsequent vegetation development proceeds slowly. The virtual absence of cover change in 30 years suggested that vegetation has expanded to the limits of available resources. Inputs of new propagates and clonal expansion may be balanced by high mortality due to the vigors of the harsh and nutrient poor site. This will result in a stable cover pattern.

As woody vegetation continues to increase in importance, the tailings environment is altered. Increased shading, litter accumulation, and the improved moisture begin to ameliorate the severe environment. Shifting dominance and increasing cover are accompanied by changes in spatial pattern.

Aggregation is highest at the outset of colonization. Individual plants are clustered together and separated by large areas of bare tailings. However, the intensity of aggregation continues to diminish with successional development. Despite the trend of declining aggregation, the pattern of vegetation cover at even the oldest mine is extremely patchy.

REVEGETATION AT THE USIBELLI COAL MINE, ALASKA

The Usibelli coal mine has been operating in Alaska for 3 decades. The company has been involved in all phases of surface coal mining from exploration to postmining reclamation. After an approximate permit area boundary (based on the identified coal resource) is outlined, environmental studies are initiated. Studies include vegetation, wildlife, surface and subsurface hydrology, soils, and cultural resources. The objectives of the vegetation studies are to determine and describe the existing vegetation and the amount of area this covers. The data are used to determine what vegetation types will be disturbed and what plant sources are available for reclamation.[6]

Vegetation inventory data include the following:

- Percentage of cover by vegetation
- Mosses, lichens, and litter
- Density of shrubs >20 cm (8 in.) tall
- Age, height, and dbh (diameter at breast height) of trees and tall shrubs
- Height and basal diameter of dominant woody species <3 m (10 ft) tall

A list of vascular plant species and moss and lichen genera is prepared. The sampling methods include reporting every plant species at 50 cm (1.5 ft) spacing along a 20-m (66-ft) transect. This is converted to percentage of cover. Stems of woody plants are counted for density estimates in belt transects (rectangles) beside the transects. Enough of these transects must be sampled to ensure that the total living vascular cover is examined with an acceptable degree of precision. The successional and ecological conditions associated with the various vegetation types are described.

The mine is located in the boreal forest zone of interior Alaska. Its coal leases contain almost 20 different vegetation types ranging from low shrub communities to dwarf tree forest types. Grasslands are dominated by bluejoints. The most important tree species in the boreal forest vegetation types include white spruce, black spruce, and paper birch.

The topography consists of flat ridges with adjacent steep slopes where streams are actively downcutting. Along the slopes, vegetation is influenced by landslides. Plant species growing under conditions of slope and aspect are identified for potential use in reclamation. Plant species that occur in steep south-facing, windblown slopes before mining are those that would succeed on the south-facing slopes after mining. Plant species occurring in early successional sites are the most likely candidates in reclamation.[7]

One of the main objectives of reclamation is to establish a diverse, self-reproducing plant community. Successful reclamation is usually among various options. Grass is usually the fastest, most reliable plant type for establishing protective cover on the mined soil. However, heavy grass cover can slow down natural succession on the site and compete with woody plant establishment. Moderate grass cover may help conserve subsurface soil moisture.

The reclamation plan is developed during the mine planning process. Site conditions such as steep slopes and permafrost are considered in both the engineering and the biological aspects of reclamation. The mined area is backfilled and graded to restore approximately the original contour. Reclamation guidelines require that topsoil (an A horizon) be replaced over the graded overburden. Some soil profiles in Alaska do not have A horizons or other material suitable for use as a plant growth medium.

Grass species are the primary plant species used for initial soil stabilization. Woody species are then planted from local transplants or cuttings or seedlings so as to provide diversity and sometimes deeper rooting, which also contributes to soil stabilization. Leaf fall from woody plants may help increase the organic matter and nutrients in the soil. Leaves of alder (*Alnus*) contain high levels of nitrogen. Woody species are usually selected based on their local availability, rapidness of growth, tolerance of various conditions, and usefulness for wildlife or other postmining land use. The reclamation plan includes plant species, planting techniques, amendments such as fertilizer or inoculation and their rates of application, and general location for different plant communities.

The reclamation plan must include a postmining bond-release standard. Before disturbing the land, the mine operator posts a bond sufficient to cover reclamation costs at any given point. If a mine operator fails to meet reclamation requirements after mining, then the state maintains funds to reclaim the existing disturbance. At Usibelli, the reclamation must meet certain standards after 10 years.

The biological objective of these bond-release standards is to ensure that a diverse, self-reproducing plant community capable of supporting postmining land use is established. Bond-release standards can be developed in several ways and possess three parts. The standards include total living vegetation cover, woody species density, and plant species diversity. A simple method of creating a standard is to develop a technical standard in which the mine operator agrees to meet a certain level of the given parameter, such as 70% (level) living plant cover (parameter). This technical standard can be justified by data from existing revegetated areas, test plots, existing native vegetation, or a combination of these. The reclamation plan must be designed to meet these standards at the end of the bond-release period.

After the mining permit is approved, the area is mined according to the mine plan. After mining is completed in an area, the pit is backfilled and graded to approximately the original contour. The slopes are sometimes furrowed for reducing runoff and for accumulating moisture and nutrition.

The site is usually aerially seeded. The seed mix consists of grasses adapted to the area, including Arctared red fescue, Manchur smooth brome, common foxtail, and reed canarygrass. Two annual species are also used to establish ground early cover. Some mines frequently are found to have seed mixes of 20 species with different mixes for different conditions.

The mine personnel regularly transplant young woody plants from nearby vegetation into the reclaimed areas to assist natural succession toward boreal forest communities. This ensures that the plants are adapted to climatic conditions and the rooting zone contains the microbial communities, especially mycorrhizae, needed for proper plant growth.

REFERENCES

1. Gordon, R. M. and J. G. Hannen. Reclamation of surface coal mines in the Hunter Valley, New South Wales, Australia, in *Innovative Approaches to Mined Land Reclamation* (Carbondale, IL: Southern Illinois University Press, 1986), pp. 525–559.
2. Sindelar, B. M. Successful development of vegetation on surface mined land in Montana, in *Ecology and Coal Resource Development,* Vol. 2 (New York: Pergamon Press, 1979), pp. 550–556.
3. Bell, T. J. and I. A. Ungar. Factors affecting the establishment of natural vegetation on a coal strip mine spoilbank in southeastern Ohio, *Am. Midl. Nat.* 105:19–31 (1981).
4. Skousen, C. A. A Chronosequence of Vegetation and Minesoil Development on a Texas Lignite Surface Mine, U.S. Bureau of Mines, IC 9184 (1988), pp. 79–88.
5. Kimmerer, R. W. Vegetation development on a dated series of abandoned lead and zinc mines in Southwestern Wisconsin, *Am. Midl. Nat.* 111:332–341 (1984).
6. Elliott, C. and J. D. McKendrick. Strip mine reclamation and Alaska's big game wildlife, *Agroborealis* 23(1):41–43 (1991).
7. Helm, D. J. From boreal forest to reclaimed site: revegetation at the Usibelli coal mine, *Agroborealis* 23(1):45–50 (1991).

CHAPTER 4

The Acid Mine Drainage Problem from Coal Mines

INTRODUCTION

Acid drainage from underground coal mines and coal refuse piles is one of the most persistent industrial pollution problems in the United States. Pyrite in the coal and overlying strata, when exposed to air and water, oxidizes, producing ferrous ions and sulfuric acid. The ferrous ions are oxidized and produce an hydrated iron oxide (yellowboy) and more acidity. The acid lowers the pH of the water, making it corrosive and unable to support many forms of aquatic life. The iron oxide forms an unsightly coating on the bottom of streams, and further limits the ability of aquatic life to survive in streams affected by acid mine drainage (AMD).

Before coal is mined, very little of the pyrite is exposed to the conditions necessary to produce acid drainage. The mining and coal cleaning process exposes the pyrite to surface or ground waters and allows pyrite oxidation to occur. A ton of coal containing 1% pyritic sulfur has the potential of producing 33 lb of yellowboy and over 60 lb of sulfuric acid. However, the rate of acid production varies, and abandoned mines and refuse piles can produce acid drainage for >50 years. The drainage, if discharged into surface streams or ponds, constitutes an extensive, expensive, and persistent environmental problem. Federal law now requires that water discharge from active coal mines have a pH between 6 and 9, and the law places limits on the total iron, manganese, and suspended solids. Controlling AMD from active mines usually requires expensive water treatment and the necessity of handling very large volumes of water.

CHEMISTRY OF FORMATION

The oxidation of iron pyrite (FeS_2) and the release of acidity into waters draining through coal mines can be represented by the reaction sequence given in Table 4.1. Fe^{2+} is released in the initiator reaction either by simple dissociation of the iron pyrite or by oxidation of the pyrite by oxygen. After the sequence has been initiated, a cycle is established in which Fe^{2+} is oxidized by oxygen to Fe^{3+}, which is subsequently reduced by pyrite, thereby generating additional Fe^{2+} and acidity.[1]

Two possible oxidants for iron pyrite are available, namely oxygen and ferric iron. The reduction of Fe^{3+} by pyrite both in the presence and in the absence of oxygen at 0.20 atm showed no difference in the rate of the reaction (the parallelism exhibited by the two straight lines is indicative of equivalent reaction rates) or in the rate of change of soluble ferrous iron. The rate of the reaction is relatively rapid; for example, at pH 1, 50 min was required for the reduction of 50% of the initial ferric iron concentration by approximately 3 m^2 of pyrite per liter of solution. In the absence of ferric iron, no oxidation of pyrite was observed even after 1 week. Further evidence for the slowness of reaction 1 is the fact that pyrite can be used as a reasonably inert electrochemical electrode. Hence, the major oxidant of iron pyrite is ferric iron, as indicated in the propagation cycle.

The rate of oxidation of Fe^{2+} by oxygen in abiotic systems is a function of pH.[2] These results were obtained in pH-buffered systems (using $HClO_4$ or CO_2 and HCO_3^-) under a constant partial pressure of oxygen and, at pH values >4.5, were found to be compatible with the kinetic relationship

$$\frac{-d[Fe^{2+}]}{dt} = k[Fe^{2+}][O_2][OH^-]^2 \tag{4.1}$$

where k = 8.0×10^{13} L^2 mol^{-2} atm^{-1} min^{-1} at 25°C. At pH values below 3.5, the reaction proceeds at a rate independent of pH, i.e.,

$$-d[Fe^{2+}] = k'[Fe^{2+}][O_2] \tag{4.2}$$

where k = 1.0×10^{-7} atm^{-1} min^{-1} at 25°C. The reaction has previously been reported to be first or second order with respect to Fe^{2+}, depending upon the ionic medium; in the presence of ligands which form relatively strong complexes with Fe^{3+} (pyrophosphate, fluoride, and dihydrogen phosphate), the rate is first order in $[Fe^{2+}]^2$. The actual rates of the reaction, however, are of the same order of magnitude as those reported here in the lower pH range. The half-time of the reaction in this acidic pH region is approximately 1000 d, reflecting the slowness of the oxygenation reaction when compared to the rapid oxidation of pyrite by ferric iron. Since reaction 2 is significantly slower than reaction 1, the oxidation of Fe^{2+} by oxygen appears to be

Table 4.1. Reactions Responsible for Pyrite Oxidation

1. $FeS_2 + \frac{7}{2} O_2 + H_2O \rightarrow Fe^{2+} + 2SO_4^{2-} + 2H^+$

2. $Fe^{2+} + \frac{5}{2} H_2O + \frac{1}{4} O_2 \rightarrow Fe(OH)_3(s) + 2H^+$

3. $Fe^{2+} + \frac{1}{2} O_2 + H^+ \rightarrow Fe^{3+} + \frac{1}{2} H_2O$

4. $FeS_2 + 14Fe^{3+} + 8H_2O \rightarrow 15Fe^{2+} + 2SO_4^{2-} + 16H^+$

Stage 1		
Mechanism	Reaction 1:	proceeds both abiotically and by direct bacterial oxidation
	Reaction 2:	proceeds abiotically, slows down as pH falls
Chemistry	pH above approximately 4.5; high sulfate; low iron; little or no acidity	
Stage 2		
Mechanism	Reaction 1:	proceeds abiotically and by direct bacterial oxidation
	Reaction 2:	proceeds at rate determined primarily by activity of *T. ferrooxidans*
Chemistry	approximate pH range of 2.5–4.5; high sulfate; acidity, and total iron increasing; low $Fe^{3+}:Fe^{2+}$ ratio	
Stage 3		
Mechanism	Reaction 3:	proceeds at rate totally determined by activity of *T. ferrooxidans*
	Reaction 4:	proceeds at rate primarily determined by rate of reaction 3
Chemistry	pH below approximately 2.5; high sulfate, acidity, total iron and $Fe^{3+}:Fe^{2+}$ ratio	

From Kleinman, R. L. P. Biogeochemistry of acid mine drainage and a method to control acid formation, *Min. Eng.* 33(3):300–305 (1981). With permission.

the rate-limiting step in the propagation cycle. It is irrelevant in this model whether pyrite or marcasite, the orthorhombic polymorph of FeS_2, is considered to be the sulfide source; reaction 2 continues to be the rate-controlling reaction.

Field investigations of the oxidation of Fe^{2+} in natural mine drainage waters were conducted in the bituminous coal regions of West Virginia. Observations of the rate of this reaction in these acidic streams (at pH values closer to 3) showed that it proceeded considerably more rapidly than the laboratory studies at pH 3 predicted.

Many agents indigenous to these natural mine streams have been cited in the literature, in various circumstances, as displaying catalytic properties in the oxidation of Fe^{2+} by oxygen. The catalytic effects of sulfate, iron(III), copper(II), manganese(II), aluminum(III), charcoal, iron pyrite, clay particles and their idealized counterparts, alumina and silica, and microorganisms were investigated and compared in synthetic mine waters in our laboratory. Of these, microorganisms appeared to exhibit the greatest effect in accelerating the oxygenation of Fe^{2+}. Comparisons between the rates of oxidation of Fe^{2+} under sterile conditions after inoculation with untreated and with sterilized natural mine water showed that microbial mediation accelerates the reaction by a factor $>10^6$.

If the reaction scheme describing the oxidation of pyrite and the formation of acidity (reactions 1 to 3) is considered, it appears that the oxygenation of Fe^{2+} is the rate-determining step and that in natural acidic systems the reaction is greatly accelerated by microbial mediation. A similar mechanism is probably applicable to acidic leaching processes in other mines, such as copper mines, where iron is also invariably present.[3] If so, copper sulfides are oxidized by ferric iron and the resultant ferrous iron is reoxygenated, again with the aid of microorganisms, to form additional iron(III).

In a coal mine, pyrite degradation may occur in the self-accelerating cycle identified in Table 4.1 with pyrite being oxidized by ferric iron. Of the two ferric iron-producing equations shown in Table 4.1 (Equations 1 and 2), only ferrous iron oxidation is a significant source of ferric iron. Assuming that influent mine waters have an average total iron concentration of 0.5 mg/L, that ferrous iron oxidation is the only ferric iron source, and that no ferric iron is hydrolyzed, 93 turns of the pyrite degradation cycle (Equations 2 and 3, in Table 4.1) must occur to result in the release of 300 mg/L total iron commonly observed in mine drainage. If the average mine water residence time is 10 months, each pyrite degradation cycle can thus take no longer than 3.2 days. Ferric iron hydrolysis and precipitation (Equation 4) will naturally increase the number of cycles required and thus decrease the theoretical time available per cycle.

The iron oxidation-pyrite degradation cycle identified in Table 4.1 will proceed slowly at pH <4.5 in the absence of catalysis. This is because at pH <4.5, the cycle is rate-limited by the ferrous iron oxidation rate. Catalysis of ferrous iron oxidation can be biologically or abiotically mediated. The effect of chemical and physical catalysts on abiotic ferrous iron oxidation is a catalysis factor of <30. The iron bacterium *Thiobacillus ferrooxidans* significantly catalyzes iron oxidation with catalysis factors >300 under laboratory conditions. However, optimal activity of this organism is at pH <3.5. Thus, no mechanism exists for rapid catalysis of the iron oxidation-pyrite degradation cycle in the pH range 4.5 to 3.5. This chapter describes the effect of a filamentous iron bacterium on catalyzing the formation of the environment necessary for optimal *T. ferrooxidans* activity and thus for pyrite degradation.

The Role of Bacteria

The major reactions responsible for pyrite oxidation are summarized in Table 4.1. These four reactions reflect the current understanding of acidification mechanisms. The actual reaction process occurs in a multistage sequence dependent upon the activity of *T. ferrooxidans* and solution Eh and pH.

During the first stage of this process, fine-grained pyrite is oxidized either by *T. ferrooxidans* or by air with equal amounts of acidity produced by the oxidation of sulfide to sulfate (reaction 1) and by the hydrolysis of Fe^{3+} (included in reaction 2). The most reactive pyrite is the framboidal form due to the presence of pyrite granules <0.5 μm in diameter. The ability of *T. ferrooxidans* to accelerate the rate of pH decline is important, for each rainfall potentially interferes with the initial buildup of acid. During this stage, it is possible to forestall acidification by adding alkalinity to the reaction system; if alkalinity exceeds acidity, the only major downstream effect is an increase in sulfate concentration. Thus, adding crushed limestone or other sources of alkalinity to freshly exposed pyritic material can stop or postpone acidification. Once acidity changes to significant alkalinity, it becomes much more difficult to return an acid-producing system to stage 1, although the fall of pH is moderated as it approaches 4.5 by a decrease in the rate of reaction 2. For example, at $[Fe^{2+}]$ = 5 ppm, and pH = 5.5, the half-time for reaction 2 is 3 d; at pH 4.5, 300 days are needed for the oxidation of half the initial Fe^{2+}. This is due to the second-order dependence of the reaction on OH^- activity, as given by rate law:

$$-d[Fe^{2+}]/dt = k[Fe^{2+}][OH^-]^2 \cdot P_{O_2} \qquad (4.3)$$

where k = 8.0 (± 2.5) · 10^{13} L^2 mol^{-2} atm^{-1} min^{-1} at 25°C.

As abiotic oxidation of Fe^{2+} slows, *T. ferrooxidans* takes on its primary role of oxidizing Fe^{2+}, thereby allowing reaction 2 to continue producing acidity and ferric hydroxide.[4] This initiates stage 2 of the reaction process. Once again, it is possible for the pH to stabilize in this region, though this will usually occur only when permeability or the amount of exposed pyrite surface area is low. The pH decline otherwise continues to the third stage, where acid production is most rapid.

At pH <3, the increased solubility of iron and the decreased rate of $Fe(OH)_3$ precipitation results in increased Fe^{3+} activity. Stage 3 begins as Fe^{3+} activity becomes significant at a pH of approximately 2.5; a vicious cycle of pyrite oxidation and bacterial oxidation of Fe^{2+} results from the combined effects of reactions 3 and 4 (Table 4.1). The rate of reaction 3 exerts primary control on the cycle by limiting the availability of Fe^{3+}, the major oxidant of pyrite. The steady-state activity of Fe^{3+} is determined by the combined effects of bacterial oxidation of Fe^{2+}, reduction of Fe^{3+} by pyrite, and the formation of associated ferric sulfate and hydroxy complexes. Stage 3 includes the oxidation of both framboidal and coarser-grained pyrite.

The transition to stage 3 can be seen in the graph of total dissolved iron and acidity vs. pH which has been compiled from a series of experiments conducted in simulations of a coal refuse pile. Acidity was determined in the laboratory by titrating with NaOH to pH 8.3 after boiling with samples of hydrogen and is expressed as milligrams per liter of $CaCO_3$ as described in the American Public Health Association Standard Methods. The advantage of small laboratory simulations is their rapid acidification as compared to the months necessary for acidification of coal refuse piles in the field. The three stages of acidification can be observed in the laboratory data and have also been observed in the field.

An Eh pH analysis indicates that pyrite and $Fe(OH)_3$ cannot theoretically coexist at acid pH. The fact that they are commonly found together at the source of acid drainage demonstrates that pyrite oxidation is controlled by kinetic processes rather than by equilibrium chemistry.

The rise in Eh that accompanies stage 2 (due to the oxidation of Fe^{2+} by *T. ferrooxidans*) controls the manner in which Eh levels off as stage 3 is reached and steady-state cycling begins between reactions 3 and 4. The Eh limit is determined by the pyritic material and the activity of *T. ferrooxidans*; in laboratory experiments, stage 3 occurred at roughly 670 mV, which represents an $Fe^{3+}:Fe^{2+}$ activity ratio of approximately 0.02. This suggests that stage 3 may be initiated at relatively low Fe^{3+} activities.[5] The pH boundary of approximately 4.5 between stages 1 and 2 does not appear because neither the mechanism nor the rate of pyrite oxidation is affected by the transition from abiotic to bacterial oxidation of Fe^{2+}.

It should be noted that relatively high Eh is also possible at near neutral pH due to the rapid abiotic oxidation of Fe^{2+} and the low concentrations of iron in solution (on the order of 10 ppb). At higher iron concentrations and lower pH, Eh rises due to the activity of *T. ferrooxidans*. In general, the ratio of ferric/ferrous concentrations for all species also stabilizes at about 1:2 as stage 3 is reached, although drainage values can be much higher as reaction 3 proceeds without reaction 4. The difference in the iron activity and concentration ratios is primarily due to the fact that most aqueous iron in this system is present as sulfate and hydroxy complexes, which are included in the iron concentrations, but not in the Fe^{3+} and Fe^{2+} activities.

Once stage 3 is reached, acid production can only be reduced by slowing reaction 3. It is traditionally done by limiting the available oxygen, but the more direct route of inhibiting *T. ferrooxidans* is also possible.

Acid mine drainage is a dilute solution of sulfuric acid and iron sulfate with iron in ferrous and/or ferric form. Treatment of the AMD involves neutralization of the acid with a suitable alkali, oxidation of ferrous iron to the insoluble ferric form, and removal of the precipitates by a sedimentation process.[6] Even though one basic treatment process is applied, many options are available in each unit process or subprocesses.

CONVENTIONAL NEUTRALIZATION PROCESS USING LIME

In the conventional process, the five basic treatment steps are

- Equalization
- Neutralization (mixing)
- Aeration
- Sedimentation
- Sludge disposal

THE ACID MINE DRAINAGE PROBLEM FROM COAL MINES

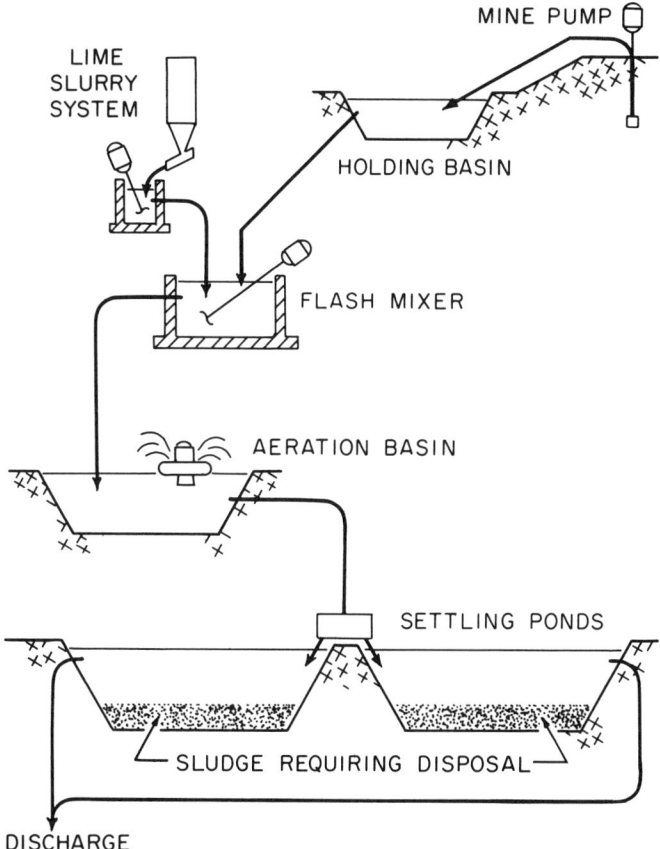

Figure 4.1. Conventional lime neutralization process.[7]

A flowsheet for a typical system is shown in Figure 4.1.

A constant flow with only small variations in quality is desired so as to minimize needed controls and operator attendance.[7] The mine drainage is collected in large holdings or equalization basins or in large sumps within the mine. The holding basins should have a capacity for 2 to 3 days' flow. Usually, 12 to 24 hr of flow is maintained in the holding basins to equalize flow and quality to the treatment facility. From the holding basin, the mine drainage flows by gravity to the treatment plant. The holding and settling basins are usually of earthen construction.

Lime is used as the alkali for neutralization of the acid in almost all large treatment facilities; however, the selection between quicklime and the hydrate is determined by cost and availability.

Aeration is used for oxidizing ferrous iron to the ferric form, which is less soluble in water. Ferrous iron is much more soluble than the ferric iron. The solubility of ferrous iron occurs in the pH range of 9.3 to 12.0. Ferric iron is much less soluble and begins to precipitate as hydroxide at a pH of 4.0, with minimum solubility occurring at about pH 8.0. An economic advantage id obvious in removing iron in the ferric form at the lower pH as less lime is needed for neutralization to the pH level needed to maintain minimum iron solubility (8.0 vs 12.0).

The oxidation of ferrous iron is commonly included in the AMD treatment processes. The oxidation is pH dependent, with the reaction proceeding rapidly at a pH above 8.0. At this pH level, iron oxidation becomes dependent upon the availability of oxygen. Theoretically, one unit weight of oxygen is required for seven unit weights of ferrous iron to be oxidized.

After neutralization and oxidation of ferrous iron, the subsequent step is sedimentation. Settling of iron hydroxide and other suspended solids is accomplished in settling basins. The basins must have a capacity of at least 12 hr of clear water detention above the sludge storage zone. When small settling basins with 12 to 48 hr detention and minimal sludge storage capacity are used, two units operated in parallel are a common practice to allow sufficient time for sludge removal. If plant site conditions allow, large impoundments that provide many years of sludge storage can be advantageous. Adequate planning for sludge handling and disposal is essential in all treatment processes. Sludge removal contributes significantly to both construction and operating costs of the AMD facilities. The simplest method for final disposal is to pump the sludge into abandoned deep mines. This practice is common, but the overall environmental effects of this disposal method have yet to be determined. Another method used is lagooning, wherein the sludge thickens naturally. Eventually the sludge must be disposed of in a more satisfactory manner, such as burial in a reclamation project. Among other methods used for sludge dewatering are drying beds, vacuum, and pressure filtration.

High-Density Sludge Process

One variation in the conventional lime neutralization process includes the sludge recirculation process. This process uses lime for neutralization and can provide a sludge with substantially less volume than is produced in the conventional process. The process is based on a high sludge recirculation rate within the system, in which the optimum ratio of solids recirculated to solids removed is in the range of 20:1 to 30:1. The sludge is transported to a reactor vessel where the lime slurry is added. The slurry is then mixed with the AMD in a neutralization reactor, and here aeration is provided for oxidizing ferrous iron.

Other Treatment Processes

Many other processes are available for the treatment of AMD wherein acid concentration is the primary problem and the flow of mine water is low. Other treatments such as soda ash, caustic soda, or limestone can be used. Portable caustic soda treatment units are common in surface mine operations. Limestone has been used in several applications for in-place treatment.

THE ACID MINE DRAINAGE PROBLEM FROM COAL MINES

Besides the conventional treatment of AMD, other methods are available to produce water of higher quality: reverse osmosis, ion exchange, and chemical softening, among others.[7]

CHEMICAL TREATMENT

Where the formation of AMD cannot be prevented or its discharge cannot be controlled, chemical treatment is necessary before the mine water can be discharged. Lime treatment is the most commonly used system. The chemistry of the lime treatment process is as follows:

$$Ca(OH)_2 + H_2SO_4 \rightarrow CaSO_4 + 2H_2O \qquad (4.4)$$

$$Ca(OH)_2 + FeSO_4 \rightarrow Fe(OH)_2 + CaSO_4 \qquad (4.5)$$

$$3Ca(OH)_2 + Fe_2(SO_4)_3 \rightarrow 2Fe(OH)_3 + 3CaSO_4 \qquad (4.6)$$

The choice of an alkali in the neutralization process is influenced by factors such as cost, suitability, relativity, availability, ease of use, and sludge volume. Each of these factors should be carefully evaluated because each alkali requires significantly different equipment, and the alkali selected may affect the design of other processes in the overall system.

Lime

Lime is a general term that, by definition, encompasses only burned forms of limestone. The two forms of lime of particular interest in AMD treatment are quicklime and hydrated lime.

Quicklime (CaO) is produced by the calcination of limestone. Limestone generally consists of 50 to 90% calcium carbonate. Based on chemical analysis, quicklime may be divided into three categories:

- High-calcium quicklime — containing <5% magnesium oxide
- Magnesium quicklime — containing 5 to 35% magnesium oxide
- Dolomitic quicklime — containing 35 to 40% magnesium oxide

Quicklime is available in different standard sizes, but the major types used in AMD treatment include ground lime and pulverized lime. The ground lime is produced by grinding the larger-sized material and screening off the fine size. A typical size is –#8 mesh and 40 to 60% –#100 mesh. Pulverized lime is produced by intense grinding to produce a size between –#20 mesh and 85 to 95% –#100 mesh. Quicklime is usually delivered in bulk carloads or trucks and then transferred to a storage silo.

For efficient use in mine drainage neutralization the quicklime must be slaked. The slaking process must be carefully controlled and requires daily attention. Well-slaked quicklime offers a low unit cost per gram of acidity neutralized. Major disadvantages include high capital investment for a slaker, grit removal, close operational control, and the danger from possible severe burn injuries to personnel.

Hydrated lime ($Ca(OH)_2$) is the most commonly used alkali for neutralizing AMD in existing treatment plants when lime consumption rates are low or when the cost of a slaking system is prohibitive.

An air classification system is used to produce the fineness necessary to meet the process requirements. A common size is 75 to 95% –#200 mesh. The commercial hydrated lime is purer than quicklime.

Storage and handling of lime in AMD treatment plants require careful attention. Most often bulk lime, either quick or hydrated, is more efficient and economical to use. The bulk lime is delivered by truck and conveyed by mechanical or pneumatic systems into weather-tight bins or silos. Bagged lime is delivered loose or palletized in truck or boxcar, and generally handled by hand truck or forklift to storage. Bagged lime should be stored in dry areas. Hydrated lime is normally packaged in multiwall paper bags. Moisture can permeate the liner and cause caking. Dry storage is important. The use of bulk lime can yield considerable savings over the use of bagged lime, not only in initial cost, but also in reduced labor costs for less handling and for other advantages including faster loading, elimination of losses from broken bags and spillage, better housekeeping by modern handling systems, and less dust hazard to employees.

Bulk lime is commonly delivered to the treatment plant by truck. The blower truck is the fastest, most common, and most economical. The lime is blown from the truck directly to silo storage via a pipeline. Conventional steel or concrete bins and silos can be used for lime storage as quicklime and hydrated lime are not corrosive. A steel silo with a cone bottom is the most popular.

The flowability of lime is an important inertia in storage bin design. The flowability varies from good, for pebble and granulated quicklime, to erratic, for hydrated and pulverized quicklime. Lime can absorb moisture quickly, forming a sticky, soft cake that can reduce flowability in the bins. Hydrated lime tends to form ratholes. In order to avoid the problems inherent to lime flowability, special design in bins, external vibrators, internal antipacking and antiarching devices, and live bin bottoms are used.

Currently AMD treatment plants utilize dry lime in a liquid suspension or slurry before feeding it into the raw water, but new practice is to feed dry hydrated lime directly to the acid water, thereby eliminating a slurry feed system. Dry lime feed has been used when the drainage streams are small and mildly acidic, the lime requirements being <0.1 kg/1000 L. For drainage systems requiring larger amounts of lime, for example, 0.36 to 0.48 kg/100 L, a special feeding arrangement such as a larger aeration or a flash mixer is used for achieving complete mixing.

The efficiency of an AMD treatment plant depends on the speed and accuracy with which the lime is fed to the process. Various types of feeders available for use include:

- Volumetric feeder
- Screen feeder
- Oscillating hopper
- Belt feeder
- Rotary paddle
- Vibrating feeder

As explained earlier, slaked lime is often used in AMD treatment plants. The term "slaking" refers to the combination of varying proportions of water and quicklime which yields milk of lime, a viscous paste of lime. Continuous slakers have largely replaced manually operated batch slaking. The variables exerting influence on slaked hydrated lime include:

- Reactivity of the quicklime
- Particle size and gradation of quicklime
- Optimum amount of water
- Temperature of water
- Distribution of water
- Agitation

Concentrations of lime solids in the milk of lime slurry vary between 5 to 20%, with 10% being very common. Lime concentrations can be checked for specific gravity by hydrometer. Automatic pH control systems for the feeding of lime solutions into the flash mix tank are commonly used.

The diluted paste or slurry is transferred to the neutralization tank. Scaling can become a serious problem in the transfer process. Lime is only slightly soluble. At saturation level the solubility of lime is 1.7 g/L. With an increase in temperature the solubility of lime decreases to 0.55 g/L at 90° to 100°C. It is economically desirable to carry lime in a much more concentrated form such as suspension. Because lime slurry has a pH higher than 12.0 it softens the water and calcium carbonate is precipitated. Calcium carbonate forms a dense, hard scale. If left unattended the scale will build up and clog pipes. No foolproof solution exists; however, some corrective measures can be taken. Scaling can be minimized by using sodium hexametaphosphate, which softens the slaking or dilution water so the calcium carbonate does not scale up in the pipes.

The feeder location should be such that the slurry flows by gravity directly into the mixing tank or the solution. Scaling at the end of the slurry pipe can be avoided by discharging slurry through an air gap into the open solution tank. Heavy-duty flexible rubber hoses are preferred over metal pipes. The components of the slurry feed system such as piping, valves, pumps, and configuration should be designed so that they provide quick-disconnect, easy-to-assemble fittings and valves; avoid confined spaces; and enhance flow patterns. Slurry feed systems have used heavy-duty plastic pipe (Schedule 80), flexible rubber hoses, and galvanized or stainless steel pipes. The pipe material choice is determined in the consideration of economics, anticipated problems, type of slaking water, and ease of assembly.

A flushing system is not essential, but highly recommended for systems with frequent shutdowns; for example, AMD plants not operating daily or operating only in wet weather. A minimum flow velocity of 1.0 to 1.2 m/sec should be maintained within the slurry loop.

The type of valves used in the slurry feed system determines the degree of maintenance. Pinch valves have proven the most successful. In fully closed or fully open situations ball or plug valves can be used.

A tight control on the metering system is essential for the bleed-feed slurry system. Material used in the tank construction should be resistive to the high pH of the slurry and suitable from a structural standpoint. The configuration is determined by the slope available. Tanks exceeding 9500 L (2500 gal) should be fitted with baffles set 90° apart to prevent vortex formation. Adequate agitation in slurry storage tanks must be provided.

AMD treatment plants do not need a sophisticated slurry feed control system. Usually the drainage flow is constant in quantity and quality because equalization basins are used. The system can be efficiently controlled by the pH of the flow leaving the neutralization tank. This pH signal can control the dry lime feeder, a lime solution feeder, or a control valve. Another type of lime slurry feeding involves proportioning the slurry flow in response to the pH signal.

Pumps used for lime slurries generally fall into two categories: controlled volume and centrifugal. Centrifugal pumps are more popular for a wide range of slurry flows as they are inexpensive and their standard design incorporates flow patterns that lend themselves to easy slurry transfer. Two configurations have been frequently used, as shown in Figures 4.2 and 4.3.

Limestone

Considerable interest has arisen in the possible use of limestone for the treatment of AMD. Limestone equals about 30% of the cost of quicklime or hydrated lime. Two types of limestone are found: the dolomitic and the high-calcium. The high-calcium type has potential application in AMD treatment.

A typical chemical composition of limestone is

	Calcium Limestone % Composition
Calcium oxide (CaO)	53.00–56.00
Magnesium oxide (MgO)	0.12–3.11
Calcium carbonate ($CaCO_3$)	92.66–98.60
Silica dioxide (SiO_2)	0.10–2.89

THE ACID MINE DRAINAGE PROBLEM FROM COAL MINES

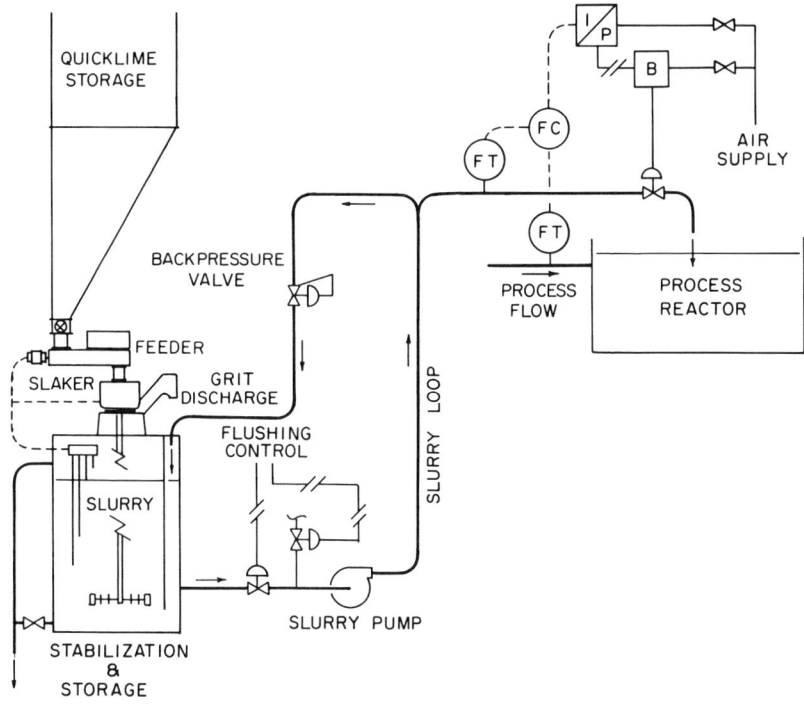

Figure 4.2. Slurry feed with pH control loop.[7]

The reaction of limestone with the components of AMD can be represented by the following equations:

$$CaCo_3 + H_2SO_4 \rightarrow CaSO_4 + H_2CO_3 \quad (4.7)$$
$$3CaCo_3 + Fe_2(SO_4)_3 + 6H_2O \rightarrow 2CaSO_4 + 2Fe(OH)_3 + 3H_2CO_3 \quad (4.8)$$
$$3CaCo_3 + Al_2(SO_4)_3 + 6H_2O \rightarrow 3CaSO_4 + 2Al(OH)_3 + 3H_2CO_3 \quad (4.9)$$

To utilize limestone effectively as a neutralizing agent, certain quality criteria must be met:

- Minimum particle size, preferably a –#325 mesh
- High calcium content
- Low magnesium content
- High specific (surface) area

For treatment of ferrous iron, limestone use is also possible, but limitations include slow reactivity of limestone and its inability to increase the pH above 7.0. At

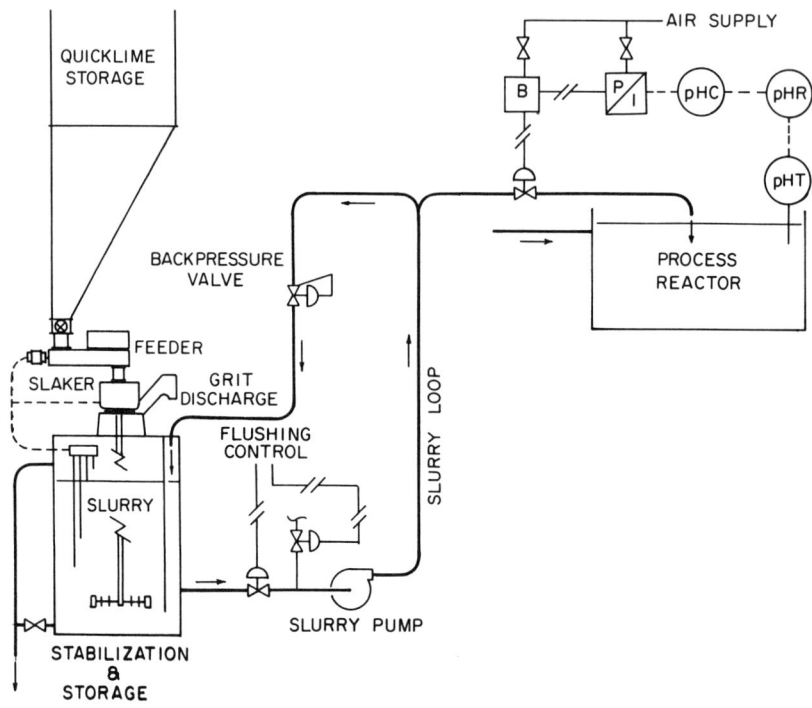

Figure 4.3. Slurry feed by flow proportioning.[7]

a pH level of 7.0, the slow rate of oxidation of ferrous iron is a deterrent. Excessive aeration time is needed for complete oxidation of the ferrous iron.

The design approaches for limestone treatment of AMD include the following:

Ferrous Iron Conc. (mg/L)	Response to Limestone Treatment
0–50	Effective treatment may be achieved without pre- or postneutralization iron oxidation
50–100	May be effectively treated, but requires postneutralization aeration and significant retention time
>100	Potential treatment is uncertain

Important factors to be considered in designing a limestone treatment process include:

- Specifications for limestone grade, size, and hardness
- Mode of operation (mixer or rotary mill)

- Supplementary reagent requirements (live, polymers)
- Operating pH and aeration requirements for ferrous iron oxidation
- Recycle volume ratio
- Sludge settleability

If iron is mostly in the ferric form (at least a 4:1 ferric:ferrous ratio) treatment with limestone is feasible, but limestone utilization efficiency is only 32%. Therefore, the economic advantage over hydrated lime treatment may not exist.

Caustic Soda

Caustic soda can be utilized to treat AMD for low-flow, mildly acidic drainage problems. Generally it is used for surface mine drainage problems in remote areas. The treatment process consists of a horizontally mounted 38 m^3 (10,000 gal) storage tank and a flume-type chemical feeder. In surface mining operations caustic soda is frequently used to treat pit water. A schematic layout is shown in Figure 4.4. The disadvantages of caustic soda use in AMD include high cost, dangers in handling, and special design requirements. Sodium hydroxide produces an excellent effluent quality that is low in suspended solids, turbidity, and iron content.

Soda ash (Na_2CO_3) is rarely used to treat large-flow AMD facilities. Due to its high cost and limited availability, soda ash has been used for treatment of low-flow drainage that contains little ferrous iron such as would occur in some surface mines. It is selected primarily for convenience rather than cost efficiency.

IRON OXIDATION

Acid mine drainage usually contains significant iron concentrations that result from the oxidation of pyritic minerals present in coal seams. The effect of the process of oxidation on pyrites produces ferrous sulfate and sulfuric acid. These salts readily dissolve in water forming the AMD.

Iron is initially present in the ferrous (Fe^{2+}) form. Ferrous iron can be oxidized to ferric iron which is much more soluble and hence can be precipitated as a hydroxide to effluent quality levels below the allowable pH of 6.0. The minimum solubility of ferric iron occurs at a pH of 8.0 (Figure 4.5), while ferrous iron does not reach minimum solubility until the pH approaches 11.0. At the maximum allowable discharge pH of 9.0, ferrous iron is soluble to about 4 mg/L, which exceeds the discharge limitations for new sources. Therefore, in most AMD treatment systems it is imperative to oxidize any ferrous iron to the ferric form and then to remove it at lower system pH. Methods used for this oxidation include:

- Mechanical aeration
- Chemical oxidation
- Biological systems

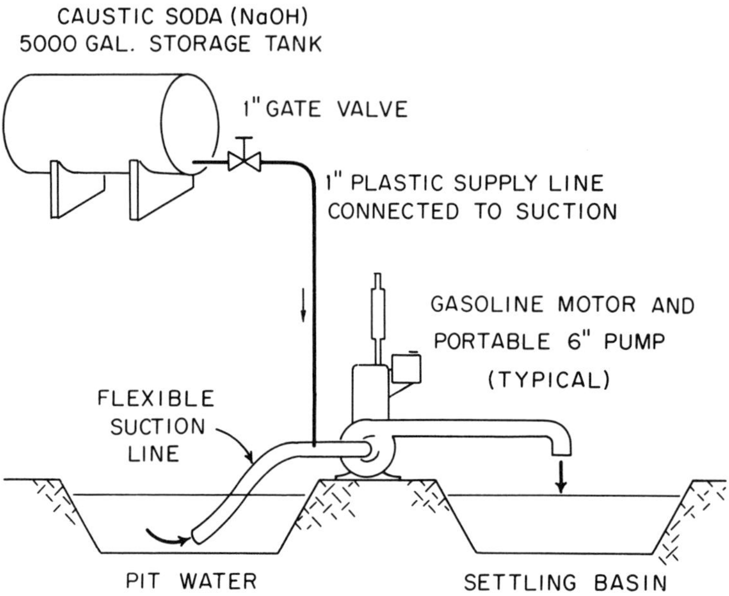

136   ENVIRONMENTAL IMPACTS OF MINING

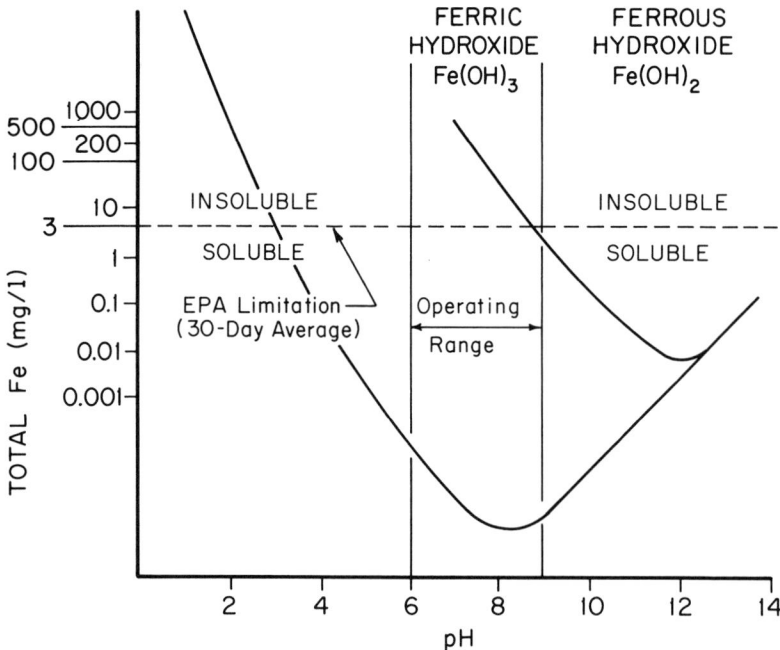

Figure 4.4. Portable caustic soda feed arrangement.[7]

Figure 4.5. Solubility of ferric and ferrous iron at various pH.[7]

Aeration Systems

Ferrous iron, when exposed to oxygen, oxidates to ferric iron. The rate of oxidation depends upon ferrous iron concentration, the dissolved oxygen concentration, and the pH of the solution. At pH values higher than 6.0, the reaction occurs according to the following rate equation:

$$\frac{-d}{df}(Fe^{2+}) = k\,(Fe^{2+})\,(O_2)\,(OH^{-2}) \tag{4.10}$$

The reaction is a first-order reaction with respect to the ferrous iron and the dissolved oxygen concentration. The oxidation rate decreases as the concentration of ferrous iron or oxygen decreases. The reaction rate is second order with respect to the hydroxyl ion (OH⁻) concentration for pH values >6.0. The reaction rate increases 100 times for each one-unit rise in pH above 6.0. The rate of ferrous iron oxidation is extremely slow at a pH of <3.0, slow in the pH range of 3.0 to 6.0, moderate to fast in the pH range of 6.0 to 8.0, and rapid above this point. At a pH level of 5.0, ferrous iron will precipitate as ferric hydroxide sufficiently to meet the effluent limits at a pH of 5.0. At this pH value, however, the oxidation rate for ferrous iron is slow. Until the pH is 8.0 or greater, the oxidation rate does not increase until the pH value reaches 8.0 or higher. When iron occurs mostly in the ferrous form, aeration processes are most efficiently operated within a pH range of 8.0 to 9.0, when oxidation takes place in a matter of minutes. At this stage the controlling parameter for the design of the aeration unit becomes a function of the oxygen transfer efficiency and not the chemical reaction of oxygen and iron. The aerator should be designed to provide dissolved oxygen saturation in the aeration basin with maximum oxygen transfer.

The capacity of the aeration system is determined by the amount of iron to be oxidated. If the oxygen requirements cannot be met the oxidation will be incomplete. The oxidation rate increases as the concentration of oxygen dissolved in water increases to its saturation point. Aeration capacity in excess of the saturation point is not beneficial.

The rate of oxygen transfer into water depends on the initial oxygen deficiency of water. It is easier to dissolve oxygen by aeration if the initial oxygen concentration is lower.

The oxidation rate of ferrous iron depends on the dissolved oxygen concentration, with the maximum rate occurring at saturation. For optimization of the aeration process, these two mechanisms must be compromised. At a pH >8.0 the oxidation rate is sufficient so that oxygen concentration near saturation level is not necessary. If aeration is performed at a pH level <8.0, then a fairly high level of dissolved oxygen should be maintained.

The chemical equations for the oxidation of ferrous iron to ferric iron and hydrolysis of ferric iron are

138   ENVIRONMENTAL IMPACTS OF MINING

$$Fe^{2+} + 1/4 O_2 + H^+ \rightarrow Fe^{3+} + 1/2 H_2O \tag{4.11}$$

$$Fe^{3+} + 3H_2O \rightarrow Fe(OH)_3 + 3H^+ \tag{4.12}$$

These chemical equations indicate that 1 kg of oxygen will oxidate 7 kg of ferrous iron under ideal conditions. In these reactions, 1 mol of acidity (as H_2SO_4) is formed for each mole of ferrous iron oxidated. Sufficient alkalinity must be added to neutralize the extra acid formed and to maintain optimum pH conditions.

The aeration system chosen must meet the oxygen demand for ferrous iron oxidation. The theoretical oxygen demand for any mine water can be calculated from the following equation:

$$O_2 = QW \times Fe \times 5.16 \times 10^{-4} \tag{4.13}$$

where O_2 = theoretical O_2 demand (kg O_2/hr)
QW = AMD flow rate (L/sec)
Fe = Initial concentration of Fe^{2+} (mg/L)

Atmospheric air contains about 21% of oxygen by volume. Only a fraction of the oxygen in the air that comes in contact with water, called oxygen transfer efficiency, is actually absorbed by water. This fraction differs for each aeration system and operating conditions. The total air needed to supply the theoretical oxygen quantity demanded can be calculated by the following equation:

$$Qa = \frac{6.324 \times O_2}{E} \tag{4.14}$$

where Qa = total air demanded (m³/min)
O_2 = theoretical oxygen demanded (kg/hr of oxygen)
E = oxygen transfer efficiency (as %)

Oxygen transfer efficiencies (E) range from 3 to 25% depending upon the type and size of aerator and the depth of submergence.

The aeration system must also be capable of keeping the ferric hydroxide solids and unreacted reagent in suspension. If the mixing is insufficient these solids will settle to the bottom of the basin. Settled solids will reduce the aeration volume and aeration time causing incomplete ferrous iron oxidation. Thus, the aerator must be designed to meet both oxygen and mixing requirements.

Aeration processes that dissolve atmospheric oxygen in mine drainage water can be classified into four types:

THE ACID MINE DRAINAGE PROBLEM FROM COAL MINES

- Mechanical surface aeration
- Submerged turbine aeration
- Cascade aeration
- Diffused air aeration

The aeration basin should be efficiently designed for complete oxidation of ferrous iron without requiring excessive aeration time. Important parameters are basin plan, depth, and inlet and outlet structures. Aeration basins are excavated in the ground and leveled with riprap or asphalt. They also can be constructed as concrete or steel structures.

Mathematical models have been developed to predict aeration times for the oxidation of ferrous iron at varying pH ranges. The nature of mine drainage is variable and the effects of the other dissolved ions on the reaction are not well known. Therefore, laboratory tests are the most reliable way to optimize the aeration system design. The tests should be conducted on a sample containing the maximum expended ferrous iron concentration and at the lowest anticipated operating pH level in the AMD.

The detention time needed for ferrous iron oxidation must be multiplied by a safety factor for the design of the operating aeration basin. The capacity of the aeration basin is determined from the following formula:

$$V = Q \cdot D_t \cdot f \tag{4.15}$$

where V =   volume (m^3)
        Q =   flow (m^3/sec)
        D_t =   detention time (sec)
        f =   safety factor

The aeration basin must be designed to efficiently accommodate the aerator, and the entire volume of the basin should be well mixed and aerated.

Besides aeration, chemicals have been used for iron oxidation. Ozone and hydrogen peroxide have been used.

Theoretically, 1.0 kg of ozone will oxidize 2.3 kg of ferrous iron. The same amount of acid is released during ozone oxidation as during aeration. At 86% ozone utilization, 1.0 kg of ozone will oxidize 2.0 kg of ferrous iron. The advantages of ozone treatment over lime aeration include:

- The oxidation reaction is efficient and quick
- Close process control needed in lime treatment process is not required
- Neutralization to pH 6.0 is all that is needed
- The sludge produced by limestone-ozone reaction is denser than lime sludge, reducing sludge handling requirements

Hydrogen peroxide can be used when specific conditions exist:

- Alkaline mine drainage (pH higher than 6.0) with 10 W oxygen requirements for iron oxidation values
- Need for a supplemental oxidizing source when the system is overloaded with iron and expansion is impossible

The chemical reaction of ferrous iron with hydrogen peroxide is given by the following equation:

$$H_2O_2 + 2Fe^{2+} + 2H^+ \rightarrow 2Fe^{3+} + 2H_2O \qquad (4.16)$$

According to the above equation, 0.45 kg of H_2O_2 will oxidize 1.5 kg of ferrous iron. A hydrogen peroxide feeding system is shown in Figure 4.6.

Biological Oxidation

Bacteria capable of oxidizing ferrous iron exist naturally in most acid mine drainages. As explained earlier, these bacteria, *T. ferrooxidans,* act as a catalyst in the formation of AMD. These bacteria obtain their carbon and nitrogen requirements from inorganic sources and their energy from the oxidation of ferrous iron. Researchers have proposed several methods to utilize their ability to oxidize ferrous iron in mine drainage. Dispersal growth systems have not been successful. Fixed growth systems, such as trickling filters and rotating biological contractors (RBC), are effective in supporting the bacterial growth necessary for oxidation in a trickling filter.

The RBC is an aerobic treatment device consisting of a series of four plastic discs mounted on a horizontal shaft. This assembly is placed in a trough through which the wastewater flows, submerging almost half of the surface area of the discs. The discs rotate slowly on the shaft, causing the biological growth on the discs to alternately contact the air and water (Figure 4.7).

Experimental data have been obtained at two peripheral disc velocities, 0.32 and 0.17 m/sec, and at five hydraulic loadings ranging between 110 and 440 e/d/m². A linear relationship has been observed between ferrous iron removal and stage retention time. A faster disc velocity produced better iron oxidation at any constant hydraulic loading. At a given disc velocity, increased hydraulic loading resulted in an increased effluent ferrous iron concentration, even though the ferrous iron oxidation rate also increased. The RBC can produce an effluent containing <10 mg/l ferrous iron at loading rates up to 88 kg Fe^{2+} applied per 1000 m² of disc surface.

Tests have been conducted on AMD flowing overland for more than 1 mi before reaching the treatment facility, thereby exposing the water to stream ecology and ambient air temperatures. Observations during the winter indicated that the effect of low temperatures on the biological oxidation process is not significant. At 0.4°C, the

THE ACID MINE DRAINAGE PROBLEM FROM COAL MINES

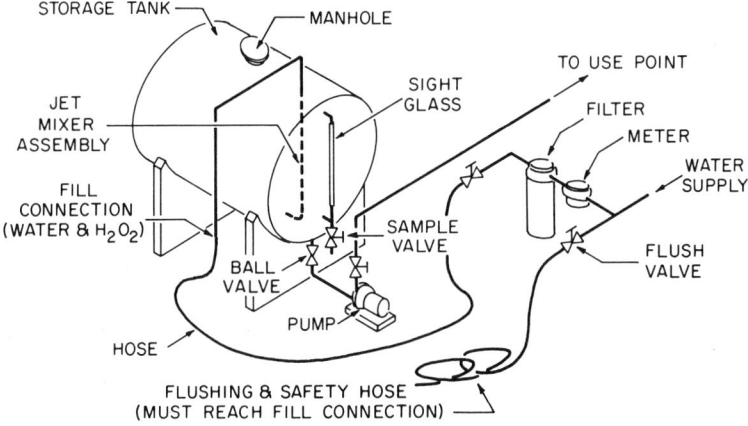

Figure 4.6. Hydrogen peroxide feeding system.[7]

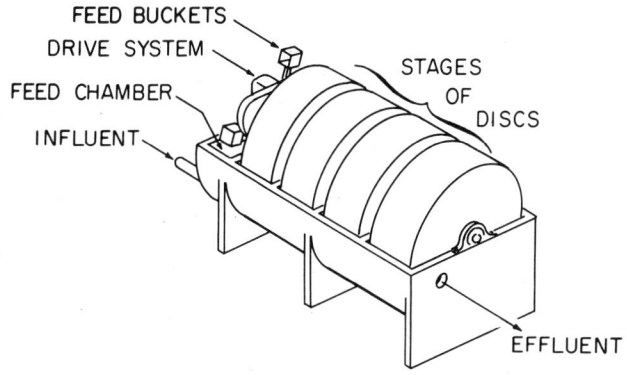

Figure 4.7. Rotating biological contactor.[7]

lowest temperature recorded, the removal efficiency decreased by only 10% below the efficiency observed at 10°C, the initial mine water temperature. This effect can be compensated for by a lower hydraulic loading.

It may not be possible to remove ferrous iron to 1.0 mg/ℓ with RBC. However, this degree of removal by a biological system is not necessary. An effluent of 10 mg/L of ferrous iron is adequate for RBC. The effluent will require neutralization which will provide enough aeration to oxidize any remaining ferrous iron.

Design of a four-stage single-shaft RBC system can be made using the following equations. The equations for disc area surface area can be determined from:

$$A = \frac{F_o Q 86.4}{L} \qquad (4.17)$$

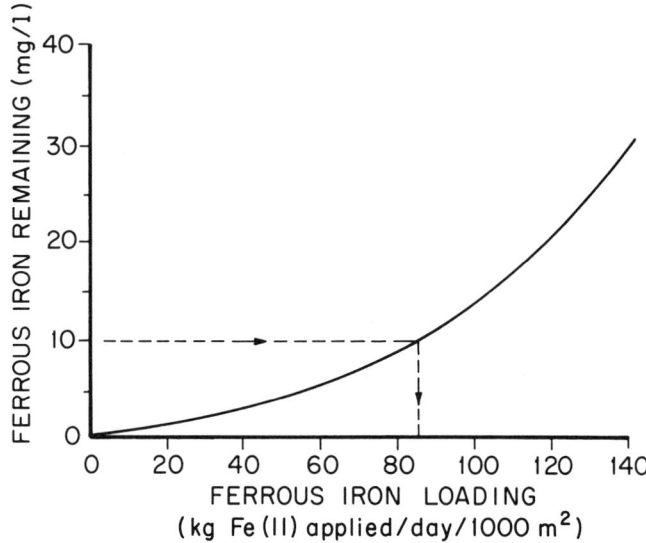

Figure 4.8. Design procedure for a four-stage rotating biological contactor.[7]

where   A   =   disc surface area (m^2)
           Fo =   initial ferrous iron concentration (mg/e)
           Q   =   flow (p/sec)
           L   =   ferrous iron loading (kg/d/1000 m^2) (calculated from Figure 4.8)

The designer chooses the desired effluent ferrous iron concentration from the RBC system listed on the vertical axis in Figure 4.8. A line is drawn horizontally to intersect the curve, and then from the point of intersection, a vertical line is dropped to the horizontal axis. This point on the horizontal axis gives the maximum ferrous iron loading (L) that will yield the desired effluent. This value of L, and the initial iron concentration (Fo) and the flow (Q) are applied to the equation, yielding the necessary disc area.

Oxidation Rate

The oxidation rate of ferrous iron is dependent upon the ferrous iron concentration at any point in time. The reaction can be expressed as:

$$\frac{-d}{dt}(Fe) = K(Fe) \tag{4.18}$$

The integrated form of the equation is

$$Fe = Fe_0 e^{-kt} \tag{4.19}$$

THE ACID MINE DRAINAGE PROBLEM FROM COAL MINES

where Fe_o is the initial ferrous iron concentration and Fe is the ferrous iron concentration at time t.

The slope of the graph, log Fe vs time, which is a straight line, determines the constant K, the reaction rate.

$$K = \text{Slope} \times 2.303 \qquad (4.20)$$

The oxidation rate test should be performed on fresh samples of mine drainage to minimize any natural oxidation of the ferrous.

A small sample is taken and carefully preserved for analysis to determine the initial ferrous iron concentration. Lime or caustic soda is added to raise the pH to the desired level. Air is supplied through an aeration stone.

Thickeners and separator units receive clarifiers underflow with a high percentage of solids. They are used to produce an acceptable overflow and to store sludge to produce denser underflows.

Tilted-plate gravity settlers are sometimes used instead of clarifiers in instances in which very limited space is available. They are inclined-plate, shallow-depth settling devices that perform the same function as a clarifier but occupy only one tenth the space.

SLUDGE DEWATERING AND DISPOSAL

Sludge is costly to handle, dewater, and dispose. In AMD treatment plants, sludge disposal presents the most recurrent and demanding problem. Environmental regulations categorize sludge as a potential pollutant. In AMD treatment plants, two effluents are produced: treated water and sludge.

The volume of sludge to be disposed depends on the chemical composition of mine drainage and the neutralization method used. Commonly, the sludge volume is 5 to 10% of the daily flow through the treatment plant.

$$V = Q t_d \qquad (4.21)$$

where t_d = detention time (min)
V = volume of sludge (m^3)
Q = design flow (m^3/min)

Additional capacity must be provided for sludge accumulation, which is estimated to be 5 to 10% of the average daily volume treated. Sludge storage volume requirements in a settling basin depend on the frequency of sludge removal. Ponds without sludge removal facilities should allow sufficient volume for sludge storage between withdrawal operations. In some cases, basins have high volumes to hold sludge for the life of the treatment plant.

Many AMD treatment plants use circular-shaped clarifiers for liquid-solids separation when limited land area is available. A clarifier is a gravitational liquid-solids separator having the primary objective of producing clean overflow regardless of underflow solids content. Sizing of the clarifier is made from the following formula:

$$V_{RR} = Q / A \tag{4.22}$$

where V_{RR} = rise rate (m/min)
  Q = design flow rate (m^3/min)
  A = cross-sectional area of clarifier (m^2)

Two types of earthen ponds are commonly used: settling ponds and impoundments. Settling ponds are small and designed permanently for settling with periodic sludge removal. The sludge is disposed of or recycled to utilize the untreated portions of the neutralizing agent. In settling ponds, surface baffles prevent short circuiting. Submerged baffles are used for sludge confinement. The two types of baffles are illustrated in Figure 4.9a and b.

Many factors are considered in calculating the size of sedimentation basins. Important factors include:

- Detention time
- Sludge removal
- Disposal method
- Mode of operation

The basin volume can be calculated, using the detention time and design flow, with the following equation:

$$V = Qt_d \tag{4.23}$$

where V = volume of settling pond, without sludge storage (m^3)
  Q = design flow (m^3/min)
  t_d = detention time (min)

Efficient design of inlet and outlet is important in providing quiescent conditions for good settling pond performance. Inflow should be uniformly distributed over as much of the pond width as possible. Multiple inlets or a continuous-width, multiple V-notch box weir offers an efficient design. This reduces the inlet flow velocity, reducing the possibility of washout or resuspension of solids. Bad inlet design can cause short circuiting, reducing the detention time and removal efficiency of the settling pond. Dead areas of noncirculating water can be created in the settling pond, creating channels of flow within the pond. Uniform distribution of influent across the width of the settling pond inhibits isolated mounding of settled particles within the

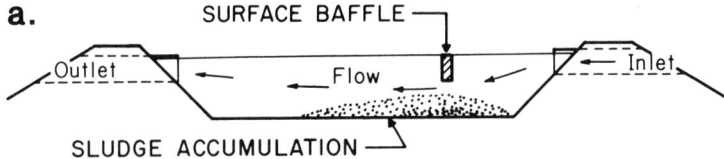

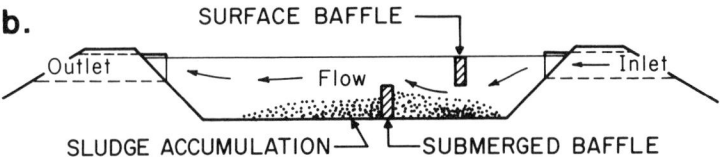

Figure 4.9. (a) Surface baffled pond.[7] (b) Combination surface and submerged baffled pond.[7]

basin, maximizing sludge storage volumes. Similarly, outlet openings should also be multiple or continuous to maintain low exit velocities. High exit velocities create turbulence that can resuspend settled solids, causing poor effluent quality.

Baffles, placed selectively in a settling pond, prevent short circuiting. Two types of baffles, surface and submerged, are used. Clays and silty clays are excellent compounds for providing an impermeable layer. Plastic membrane liner can be placed in areas where fill material is unsuitable. Soil conditions must provide stable support for pond embankment foundation, along with necessary imperviousness to passage of water. Mixtures of coarse- and fine-textured soils like gravel-sand, sand-clay, and sand-silt provide good stability and resistance to the passage of water. Layers of clay and silty clay are often used to ensure tight contact between embankment and foundation. The basin bottom should not be founded on bedrock or on stony, rocky soils. A thick layer of relatively impervious consolidated material can form the most suitable foundation.

The embankments forming the sidewalls of the basin should be constructed of impervious materials similar to those used in the basin's liner. The design of the embankment elements, such as slope of the sidewalls, height, crown width, etc., should provide stability. The inside slope of the embankments should be protected against erosion caused by wave action. Piling of a 0.61-m wide collar of riprap at the expected upper level on the embankment is an effective method of erosion protection.

In AMD treatment plants, settling units vary in size, configuration, and method of solids removal. Excavated settling ponds in the ground are most popular for economic reasons. Mechanical clarifiers or thickeners may be preferred as they enable the operator to exercise control over the treatment process and improve sludge densities.

Before sizing and designing settling ponds or clarifiers, tests should be conducted to determine the behavior and characteristics of the sludge and quality of the incoming fluid. The tests determine sludge settling velocity, optimum pH, best neutralizing agent, dosage rate, sludge density, and volume.

Earthen ponds are commonly used as settling basins. Earthen basins should not be located in swamps, marshes, on floodplains, or over abandoned wells or mine workings. A layer of impervious material about 0.6- to 0.9-m thick should be placed at the bottom of the pond to prevent seepage.

The performance of the separator is related to other variables:

- Flow turbulence in the basin
- Velocity distribution throughout the basin
- Particle interaction
- Particle resuspension

Expressed empirically:

$$\frac{\text{Depth}}{\text{Settling velocity}} = \text{Detention time} \tag{4.24}$$

$$\frac{D}{V_s} = \frac{L \times W \times D}{f} \tag{4.25}$$

or

$$V_s = \frac{F}{L \times W} = \frac{\text{Flow rate}}{\text{Area}} \tag{4.26}$$

where V_s = settling velocity (m/min)
  D = depth (m)
  L = length of pond (m)
  W = width of pond (m)
  F = flow rate (m³/min)

Therefore, all particles with settling velocity higher than or equal to hydraulic surface loading (flow/area) are removed.

Settling velocity is influenced by many other factors as well:

- Inlet and outlet devices
- Wind-induced turbulence
- Nonquiescent flow current
- Flocculation of particles

Settling performance is dependent on hydraulic surface loading, which can be calculated by dividing the design flow (L/d) by the surface area of the pond or clarifier (1 m^2), with the resulting hydraulic surface loading of L/d/m^2. Common values lie in the range of 175 to 350 L/d/m^2. The AMD sludge exhibits three types of settling behavior. The type 1 sludge settles rapidly with a clear liquid-solid interface. Figure 4.10 illustrates this settling phenomenon. In the first part of the figure the precipitated sludge forms a homogeneous mixture. In the second part the stratification begins and the particles begin settling onto already settled particles. Water trapped inside forms a gelatinous mass. The adjacent upper layer (zone C) is a transition zone characterized by a suspended solids concentration lower than zone D, but greater than zone B. The supernatant zone A develops as the liquid-solids separation is completed. Settling continues in part III, and zones A and D increase in depth while B and C decrease. In IV, only zones A and D remain, with most of the solids present in zone D. At this point, zone D begins to compress. Compaction forces of the sludge displace the trapped water.

Type 2 settling involves sludge generated from mine drainage containing low concentrations of iron and aluminum with a pH of 6.5 to 7.5 and little acidity. Lime addition is not always required. Because no natural nuclei exist for flocculation, the sludge has poor settling rates and most remain in solution. The settled particles are light and fluffy in character and produce sludges with only 0.5% solids. Satisfactory liquid-solids separation does not occur if the treatment does not involve neutralization. The volume of sludge is determined by iron content, suspended solids, and other precipitable elements present in the water. Polymer or flocculent addition can be useful to enhance settling rate.

Type 3 settling behavior does not depend on solids content and pollutional loadings. It is produced by limestone or sodium carbonate neutralization. A two-phase separation system is observed. Most of the solids settle rapidly, with a distinct liquid-solids interface and a turbid, cloudy supernatant.

Sludge production increases as the iron and aluminum concentrations increase. Sludge formed in the pH range of 6.4 to 7.2 possesses the best settling properties, but the iron oxidation rate is low. The ideal pH for iron oxidation is around 8.5.

Sludges produced from limestone treatment have a higher density than those generated by other lime treatment processes. This results in a smaller volume required for sludge disposal. Sludges produced from highly acidic waters neutralized with caustic soda (NaOH) have very low densities with long detention times. Sludges produced from soda ash (Na_2CO_3) treatment possess densities somewhere between those generated by lime and limestone.

Polyelectrolytes may be added to improve sludge settling rates. These are water-soluble, high molecular weight, organic polymers that may be cationic or anionic. These changed polymers adhere to sludge particles and improve settling. Determinations of polymer dosages and whether cationic or anionic treatment should be used are made from treatable tests.

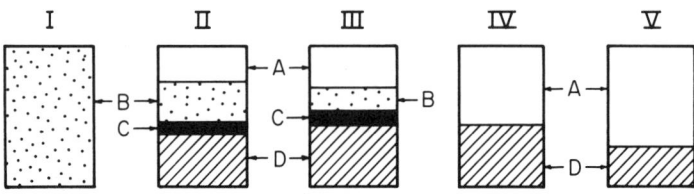

Figure 4.10. Type 1 settling.[7]

Sludge generated in the AMD treatment process is generally composed of hydrated ferrous or ferric oxide, gypsum, hydrated aluminum oxide, unused lime, sulfates, calcium carbonates, bicarbonates, and trace amounts of silica, phosphate, manganese, copper, titanium, and zinc. The characteristics of the sludge vary with mine drainage quality and neutralization method. Important characteristics include settleability, density, dewaterability, particle surface properties, and viscosity. Sludge settleability is determined by settling rate and final sludge volume. Sludge density is reported as percent solid by weight. Sludge dewaterability is determined by the ability of a sludge to be concentrated into a more manageable and less voluminous form by centrifuging, filtering, or lagooning. The electrostatic charge or particle surface property influences flocculation characteristics of the particles. The viscosity of sludge measures flowability when pumping sludges.

Neutralization of mine drainage with lime produces hydrated sludges which are light, gelatinous, and very voluminous. The ferric hydroxide sludges from AMD have high iron concentrations and are generally a fluffy mass with very low solids.

The chemical characteristics of AMD sludge (called yellowboy) vary with raw drainage and method of treatment. Sludge is generally composed of hydrated ferrous or ferric oxides, gypsum, hydrated aluminum oxide, varying amounts of sulfates, calcium carbonates, bicarbonates, and trace quantities of silica, phosphate, manganese, titanium, copper, and zinc.

Table 4.2 presents the primary composition of some typical mine drainage sludges generated from different mine waters and neutralization processes. The chemical quality of mine water, the type of neutralization process, and the chemical agents applied influence sludge composition and its characteristics.

The important physical properties that influence sludge handling and treatment processes include its density, settleability, viscosity, particle size and surface properties, and dewaterability.

Figure 4.11 illustrates the relationship of settling time to sludge volume for various neutralizing agents.

Main sludge disposal methods include lagooning, deep mines, underground disposal, filtration, bed drying, and centrifugation.

When land is available, lagooning offers one of the cheapest methods of sludge storage and dewatering. Mine drainage treatment plants are usually located in remote or isolated areas where land is readily available. Lagoons function as settling ponds

Table 4.2. Chemical Analyses of Sludges

		Weight, % (Dry Basis 105°C/24 hr)						
Alkali Used:		Hydrated Lime-Air Oxidation		Hydrated Lime Bio-Oxidation	Dolomite		Calcined Dolomite	
Mine Water:	Bennett's Branch	Proctor 1	Proctor 2	Proctor 2	Proctor 2	Tyler	Proctor 1	Proctor 2
Component								
Al	3.8	4.7	3.1	8.0	2.8	5.5	4.5	4.8
Fe	19.5	17.7	23.1	24.3	13.0	7.4	13.5	23.2
Ca	6.9	5.8	5.2	4.8	17.2	10.7	6.7	5.2
Mg	6.6	4.3	5.1	1.0	3.8	11.8	9.8	5.8
SO_4	5.7	6.8	5.8	11.5	4.4	1.6	2.3	5.5
H_2O at 180°C	12.5	15.8	14.8	—	10.2	8.7	11.7	14.7
			Compound Composition					
$Al(OH)_3$	11.1	13.7	8.9	23.1	8.2	16.0	13.2	13.9
$Fe(OH)_3$	37.6	34.0	44.3	46.6	24.9	14.2	25.9	44.7
$CaCO_3$	11.4	7.5	7.0	0.0	38.3	25.0	14.4	7.2
$MgCO_3$	—	—	—	—	—	21.0	12.0	6.1
$3MgCO_3 \cdot Mg(OH)_2 \cdot 3H_2O$	25.2	16.4	19.4	5.1	14.4	22.4	24.6	15.6
$CaSO_4 \cdot 2H_2O$	10.7	12.7	10.9	20.7	8.4	3.1	4.4	10.4
Total	96.0	84.3	90.5	95.5	94.2	101.7	94.5	97.9

From Penn Environmental Consultants. Design Manual, Neutralization of Acid Mine Drainage, U.S. EPA Rep.-600/2-83-001 (1983).

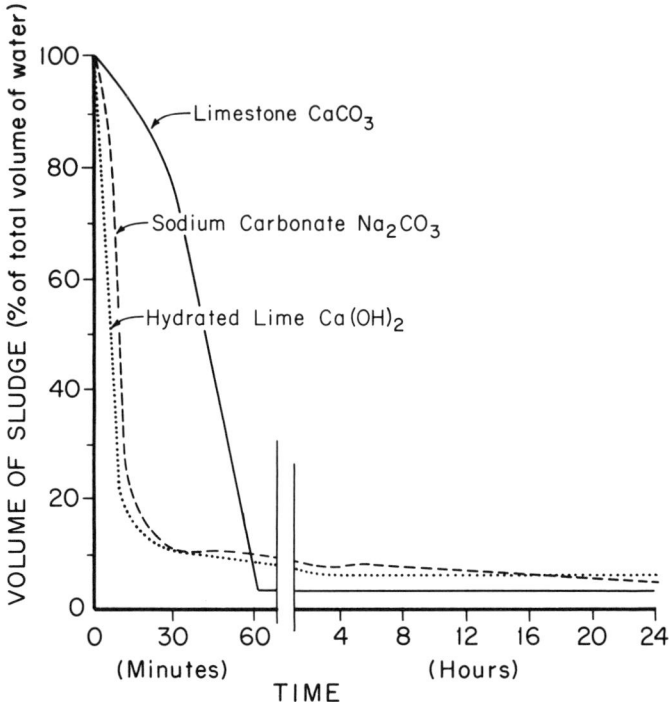

Figure 4.11. Treatability test settling curves.[7]

or impoundments for collecting and dewatering the sludge, and as permanent storage and disposal facilities.

In mine drainage treatment, settling ponds are used in series in which the first pond serves as the primary settling unit and the second serves as a polishing pond. The system isolates the majority of the solids in the first pond, but little sludge dewatering occurs. The primary pond is equipped with a sludge removal device that transfers the sludge to a disposal lagoon in which atmospheric dewatering can occur.

A dual or parallel arrangement of settling ponds together with an isolated dewatering lagoon offers the most effective natural method of sludge dewatering. Two settling ponds, each with volume sufficient to treat the designed flow, are best suited. The ponds can be used alternately so that the inactive pond can undergo dewatering. Its contents are then transferred to the final disposal lagoon for further dewatering. In both ponds, the supernatant, or the surface water, must be decanted to allow the sludge to dry. After most of the water is decanted from a lagoon, shrinkage cracks appear over the entire pond. Sludge lagoons are difficult to cover with soil. Sludge exhibits thixotropic tendencies; i.e., it will tend to liquefy upon vibration and then exhibit low compressive strength, which presents difficult problems for the weight and movement of the machinery employed in covering operations using soil.

THE ACID MINE DRAINAGE PROBLEM FROM COAL MINES

Spread burial of the sludge utilizing the benefits for soil conditioning (i.e., alkalinity, minerals) can be economical if land area is available. Controlled amounts of sludge are applied to an area and the area is then covered by soil in layers. The heavy metal content and other toxic constituents of the sludge should be evaluated for their long-term effect on vegetation.

Disposal of mine drainage sludge in abandoned deep mines through boreholes has been practiced. Sometimes this means of disposal offers the most economical and simplest method for sludge disposal. However, all legal and environmental ramifications must be considered. For underground disposal, the sludge must have a pH of 7.0 or above, and all of the iron must be in the ferric form. For underground disposal, the following environmental factors are considered:

- Mine hydrology — presence of water in the mine, water level, location, and quantity of discharge from the mine
- Connection to nearby workings and other mines
- Quality of mine water, pH, acidity, and iron
- Sludge input — volume of sludge, mine volume, leachate quality from the sludge, sludge concentration
- Effect of sludge on discharge from the mine
- Geology — structure of the coal beds, strike, dip, and presence of faults

Vacuum filtration can be applied to mine drainage sludge for dewatering. Operational variables include amount of vacuum, drum speed, drum submergence, filter media, and sludge conditioning before filtration.

Pressure filtration is an accelerated process of vacuum filtration. This process has not often been used on mine drainage sludge because of high capital, labor, and maintenance costs.

Porous bed drying of mine sludge has been employed. A variety of filter media such as sand, crushed limestone, coal, red dog, and gravel has been used. Water is removed by decanting the surface water in ponds, by percolation through the bottom of the bed, and by evaporation. Figure 4.12 illustrates a typical drying bed. The method was found to be impractical in the eastern United States, which experiences freezing weather conditions in some seasons.

The high-density sludge process developed by Bethlehem Steel Corporation is illustrated in Figure 4.13. This process achieves settled sludge concentrations between 15 to 40% compared to a maximum 15% from the conventional lime neutralization process. The resulting sludge storage or disposal volume is reduced significantly.

REVERSE OSMOSIS

Reverse osmosis (RO) can be highly effective in removing most of the dissolved solids in AMD. The product water is low in dissolved solids, usually <100 mg/L, but

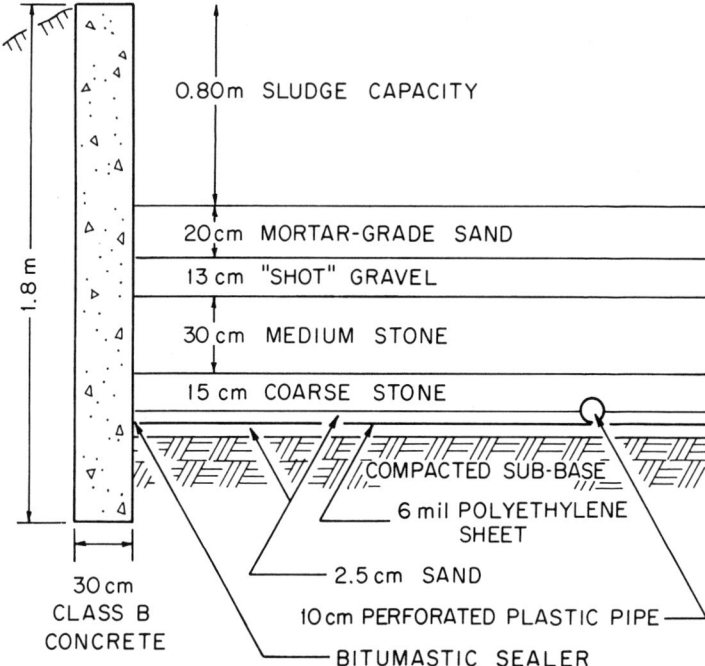

Figure 4.12. Cross-sectional view of the drying bed construction.[7]

may contain chemical or bacterial constituents that exceed drinking water standards. The product water can be used for other industrial purposes.

Osmosis occurs if two solutions of different concentrations in the same solvent are separated from one another by a membrane. If the membrane is semipermeable (i.e., permeable to the solvent and not to the solute), then the solution will flow from the more dilute solution to the more concentrated solution until equal concentration results. In RO, the direction of solvent flow is reversed by the application of pressure to the more concentrated solution. As a result, the concentrated solution termed "brine" becomes more concentrated. The solvent, termed "the permeate," is the product of the process.

Tubular, hollow-fiber, and spiral-wound membrane types have been tested for use in treating AMD. The spiral-wound configuration with a formamid-modified cellulose acetate membrane may be slightly superior to others with respect to the average flux (permeate flow rate), long-term flux stability, and dissolved solids rejection.

Problems with membrane fouling can occur as the concentrations of various compounds increase during the process. Most important is the possibility of iron fouling and calcium sulfate formation. Iron fouling can be minimized by lowering the feed water pH or flushing the RO membrane by operating at lower pressures for short periods. When the raw AMD contains high concentrations of sulfates, gypsum ($CaSO_4$) can form if its solubility is exceeded. In this case, this process may not be applicable.

THE ACID MINE DRAINAGE PROBLEM FROM COAL MINES   153

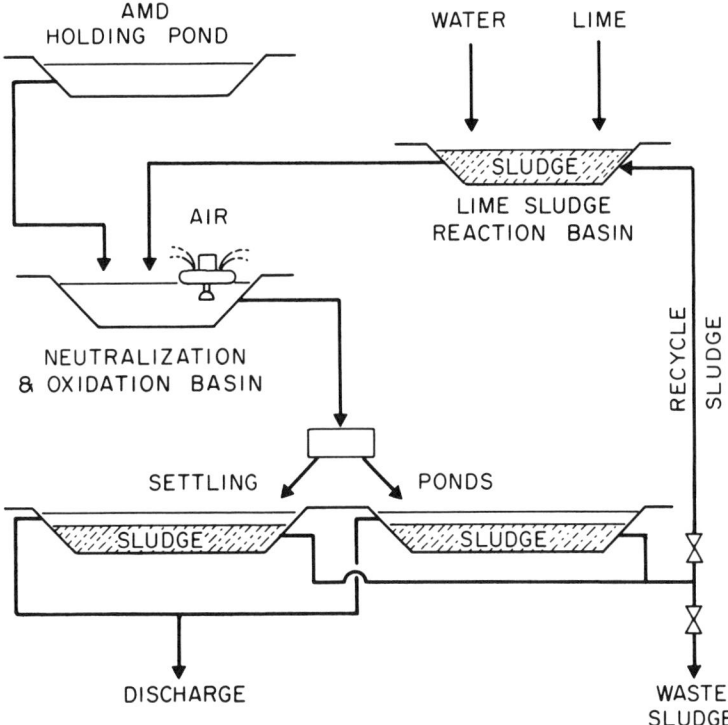

Figure 4.13. High-density sludge neutralization process.[7]

Bag filters or cartridges should be used to minimize fouling by suspended solids with feed water. Membrane life can be increased and rejection levels improved. The filters should have the capability to remove particles larger than 20 μm. The filters are placed in the suction side of the RO feed pumps.

In AMD treatment, the pH of the feed should be maintained between 2.8 and 3.0. At a pH lower than 3.0, ferric iron remains dissolved. When pH values exceed 3.0, ferric hydroxide may begin to precipitate on the membrane surface.

Although a low pH is necessary to improve operating conditions when treating AMD, it is lower than the optimum range of 5.0 to 6.5 for the cellulose acetate membranes. The life of this type of membrane is reduced. New, pH-resistant membranes are now available.

Disinfection can be used to reduce iron fouling problems and to inhibit microbial activity in the raw mine drainage feed. Ultraviolet (UV) light, proven to be an effective bactericide, can be used to prevent accumulations of iron-oxidizing bacteria on the membrane surface.

One of the design parameters critical to a successful RO application is an accurate permeation rate (flux rate) over the membrane life. This is essential to estimate the quantity of installed surface area, cleaning cycles, and membrane replacement. With a constant feed rate, a decline of permeate flow will occur due to membrane fouling. Even without fouling, a slight flux decline will occur because of the membrane compaction.

The system is designed to produce a constant permeate output based on the daily design flow. Use of pressure-compensating flow controls that automatically adjust the operating pressure for flow variations can accomplish a constant permeate output.

The system is designed to operate at 281 kg/cm^2 because, at this operating pressure, minimum membrane compaction is experienced, while maintaining adequate flux ratios to assure high effluent quality.

As the flux declines, the pressure control system compensates for the decreased flow by increased operating pressure, thereby maintaining a constant product flow. Due to compaction, a loss of flux is typically offset by appropriate sizing and start-up at a reduced pressure with a gradual increase in operating pressure over the life of the modules. To reduce fouling, each module should operate with a 10:1 brine:product flow ratio. This brine velocity prevents "boundary layer" development, a stagnant layer of water against the membrane surface.

Module configuration is an important design item. The type of pressure vessel manifolding arrangement is determined by the desired recovery level and the need to maintain an adequate brine:product flow ratio. Pressure vessel arrangements and modules are designed so that the raw feed enters a parallel bank of pressure vessels, and the concentrate from this bank enters as the feed for the next parallel arrangement of vessels.

In high-recovery continuous flow systems, only a small number of modules should process the most concentrated portion of feed stream so as to confine any fouling due to chemical precipitation to a minimum number of modules.

Two controlling factors limit this overall recovery of water from the treatment process. The first factor is precipitation of calcium sulfate (CaSO4). Acid mine water typically contains high concentrations of calcium and sulfate ions. With the progress of the RO process, calcium and sulfate concentrations increase. Once the solubility is exceeded, calcium sulfate precipitates. The empirical limit for RO recovery levels with AMD is given by:

$$R = 100 - 0.55(Ca \times SO_4)^{-1/2} \qquad (4.27)$$

where Ca and SO4 concentrations (mg/L) are those used in the raw feed. Gypsum (CaSO$_4$) is only slightly soluble. When concentrated, it precipitates and forms a hard, tenacious scale on tanks, piping, and membranes.

The second factor influencing recovery rates is the desired quality of the permeate. As the drainage is processed, the concentration of fatal dissolved solids in the permeate increases linearly with the total dissolved solids. Increasing recovery increases the concentration of pollutants in the waste (reject) stream and in the

THE ACID MINE DRAINAGE PROBLEM FROM COAL MINES

Table 4.3. Anticipated Permeate Water Quality

Parameter[a]	Raw Water Quality	Product Water Quality
pH (units)	3.4	4.3
Specific conductance (μmhos)	1020	32
Acidity	210	32
Calcium	150	1.2
Magnesium	115	1.4
Iron, total	110	1.2
Iron, ferrous	71	0.8
Aluminum	15	0.8
Manganese	43	0.4

[a] All values expressed as milligrams per liter unless otherwise noted.

From Penn Environmental Consultants. Design Manual, Neutralization of Acid Mine Drainage, U.S. EPA Rep.-600/2-83-001 (1983).

product water. The maximum recovery of the process is determined by the final use of the permeate.

In most cases the spiral-wound cellulose acetate membrane rejects 99% or more of the dissolved salts in the raw AMD feed. Table 4.3 shows expected permeate water quality.

Membrane life is affected by pH level, temperature, and operating pressure. The useful membrane life is the time taken to lose 40% of its initial flux.

The RO process removes dissolved solids from the AMD. The process generates a highly concentrated waste stream that requires treatment before disposal. The exact volume and salt content of the concentrate stream depends on the influent quality as well as the recovery rate. The possible treatment and disposal methods include:

- Lime neutralization
- Evaporation — mechanical or atmospheric
- Contract disposal

Lime neutralization of the waste stream is a possible treatment and disposal method.

Evaporation techniques include a mechanical, wiped-filmed unit, which can reduce the volume by 75% or more.

Contract hauling and disposal by a waste disposal firm can be another alternative.

ION EXCHANGE

The ion exchange process can be used to remove unwanted dissolved ions in AMD to produce a water of excellent quality for many industrial uses. Ion exchange can produce drinking-quality water, but it will additionally require use of filtration and disinfection so as to comply with public health regulations.

Ion exchange involves the reversible interchange of ions between a solid medium and the aqueous solution. To be effective, the solid ion exchange medium must contain ions of its own, be insoluble in water, and have a porous structure for the free passage of the water molecules. Within the solution and the ion exchange medium, the number of charges (charge balance) must stay constant. Ion exchange materials show affinity for multivalent ions, therefore, they tend to exchange their monovalent ions. This reaction can be reversed by increasing the concentration of monovalent ions. Thus, the ion exchange material can be regenerated once its capacity to exchange ions has been depleted.

Ion exchange is commonly used for softening of hard water. The ion exchange material is charged with monovalent cations, usually sodium (sodium chloride). The hard water is passed through a bed of ion exchange material, and the divalent calcium and magnesium cations are exchanged for sodium ions as follows:

$$Ca^{2+} + 2Na^+ \text{ (resin)} \rightarrow Ca^{2+} \text{ (resin)} + 2Na^+ \qquad (4.28)$$

Ion exchange processes which have been developed for treatment of AMD include the sul-bisul process, the modified desal process, and the two-resin process.

Sul-Bisul Process

The sul-bisul process (Figure 4.14) employs a two- or three-bed system, depending upon the mine drainage quality. Cations are removed by a strong acid resin in the hydrogen form, or by a combination of weak acid and strong acid resins. The AMD feed is first passed through the cation exchanger which removes the metal cations and exchanges these for hydrogen protons, or (H$^+$) ions. This reaction is expressed by the chemical equation:

$$Fe^{2+}SO_4 + 2HR \rightarrow MR_2 + H_2SO_4 \qquad (4.29)$$

Where R represents the strong-acid exchange groups on the resin, and M represents a divalent metal cation, such as iron (ferric), calcium, or manganese.

The water produced from the first stage contains additional sulfuric acid. The water is decarbonated to remove carbon dioxide formed during the cation exchange process. Then, a strong base anion resin (R') operating in the sulfate-to-bisulfate cycle removes both the sulfate and hydrogen ions during this exchange reaction:

$$R'SO_4 + H_2SO_4 \rightarrow R'(HSO_4)_2 \qquad (4.30)$$

The sulfate ions in the solution and on the resin are converted to the bisulfate form by the high acid content in the feed. This conversion of bivalent sulfate to mono-

THE ACID MINE DRAINAGE PROBLEM FROM COAL MINES 157

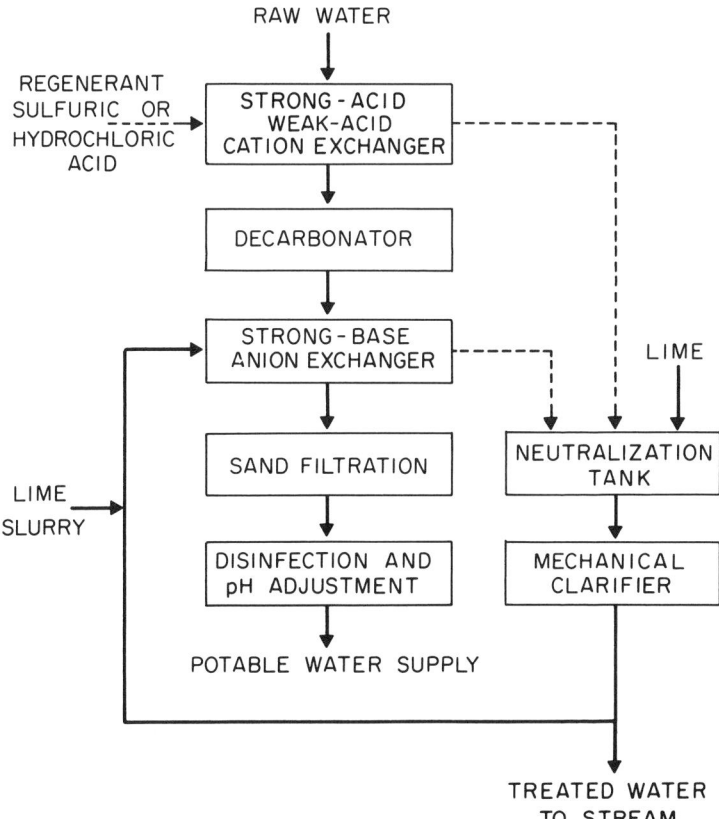

Figure 4.14. Sul-bisul process flow sheet.

valent bisulfate provides for twice the amount of sulfate to be stored in the resin. Removal of sulfur produces good quality water. Regeneration of the cation exchange bed is accomplished via either hydrochloric or sulfuric acid. In the regeneration process of the anion bed bisulfate ions are converted back to the sulfate form by the feed water. Addition of lime slurry to the regenerant speeds this reaction. The product water must be filtered and chlorinated according to public health regulations before use as potable water. Wastes from the regeneration process would have to be treated before discharge.

The sul-bisul process can be used to demineralize brackish water containing predominantly sulfate anions with a dissolved solids content up to 3000 mg/L. The raw water should have an alkalinity content of about 10% that of the total anion concentration and a sulfate:chloride ion ratio of at least 10:1. The process is especially suitable for alkaline waters containing calcium sulfate, such as mine drainage waters.

158 ENVIRONMENTAL IMPACTS OF MINING

Limitations of the process include low exchange capacity of the anion exchange resin and its inefficient method of regeneration. The expended anion resin can be regenerated by the raw water itself, requiring a considerable volume of water over a significant length of time if the sulfate content is low.

The large volume of regenerants requires disposal. This water must be sufficiently alkaline and abundant so that it can be used as the anion bed regenerant. Otherwise, other alkalis must be employed. When this becomes necessary, the process may not be economically competitive.

Modified Desal Process

The modified desal process (Figure 4.15) is another ion exchange process that has been investigated for purification of AMD in order to produce potable water. The process uses a weak base anion resin in the freebase form, which is converted to the bicarbonate form to treat the raw AMD. The weak base resin exchanges sulfates for bicarbonates, allowing the cations to pass through the bed according to

$$MSO_4 + 2R''HCO_3 \rightarrow R''_2SO_4 + M(HCO_3)_2 \qquad (4.31)$$

where R'' is the weak-base exchange group on the resin matrix and M is a divalent metal ion.

Aeration of the solution of metal bicarbonates is required to oxidize ferrous iron to ferric iron and to purge the carbon dioxide gas. The effluent is then treated with lime to precipitate metal hydroxides, settled to remove suspended solids, and then filtered and chlorinated before use as potable water.

Ammonia is used as the alkaline regenerant for displacing sulfate from the exhausted resin. Lime is used to precipitate the ammonia regenerant for reuse. Thus, ammonia is recycled. Lime and carbon dioxide used in the process can be recovered by roasting the lime sludge wastes in a kiln. All the principal chemicals used in the process can be recycled so that the net discharge of the process would approach zero. The products of the process include potable water, iron hydroxide, and calcium sulfate.

Use of the modified desal process is not limited by total dissolved solids or pH levels, but large quantities of carbon dioxide are needed to achieve good resin utilization for high total dissolved solids or alkaline feed waters. Application of this process is limited to waters containing <2200 mg/L of sulfate. Mine waters containing iron in the ferric form may cause fouling of the anion bed due to precipitation of ferric hydroxide.

Two-Resin Process

The two-resin process (Figure 4.16) involves the use of a strong acid cation exchanger in the acid form (H+) followed by a weak base anion exchanger in the

THE ACID MINE DRAINAGE PROBLEM FROM COAL MINES

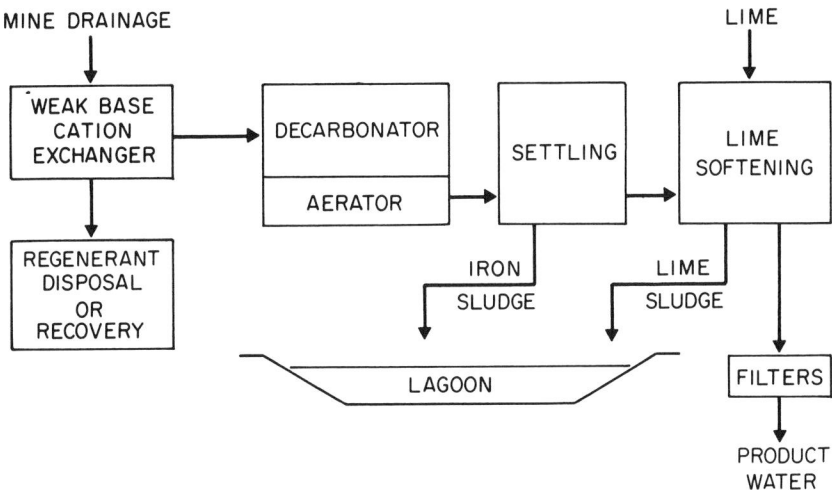

Figure 4.15. Modified desal process.[7]

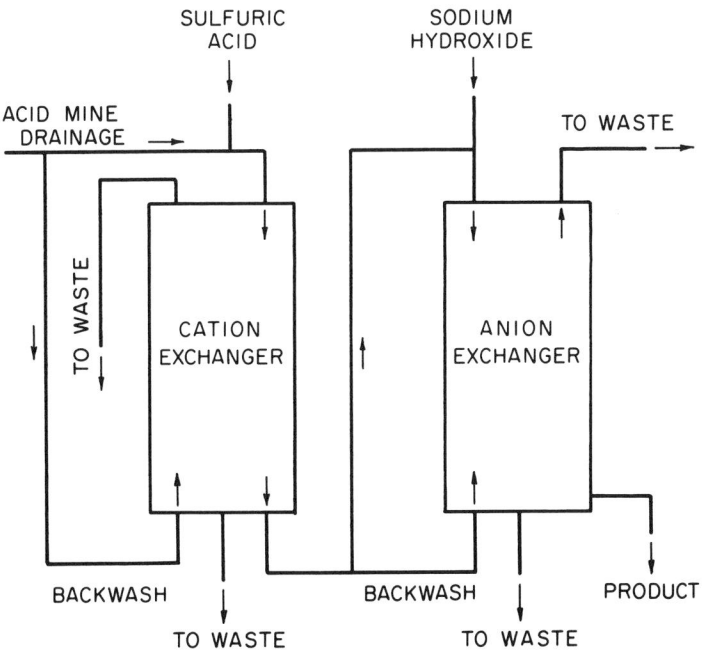

Figure 4.16. Two-resin ion exchange system.[7]

freebase (DH-) form. In the cation column, proton (H+ ions) are exchanged for the metal ions in the AMD. Following cation exchange, the anion column feed is rich in sulfuric acid.

The removal of total metal cation greatly increases the regenerant dosage and the operating cost of the system. The concentration of the residual metals in the cation exchanger effluent can be optimized by controlling the dosage of regenerant. The anion exchange is fed predominantly by sulfuric acid, which is totally absorbed by the resin. A weak base anion exchange resin absorbs only acids. It cannot split neutral salts. The anion exchange effluent is alkaline and some precipitation of residual iron and aluminum ions can be expected.

Sulfuric or hydrochloric acids may be used for regenerating the cation exchanger. Because of its lower cost, sulfuric acid is usually preferred; however, gypsum may be formed. Treatment for disposal of both regenerant streams is required.

Chemical Softening

The chemical softening process can be used for treatment of AMD when the effluent water is to be used as potable water or for drinking purposes. Two processes that merit application to AMD treatment are lime-soda and alumina-lime-soda.

Lime-Soda Process

The lime-soda process takes advantage of the low solubilities of calcium and magnesium components and removes these precipitates by sedimentation. Calcium is precipitated as calcium carbonate and magnesium is precipitated as magnesium hydroxide. Lime and soda ash are the chemicals most often used to bring about these chemical reactions.

The total dissolved solids in the water are not affected by this process. Calcium and magnesium ions are replaced by sodium ions while the sulfate concentration remains constant. In the lime-soda process, the first four unit processes (equalization, neutralization, iron oxidation, and solids removal) are the same as for the conventional lime neutralization process. Then, the effluent from the solids removal unit enters a flash in X tank for chemical addition, the first softening process. The next step is the softening reaction (flocculation) tank, settling basins, a re-carbonation chamber, filters, and chlorination.

The neutralization and iron oxidation processes are similar in any typical AMD plant: pH adjustment and iron and manganese removal. At this stage, lime is added for the neutralization of the mine drainage ability and the precipitation of iron, manganese, and aluminum as hydroxides.

Lime is again added to the sedimentation basin effluent as the first step in the softening process. Lime is also required for manganese and magnesium removal, both of which are precipitated as hydroxides according to the equations:

$$MgSO_4 + Ca(OH)_2 \rightarrow Mg(OH)_2 + CaSO_4 \qquad (4.32)$$

THE ACID MINE DRAINAGE PROBLEM FROM COAL MINES

$$MnSO_4 + Ca(OH)_2 \rightarrow Mn(OH)_2 + CaSO_4 \qquad (4.33)$$

Soda ash is then added to remove noncarbonate calcium hardness or calcium sulfate:

$$CaSO_4 + Na_2CO_3 \rightarrow CaCO_3 + Na_2SO_4 \qquad (4.34)$$

The pH must be maintained at 9.5 or higher for precipitation to occur.

Alumina-Lime-Soda Process

This process is suited for waters in which the principal source of salinity is sulfate. Heavy metals and hardness can be removed as well. The process is most suitable for treating AMD with sulfate concentrations between 400 and 1200 mg/L. The lower limit is 100 mg/L, below which sulfate removal is not economically feasible. The drinking water standards require a maximum limit on sulfate concentration to 250 mg/L.

The process is applied in two stages. The raw AMD is split into two streams. The larger stream is treated with lime and sodium aluminate ($NaAlO_2$). The proportion of the raw feed treated in stage 1 depends on sulfate concentration desired in the total plant effluent. The effluent from stage 1 typically contains approximately 100 mg/L sulfate, while the sulfate concentration bypassed to stage 2 remains constant. The sulfate concentration in final blended flow can be calculated by a mass balance method. Effluent from stage 1 is mixed with the smaller AMD stream and carbon dioxide is added for pH control. In both stages, the produced solids are removed by filtration.

The key reactions take place in stage 1. The sodium aluminate and lime neutralize the raw acidity, precipitate the heavy metals and magnesium, and remove calcium sulfate. Sulfate is removed by the sodium aluminate, and to a lesser degree, by the iron and aluminum present in the raw AMD. The reaction between AMD, the sodium aluminate, and lime produces insoluble calcium sulfa-aluminates which dominate the sludge produced.

After the stage 1 reaction, precipitating solids are removed in a mechanical settling unit. The resulting sludge is further dewatered by filtration.

The water produced in stage 1 contains excess lime which is removed in stage 2 by mixing with the smaller stream of raw AMD. Carbon dioxide is added to lower the pH to 10.3, the minimum solubility of calcium carbonate. Calcium carbonate is formed by the reaction:

$$2OH^- + Ca^{2+} + CO_2 \rightarrow CaCO_3 + H_2O \qquad (4.35)$$

Excess carbon dioxide redissolves calcium to form calcium bicarbonate. Precipitates of calcium carbonate and metal hydroxide precipitates are removed by sand filtration. The resulting filtrate will have a pH of 10.3 and will contain about 35 mg/L of dissolved calcium carbonate at its minimum solubility level. Additional carbonation can reduce the pH to an acceptable level for potable water.

BACTERICIDES IN AMD CONTROL

Inhibiting or destroying thiobacilli can significantly slow the rate of acid production. The controlled release of bactericides over a long period of time has been developed. The ProMac system, developed by BF Goodrich, is an example. A case study using the ProMac system is discussed below.[8]

The huge pile of refuse that accumulated during decades of coal mining operations spewed its nasty toxic acids into the nearby West Fork River at Dawmont, West Virginia.

The abandoned coal tipple and preparation plant site a few miles northwest of Clarksburg proved to be an eyesore, in addition to causing a detrimental effect on the West Fork, which contributes to the water supply of Fairmont to the north.

That was the situation at Dawmont in 1986.

Today, the site is covered with a lush growth of birds-foot trefoil, a legume vegetation that is high in nitrogen and has a strong root system that grips itself firmly to the soil. And the acids that once found their way into the West Fork have been brought under control.

The 36-acre reclamation project experienced such a quick and strong comeback that it earned the Grafton Coal Company a 1988 Reclamation Award from the West Virginia Mining and Reclamation Association and West Virginia Department of Energy.

Grafton Coal Company, contractor on the reclamation, was recognized for the "timely, efficient, and successful restoration of 36 acres known as the Dawmont Project. The company's innovative techniques in dealing with highly acidic soil and water conditions transformed an eyesore and environmental liability into an aesthetically pleasing and potentially commercial asset, thus representing the highest potential of the Abandoned Mine Lands program."

The West Virginia DOE decided to use the BF Goodrich Company's ProMac system in the reclamation at Dawmont.

A technology developed in the 1980s, the ProMac System involves controlled-release bactericides designed to substantially reduce acid drainage problems, improve water quality, and reduce associated treatment costs in active mining, at coal preparation plants, in active coal piles, active refuse piles, and mine tailings.

For use in mine land reclamation the ProMac system requires a one-time application that sets up a recovery cycle to help nature permanently restore a site with healthy vegetation and to provide assurance against postreclamation acid discharge.

Reclamation of the Dawmont site began in September 1986 and was completed 1 year later by Grafton Coal. It was necessary to move about 250,000 yd of coal refuse, or gob as it is called in the coal industry, and spread it over the 36-acre site.

THE ACID MINE DRAINAGE PROBLEM FROM COAL MINES

The refuse was the waste material that came from the underground mines around Dawmont, after separation of the usable coal. The gob pile built up from the day the mines began operation in the 1920s until they were closed in the 1960s.

Since the mine closings, the gob pile continued to pour out its toxic acids into the nearby West Fork River and even occasionally experienced minor fire flare-ups. It remained an eyesore.

After regrading the pile to stable slopes, treatment with the ProMac System was begun during the summer of 1987 and finished in September. Application of the chemicals was accomplished with a hydroseeder.

Included in the treatment was a powder that reacts immediately and three different time-release pellets. The time-release bactericides are encased in plastic pellets, with one formula releasing beneficial ingredients for the first 2 years, a second formula doing its job up to 5 years, and the third pellet continuing for 7 years or longer.

By the time the chemicals have released over the 7 years, the natural life cycle will have taken over to maintain vegetative growth and eliminate acid generation.

Following application of the powder and pellets at Dawmont, a small quantity of lime was spread over the site, soil cover of 12 to 18 in. was provided, and the reclamation completed by seeding, fertilizing, and covering with straw.

With the ProMac system still being considered to be "under the microscope" by the West Virginia DOE, 1 acre at the Dawmont site was left untreated as a control area. The treated and untreated areas are tested on a monthly basis.

In the 2 years since the site was finished, the Dawmont Project has experienced a rapid turnaround. The refuse pH was 3 or less before reclamation and currently is close to 5 on the treated portion. In the 1-acre control area, the pH has never gone above 3.

Reduction of the refuse acidity has shown significant improvement, with the control area still measuring about 1000 ppm as compared to <50 in the treated area. The generation of iron and manganese also provides evidence that the treatment is achieving results greater than those in the control area.

DETERMINATION OF ACID-GENERATING POTENTIAL

Acid potential (AP) is defined as the amount of acid that would be produced if all pyrite in the sample reacted according to the following stoichiometry:

$$FeS_2 + \frac{15}{4}O_2 + \frac{1}{2}H_2O = Fe(OH)_{3(s)} + 2SO_4^{2-} + 4H^+ \quad (4.36)$$

Pyrite, 1 mol, containing 64 g sulfur, produces 4 mol of H^+ acidity, equivalent to 200 g $CaCO_3$ acidity. Thus, the AP of pyritic sulfur is 3.125 g acidity (as $CaCO_3$) per gram of sulfur; 1% pyritic sulfur has the potential to produce 31.25/1000 tons acidity. Total sulfur concentrations can be measured with an automated Fisher sulfur analyzer, consisting of a furnace and SO_2 titrater, according to the instrument instructions. Sulfur speciation can be analyzed for samples containing >0.5% sulfur by hot extraction with hydrochloric and nitric acid to remove sulfate and pyritic sulfur, respectively.

Organic sulfur is defined as the sulfur extraction remaining after acid extraction. AP can be calculated from both total sulfur and, when analyses are available, pyritic sulfur. Total sulfur is often used in mine permit applications to minimize analytical costs.[9]

Neutralization potential (NP), defined as the ability of the stratum to neutralize strong acid, can be determined by treating a 2-g sample with 20 to 80 mL of 0.1 M HCl, heating nearly to boiling, and swirling periodically until no gas evolution is observed. The samples are made up to 125 mL with distilled water, boiled for 1 min, and cooled to room temperature. The treated sample is then titrated with standard NaOH (0.1 or 0.5 M) to pH 7. NP is calculated as the amount of HCl consumed by the sample and converted to the units of tons of $CaCO_3$/1000 tons material:

$$\text{NP, tons} / 1000 \text{ tons} = \frac{\text{g HCl consumed}}{\text{g sample}} \times \frac{50 \text{ g CaCO}_3}{36 \text{ g HCl}} \times 1000 \quad (4.37)$$

Net neutralization potential (NETNP) is calculated for the stratum by subtracting AP from NP. A positive NETNP indicates an excess of neutralizers, while a negative NETNP indicates a deficiency of neutralizers in the stratum.

Geochemically, a number of fundamental concerns is found in the theoretical basis on which the acid-base or NAPP technique is based. The major concerns are

1. Assumes all sulfur occurs as pyritic sulfide and is acid forming;
2. Assumes all ANC is available to neutralize acid; and
3. Does not consider the kinetics of either the acid-forming or acid-neutralizing processes.

The first two concerns may be considered to balance each other; however, failure to consider the site-specific mechanisms and kinetics when assessing the likely field geochemistry of a particular waste can result in significant error.

The NAPP procedure is an essential tool in waste characterization, but the interpretation and identification of the implications for waste management require a more detailed determination of waste geochemistry. Once the broad principles of the site-specific processes have been identified, a monitoring and management program can normally be developed which is based on the NAPP procedure.

REFERENCES

1. Kleinmann, R. L. P. Biogeochemistry of acid mine drainage and a method to control acid formation, *Min. Eng.* 33(3):300–305 (1981).
2. Singer, P. C. and W. Stumm. Acid mine drainage: the rate-determining step, *Science* 167:1121–1123 (1970).
3. Nordstrom, D. K. Hydrogeochemical and Microbiological Factors Affecting the Heavy Metal Chemistry of an Acid Mine Drainage System, Ph.D. thesis, Stanford University, Stanford, CA (1977).
4. Dugan, P. R. Bacterial ecology in strip mine areas and the relationship to the production of acidic mine drainage, *Ohio J. Sci.* 75(6):266–279 (1975).
5. Kleinmann, R. L. P. *Thiobacillus ferroxidans* and the formation of acidity in simulated coal mine environments, *Geomicrobiol. J.* 1(4):373–388.
6. Walsh, F. and R. Mitchel. A pH dependent succession of iron bacteria, *Environ. Sci. Technol.* 6(9):809–812 (1972).
7. Penn Environmental Consultants. Design Manual, Neutralization of Acid Mine Drainage, U.S. EPA Rep.-600/2-83-001 (1983), pp. 1–183.
8. Sobek, A. A. Successful Reclamation Using Controlled Release Bactericides, American Society for Surface Mining and Reclamation Conference, Charleston, WV (1990).
9. Erickson, P. M., and R. S. Hedin, Evaluation of Overburden Analytical Methods as a Means to Predict Post Mining Drainage Quality, U.S. Bureau of Mines Information Circular 9183 (1988), pp. 11–19.

CHAPTER 5

Acid Rock Drainage and Metal Migration

INTRODUCTION

Acid rock drainage (ARD) can be produced from locations in which sulfidic rock has been exposed as a result of mining, construction, or other activities. The sources of ARD from mining operations include:

- Drainage from underground workings
- Runoff from open pit workings
- Waste rock dumps from mining activities
- Mill tailings
- Ore stockpiles
- Spent ore piles from heap leach operations

The ARD from underground workings has been known since earlier times because it generally occurs as a point discharge of substantial flows of low pH water. Many of the ARD sources have been drainage tunnels or mine adits.

The environmental effects of ARD from open pit mines have been of more recent concern because many of the ARD producers or potential producers are still in operation and the ARD is not being treated. Many of these mines have been developed without plans to deal with ARD. The large areas of exposed rock in open pits can produce large volumes of ARD. Slope deterioration in the long term can cause continuous exposure of fresh rock to the natural elements, producing ARD.[1]

Waste rocks produced from mining operations are exposed to precipitation, runoff, and possibly seepage. Waste rocks containing sulfides are potentially large sources of ARD. The chemical and physical properties of these waste rocks significantly affect the quality of ARD and the change in its quality over time.

Sulfide-rich tailings are well known as potential sources for acid generation. The low permeability of the mill tailings and the flooding occurring in both operating and abandoned tailings impoundments limit the role of ARD generation and release. Consequently, the full potential effect of very large tailings dumps of more recent origin has not yet developed. Many of these tailings dumps are in active use and their potential impacts with ARD are controlled. The ARD may be produced after abandonment.

Low-grade ore stockpiles are often of particular concern as they can become concentrated sources of ARD. Similarly, the spent heap-leach piles can be sources of ARD, particularly those associated with low pH leachates.

THE ACID GENERATION PROCESS

Acid rock drainage is produced by the exposure of certain sulfide minerals, most commonly pyrite, to air and water, resulting in the production of acidity and elevated concentrations of metals and sulfate. The sulfur in the mineral is oxidized to a higher oxidation state, and aqueous iron, if present, is precipitated as ferric iron. Sulfide minerals are often found in rock that lies below a mantle of soil beneath the water table. Under natural conditions, the overlying soil and groundwater allow very little contact with oxygen, therefore acid generation proceeds at a slow rate and its effect on groundwater quality is negligible. When the rock is exposed to air and water by the mining process, the rate of acid generation is accelerated.

The ability of a particular rock formation to generate acid is determined by the relative content of acid-generating and acid-containing minerals. The process by which acid is consumed is known as "neutralization." Acid waters produced by the oxidation of sulfides in a rock may be neutralized upon contact with acid-consuming minerals. As a result, the water flowing from the rock may have a neutral pH and negligible acidity. However, if the acid-consuming minerals are dissolved, washed out, or coated by other minerals through encapsulation then, as acid generation continues, acid water will eventually drain from the rock. Whenever the acid-generation capability exceeds the acid-consuming capacity of the rock, it is generally expected that the water draining from the rock will not be of neutral pH.

It is important to consider the scale of examination when addressing acid drainage. A rock that produces pH-neutral conditions in water passing over it may experience acid generation in microenvironments around sulfide grains. The resulting acidic water may be neutralized by the remainder of the sample as it leaves the microenvironment. If an acid-generating rock has no neutralization potential, then acidic water will be present on the scale of both the sulfide grains as well as the rocks as a whole. A more complex situation occurs when pH-neutral water flowing over a rock invades and flushes the microenvironments of the sulfide grains, resulting in a slower rate of acid generation and a slower consumption of neutralization potential. To clarify terms, "acid generation" generally refers to the reaction in the microenvironment around a sulfide grain, whereas "acid mine drainage" refers to the chemical composition of water emanating from a rock or waste rock pile on a larger scale.

The time interval between the initial disturbance of rock and the peak rate of acid generation may range from days to years, depending on a number of environmental factors and the neutralization potential of the rock. In addition, the rate at which acid generation occurs through time will vary depending on several environmental factors and geochemical characteristics of the sulfide minerals. As a result, acid generation is not a simple process; rather, it is a complex set of chemical reactions changing through time, which are currently the topic of a large amount of scientific research.

SULFIDE MINERALS

Crystalline substances that contain sulfur combined with a metal (e.g., iron) or semi-metal (e.g., arsenic) but no oxygen are called sulfide minerals (Table 5.1). If a metal or semi-metal are both present in a mineral (e.g., arsenopyrite, FeAsS), the semi-metal substitutes for sulfur in the crystal structure. These minerals form in strongly anoxic (i.e., chemically reducing) environments, as indicated by sulfur, which is present in its lowest natural oxidation state. In oxygenated environments, sulfur exists in higher oxidation states such as $S_2O_3^{2-}$, SO_3^{2-}, and SO_4^{2-} (sulfate) and forms minerals with oxygen (e.g., gypsum, $CaSO_4$–$2H_2O$).

Under certain geological conditions, most notably near-surface, low-temperature deposits (bogs and swamps, etc.) and/or rapid deposition (e.g., mid-oceanic ridge sulfide deposits), sulfide may be precipitated in amorphous (noncrystalline) or poorly crystalline forms. For iron-sulfide minerals, amorphous FeS or greigite may form initially and then alter to pyrite via sulfurization. This process may lead to the formation of raspberry-like balls or "framboids" of fine-grained pyrite crystals. This framboidal pyrite has a significantly higher rate of acid generation when exposed to an oxidizing environment than coarsely grained, euhedral pyrite.

Marcasite (Table 5.1) is a low-temperature iron-sulfide mineral that may form instead of pyrite and which reportedly has a higher rate of acid generation under oxidizing conditions than crystalline pyrite. Marcasite may also be found in higher temperature paleoenvironments where it is metastable with respect to pyrite at temperatures >157°C.

At elevated temperatures, sulfide may be mobile, leading to recrystallization as a massive sulfide sometimes found at metal mines. The rate of acid generation from massive sulfide may be relatively slow, but the rate may be accelerated during the mining process through blasting and grinding. In general, the relative rates of oxidation for sulfide minerals under typical environmental conditions are unclear and detailed experimentation is recommended.

Table 5.1. Summary of Common Sulfide Minerals and Their Oxidation Products

Minerals	Composition	Aqueous End Products of Complete Oxidation[a]	Possible Secondary Minerals Formed at Neutral pH after Complete Oxidation and Neutralization[b]
Pyrite	FeS_2	Fe^{3+}, SO_4^{2-}, H^+	Ferric hydroxides and sulfates; gypsum
Marcasite	FeS_2	Fe^{3+}, SO_4^{2-}, H^+	Ferric hydroxides and sulfates; gypsum
Pyrrhotite	$Fe_{1-x}S$	Fe^{3+}, SO_4^{2-}, H^+	Ferric hydroxides and sulfates; gypsum
Smythite, greigite	Fe_3S_4	Fe^{3+}, SO_4^{2-}, H^+	Ferric hydroxides and sulfates; gypsum
Mackinawite	FeS	Fe^{3+}, SO_4^{2-}, H^+	Ferric hydroxides and sulfates; gypsum
Amorphous	FeS	Fe^{3+}, SO_4^{2-}, H^+	Ferric hydroxides and sulfates; gypsum
Chalcopyrite	$CuFeS_2$	$Cu^{2+}, Fe^{3+}, SO_4^{2-}, H^+$	Ferric hydroxides and sulfates; copper hydroxides and carbonates; gypsum
Chalcocite	Cu_2S	Cu^{2+}, SO_4^{2-}, H^+	Copper hydroxides and carbonates; gypsum
Bornite	Cu_5FeS_4	$Cu^{2+}, Fe^{3+}, SO_4^{2-}, H^+$	Ferric hydroxides and sulfates; copper hydroxides and carbonates; gypsum
Arsenopyrite	FeAsS	$Fe^{3+}, AsO_3^{3-}, SO_4^{2-}, H^+$	Ferric hydroxides and sulfates; ferric and calcium arsenates; gypsum
Realgar	AsS	$AsO_3^{3-}, SO_4^{2-}, H^+$	Ferric and calcium arsenates; gypsum
Orpiment	As_2S_3	$AsO_3^{3-}, SO_4^{2-}, H^+$	Ferric and calcium arsenates; gypsum
Tetrahedrite and tennenite	$Cu_{12}(Sb,As)_4S_{13}$	$Cu^{2+}, SbO_3^{3-}, AsO_3^{3-}, SO_4^{2-}, H^+$	Copper hydroxides and carbonates; calcium and ferric arsenates; antimony materials; gypsum
Molybdenite	MoS_2	$MoO_4^{2-}, SO_4^{2-}, H^+$	Ferric hydroxides; sulfates; molybdates; molybdenum oxides; gypsum
Sphalerite	ZnS	Zn^{2+}, SO_4^{2-}, H^+	Zinc hydroxides and carbonates; gypsum
Galena	PbS	Pb^{2+}, SO_4^{2-}, H^+	Lead hydroxides, carbonates, and sulfates; gypsum
Cinnabar	HgS	Hg^{2+}, SO_4^{2-}, H^+	Mercuric hydroxide; gypsum
Cobaltite	CoAsS	$Co^{2+}, AsO_3^{3-}, SO_4^{2-}, H^+$	Cobalt hydroxides and carbonates; ferric and calcium arsenates; gypsum
Niccolite	NiAs	$Ni^{2+}, AsO_3^{3-}, SO_4^{2-}, H^+$	Nickel hydroxides and carbonates; ferric, nickel, and calcium arsenates; gypsum
Pentlandite	$(Fe, Ni)_9S_8$	$Fe^{3+}, Ni^{2+}, SO_4^{2-}, H^+$	Ferric and nickel hydroxides; gypsum

[a] Intermediate species such as ferrous iron (Fe^{2+}) and $S_2O_3^{2-}$ may be important.
[b] Depending on overall water chemistry, other minerals may form with, or instead of, the minerals listed here.

From *Draft Acid Rock Drainage Technical Guide*, Vol. 1 (Vancouver, Canada: BiTech Publ., 1989). With permission.

Oxidation of these minerals may lead to the formation of secondary minerals after some degree of pH neutralization or when pH is maintained near neutral during oxidation. Some of these minerals are listed in Table 5.1; other minerals may form in addition to, or instead of, these minerals depending on water chemistry, extent of oxidation, and the presence of other compounds such as aluminosilicates. These secondary minerals may encapsulate the sulfide mineral and/or any neutralizing mineral, slowing the reaction rate.

CHEMICAL AND BIOLOGICAL REACTIONS RELATED TO ACID GENERATION

Acid generation, as well as acid consumption, is the result of a number of interrelated chemical reactions. The primary ingredients for acid generation are

- Sulfide minerals
- Water or a humid atmosphere
- An oxidant, particularly oxygen, from the atmosphere or from chemical sources

Total exclusion of moisture or oxidant will stop acid generation. In most cases, bacteria play a major role in accelerating the rate of acid generation, and the inhibition of bacterial activity in these cases will lessen the rate of acid generation.

The reactions of acid generation are best illustrated by examining the oxidation of pyrite ($FeS2$), which is one of the most common sulfide minerals. The first important reaction is the oxidation of the sulfide mineral into dissolved iron, sulfate, and hydrogen (H^+):

$$FeS_2 + 7/2\, O_2 + H_2O \rightarrow Fe^{2+} + 2SO_4^{2-} + 2H^+$$
$$(\text{Solid} + \text{gas} + \text{water} \rightarrow$$
$$\text{ferrous iron, sulfate, and } H^+ \text{ in water}) \quad (5.1)$$

The dissolved Fe^{2+}, SO_4^{2-}, and H^+ represent an increase in the total dissolved solids and acidity of the water, and, unless neutralized, the increasing acidity is often associated with a decrease in pH. If the surrounding environment is sufficiently oxidizing, much of the ferrous iron will oxidize to ferric iron:

$$Fe^{2+} + 1/4\, O_2 + H^+ \leftrightarrow Fe^{3+} + 1/2\, H_2O \quad (5.2)$$

At pH values above 2.3 to 3.5, the ferric iron will precipitate as $Fe(OH)_3$, leaving little Fe^{3+} in solution while lowering pH at the same time:

$$Fe^{3+} + 3H_2O \leftrightarrow Fe(OH)_3 \text{ (solid)} + 3H^+ \quad (5.3)$$

Any Fe^{3+} that does not precipitate from solution may be used to oxidize additional pyrite:

$$FeS_2 + 14Fe^{3+} + 8H_2O \rightarrow 15Fe^{2+} + 2SO_4^{2-} + 16H^+ \quad (5.4)$$

Based on these simplified basic reactions, acid generation that produces iron which eventually precipitates as $Fe(OH)_3$ may be represented by a combination of reactions:

$$FeS_2 + 15/4\, O_2 + 7/2\, H_2O \rightarrow Fe(OH)_3 + 2SO_4^{2-} + 4H^+ \quad (5.5)$$

On the other hand, the overall reaction for stable ferric iron that is used to oxidize more pyrite is

$$FeS_2 + 15/8\, O_2 + 13/2\, Fe^{3+} + 17/4\, H_2O \rightarrow$$
$$15/2\, Fe^{2+} + 2SO_4^{2-} + 17/2\, H^+ \quad (5.6)$$

Equations 5.3 to 5.6 assume the oxidized mineral is pyrite and the oxidant is oxygen. However, other sulfide minerals such as pyrrhotite (FeS) and chalcocite (Cu_2S) have other ratios of metal:sulfide and metals other than iron. Other oxidants and sulfide minerals also have different reaction pathways, stoichiometries, and rates, but research on these variations is limited.

The primary chemical factors that determine the rate of acid generation are

- pH
- Temperature
- Oxygen content of the gas phase, if saturation is <100%
- Oxygen concentration in the water phase
- Degree of saturation with water
- Chemical activity of $Fe3+$
- Surface area of exposed metal sulfide
- Chemical activation energy required to initiate acid generation

Certain bacteria may accelerate or decelerate the rate at which some of the above reactions proceed, thereby increasing or decreasing the rate of acid generation. *Thiobacillus ferrooxidans* in particular is known to accelerate these reactions through its enhancement of the rate of ferrous-iron oxidation. *T. ferrooxidans* may also accelerate the reaction through its enhancement of the rate of reduced sulfur oxidation. Experimental testing of the many other bacterial species capable of oxidizing iron and sulfur is generally limited. In addition, several species are known to reduce sulfur and iron, potentially counteracting acid generation (Table 5.2).

Most testing of *T. ferrooxidans* has involved oxidation of pyrite (FeS_2); however, the bacterium may accelerate the oxidation of sulfides of antimony, gallium, molybdenum, arsenic, copper, cadmium, cobalt, nickel, lead, and zinc.

For bacteria to thrive, environmental conditions must be favorable. *T. ferrooxidans*, for example, is most active in waters with a pH around 3.2. If conditions are not favorable, the bacterial influence on acid generation will be minimal. This apparent importance of environmental conditions explains the contradiction in reported experimentation that shows bacterial influence ranges from major to negligible. Experimental laboratory-based and in-field tests with bactericides indicate bacterial activity enhances the rate of acid generation (as indicated by sulfate and acidity) by a factor of up to 5, with one extreme measurement of a factor of 20.

In situations in which bacterial acceleration is significant, additional factors determine the bacterial activity and the associated rate of acid generation:

- Biological activation energy
- Population density of bacteria
- Rate of population growth
- Nitrate concentration
- Ammonia concentration
- Phosphorus concentration
- Carbon dioxide content
- Concentrations of any bacterial inhibitors

The bacterial contribution to sulfide and iron oxidation can be complex through contributions from many species. However, research indicates that *T. ferrooxidans* often plays a major role in organic-enhanced oxidation in natural environments. This justifies the attention paid to this bacterium, but does not always justify the exclusion of other bacteria from consideration.

Following the oxidation of a sulfide mineral, the resulting acid products may either be immediately flushed away by water moving over the rock or, if no water movement occurs, accumulate in the rock while remaining readily available for flushing. If the acid products are flushed away from the sulfide mineral, they may eventually encounter an acid-consuming mineral; the resulting neutralization will remove a portion of the acidity and iron from solution and will neutralize the pH. Sulfate concentrations are not usually affected by neutralization unless mineral saturation with respect to gypsum is attained. Consequently, sulfate sometimes may be used as an overall indicator of the extent of acid generation, even after neutralization by acid-consuming minerals has occurred.

The most common acid-consuming mineral is calcite ($CaCO3$), which consumes acidity through the creation of HCO_3^- or $H_2CO_3^0$:

Table 5.2. Bacterial Species that Influence Rate of Sulfur and Iron Oxidation

Bacterial Species	Type	Optimal Growth of Chemical Environment
Thiobacillus ferrooxidans	Sulfur oxidizing Iron oxidizing	pH = 2.5–3.5
T. novellus	Sulfur oxidizing	pH = neutral to alkaline
T. thioporus	Sulfur oxidizing	
T. denitiricans	Sulfur oxidizing	pH = neutral to alkaline Nitrate supply for reduction to N_2
Arthrobacter sp.	Sulfur oxidizing	—
Bacillus sp.	Sulfur oxidizing	—
Flavobacterium sp.	Sulfur oxidizing	—
Pseudomonas sp.	Sulfur oxidizing	—
Desulfavibrio sp.	Sulfur reducing	—
Desulfotomaculum sp.	Sulfur reducing	—
Salmonelis sp.	Sulfur reducing	—
Proteus sp.	Sulfur reducing	—
Suffolosu sp.	Sulfur reducing	—
Metalloginium sp.	Iron oxidizing	—
Sidenocapsa sp.	Iron oxidizing	—
Leptothrax sp.	Iron oxidizing	—
Gallionella sp.	Iron oxidizing	—
Vibrio sp., Bacillus sp.	Iron oxidizing	—
Aerobacter aerogonus	Iron oxidizing	—

From *Draft Acid Rock Drainage Technical Guide,* Vol. 1 (Vancouver, Canada: BiTech Publ., 1989). With permission.

$$CaCO_3 + H^+ \rightarrow Ca^{2+} + HCO_3^- \tag{5.7}$$

and

$$CaCO_3 + 2H^+ \rightarrow Ca^{2+} + H_2CO_3^0 \tag{5.8}$$

There are also other acid-consuming minerals such as $Al(OH)_3$:

$$Al(OH)_3 + 3H^+ \rightarrow Al^{3+} + 3H_2O \tag{5.9}$$

It is not unusual for a rock to contain both sulfide minerals and acid-consuming minerals. The balance between the two types will determine whether the rock will eventually produce acid conditions in the water passing over and through it, and this balance forms the basis of the experimental procedure used in static tests.

Metal Leaching and Migration Processes

Acid-generation processes in rocks which produce low pH water are capable of dissolving heavy metals contained within the rocks. This water then migrates from the generation site and enters the receiving environment. High-metal loadings in the water are the most harmful to the environment.

A series of reactions occurs along the path as the low pH water migrates from the source to the receiving environment. The resulting quality of the water is determined by the following factors:

- Nature of the sulfides
- Availability and type of soluble constituents
- Nature of alkaline reactants
- Physical properties of the waste

Some examples of quality of ARD are given in Table 5.3.

Several naturally occurring physical, chemical, and biological properties of mine waste affect metal solubility and contaminant migration. The mobilization of metals is mainly controlled by chemical factors, while the processes that occur along the migration route are controlled by physical and chemical factors.

Physical properties that influence metal solubility include waste particle size and shape, and temperature and pressure of pore gases. However, chemical factors are more predominant than physical properties in the metal mobilization process. Physical properties are important in the migration rate of ARD and in the reactions that occur along the migration path. Important characteristics include:

- Climatic conditions
- Waste permeability
- Availability of pore water
- Pore water pressure
- Movement mechanism, whether by stream flow or diffusion

These factors control the rate of movement of contaminant fronts, the amount of dilution, and the degree of mixing that occurs as the ARD moves from the source to the environment. The physical properties of the subsurface are different than those of waste, so a number of contaminant fronts may develop, all moving at different rates. Surface water yields tend to occur before groundwater yields in hard-rock waste dumps because of lower retardation and the resulting rapid migration through the waste rock. The quality of the yields, whether surface or groundwater, is a function of the dilution and buffering reactions that occur en route.

The solubility of the metals is generally determined by the pH of the leachate. Other chemical factors include the specific metal being dissolved, Eh, adsorption characteristics, and the chemical composition of the leachate. As the ARD moves

Table 5.3. Examples of Acid Rock Drainage Quality

Parameter[a]	Seepage from Abandoned Uranium Mine Tailings Pond in Ontario	Waste Rock Dump Seepage from Active Silver Mine in British Columbia	Mine Water from Underground Copper Mine in British Columbia
pH	2.0	2.8	3.5
Sulfate	7,440	7,650	1,500
Acidity	14,600	43,000	—
Iron	3,200	1,190	10.6
Manganese	5.6	78.3	6.4
Copper	3.6	89.8	16.5
Aluminum	588	359	—
Lead	0.67	2	0.1
Cadmium	0.05	0.5	0.143
Zinc	11.4	53.2	28.5
Arsenic	0.74	25	0.05
Nickel	3.2	8.0	0.06

[a] Units are mg/L except pH.

From *Draft Acid Rock Drainage Technical Guide,* Vol. 1 (Vancouver, Canada: BiTech Publ., 1989). With permission.

away from the sulfide source through the waste material, more acid-generating material may be encountered, causing a reduction in pH. The drainage over alkaline material may also cause complete or partial neutralization.

With gradual lowering of the pH level the dissolved metal load generally increases. However, a combination of chemical conditions could cause increasing mobilization of metals even at neutral or alkaline conditions. During the neutralization process of the drainage, precipitation of many of the soluble metals may occur, and the resultant drainage will contain the residual metals.[3]

An interesting phenomenon that has been observed in copper and massive sulfide ores is elevated zinc loadings in neutral drainage. Dissolved copper precipitates out as the pH is raised; the zinc remains at relatively high concentrations until the pH is raised to values above 9.5. As the ARD front moves through the waste or subsurface strata, the chemical composition of the front undergoes continuous change.[2]

Biological activities along the route may influence metal dissolution. Metal leaching occurs where iron-oxidizing bacteria are present with iron and metal sulfides. Biological species can also attenuate the mobility of metals by absorption and precipitation.

Prediction of Acid Drainage

The prediction process for determining potential acid generation from metal mining operations includes:

ACID ROCK DRAINAGE AND METAL MIGRATION

- Comparison with similar and neighboring mines
- A systematic sampling program to collect representative samples
- Static tests on the samples
- Kinetic tests using anticipated on-site conditions using potentially acid-generating samples
- Modeling

In the exploration stage, samples of the ore and waste rock should be collected for acid-base accounting analyses. These results early in the mine planning stages would indicate whether acid drainage might be a concern.

The geological units of ore and waste rocks should be identified based on lithology, mineralogy, and continuity of units. Comparison should be made with neighboring mines and similar geological and paleoenvironmental areas for obtaining an initial indication of potential acid generation. A sampling program for each geologic unit should be implemented. These samples should be subjected to acid tests to determine the potential for net acid production.

If the static test results give an uncertain indication of acid generation, then kinetic tests could be of value in determining acid production potential. If the potential for net acidity is identified for any geological unit, the mine plan should be revised. Mathematical models should be utilized along with kinetic tests to predict acid generation over a longer time period.

A simple approach to assess acid-generation potential involves geological comparison with nearby mines. This approach assumes that all factors influencing the acid-generation process are identical for the mines. This is rarely the case in vein deposits as the host rocks, alterations, and mineralogies are often dissimilar. On a larger scale, comparison over a wider geographical region is likely to be unreliable as the nongeological factors that affect acid generation, like climate and physiography, will vary.

A basic approach for the assessment of acid-generation potential is to compare paleoenvironmental and geological characteristics. In this process it is necessary to classify both deposits from the standpoint of acid generation. Some of the existing mineral deposit classification models can be useful in the prediction of ARD.

The geological factors controlling generation of ARD include:

- Oxidation state of minerals
- Sulfide mineral compositions
- Texture and crystal development in sulfides
- Presence of acid-consuming minerals
- Presence of rock structures that increase permeability

Any available database to refine a geological classification in terms of ARD potential should also be helpful.

In the prediction process, a reliable sampling program is the initial step. Beginning a sampling program is complicated because static tests on samples from a defined geologic unit may indicate significant variability in acid-generation potential. This variability may indicate that the geological unit actually consists of two or more units from the standpoint of acid generation. The sampling plan should be revised to define the additional units. Such an iterative sampling program may be necessary to clearly define acid-generating units.

The sampling program should be directly based on the mine plan. The samples should be taken from different areas in the mine plan. This approach will help predict the timing of acid generation as mining progresses.

The design of a sampling program is initiated at the exploration stage when geological units are identified. In the next stage a more detailed sampling program is required to define more reliably the potential for net acid generation. The sampling program should also respond to any change in mine plan. A minimum number of samples to characterize each geological unit in terms of its potential to generate net acidity will be needed.

The potential sources of samples are outlined in Table 5.4.

For large-scale kinetic tests such as on-site rock piles large amounts of a specific unit are required. For proposed mines such large volumes of specific units may not be available.

Static Tests

A static test defines the balance between potentially acid-generating minerals (potential acidity) and acid-neutralizing minerals (neutralization potential) in a sample. In particular, acid-generating compounds include reactive sulfide minerals and acid-neutralizing compounds include carbonate minerals. A sample will theoretically generate net acidity at some point only if the potential acidity exceeds the neutralization potential (NP); otherwise the sample will not produce net acidity as long as the NP is not dissolved more quickly than the generation of acidity.

Despite the theoretical simplicity, static tests cannot be used to predict the quality of drainage emanating from waste materials at any future time. Acid-generation processes and therefore drainage quality are time dependent and functions of a large number of complex factors such as mineralogy, rock structure, and climate. For this reason, static tests should be treated as a qualitative predictive method; i.e., they can only indicate whether a potential exists for the generation of net acidity at some unknown time.

Several types of static tests are available such as acid-base accounting and APP:sulfur ratio. However, all of these tests are simply variations on a basic procedure and all require variations of the same basic analyses for determining the balance between potential acidity (AP) and NP. Consequently, the basic, common procedure will be presented and the names of the variations will be deemphasized.

The initial step in defining the acid-generating/acid-neutralizing balance in a sample begins with a measurement of total sulfur in a sample, commonly performed with a Leco furnace/analyzer. The measurement of total sulfur allows the calculation

Table 5.4. Potential Sources of Samples for Acid-Generation Prediction

Mine Component	Existing Mines	Proposed Mines
Pit walls	Drill core Pit walls	Drill core Underground exploration passages Trenches
Underground workings	Drill core Walls Excavated rock	Drill core Underground exploration passages
Waste rock/overburden piles	Waste rock piles Drill core	Drill core Underground exploration passages
Tailings	Tailings Impoundments	Pilot plant for mill process
Ore stockpiles	Ore stockpiles	Drill core Underground exploration passages
Spent ore	Heap leach	Pilot plant for heap leach

From *Draft Acid Rock Drainage Technical Guide*, Vol. 1 (Vancouver, Canada: BiTech Publ., 1989). With permission.

of "maximum potential acidity," which may overestimate the potential for acid generation if all sulfur in a sample is not acid generating. Therefore, additional analyses may be performed to refine the potential acidity. The analyses, which have not yet proven to be as reliable as total sulfur, are

- Sulfur species, which defines short-term leachable sulfate and leachable sulfide using acid extractions
- Reactive sulfur, which defines short-term oxidizable sulfide using hydrogen peroxide

The unproven nature of these additional analyses makes them options in a static test.

Following the delineation of AP, the next parameter, NP, is defined. The measurement of NP provides a gross value for neutralization; however, this value may overestimate the capacity of the sample to neutralize the pH to an environmentally acceptable level above 6. An analysis of carbonate content will provide a more meaningful measure of NP from the perspective of pH neutralization. The carbonate analysis is recommended as an optional portion of static tests.

Paste pH is measured in a paste, formed by water and the ground sample. The pH value will indicate the immediate reactivity of neutralizing minerals in the sample and will indicate whether significant acid generation occurred prior to the measurement.

Following these analyses, the potential for net acidity is calculated by subtracting AP from NP with a negative value indicating the potential for net acidity. Alternatively, a ratio of NP to AP can be used (APP:sulfur ratio), but the subtraction method (acid-base accounting) is adopted here.

The subtraction of maximum AP (based on total sulfur) from the gross NP yields the "net neutralization potential" (NETNP). Theoretically, a sample can be expected to generate net acidity at some point if the NETNP is <0. However, based on general experience, values of NETNP in the range of –20 and +20 tons of CaCO3/1000 tons of sample (-2 to +2% CaCO3) may be considered to have the ability to generate net acidity. This range of uncertainty is attributed to the sources of error in:

- Obtaining the objective of defining true potential acidity and neutralization
- Converting total sulfur to acidity using a restricted conversion factor
- Analysis

The subtraction of AP (based on reactive sulfide) from carbonate content yields the "net neutralization potential from species" (NETNP(S)). This value will presumably reflect the actual NETNP due to the narrower range of uncertainty and, thus, provide more reliable predictions, although no database exists to confirm these conclusions. The primary sources of error are similar to those for the NETNP (above), except that estimating long-term reactive sulfide from a short-term test may result in some uncertainty.

In the event the samples from a geologic unit indicate the unit has or may have the potential for net acid generation, kinetic tests should be conducted.

Kinetic Tests

Static tests identify the geologic units at a site that may have the potential to generate net acidity. Geochemical kinetic tests involve weathering (under laboratory-controlled or on-site conditions) samples of these units in order to confirm the potential to generate net acidity; determining the rates of acid generation, sulfide oxidation, neutralization, and metal depletion; and test control/treatment techniques. This information is critical because, for example, the rate of acid generation may be negligible, or in extremely rare cases, may be severe for only a short period of time so that long-term control or treatment techniques may not be necessary. Based on the results of kinetic tests, the optimization of treatment and control techniques to address the specific severity and duration of acid drainage from a geologic unit will minimize overall costs of acid-generation abatement.

Whereas static tests provide some information on overall potential acid generation independent of time, kinetic tests explicitly define reaction rates through time under specific conditions. As a result, kinetic tests are significantly more expensive and continue for months or years. Laboratory kinetic tests conducted in the short term only provide semiquantitative information on drainage water quality because they do not reproduce site conditions. In order to provide quantitative data on water quality at the site, waste material test pads can be monitored for several years. Ultimately, true prediction of long-term drainage quality will only be possible through quantitative mathematical models that can reliably extrapolate results beyond the time of the tests.

The initial step in a kinetic test is the definition of material characteristics in addition to those measured in static tests, specifically surface area, mineralogy, and total metals. These characteristics are important to the interpretation of the results from kinetic tests as they can affect the acid-generation process or overall water quality.

The particle size of a material can affect the acid production and acid-consumption results. Smaller grain-sized materials have a greater surface area per unit weight and a greater density of broken crystal bonds.

The mineralogy of a sample may also be directly related to reaction rates. Both the chemistry and crystal form of the minerals in a sample control the rates of acid generation and neutralization. For example, poorly crystalline minerals react faster than their crystalline counterparts, and some sulfide minerals oxidize faster than others. Additionally, the mineralogy of a sample may determine the metals that could be leached during acid generation and the extent to which pH may be neutralized by the sample.

Total metal analysis assists in the evaluation of the water quality from the tests. First, total metal analysis indicates any metals present in high levels that may warrant attention. Second, the leaching rates of a metal when compared to the total metal content will suggest when a metal may be depleted within the sample, resulting in negligible leach concentrations, even though it is difficult to extrapolate laboratory test concentrations to field leaching conditions.

Once the material characteristics have been determined, which include the sulfur and carbonate content determined during static tests, the overall program objectives must be defined before the selection of a kinetic test. The program objectives should be based on the mine plan and the proposed handling of acid-generating rock. Program objectives could include one or more of the following:

- Selection or confirmation of disposal options
- Determination of the overall water quality impact
- Determination of the effect of the flushing rates through a sample on water quality
- Determination of the influence of bacteria on the acid-generation sample

Kinetic tests are selected for each acid-generating component based on the information required to meet program objectives.

Both small-scale controlled tests (e.g., humidity cells) and large-scale on-site weathering trials have been used in assessing acid-generation reactions. The controlled tests have the advantage of simulating specific climatic and weathering conditions. On-site tests may be considered more representative than controlled tests because of the natural conditions under which the tests are conducted; however, because results vary as climatic conditions change, the interpretation and extrapolation of the test results are more complicated.

The data from kinetic tests are evaluated to define the rate and temporal variation of acid generation and water quality of a sample or a treatment/control technique.

The results are assessed to determine if they are environmentally acceptable with respect to the proposed mine plan. For example, if the proposal is made to mix waste rock with limestone and the tests indicated that acidic drainage occurred in a kinetic test, then the results would not be environmentally acceptable.

If the results are not environmentally acceptable, then the mine plan and the program objectives must be redefined. The mine plan must be redefined to ensure that the appropriate acid-generation control and treatment techniques are used. The program objectives may have to be redefined to incorporate the changes in the mine plan, and to test for additional appropriate acid-generation control and treatment techniques.

Additional tests would not be necessary if the existing data (through extrapolation) were sufficient to evaluate environmental acceptability concerning the new mine plan. If the existing data are not sufficient, then additional kinetic tests should be conducted to meet the new objectives.

When the results are environmentally acceptable, experimental results can be extrapolated to other conditions or into the future using mathematical models.

CONTROL OF ACID GENERATION

The ARD control can be divided into three stages:

- Control of acid-generation process
- Control of acid-generation migration
- Collection and treatment of ARD

Controlling ARD by preventing or inhibiting acid generation is the most preferable stage of control. The objective of acid-generation control is to prevent or reduce the rate of acid formation at the source by inhibiting sulfide oxidation. This can be achieved by excluding one or more of the principal ingredients or by controlling the environment around the sulfides.

The primary components in the acid-generation process are

- Wastes containing reactive sulfide
- Oxygen
- Water

Factors that influence acid generation include:

- Bacterial activity
- Temperature
- pH

Acid generation can be controlled by eliminating or reducing one or more of the essential components, or by controlling the environmental factors at the source in order to retard the rate of acid generation. This control can be achieved in one of the following ways.

1. Sulfide removal or isolation: If sulfide minerals in waste rock and tailings are removed, reduced, or isolated by coating or through some other means, then the sulfide-oxidation-producing acid will not occur. High sulfide content in waste can be concentrated and separated from the bulk of mine waste. The procedures used to concentrate, remove, or isolate the sulfides are termed "conditioning" of tailings and waste rock.
2. Exclusion of water: Total exclusion of water to prevent acid generation may not be practical. Water includes surface water, infiltration due to precipitation, and groundwater seepage. The main source of water depends on the type and location of the waste facility. In underground mines, groundwater seepage is prominent, while for waste rock dumps and tailings deposits, surface water and infiltration are important. Water can be excluded with impermeable barriers such as a synthetic membrane cover, but in the long term, degradation would result from water penetrating the barrier to facilitate acid generation.
3. Exclusion of oxygen: The elimination of oxygen from waste rocks would prevent the oxidation of sulfide minerals or reduce the rate of contaminant production. Although it is possible that acid can be generated under anaerobic conditions, this has not been a significant factor in mining wastes. Significant reduction in oxygen level can be achieved through the placement of a cover with an extremely low oxygen diffusion characteristic. Appropriate cover materials include soil, water, and synthetic materials.
4. pH control: If the pH of the water can be maintained within the alkaline range, acid generation can be inhibited. The pH may be controlled by the addition of alkaline materials to potentially acid-generating wastes. Blending of acid-consuming waste with acid-producing waste to achieve a net acid-consuming mixture can be a successful approach. Adding and mixing imported alkaline material such as ground limestone can be an efficient procedure.
5. Control of bacterial action: When the pH within a reactive waste pile drops below 4, the rate of acid generation increases fivefold or more by the presence of the bacteria *T. ferrooxidans*. The use of bacterial control compounds such as anionic surfactants (sodium lauryl sulfate), organic acids, and food preservatives can control bacterial action.

Available Control Measures

The principal objective in the selection of ARD control measures is to achieve the necessary environmental control in the most cost-effective way. The effectiveness of any control measure is determined by several site-specific factors:

- Degree of acid-producing potential of the mine waste, including the nature, quantity, and reactivity of sulfide minerals present, neutralizing potential of the rock, etc.
- Physical characteristics of the waste
- Climate, topography, and surface and groundwater hydrology

184 ENVIRONMENTAL IMPACTS OF MINING

- The expected time period over which the measure will be effective
- The sensitivity of the receiving environment to AMD

CONDITIONING OF TAILINGS/WASTE ROCK

The generation of ARD may be reduced by placing tailings and rock dumps in a condition that is favorable for ARD prevention. The sulfide content of tailings may be reduced by means of bulk sulfide flotation prior to placement. This process will produce a sulfide concentrate and flotation tailings. The latter will contain residual sulfide, and hence may still be potentially acid generating; however, control of acid generation will be easier to achieve for this, the bulk of the waste. The disposal of sulfide concentrate remains a consideration. An option that has been identified is pressure leaching of the concentrate to produce acid, filtering the acid off, and disposing of the remainder of the concentrate. The cost of flotation and disposal of sulfide concentration will influence the feasibility of this method. Another approach to tailings disposal may be to utilize the "dry" tailings disposal technique combined with an additive such as cement or bentonite, for example. These procedures could be used to produce a compacted soil cement with the intention of reducing oxygen and water access to the sulfide minerals. No records are available of these methods having been used to control acid generation.

The potential for isolation of pyrites by developing a coating of some form has been evaluated; however, these methods are still experimental and do not yet indicate adequate, economical control of ARD.

Placing tailings in a systematic managed manner in order to achieve a uniform deposit with maximum density and minimum segregation results in the minimum permeability to both air and water. Layered tailings placement, with minimized pool areas and maximized discharge densities, is a placement method often adopted. This technique is often referred to as "subaerial" or "semidry" placement. While this technique may have advantage under certain conditions, abatement of ARD does not necessarily occur. If the tailings remain in a saturated state, reductions in acid generation due to oxygen exclusion and reduced infiltration (due to reduced surface permeability) are noticeable but still comparatively small. However, once the tailings are allowed to dry (which is the case in this method), shrinkage cracks extending from the surface into the tailings deposit may cause a dramatic increase in permeability. Evidence indicates that this secondary permeability permits both oxygen and water entry into the tailings and continued acid generation. While underdrainage is maintained this may increase the rate of both oxygen entry and ARD. Thus, the direct beneficial effect of layered tailings on ARD abatement is small and, in some instances, may be detrimental. Of greater importance is the improved consolidation characteristics and surface trafficability, which permits easier cover placement.

The relatively poor control of ARD provided by layered or subaerial deposition is demonstrated by the experience with South African gold tailings, in which layered tailings deposition is practiced extensively. Oxidation and acid generation have penetrated many meters, in some cases tens of meters, into these tailings.

For acid-generating waste rock, merit may be found in segregating and isolating the high sulfide wastes during mining. This may serve to concentrate the high sulfide wastes in one location. While processes that concentrate high sulfide wastes may have definite potential in terms of waste management aspects, some form of ARD control is still required.

WASTE SEGREGATION AND BLENDING

Waste segregation involves the careful removal and separate handling of various geologic units at a mine site. Mines with acid-generating geologic units may also have other geologic units with excess acid-consuming capability. As a result, the segregation and separate handling of each unit provide two primary benefits. First, the volume of rock that may generate acidity and require treatment or control is minimized. Second, if acid-consuming units contain carbonate, which readily and reliably reacts to acidic pH conditions, these units can be blended with the acid-generating units in experimentally defined proportions for pH control. This is practiced effectively at some coal mines in the eastern United States.

The blending of acid-generating and acid-consuming rock units is similar to the alternative control technique of adding limestone or other neutralizing additives to the acid-generating waste. Consequently, successful blending is primarily dependent on the same factors as limestone addition:

- The movement of water through the system
- The nature of contact of acidic waste/water with the acid-consuming rock/water
- The proportion of excess acid-consuming rock
- The type and reactivity of the acid-consuming minerals

These factors determine the required procedure for blending.

Because acid-consuming rock units rarely contain $CaCO_3$ and other highly reactive carbonates in high proportions, overall costs/benefits of transporting and adding low-volume, highly neutralizing additives may be less than blending with higher volume, less reactive acid-consuming rock. The difference in volume of the mixture, associated catchment area, and monitoring requirements are also factors in determining the overall cost/benefit analysis of the alternatives. However, situations occur in which a mine plan may require similar handling, transportation, and disposal for both acid-generating and acid-consuming rock and, if blending is experimentally demonstrated to be successful and reliable, the sole cost of blending of the rock units may be more economical.

The costs of segregation and blending are site specific and dependent on the mine plan, the handling and transportation of the material, and the technique of blending.

BACTERICIDES

The rate of sulfide oxidation and acid generation is enhanced in some environments by microbiologic activity, particularly that of *T. ferrooxidans*. This bacterial activity can accelerate the oxidation both of ferrous iron to ferric iron in water and of reduced sulfur in the sulfides to a higher oxidation state. The purpose of bactericides is to create a toxic environment for bacteria so that the inorganic rate of acid generation cannot be enhanced. This does not imply that acid generation will cease.

The most popular bactericides for acid-generating materials include benzoate compounds, sorbate compounds, anionic surfactants such as sodium lauryl sulfate, and phosphate compounds. Laboratory and field experiments indicate bactericides reduce the rate of acid generation (as indicated by sulfate and acidity) as well as concentrations of certain metals generally by factors of up to 5, with one reported case of a 20-fold reduction. The overall effectiveness of each bactericide compound appears to be similar, generally between 50 and 95% effective in the short term. It should be noted that the period during which bactericides remain effective is limited due to the fact that they degrade and are removed by infiltrating and percolating water. The results quoted in Table 5.5 do not take into account degradation and depletion of bactericides.

Available studies do not discuss the impact of bactericides on pH, presumably because there is little effect on pH is seen, or effects are unpredictable. A 10-fold decrease in acidity theoretically changes the pH by only 1 pH unit (e.g., from pH 2 to pH 3), although aqueous buffering by sulfate and metals would further limit the pH change. The one detailed set of pH data in the referenced studies indicated the successful inhibition of *T. ferrooxidans* resulted in a maximum pH increase of only 0.3 units from pH 2.9 to pH 3.2.

Based on the laboratory and field data, bactericides may reduce the rate of acid generation but will not eliminate AMD. Consequently, bactericides must be used in conjunction with other control techniques for proper environmental protection. Additionally, some concern has been expressed over environmental toxicity of the bactericides that must be applied in strong concentrations to eliminate bacteria.

Bactericides are applied to surfaces of piles and fields using sprayers or hydroseeders. The cost is often several thousands of dollars for each hectare of surface. Because bactericides degrade and are removed by infiltrating water, occasional reapplication is necessary. The timed release of bactericide from rubber pellets is reported to extend the lifetime of one application.

BASE ADDITIVES

In mining environments with sulfide-rich rock, the potential for acid drainage is based on the relative proportions of acid-producing and acid-consuming materials. Acid-consuming minerals are also known as "alkaline," "basic," or "neutralizing"

Table 5.5. Effectiveness of Bactericidal Methods

Method	Results
Spray and controlled-release pellets	80% reduction in acidity, sulfates, irons, manganese, and aluminum
Sodium lauryl sulfate potassium benzoate, potassium sorbate on silver mine waste rock	Complete inhibition of *T. ferrooxidans*; 92 to 84% lower acidity
As above on coal mine waste	Short-term reduction in acidity
Sodium lauryl sulfate controlled release from rubber pellets	50 to 95% reduction in acidity
Sodium lauryl sulfate	60 to 95% reduction in acid production
BF Goodrich ProMac System Co.	58 to 72% reduction in acidity, 58 to 68% reduction in sulfate

From *Draft Acid Rock Drainage Technical Guide*, Vol. 1 (Vancouver, Canada: BiTech Publ., 1989). With permission.

material. If the potential for acid drainage exists through excess acid-producing material, one potential control technique is the addition of excess neutralizing material, particularly carbonate and hydroxide compounds, which produce a neutral to alkaline pH in the associated water.

The common additives are limestone ($CaCO_3$), lime (CaO or $Ca(OH)_2$), and sodium hydroxide (NaOH). These additives are usually used in solid rather than dissolved form because the liquid represents a less concentrated source through solubility constraints. For example, 1 m^3 of $CaCO_3$-saturated water provides approximately 1 kg of $CaCO_3$, whereas 1 m^3 of high-purity limestone provides around 3000 kg of $CaCO_3$.

The success of base additives to control acid drainage depends primarily on

- The movement of water through the system
- The nature of contact of acidic rock or water with neutralizing additives or water
- The proportion of excess neutral material
- The type and purity of neutralizing additive

The movement of water can affect the success of this technique in several ways. For example, the movement of water can affect the rate of acid generation, particularly in a saturated system in which the sole source of oxygen is dissolved in and carried by the water. If background groundwater with little oxygen moves upward into a saturated acid-generating rock pile, the rate of acid generation can be expected to decrease as soon as the oxygen added by disturbance and transport is depleted. Such a situation would lessen the severity of the acid drainage and minimize the necessary quantity of additive for pH control.

An occasionally overlooked complication related to water movement is the consumption of neutralizing additives by pH-neutral water. For example, rainfall passing through a surficial layer of additive and upwelling groundwater passing through a basal layer of additive can usually dissolve some of the additive in excess of the quantity needed to control acid drainage. The rate of water flow will determine the amount of required excess additive required to avoid early depletion of the additive.

The rate and direction of water movement provide the primary connection between acid generation and acid neutralization because these processes cease in the absence of water. This is closely related to the second factor determining successful control, the nature of contact of acidic rock or water with neutralizing additives or water. The nature of the contact can be divided into three basic scenarios:

- The additive lies above or "upstream" of the sources of acid generation
- The additive is mixed with the acid source
- The additive lies below or "downstream" of the acid source

The first scenario of acid/neutralizing contact involves the dissolution of additive into ambient water, followed by the movement of this water into acid-generating material, such as with rainfall moving downward through a surficial layer of additive into an acid-generating rock pile. Unless the rate of acid generation is low, this scenario is not effective because sufficient acidity can often be released into the water to overcome the alkalinity of the water contributed by the additive.

The second scenario of acid/neutralizing contact involves the mixing of additive with the acid-generating material. This provides a continual dissolution of additive in response to the acid generation, preventing the development of acid drainage outside the microenvironment around sulfide grains. In fact, laboratory experiments have demonstrated that the rate of acid generation can be slowed significantly, leading to decreased additive consumption, if alkaline water invades the sulfide microenvironment. As the thoroughness of mixing of additives and acid-generating material decreases, the potential success decreases as the first scenario comes into play and "hot spots" of unhindered acid generation arise. This accounts for the recognized poor success of layers of additive within acid-generating material.

The third scenario of acid-neutralizing contact has the additive downstream of the acid source, such as in a basal layer or a collector trench. In this case, the optimum use of additive requires the flow of acid water through the additive rather than over the top of the layer, which may not occur in a collector trench lined with a base additive. The elimination of hydraulic short circuits, which would allow water to flow around the additive, is critical to this scenario. Unlike the second scenario, no opportunity exists for in situ control of the rate of acid generation.

The third factor determining the potential success of neutralizing additives is the amount of excess additive. The deficit of natural neutralization potential as defined by static and kinetic tests indicates the minimum required quantity of additive as $CaCO_3$ equivalent. This quantity must then be increased to account for the dissolution by ambient precipitation and groundwater, and for the encapsulation of additive by precipitates. As acidic water comes into contact with additive and is neutralized, metal compounds such as hydroxides precipitate from the water and may encapsulate the additive, slowing or preventing further neutralization. This is a recognized problem with base additives that has not yet been solved, requiring a significantly higher quantity of additive than would otherwise be required. The alternative of forcing alkaline water into the microenvironments around sulfide grains, thereby encapsulating the sulfide minerals, has not been addressed experimentally in detail, but may warrant attention.

The fourth factor determining the potential success of neutralizing additives is the type and purity of additive. The common additives of limestone, lime, and sodium hydroxide differ in their solubilities and, thus, differ in the pH they create upon dissolution. Limestone, a common natural mineral, often raises aqueous pH to around 7.0 to 8.5. However, dissolution of limestone into water is restricted by high calcium concentrations, such as are found in gypsum-saturated waters (like most acid drainage), through the "common-ion effect." The common-ion effect may limit pH neutralization to pH 5 to 7. Lime and sodium hydroxide, which are not found in surficial natural environments because of their highly alkaline character, create aqueous pH values approaching 10 and above upon dissolution. Like limestone, their dissolution can also be restricted by the common-ion effect. Still, lime and sodium hydroxide provide greater neutralization than limestone per unit weight, but the greater unit cost and the environmentally unacceptable alkaline pH detract from the benefit.

The purity of the additive is important for successful control in that the lower the purity, the greater the quantity needed for equivalent neutralization. Lime and sodium hydroxide are manufactured and can be obtained in essentially pure form. On the other hand, limestone is quarried and its purity is often <100% as $CaCO_3$. The impurities are other carbonates and other minerals that may not contribute to neutralization. For any choice of additive, a neutralization-potential test identical to the tests carried out on acid-generating material should be carried out so that consistency is maintained.

After the additive is adjusted to a proper grain size to maximize reactivity and geotechnical stability, standard earth-moving equipment is required for the application of base additives as layers at locations upstream or downstream of the acid-generating material. For mixing of additive and acid-generating material rather than layering, the additive must be brought to the proper grain size to maximize reactivity and ease of application, as well as to minimize the potential for additive migration after application. Viable application procedures include slurry spraying, mechanical mixing during disposal, and slurry injection into boreholes following disposal.

The purchase, transport, and application of base additives may cost tens to hundreds of dollars for each cubic meter. These costs will be relatively low if a mine site is located near a limestone deposit that can be easily quarried and transported.

COVERS AND SEALS TO CONTROL ACID GENERATION

Covers and seals offer the ability to restrict the access of oxygen and water to reactive wastes. The restriction of water can serve to limit both the formation of acid and the subsequent transportation of the oxidation products into the environment. The exclusion of oxygen is more practical than the exclusion of water for the purpose of acid-generation control.

To limit oxygen and water entry, the cover must itself have a low permeability to either air or water, and it must not have holes or imperfections where entry can occur. If holes or cracks occur in, for example, a cover on a waste rock dump, then oxygen entry takes place as a result of convective flow of air into and out of the dump in response to natural barometric pressure changes and thermal currents through the dump. Cracking of a dump surface, as a result of the large settlements to which dumps are prone over the long term, may result in inflow of surface runoff. The resistance of the cover to cracking, the burrowing effects of roots and animals, and erosion and degradation due to weathering and frost action determine the long-term effectiveness of the cover.

A variety of materials may be used to provide surface covers depending on local availability and site conditions. These include different types of soils, synthetic membranes, water and a combination of soil and water, which result in saturated soil or bog conditions, and various other materials such as concrete, asphalt, etc. Alternative cover materials (other than water) and their permeability to water are shown in Table 5.6.

The most effective means of excluding oxygen is by means of a water cover. The other cover materials are generally more effective as inhibitors of infiltration in the control of ARD migration.

Soil Covers

Soil covers show promise as oxygen inhibitors as can be seen in the published data from the Rum Jungle site in Australia. The effectiveness of soil covers as oxygen barriers is influenced by the moisture content maintained in the cover. A cover that can be maintained in a saturated condition will be more effective, primarily due to the low diffusivity of oxygen in water, and to the absence of desiccation cracking. In this regard, composite covers with layers of different soil types to prevent desiccation of clay or till have been suggested as potentially beneficial. The reduction in oxygen entry due to soil covers is greatest for very coarse waste rock dumps where oxygen entry by convection is large. Convective transport of air into coarse waste rock is driven by changes in both temperature and barometric pressure and can be large through small holes or cracks in a cover. The long-term permanence of soil covers in resisting disruptive forces such as erosion, cracking, frost action, root action, and burrowing animals has yet to be proven in the field.

Table 5.6. Alternative Cover Materials

Cover Material	Permeability to Water (m/sec)	Advantages/Disadvantages
Compacted clay	10^{-9}–10^{-11}	Availability of large quantities problematic in many areas; subject to erosion, cracking, and root penetration; good sealing if protected and maintained
Compacted till	10^{-7}–10^{-9}	As above, but generally more permeable
Compacted topsoil	10^{-5}–10^{-8}	As above, but less robust, more permeable; questionable longevity
Peatland bog	10^{-5}–10^{-6}	Need to maintain in saturated condition; normally impractical for elevated waste dumps and sideslopes
Concrete	10^{-10}–10^{-12}	Subject to cracking, frost, and mechanical damage
Asphalt	10^{-20}	As above
HDPE synthetic	Impermeable	Requires proper bedding and protective cover; highly impermeable; life span unlikely to exceed 100 years; subject to root and mechanical penetration

From *Draft Acid Rock Drainage Technical Guide*, Vol. 1 (Vancouver, Canada: BiTech Publ., 1989). With permission.

While soil covers inhibit the access of oxygen to the waste and may control acid generation, they are generally more effective in controlling infiltration and the migration of ARD.

Synthetic Membrane Covers

Synthetic membranes such as polyvinyl chloride (PVC), high-density polyethelene, etc., have the potential to provide covers with an extremely low conductivity to air and good oxygen exclusion. These covers may be installed to provide good medium-term control; however, long-term degradation, loss of plasticity, and ultimate cracking limit the long-term effectiveness of such membranes.

Because of their vulnerability to puncture, membrane liners must be installed with adequate bedding preparation and protective surface covers. They are of low permeability and offer the potential of acting both as oxygen and infiltration barriers. Thick (2 mm) high-density polyethelene (HDPE) membranes are less susceptible to the disruptive forces affecting soil liners, except for the likelihood of tearing under differential settlement and long-term weathering. To allow for long-term degradation it is probably necessary to provide for liner replacement in 50 to 100 years.

Although the present value of the replacement cost of a synthetic cover on a 100-year basis may be quite small, the monitoring of the condition of the cover, administration of the fund, and actual implementation of the replacement are impediments to this approach, despite the fact that it can provide a relatively positive seal to moisture and oxygen penetration.

Other synthetic materials include geopolymers, asphalts, and cements such as high-volume polypropylene fiber-reinforced, sulfate-resistant shotcrete. The cost of such synthetic covers is often prohibitively high, and these often suffer from cracking and disruption over a long period of years.

As in the case of soil covers, synthetic membrane covers are more effective in the control of ARD migration.

Water Cover

Water cover is currently the most promising oxygen-inhibiting technique, and hence the most promising acid-generation control measure. The solubility of oxygen in water and the diffusion rate of oxygen through water are both very low. Thus, in the absence of convective transport, the rate of oxygen transport through water is sufficiently low to be of no concern in terms of acid generation. Evidence is steadily accumulating that underwater disposal of potentially reactive wastes reduces acid generation to negligible levels. Although oxidation of sulfide and resultant acid generation may not be halted entirely by placing wastes underwater, the rate of acid generation is generally reduced sufficiently to make the impact negligible. Care must be taken when considering placing old wastes that have previously generated acid below water because the solution of acid products contained in the waste may occur. The availability of water and the cost of maintaining this cover over the long term are obvious and important site-specific criteria that would influence its use as a water cover. Water cover may be achieved by the disposal of waste into natural waters, into manmade impoundments, or into flooded underground mine workings and open pits. However, it may not be feasible to achieve water cover on existing deposits and some types of waste facilities.

Saturated Soil or Bog

The effectiveness of a saturated soil layer for the exclusion of oxygen has been demonstrated and may be suitable for some categories of waste. Bog conditions can be achieved by the combination of a shallow soil cover with a shallow water cover provided by a water-retaining structure. Under these circumstances the waste will be effectively underwater. The soil helps to prevent total loss of coverage when the water depth reduces during dry periods, and it also prevents convective currents and wave action. Vegetative accumulation is also believed to have a marginal but beneficial effect on ARD abatement.

SUBAQUEOUS DEPOSITION

It has already been identified that the disposal of acid-generating waste underwater is currently the most promising abatement measure. Field evidence is growing that the disposal of reactive mine wastes underwater curtails oxidation to negligible levels. This is due to the very low diffusivity of oxygen through water (approximately 2×10^{-6} cm^2/sec). This concept can include disposal of waste into natural waters, flooded mine workings, or manmade reservoirs. A number of factors and possible limitations associated with underwater disposal are to be considered, however.

Current legislation and the politically controversial nature of lake and marine disposal of mine tailings (acid generating or otherwise) suggest that gaining approval for an application for this method of disposal for tailings will be difficult to achieve. For this reason it is probably beneficial to fully investigate any "on-land" means to achieve water cover. Lake and marine disposal of tailings should generally only be considered when all on-land options have been exhausted. This may not necessarily be the case for other waste types, for example, waste rock.

Disposal into Manmade Impoundments

Because available evidence indicates that water cover provides the most secure method of acid-generation control, consideration should be given to the construction of a water retention facility. The practicality and cost of a manmade reservoir relative to alternative measures are clearly dependent on site-specific criteria, for example, topography and volume of waste to be stored. The cost of flooding existing waste facilities is likely to be very high.

The design of a facility to provide water cover for combined tailings and waste rock may prove beneficial and cost effective for proposed developments. Combined tailings and waste rock disposal may have definite advantages in terms of acid-generation control, particularly if the tailings are not acid generating and are discharged at elevated pH (>7). If intimate mixing of tailings and waste rock can be achieved, the permeability of the coarse waste rock would be significantly less than if the rock were placed alone. This has the advantage of reducing potential water movement through the waste rock.

There are, however, limitations and design considerations that place manmade water cover facilities at a disadvantage, including:

- Water retention dams require detailed design of embankment and spillway facilities, careful construction control, and maintenance over the long term. Depending on the site-specific conditions, this may not prove economical.
- Reliable water sources must be available to provide a continuous water cover of sufficient depth to avoid exposure of the waste and erosion due to wave action or water flow.
- Minimum water cover needs to be maintained in low precipitation and drought periods.
- Water reservoirs may induce unacceptable seepage. If other soluble deleterious products are found in the wastes, these may result in increased contaminant loading of the environment. Whether these are significant for the specific project and site conditions needs to be determined.

Disposal into Flooded Mine Workings

Flooding underground mine workings and open pits is a means of controlling acid generation from their exposed rock faces. This method also provides a potential disposal area for acid-generating waste.

Flooding of worked-out coal mines has been successful in the control of AMD in several instances, with acidity reductions of 45 to 99% being reported. The potential benefits from flooding underground mine workings have been reported for several anthracite coal mines in eastern Pennsylvania. Field investigations at the mines, which were allowed to flood some 14 to 20 years ago, revealed that the mine waters, which were formerly highly acidic, are now slightly alkaline. Sulfate reductions of approximately 54 and 74% in mine waters were seen in comparison to 1960s data. In addition, marked decreases in the iron, aluminum, manganese, calcium, and magnesium levels were observed.

The disadvantages associated with storage of waste in flooded mine workings are

- At single pit operations it is necessary to store reactive mine waste rock for the life of the mine and to then incur rehandling costs in moving the material back to the pit when the operation is producing no revenue
- If all waste removed from the pit is reactive, the swell factor (usually about 30%) will generally result in an excess volume of reactive waste to available underwater storage, particularly as the pit will likely flood to a less than full point
- Any sulfides in the pit walls above the final water elevation will oxidize, causing a deterioration in water quality unless preventative measures can be applied
- Backfilling and flooding preclude future underground development that might be associated with the ore body

Lake Disposal

Lake disposal is a subaqueous method of the disposing of mine waste into an existing natural system. Lakes have been used in the past for the disposal of both acid-generating and non-acid-generating tailings. Available evidence indicates that the deposition of acid-generating rock underwater, such as into a lake, will effectively control the rate of acid generation. However, lake disposal of tailings also presents other problems related to turbidity and metal mobilization, which may affect the biological communities in the lake. Consequently, the environmental concerns related to lake disposal include:

- Toxicity of reagents and heavy metals from the mill process
- Excessive nutrient additions from the use of explosives
- Increased turbidity due to suspended solids causing a reduction in light penetration
- Direct physical impact from the placement of waste on the habitat

Therefore, in selecting a lake for the disposal of tailings or waste rock, the relationships between the various site, physical, chemical, and biological characteristics should be investigated. As an example, the overall climatic conditions and depth (from bathymetry) will determine the development of density and temperature stratification, which in turn may lead to an annual lake turnover. Mixing and stratification control the chemical characteristics (baseline conditions) and therefore the ability of the lake to assimilate the loading of acidity from the waste material. Finally, all three nonbiological characteristics combine to influence the lake ecosystem and its productivity and the tolerance of the ecosystem to mine materials disposal. Table 5.7 outlines some of the characteristics that should be determined to define the suitability of a lake location.

The characteristics of the tailings or waste rock that should be assessed are acid-generation potential, leaching characteristics, and settling properties of tailings. These characteristics will determine the physical and chemical impact of the material on the lake environment. For example, the leaching characteristics will indicate which metals, if any, could dissolve in lake water.

Marine Disposal

Marine disposal is another subaqueous method for disposing of mine waste into an existing natural system, thereby achieving water cover. This option is available for mines situated in proximity to a marine water body and is popular in countries such as Norway. As with lake disposal, very limited research has been carried out at the sites related to acid-generation control, and most of the impact observed is due to suspended solids and metal leachings from tailings solids.[5]

Many of the site, physical, chemical, and biological characteristics of importance in lake disposal must also be evaluated at marine disposal sites; in addition, several special factors should be considered. In particular, seawater has a greater buffering capacity and higher alkalinity than freshwater, which will affect the chemical interaction between water and tailings. Further, strong tides and currents will disturb the tailings, carrying them to other sites and affecting fisheries resources over a wider area.

Some of the waste material's characteristics that should be assessed are acid-generation potential, settling properties of tailings, and metal leaching characteristics. In particular, metal leaching in a saline environment should be considered.

Because many nearshore coastal environments frequently have fisheries value, reliable evidence for minimal environmental impact of marine disposal would be difficult to obtain. However, one may find local inlets and other coastal areas in which fisheries value is minimal and waste disposal may be environmentally acceptable. Otherwise, this option may only be acceptable if long pipelines or barges carry the waste far offshore, but the associated costs would be relatively high.

Migration Control of ARD

Where acid generation is not prevented, the next level of control is to prevent or reduce the migration of ARD to the environment. Because water is the transport

Table 5.7. Important Characteristics in the Assessment of Lake and Marine Disposal Sites

Site characteristics	Proximity of site to mine
	Route to lake from mine for tailings transport
	Regional climatic conditions
	Water/recreation use
Physical characteristics	Bathymetry
	Thermal stratification
	Hydrology — turnover and flushing events
	Hydrogeology — recharge/discharge characteristics
Chemical characteristics	General water quality including pH, buffering capacity, metal concentrations, and alkalinity (seasonal variations); suspended solids loading
Biological characteristics	Identification of resident fish
	Identification of salmonids
	Identification of benthos communities
	Identification of salmon, crab, and shrimp fisheries (marine)
	Productivity
	Unique systems associations

From *Draft Acid Rock Drainage Technical Guide,* Vol. 1 (Vancouver, Canada: BiTech Publ., 1989). With permission.

medium for contaminants, the control technology relies on the prevention of water entry to the ARD source. Control of water exit is of little value because in the long term all water entering the ARD source must exit, long-term storage being negligible. Water entry may be controlled by:

- Diversion of all surface water flowing toward the ARD source
- Prevention of groundwater flow into the ARD source
- Prevention of infiltration of precipitation into the ARD source
- Controlled placement of acid-generating waste

Diversion facilities usually consist of ditches. Diversion of surface flows, while easily implemented, is often difficult to maintain over many years. The best long-term solution to such surface flows is to select a disposal site that minimizes the need for diversion. Site selection is generally not an option in the case of open pits and underground operations, while factors other than ARD control may take precedence in site selection for other facilities.

If the ARD source is located over a groundwater discharge area, interception and isolation of the groundwater are very difficult to achieve and maintain over a long time. While measures such as underdrains and sealing layers can be employed, their long-term performance is questionable. The most effective solution is to select a site that is not located on a groundwater discharge area. Site selection may not be a practical solution.

The long-term secure prevention of infiltration is the most difficult to achieve. Covers of different types may be considered.

Controlled placement of acid-generating wastes includes cellular construction of dumps and tailings deposition methods that increase density and reduce permeability.

Diversion of Surface Water

Surface water can be prevented from entering acid-generating facilities by: (1) construction of drainage ditches and berms, and/or (2) site selection to avoid high runoff. The construction of ditches and berms is generally considered to be a short-term control because of high maintenance costs for long-term usage. However, maintenance requirements can be reduced for long-term structures by designing for extreme flow conditions with consideration for debris accumulation and providing appropriate erosion protection. While maintenance requirements can be substantially reduced, some form of periodic inspection and maintenance will nevertheless be required. Design flows for diversions can be reduced for facilities such as stockpiles, waste and spoil sites, and tailings deposits by locating these facilities near catchment watersheds. Site selection to reduce runoff, however, may have unfavorable implications to construction and operating costs.

Underground Mines

Surface water inflow is generally not the main source of water in underground mines. Surface water may flow into underground mines through portals, ventilation shafts, or possibly through cracks in the rock that develop by mining-induced settlement. At the Balaklala, Mammoth, and Walker mines in California, subsidence has created caved areas that channel surface water into underground workings. Diversion can be achieved by ditches and berms in the short term and concrete plugs in the long term.

Open Pits

Diversion of surface water around open pits and strip mines can be achieved with ditches and berms. Creeks and streams may have to be rerouted around these operations. These structures require periodic inspection and maintenance in the long term. The period between inspections depends on the design criteria and level of design adopted. Long-term facilities should be designed for severe flows, and possibly with allowance for flow blockage. Erosion protection, such as drop structures or riprap, should also be provided. Design flows for short-term structures are generally less severe due to the shorter return period.

Diversion of surface water may also be conducted within the pH limits in which one portion of the pit is not an ARD source. This can be achieved with in-pit ditches or berms and by sloping of the pit floor. This is a short-term control and may be expensive to operate because of practical problems and restrictions on mining activities.

Waste Rock Dumps and Spoil Piles

Diversion of surface water around waste rock dumps and spoil piles is achieved with ditches or berms and rerouting of creeks. Requirements for diversion can be reduced by site selection. However, other factors such as haul distance to the dump location influence site selection. Favorable sites are at the crest of slopes, or on small plateaus and near the upstream end of a watershed; however, unfavorable cost implications may be found with these locations. Long-term facilities will require more stringent design than short-term facilities and will require some form of maintenance.

Diversion of surface water around acid-generating dumps cannot be avoided by the use of underflow through base drains. Even if the drain is composed of non-acid-generating waste, the water flowing through the drain will become contaminated by water that has infiltrated through the overlying waste. Separation of these waters with barriers in the dump is not considered practical. Differential settlement within the dump would damage or destroy such a barrier.

Tailings Deposits

The approach and methods for the diversion of surface water around tailings deposits are the same as those described for waste rock dumps. However, natural topography is important in site selection for tailings dams. The most suitable location for a tailings dam, in terms of minimizing construction materials, is most often one that requires a substantial diversion of surface water. Construction cost and the cost of long-term maintenance of diversion facilities need to be optimized in site selection. Other forms of ARD control also need to be considered. For example, the control of acid generation by means of water cover eliminates the need for surface water diversion. In the case of acid heap-leach operations, the inherent low pH of the material and pore water is an important consideration in the design of control measures.

Stockpiles and Spent Heap-Leach Piles

The methods for the diversion of surface water around stockpiles and spent heap-leach piles are the same as those described for waste rock dumps. Proximity to the mill site will usually be a principal factor in site selection for stockpiles. The presence of chemicals from the leachate used for mineral extraction in heap-leach operations may require a more stringent design than is required for stockpiles.

Groundwater Interception

When groundwater enters or comes into contact with acid-generating waste, this provides a transport medium for the contaminants. The entry of groundwater into waste facilities may be prevented by interception or isolation of groundwater before it enters the waste, or site selection to avoid groundwater discharge into the waste. The objective is to minimize groundwater contact with acid-generating material or other water that has become contaminated by acid generation.

All collection and interception methods are prone to failure over the long term. Therefore, selection of a site that avoids groundwater discharge is the best method of control. Site selection is not an option for underground mines and open pits, but should be considered for waste dumps, spoil piles, tailings deposits, stockpiles, and spent ore piles. Any potential storage or waste disposal site will be located in either a groundwater recharge area, a groundwater discharge area, or an area with sufficient seasonal variation in groundwater level to be both a recharge and discharge area, depending on the time of year. The latter case may be the most severe in terms of environmental impact. The performance and cost of different groundwater interception and isolation methods vary over a wide range, depending on hydrogeologic and other site-specific parameters.

Underground Mines

Underground interception or control can be achieved by various methods for underground mines, depending on the site geohydrology. Regional dewatering around underground openings can be achieved with wells from surface or perimeter drainage galleries. Localized control of groundwater in conductive aquifers encountered by mining can be achieved by grouting the aquifer or dewatering with drainholes. In any method in which the water is collected within the mine, effort will be required to prevent it from contacting acid-generating material. Exploration drill holes can be a large source of groundwater flows into underground workings. Inflow through boreholes can be controlled by properly grouting and sealing the holes after drilling.

Open Pits

Dewatering around open pits is a common practice for improving slope stability. Groundwater interception and removal can be achieved with perimeter wells, drainage adits, and horizontal drainholes. However, an important consideration is that ARD could develop in wells and adits if these pass through acid-generating rock and oxygen is present. A means of preventing this acid generation is to keep the adit, well, or drain flooded, hence, minimizing oxygen access.

Horizontal drainholes are not well suited to groundwater interception for the prevention of ARD if these drainholes flow into the pit where contact with acid-generating rock is likely to occur. A pipe collection system can be installed to prevent ARD contamination. However, a low level of performance should be anticipated because of damage caused by mining activities, and bench and pit slope failures. Groundwater isolation by grouting may be effective for areas of high inflow.

All of these methods are short-term controls, although drainage adits that drain away from the pit could provide long-term isolation.

Waste Rock Dumps and Spoil Piles

Groundwater interception or isolation under waste rock dumps and spoil piles will generally be difficult to implement and maintain. Site selection that avoids groundwater discharge is the best method of preventing groundwater from flowing into dumps. If a waste facility is located in a groundwater recharge area, and acid generation and migration are not controlled, there is a risk of groundwater contamination occurring. In some cases it may be necessary to situate the dump on a groundwater discharge area and provide measures for acid-generation control or facilities for collection and treatment.

Tailings Deposits

Groundwater interception or isolation at tailings impoundments is generally not practical. Avoiding groundwater discharge by site selection is difficult because the best disposal sites are usually valleys, which are often also groundwater discharge areas.

Stockpiles and Spent Heap-Leach Piles

Groundwater isolation under stockpiles and heap-leach piles can be achieved with engineered barriers, though these are not likely to provide long-term control. Isolation can be achieved by placing the stockpile on an impermeable layer overlying a drainage layer to remove groundwater inflow. Integrity of the impermeable layer is critical for successful isolation. If damaged, the impermeable layer is very difficult to repair without removing the stockpile.

Covers and Seals to Control Infiltration

The transport medium for contaminants is water and the principal source of this water is infiltration of precipitation. The control of infiltration is therefore important in controlling ARD migration. The most practical way of controlling infiltration of precipitation is by means of low-permeability covers or seals. Soil and synthetic materials are commonly used to construct covers. These can be applied to rockfaces in open pits or underground mines, waste rock dumps, spoil piles, tailings deposits, stockpiles, and heap-leach piles. The length of time during which control is required is an important consideration in selecting the most appropriate cover material or combination of materials.

Soil Covers

Soil covers show promise as oxygen inhibitors, but they are generally more effective in controlling infiltration of precipitation. The effectiveness of soil covers as inhibitors of infiltration depends on factors such as climate, cover design, and construction.

ACID ROCK DRAINAGE AND METAL MIGRATION   201

For minimizing cost, a simple, single-layer soil cover is preferred. A fine-textured soil, such as clay or silt, is required to limit infiltration. To effectively limit oxygen transport, it is necessary to maintain the layer at a high moisture content. A single soil layer, however, is limited in its effectiveness for the following reasons:

- Without capillary barriers, a simple soil cover is prone to large seasonal variations in moisture content. This could result in desiccation cracking and an increase in permeability. In addition, decreasing the moisture content of the soil increases the rate of oxygen diffusion. These seasonal variations are greatest near the surface and their effect is therefore greatest on thin covers. For single-layer soil covers to be effective, they need to be relatively thick to maintain a saturated zone during the dry season. The cover thickness required is probably a function of the climate.
- The fine-grained soils required to limit infiltration may be susceptible to frost. Ice segregation may result in degradation of the cover and increased permeability. Frost heave may also cause the surface of the cover to become irregular, allowing ponding and increased infiltration.
- A simple soil cover does not have the ability to prevent moisture from being sucked up from underlying tailings by capillary action. Likewise, it does not limit the migration of salts from the tailings to the surface due to surface evaporation and transpiration.
- A simple, single-layer fine-grained soil cover may not be able to adequately withstand wind and water erosion or burrowing and root action. Some form of erosion protection, such as vegetation or riprap, is normally required.

These limitations on the effectiveness of a single soil layer can be overcome by using complex covers.

The effectiveness of a soil cover is greatly improved by adopting a complex cover design consisting of several layers, each performing specific functions to improve water and oxygen exclusion and long-term stability. These layers and their specific functions are described below.

Erosion protection can be provided by vegetation or by a layer of coarse gravel or riprap. The establishment of vegetation on the waste dumps is desirable for aesthetic and land use reasons. Therefore, revegetation is usually the most desirable method of providing erosion control. However, where revegetation is not practical or will not sufficiently control erosion, coarse gravel or riprap may be required.

Studies for uranium tailings deposits in Canada indicate that forest cover would adequately control sheet and rill erosion and wind erosion, but no methods of analysis are available to assess the effectiveness of vegetation on gully erosion.

A special study on vegetative covers was recently carried out as part of the Uranium Mill Tailings Remedial Action Project (UMTRAP) in the United States (U.S. Department of Energy, 1988). This study investigated the use of vegetation to stabilize uranium tailings and specifically includes the use of vegetation to intercept infiltration. The principal finding of the study is that properly developed plant communities on complex soil covers can be effective in stabilizing covers and controlling infiltration on topslopes of waste piles. The study showed that the appropriate vegetative cover will adapt to climatic change, will repair itself after severe disturbances such as fires and droughts, and will persist indefinitely with little or no maintenance. The plants were found to protect topslopes against sheetwash erosion; however, resistance to gully erosion depends more on the overall pile configuration than on the vegetation and soil.

Certain physical, chemical, and vegetative stabilization methods have been evaluated for purposes of mine reclamation by the U.S. Bureau of Mines. This study incorporated field testing of these different methods and costs for the various stabilization procedures.

The purpose of the moisture retention zone is to provide a zone for moisture retention to limit desiccation of underlying layers. It also provides a growth medium to support vegetation. Moisture retention is therefore desirable for two reasons:

- It helps to keep the infiltration/oxygen barrier moist. This helps prevent desiccation cracking and reduces oxygen diffusion.
- By retaining moisture after a precipitation event, it supports vegetation and allows time for evapotranspiration to occur, thus reducing infiltration.

The soil used to construct the moisture retention zone would generally be a loam soil with a substantial sand fraction.

The upper drainage/suction break layer serves two primary purposes:

- To drain water laterally from the surface of the infiltration barrier, preventing ponding
- To prevent moisture loss from the infiltration barrier due to upward capillary suction

Prevention of ponding reduces infiltration. Keeping the infiltration barrier moist helps to reduce oxygen diffusion and prevents desiccation cracking. This layer can also be designed to prevent intrusion by burrowing animals if it incorporates large gravel. For drainage to be effective it must be constructed with a cross fall of 1% or greater.

The effectiveness of this layer would be expected to decrease with time as it becomes clogged with roots and organic debris and fines, and as the drainage slope is modified by long-term settlement of the underlying tailings or rock waste.

The infiltration barrier is a low-permeability layer consisting of fine-grained soil or synthetic materials (or a combination of both). Its purpose is to prevent the downward infiltration of moisture and the diffusion of oxygen into the waste. The lower the permeability of this material, the more effective it is as a barrier to infiltration. The objective of this layer is to provide a sufficient barrier to enable the overlying coarse-grained layer to drain infiltration and prevent ponding.

Capillary barriers are used beneath the infiltration barrier to reduce infiltration. The principle is that if negative pore-water pressure is maintained in the low-permeability material at the interface with the underlying coarse-grained capillary barrier, infiltration into the lower layer would be prevented. It was found that this would only be effective if ponding on the low-permeability layer does not occur which, in practice, would be difficult to achieve. However, for soil covers over fine-grained waste deposits such as tailings, a capillary barrier beneath the infiltration barrier may be useful in preventing suction of contaminated pore water from tailings up into the cover during dry periods.

The long-term performance of a complex soil cover could be greatly reduced if fine-grained materials are allowed to migrate into the coarse-grained layers. Filter layers should be considered.

A basic layer could be incorporated into the design to reduce the pH of infiltrating water and therefore acid-generation rates. Alkaline materials such as limestone could be spread over the surface of the waste before placing the cover or mixed into the cover layers.

Limestone is commonly mixed with waste rock during placement at coal mines with great success, and research is being done on the addition of phosphate rock. The potential for acid-generation control by surface applications of alkaline materials is less attractive than mixing them with the waste. Limestone has a low solubility in near-neutral water, and the resulting alkaline charge is therefore small and may be insufficient to control ARD. Surface inflows tend to be concentrated at isolated locations such as depressions, cracks, permeable zones, etc. At these locations the available alkaline materials are quickly exhausted. The addition of a basic layer would not significantly reduce AMD where unsaturated conditions predominate, such as in waste piles. It would be more beneficial in saturated tailings and might be usefully employed in tailings impoundment covers.

Information on the relative effectiveness of soil covers in controlling AMD may be obtained from the results of mathematical model simulations of covers, and from the results of monitoring a limited number of actual covers. The results of the infiltration modeling runs are discussed below and illustrated in Figure 5.1a and b.

Bare tailings can be expected to have high runoff rates, modest evapotranspiration losses, and substantial net infiltration or seepage. With an unvegetated surface, the runoff can be expected to be quite high. In the example model runs, runoff, evaporation, and seepage rates account for 19, 63.5, and 17.5% of the annual precipitation, respectively. When the tailings permeability was increased by a factor of 3, runoff rates decline by 3% and seepage rates increase by 3%.

Vegetation has a marked effect on the water balance at a tailings site. With the growth of a "moderate" vegetation cover, runoff rates decrease from 19 to 9% of the annual precipitation, while evapotranspiration increases from 63.5 to 73.6%. The major finding is that seepage rates are not changed. With a good vegetative cover, runoff rates are again reduced further, to 3.3% of the annual precipitation. Although evapotranspiration rates are increased, this may not offset the reduced runoff. One should not conclude from this that this phenomenon is universally applicable to all sites.

Direct application of soil to the tailings area surface may have mixed effects. If the soil retains its low permeability, runoff will increase substantially and seepage rates will

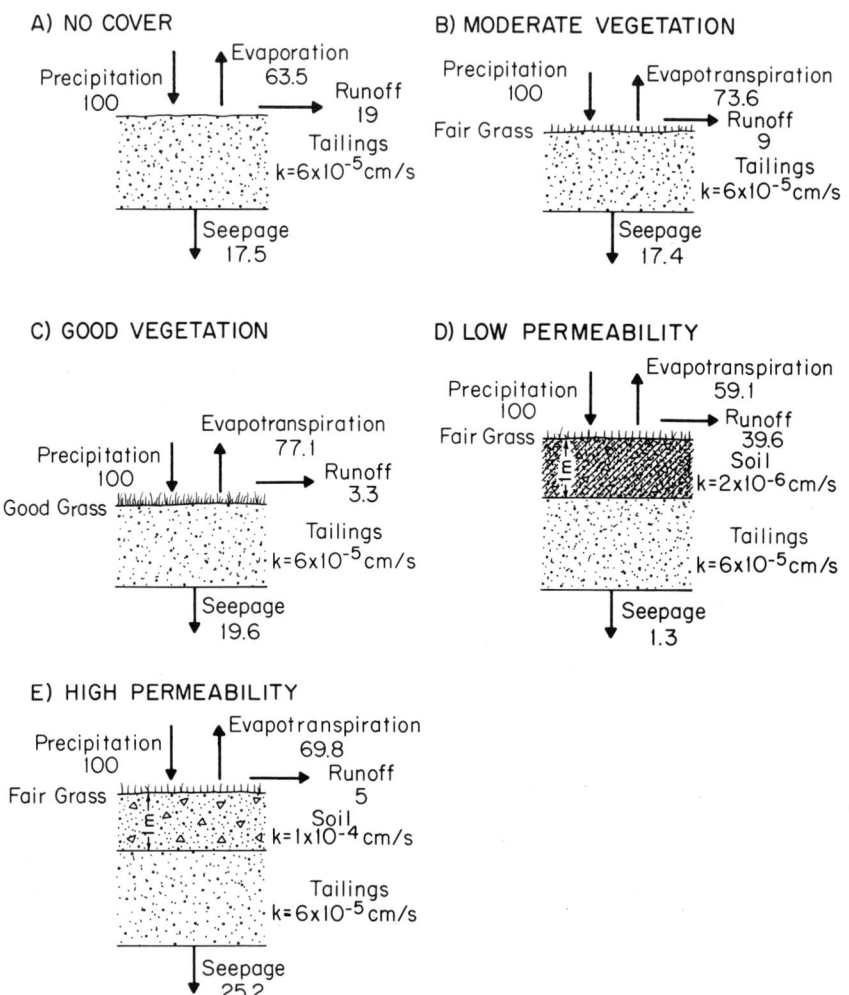

Figure 5.1. (A–I) Effect of cover types on infiltration rate· (From *Draft Acid Rock Drainage Technical Guide,* Vol. 1 (Vancouver, Canada: BiTech Publ., 1989.)

be greatly reduced. A compacted till cover with a permeability of 2×10^{-8} m/sec will reduce seepage rates to <2% of the annual rainfall. If this cover cracks and weathers (as is expected), infiltration rates increase substantially. The example indicates that if the effective permeability of the cover increases to 1×10^{-4} m/sec, seepage rates exceed those for bare tailings. The increased permeability results in a major reduction in the rate of surface runoff.

Rock or gravel is often applied to stabilize the surface of a tailings area. This pervious layer effectively eliminates runoff and therefore can substantially increase

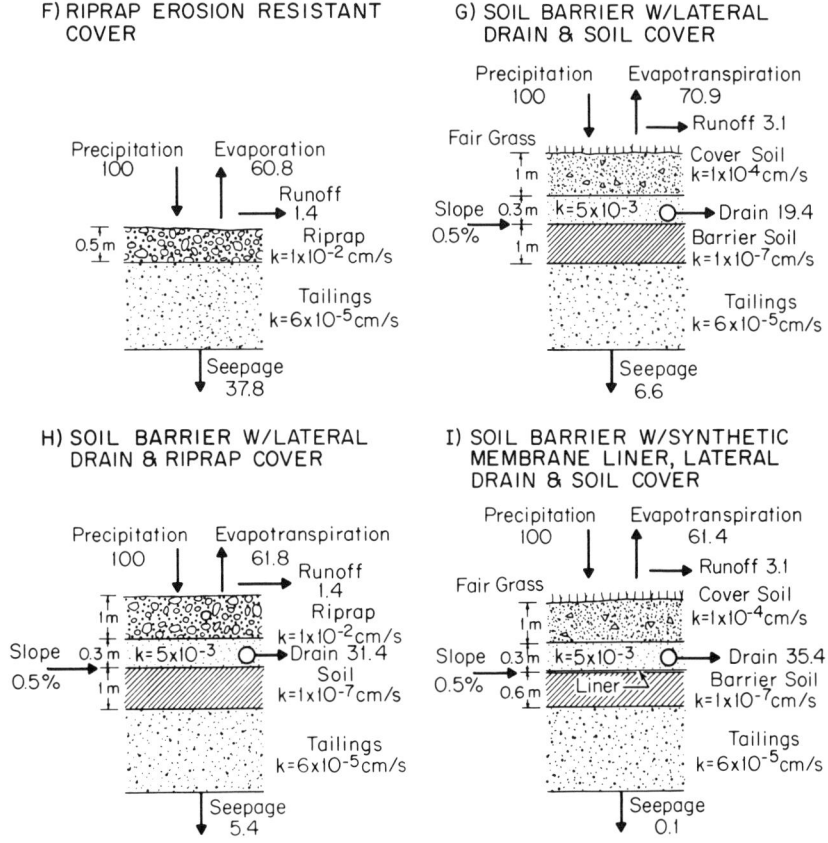

Figure 5.1 (Continued).

infiltration rates. For the modeled case, the rock/gravel cover increases seepage rates from 17.5 to 37.8%. This is more than a factor of 2 and further demonstrates how the permeability of the surface layer can affect the overall amount of seepage produced.

A properly constructed engineered cap can greatly reduce infiltration rates. The example modeled includes a cap with 1 m of soil for frost protection and vegetation, 0.3 m of lateral drainage layer, and 1 m of a low-permeability seepage barrier. This cap reduces surface runoff to 3.1% of the annual rainfall. The lateral drain intercepts 19.4%, while evapotranspiration accounts for 70.9%, leaving 6.6% as seepage. This is a 62% reduction in the total seepage, as compared to bare or vegetated tailings. The major finding is that these layers are effective but not 100% efficient in limiting seepage. At 6.6% infiltration this represents approximately 60 mm of precipitation or 60,000 m^3/year from one 100-ha disposal site at the Elliot Lake project.

Rock/gravel surfaces layers have a major effect, increasing the infiltration. With a pervious surface zone, the lateral drains become more efficient, reducing the seepage rates.

An engineered cap with a synthetic membrane liner is by far the most effective infiltration barrier. The seepage rates predicted for an engineered cap with a liner that was 99% efficient are 0.1% of the annual rainfall. The life of the liner, however, needs to be considered.

Although theoretical simulations are useful in comparing alternative cover types, the true effectiveness of covers in controlling AMD can be determined only from monitoring the performance of actual covers in the field. Unfortunately, monitoring results are limited.

The best documented case of a soil cover in use on an actual mine waste dump is that of the Rum Jungle uranium and copper mine in Australia. Composite covers were placed on three acid-generating overburden heaps. The top surface covers consisted of a 225-mm compacted clay liner, overlain by a 250-mm sandy clay loam retention zone layer, which was overlain by a 150-mm gravelly sand erosion layer. Rehabilitation of the heaps also included reshaping their surfaces and providing surface drainage systems. A typical cross section of the rehabilitated heaps is shown in Figure 5.2.

Measurements of oxygen concentrations in the pore gas in the heaps show a marked reduction in oxygen concentrations after installation of the compacted clay cover. Although measurements indicate that the transition rate of gas through the seal has increased since its initial placement, due to desiccation cracking in the dry season, the oxygen concentrations in the heaps are still much less than they were before rehabilitation. The effect of this reduction of oxygen concentration on oxidation rates has not been quantified.

Pre- and postrehabilitation measurements on and near the heaps indicate that the cover has provided some reduction in infiltration. However, the amount of that reduction remains questionable. Estimates based on lysimeter measurements indicate reductions >90%, while others based on groundwater estimates indicate only a 50% reduction.

The final measure of the effectiveness of the covers, however, is reduction of metals loads in the local river system. Precipitation, flow, and metal load data are summarized in Tables 5.8 and 5.9. The four wet seasons during and after rehabilitation (1984 to 1985) have all experienced below-average rainfall and well below-average runoff. The reductions in postreclamation metal loads are indicated in Table 5.10. Although samples taken from the East Finniss River show large reductions in metals loading, it is unclear whether these reductions are due to covering the heaps, or due to variations in precipitation rates; i.e., it is not clear whether the reduction in the loads is due to reduced acid generation and migration resulting from cover placement or merely due to reduced migration resulting from low precipitation in the years following cover placement.

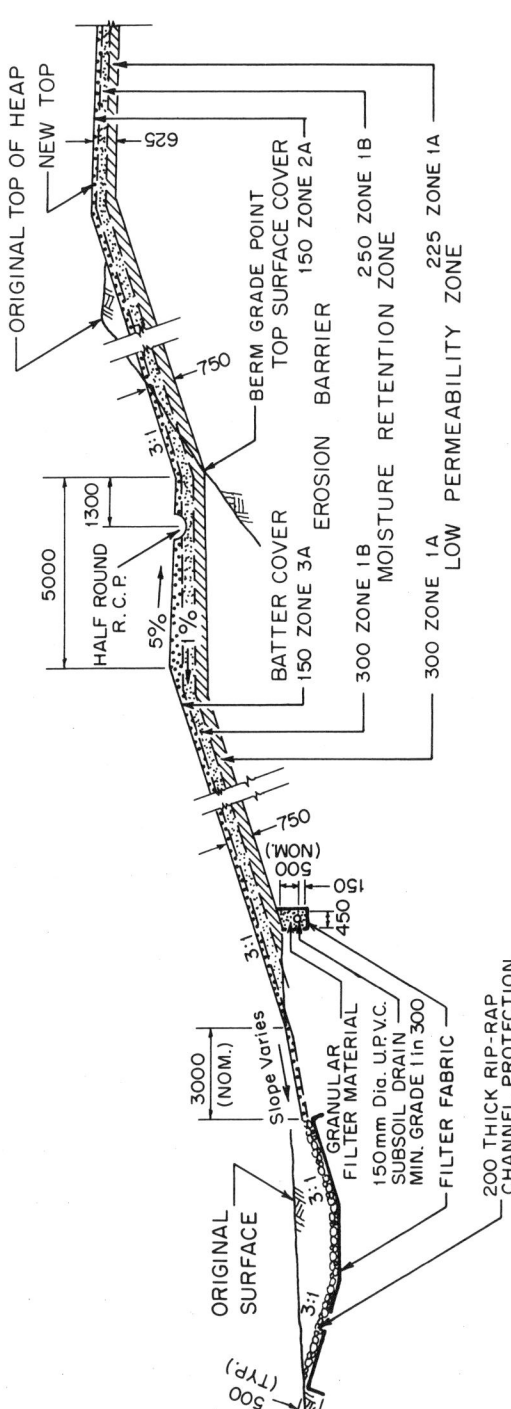

Figure 5.2. Cross-section of rehabilitation heap showing cover details. (From *Draft Acid Rock Drainage Technical Guide*, Vol. 1 (Vancouver, Canada: BiTech Publ., 1989.)

Table 5.8. East Branch of the Finniss River Revised Pollution Loading Values

Season	1971/72	1972/73	1973/74
Rainfall (mm)	1542	1545	2000
Total flow $m^3 \times 10^6$	31	22	69
Metal load (t)			
Copper	77	67	106
Manganese	84	77	87
Zinc	24	22	30

From *Draft Acid Rock Drainage Technical Guide,* Vol. 1 (Vancouver, Canada: BiTech Publ., 1989). With permission.

Table 5.9. Summary of Monitoring Results for the East Branch of the Finnis River

Season	1982/83	1983/84	1984/85	1985/86
Rainfall (mm)	1121	1704	1112	910
Total flow $m^3 \times 10^6$	9.5	48	11.7	11.4
Metal load (t)				
Copper	23	28	9	4
Manganese	6	21	7	8
Zinc	5	9	4	3

From Lapakko, K. Prediction of AMD from Duluth complex mining waste in North Eastern Minnesota, in *Proc. Acid Mine Drainage Seminar/Workshop, Halifax, Nova Scotia* (Ottawa: Environment Canada, 1987), pp. 187–221. With permission.

Table 5.10. Percentage of Reduction in Pollution of the East Branch of the Finniss River

Season	1983/84 (Stage 1 & Part Stage 2 Complete)	1984/85 (Stage 3 Complete)	1985/86	Target
Metal				
Copper	70%	80%	91%	70%
Manganese	76%	88%	86%	56%
Zinc	67%	73%	79%	70%

From Lapakko, K. Prediction of AMD from Duluth complex mining waste in North Eastern Minnesota, in *Proc. Acid Mine Drainage Seminar/Workshop, Halifax, Nova Scotia* (Ottawa: Environment Canada, 1987), pp. 187–221. With permission.

One interpretation of the results of surface water monitoring of the East Finniss River indicates a decline in annual copper loads since placement of the waste dump covers.

Measures of groundwater quality criteria beneath rehabilitated dumps at Rum Jungle indicate that the groundwater quality has not changed significantly during the 4 years since cover placement. It would appear that there is a store of contaminants in the groundwater and within the dump. The response of downstream surface-water quality to cover placement may well be affected by the release of these stored contaminants. Ongoing monitoring is essential until trends in surface and groundwater quality are established.

Synthetic Covers

The use of synthetic membranes as liners for tailings impoundments has been investigated.

Flexible membrane liners are commonly referred to as geomembranes. The common types are

- Polyethelene (PE)
- High-density polyethylene (HDPE)
- Chlorinated polyethelene (CPE)
- Chlorosulfonated polyethylene (CSPE, commonly known by the Dupont trademark Hypalon®)
- PVC
- Ethylene propylene diene monomer (EPDM)
- Butyl rubber

Occasionally neoprene and polyurethane are also used. Collectively, synthetic membranes display a number of advantages and disadvantages that may be summarized as follows:

Advantages

- They can contain a wide variety of fluids with minimum seepage due to low reported permeabilities of typically 1×10^{-1} cm/sec or less
- They have a relatively high resistance to chemical and bacterial deterioration
- They are readily installed for many applications
- They are relatively economical to install and maintain

Disadvantages

- They are relatively vulnerable to attack from ozone and ultraviolet (UV) light
- They have a limited ability to withstand stress from heavy machinery
- They have not been in service long enough to evaluate long-term performance
- They are comparatively susceptible to laceration, abrasion, and puncture
- Some materials are prone to cracking and creasing at low temperatures or stretching and distorting at high temperatures
- Although readily installed, there are often difficulties associated with the material's seams

Polymeric membranes offer wide-ranging chemical resistance and may be readily inspected. However, they are susceptible to damage during installation, largely due to improper subgrade preparation and vehicular traffic. They require very careful installation, and their performance is dependent on careful and successful field seaming. Field seaming is, in general, a detailed and sensitive operation. Weather, including temperature and precipitation, is generally the governing factor. In this regard, the elastomeric liners, namely butyl, polychloroprene, and EPDM, would appear to present the most problems in field seaming. Of the remaining liner types considered, successful field seaming has been demonstrated with HDPE, PE, CSPE, CPE, and PVC. It is noted, however, that serious concerns have been expressed about the long-term weatherability of PVC and PE.

Proper subgrade preparation and construction are crucial for a successful liner installation and would typically consist of the excavation of compressible materials; sterilization of the subgrade; removal of all roots, sticks, stones, and debris; grading and proof-rolling; and installation of the sand cushion, liner, and soil cover. Installation of the liner and field seaming should be carried out by approved installers meticulously following liner supplier instructions. Soil cover is desirable but will require liner inclinations flatter than about 3 horizontal to 1 vertical. This is a severe limitation when applied to waste dumps.

With the exception of polyurethane, the base polymeric resins and asphalt show promise for long-term resistance to the major anticipated constituents of AMD. However, CPE may be affected by weak sulfuric acid solutions.

Thin, flexible membrane liners are susceptible to overstressing by strains associated with large differential deformations in the subgrade. It may be necessary to subexcavate and replace compressible materials encountered over the subgrade prior to liner installation. Similar concerns exist for liners placed on slopes and where a potential exists for excess hydrostatic or gas pressure buildup beneath the liner.

Seepage through liners occurs primarily through liner defects. A rational approach to evaluating apparent or field liner permeabilities is via detailed monitoring of existing installations. Test results show that the type and thickness of the membrane liner have a relatively small influence on leak rates, while the low-permeability subbase is important in restricting flow through a flawed liner.[1]

Estimates of liner release rates indicate that an asphaltic membrane would reduce seepage to about 50% of an unlined basin for a field liner permeability of 1×10^{-8} cm/sec. Polymeric liners with an effective permeability of 1×10^{-10} cm/sec would reduce seepage to <10% of an unlined basin.

In West Virginia, a PVC liner was used to cover a 45-acre backfilled site to prevent seepage into acid-producing materials. Results showed substantial decreases in flow and acidity from associated seeps.

Used as the barrier layer in combination with soil material layers in a complex cover, geomembranes should prove to be very effective in limiting oxygen and water transport.

Asphaltic and spray-on surface sealants can be applied to the surface of the waste to form a barrier to infiltration and oxygen diffusion. A number of products are available, including:

ACID ROCK DRAINAGE AND METAL MIGRATION 211

- Alkyd
- Asphalt
- Concrete
- Epoxy
- Polyester
- Polysulfide
- Polyurethane
- Silicone
- Synthetic rubber
- Thermoplastic molten sulfur
- Vinyl

These materials have, in general, been developed for applications such as caulking sealants, soil stabilizers, waterproof barriers, and corrosion protective coatings. To date their application in mine waste covers is limited.

Surface sealants can be formulated to produce either flexible or rigid linings for covers. Surface sealants can be installed with three basic techniques:

1. In *situ chemical cure*. The materials chemically cure or harden after being applied to the surface. These materials usually involve more than one specific chemical.
2. *Heat application*. Materials that are solid in the desired operating temperature range are applied at elevated temperatures to improve ease of application.
3. *Surface drying*. The material is formulated in a water emulsion or diluted in a solvent carrier for application. The carrier evaporates, leaving a solid coating.

Combinations of the above techniques are also feasible in many cases. The object is to prepare the material for ease of application, usually with conventional spraying equipment. The actual technique for application is a function of the specific material.

The primary advantages and disadvantages of surface sealants are

Advantages

- Either sufficient flexibility to conform with or sufficient strength to support the design load (pedestrian or vehicle traffic, for example)
- Good weatherability and service life
- Compatibility with the stored product
- Immunity to biological attack
- Sufficient puncture and abrasion resistance
- Capability of being placed with minimal defects
- Easily reparable
- Ease of application and production of an integral liner with no joints

Disadvantages

- Relatively difficult to regulate the rate of application and thus the thickness and uniformity of the sealant.

- As a class, these materials are relatively expensive. The high initial cost vs relative ease of application for the spray-ons should be considered for specific applications.

"Geopolymer" is the term given to a compound of minerals, principally containing silica, phosphate, and oxygen, that bond to form a ceramic-type product. The suitability of using this product as a control measure for acid generation is currently being investigated. It is anticipated that geopolymers may be mixed with tailings to form a solid mass, preventing oxygen and/or water access to sulfides. A possible alternative is to mix the geopolymer with soil or other material and apply this as a cover to the waste. The behavior of geopolymers when mixed with different waste materials is not fully known at this stage, nor is their resistance to natural processes such as freezing and thawing. Geopolymers are still in the development stage and require extensive research to establish their suitability as a control measure.

Shotcrete is the name given to concrete pneumatically delivered through a hose and applied to a surface at high velocity. Shotcrete may be effective in the control of acid generation when applied as a cover to certain wastes. The advantage of shotcrete is that it can be applied to steep rock slopes or other surfaces that may be difficult to cover using other methods.

The effective use of shotcrete as a cover is dependent on the stability of the underlying material. This method has been used very successfully on rock faces and on compacted materials. However, if the material to which the shotcrete is applied undergoes consolidation or settlement, causing relative displacement at the surface, cracking of unreinforced shotcrete will occur. Once the shotcrete liner has cracked, the effectiveness of the cover is lost. Experience has shown that displacement of uncompacted waste dumps often occurs and for this reason unreinforced shotcrete is not appropriate as a cover for these materials. High-temperature-induced stresses in a shotcrete cover may also result in cracking.

The resistance of shotcrete to cracking may be increased by providing reinforcement. Conventional steel-mesh reinforcing is expensive, difficult to handle, and subject to corrosion in the long term. Steel-fiber reinforcement is easier to apply; however, is also vulnerable to corrosion. A method of reinforcement using high-volume polypropylene fiber reinforcement, which is corrosive resistant and relatively flexible, shows promise. A benefit of the fiber-type reinforcement is that it reduces crack widths in the shotcrete. The shotcrete cover is then given the facility to accommodate movements larger than mesh-reinforced shotcrete.

Placement of Covers

The establishment of covers on mine waste and tailings is complicated by the difficulties of access, trafficability, and stability of the surfaces onto which the cover is to be placed. These difficulties often render a particular cover type impractical or prohibitively expensive. The placement of some cover types requires access by wheeled vehicles working on fairly flat surfaces (asphalt covers). Others require careful bed preparation and moderate slopes (synthetic membranes), while others a firm surface to compact against (clay layers).

Rock waste surfaces are conveniently subdivided into the dump surface and the dump slopes, with different conditions applying to each.

The upper surface of a rock waste dump is usually readily accessible, trafficable, and nearly flat. The placement of any type of surface cover, except a water cover, is usually not difficult.

During dump development the material on the upper surface of dumps placed by trucking is often broken down and compacted under the wheel traffic of the dump trucks. This results in a fairly compact, lower permeability upper surface. This surface reduces infiltration, and ponding is often experienced on such surfaces. Despite the initial coarse nature of the material in such dumps, it may be necessary to install a suction-breaking layer to prevent downward suction on low-permeability cover layers.

Dumps are subject to long-term consolidation and settlement under their own weight, and as the dump rock weathers. These settlements are large (a few percent of the dump height) and uneven, reflecting the natural variation of the waste rock and dumping procedures. Differential settlements result in disruption of the drainage pattern on the dump surface and cracking of cover materials. Settlement and crack patterns are often such that drainage is toward cracks, resulting in considerably increased infiltration.

Dump slopes are usually placed at their angle of repose. At this angle slopes are inaccessible, untrafficable, and marginally stable. Cover placement on such steep slopes is essentially impractical. Crest dumping of cover materials has been attempted at some sites, creating uncompacted (permeable), uneven layers of questionable stability.

For dump slopes to be accessible, it is necessary to first reslope. At a slope of 3 horizontal to 1 vertical (3:1), the slopes are trafficable by tracked vehicles and it is possible to place soil-type cover materials. At this slope it is possible to also place synthetic membrane liners but the stability of cover layers over such membranes is questionable. The cost of resloping large dumps to 3:1 is very high, requiring large expenditures of dozer time, unless the resloping had been planned for and the dumps constructed with a staggered dump slope.

Dump slopes are subject to the same concerns regarding different settlement as are the dump surfaces. Erosion on the steep dump slopes is a major long-term concern.

Wet, unconsolidated tailings always present difficult access conditions. Access improves as the tailings are drained and consolidate. Where tailings have been spigotted onto beaches, the sand fraction is deposited near the spigots and drains more freely than the slimes, which accumulate nearby and in the pond. Drained sandy beaches may be trafficable within days of deposition, while pond areas may never achieve this condition. Thus, it is possible to place and compact covers on beach areas with little preparation. In pond areas it may be necessary to apply drainage measures to remove free and near-surface water, and to use geofabrics on slimes, followed by thin layers of the cover. These techniques have been successful in placing a cover over slimes that could not support foot traffic at the start of cover placement. Covers may also be placed during winter when freezing conditions allow access, as was done for the cover placement over wet tailings at the Beaverlodge mine.

To prevent capillary suctions in covers, it may be necessary to utilize capillary barriers over tailings fines and slimes.

Prior to placing covers on tailings it is necessary to first develop a tailings surface that has an adequate slope and drainage pattern. Much can be done to achieve such slopes by adopting an appropriate tailings placement and management method. Reshaping of tailings surfaces after mine closure with earth-moving equipment may be difficult and prohibitively costly.

After closure, tailings continue to consolidate and settle as a result of dissipation of pore pressures and thawing of included ice. These settlements can be a substantial portion of the total tailings depth and result in disruption of the drainage pattern, leading to extensive ponding on the tailings surface and cracking of covers.

The placement of covers on steep rock surfaces, such as pit walls, poses a particular problem. Two approaches can be used. The first requires the construction of thick self-supporting covers. The second requires adherence of the cover to the rock face and relies on the rock face to support the cover. The use of gunite or shotcrete methods is appropriate for the second. Both asphalt and concrete materials can be considered. Because of the corrosive nature of ARD upon cement and steel, the use of synthetic fibers and silica fume concrete is appropriate.

Waste Rock and Tailings Placement Methods

Control of ARD migration in waste rock dumps can be assisted by engineered placement methods such as cellular dump construction, compacting, mixing with low-permeability material, etc.

Cellular dump construction, when used in conjunction with a cover layer, can significantly reduce the area exposed to precipitation. Cellular construction may utilize a layer construction or lateral cell construction. This method is a short-term control as it applies only to the construction period.

Control of ARD migration by compacting or mixing with low-permeability material are both intended to reduce the bulk permeability of the waste rock dump, which will reduce infiltration. They can be used together or separately.

Compacting of waste rock will require dump construction in thin layers and is only suitable for soft rocks. This approach will also reduce settlement and increase dump stability. The result is long-term integrity and improved cover performance. Compacting will be very expensive.

Mixing low-permeability material into the waste rock will reduce infiltration. A non-acid-generating waste material or, preferably, a net alkaline waste, can be mixed during dumping. Dump stability must be addressed. This approach should provide good long-term reduction of infiltration.

Control of ARD migration in tailings impoundments is limited to methods that reduce the permeability of the tailings with density control. These methods include predisposal thickening, subareal deposition, flocculant addition, and installation of a dewatering system such as wick drains. It may be possible on some sites to reduce tailings permeability with the addition of clay to the tailings line. All of these methods will provide long-term control of ARD migration. Methods that increase the density of tailings may be required for stability in earthquake-prone regions.

MONITORING

In general terms, environmental monitoring in and around a mine site is intended to define baseline conditions and to identify changes in conditions during and after mining. This information is generally used for decision making regarding mitigation and reclamation strategies.

The environmental conditions monitored typically include physical processes such as water flow and geotechnical stability, chemical characteristics such as water quality, along with biological response and impacts such as productivity. The major objective of a monitoring program in the ARD context is to monitor the effectiveness of the prevention/control/treatment techniques and to detect at the earliest point in time whether the techniques are unsuccessful.

In the preoperational phase, baseline monitoring defines existing environmental conditions of the physical, chemical, and biological aspects of the area. This information leads to the identification of areas that are particularly sensitive to changes in environmental conditions and provides essential data to allow the assessment of changes or impacts caused by each component of the mine and mining activities.

In terms of acid generation, test work is conducted in the preoperational phase to determine the potential of each waste material from the various mine components to generate net acidity and acid drainage. Each of the potentially acid-generating materials may be further tested to determine the rate and duration of acid generation and its associated water quality. The design and testing of the required control and treatment techniques may also be conducted in this phase. The mine plan is adjusted in order to reliably implement the control/treatment techniques, to reliably eliminate acid drainage, and to minimize combined costs of environmental protection, mine construction, and operation.

Using the baseline information and the final mine plan, a monitoring program is established for mine operation. Two types of monitoring stations have been defined for this approach: an effluent discharge point and the receiving environment. An effluent discharge point is generally but not necessarily located on the mine property (onsite), while the receiving environment stations will generally but not necessarily be located offsite.

In the proposed program, monitoring stations, including both surface water and groundwater, are established in or near all environmentally sensitive areas potentially affected by the development.

A minimum of one surface water and/or groundwater monitoring station should be selected at a defined discharge point for each component to be an "advance warning" station. This would provide early warning of potential failure of acid prevention/control/treatment techniques. The advance warning stations are located at a point of direct discharge from the mine component into the receiving environment. Each mine component usually has at least one discharge point that can be selected as an advance warning station. These stations should be monitored at least monthly for pH, sulfate, alkalinity, acidity, iron, and electrical conductance. A significant decrease in pH or alkalinity and an increase in sulfate may be an indication of the onset of acid drainage. However, extreme care must be used in separating site-specific trends in water quality, such as seasonal variations in pH, from the onset of acid drainage.

Monitoring stations in the receiving environment in the vicinity of each mine component can be established within, upgradient of, and downgradient of the component for surface and groundwater flows. Downgradient stations, as defined by the movement of surface water or groundwater from the component, should be placed at various distances from the mine component in the receiving surface water and groundwater. Upgradient stations provide data for comparison with downgradient stations to determine the degree and spatial extent of impacts due to each component.

All stations should be monitored at least semiannually for a full set of water quality analyses: pH, sulfate, alkalinity, acidity, iron, electrical conductance, major cations and anions, nutrients, and a suite of metals. Station monitoring may also be used for monitoring water flow rates.

Biological monitoring is not considered to be as reliable, as rapid, or as consistent as water quality measurements and visual observations for the detection of acid drainage. Consequently, biological monitoring is not emphasized here for detecting acid drainage (although certain mines such as those located near important fisheries may be required to monitor productivity, species diversification, or metal levels in fish tissue). Nevertheless, an annual biological survey of the mine site and surrounding offsite region is recommended as contingency monitoring to check for changes in vegetation or fisheries that may indicate the migration of acid drainage undetected by an established monitoring network. A visual inspection to identify changes in color due to iron staining in seeps and streams in the area is also recommended. Color changes, for example, could arise during a first-flush event where acid products are released between sampling periods of the monitoring stations.

The monitoring program implemented during the operational phase of the mine should detect changes in environmental conditions at any station. If these changes are significant, more frequent monitoring should be performed at that station and at other stations to confirm the presence and spatial extent of the change. If an adverse impact is determined, alternative control or treatment techniques should be designed, tested, and implemented. The monitoring program should be revised to monitor the success of the new techniques.

If no unacceptable impact exists toward the end of the mine life, a long-term monitoring program for closure would be defined. Each of the steps should be performed because operational conditions at a mine ("baseline" for closure) will be different than preoperational conditions.

The long-term monitoring program implemented during and after closure would decrease in frequency of sampling as time after closure increases. If significant changes in environmental conditions are detected at any station, additional monitoring should be performed at that station and at other stations to confirm the presence and spatial extent of the change. If the adverse impact is confirmed, alternative control or treatment programs must be designed, tested, and implemented. The monitoring program must then be revised to monitor the success of the new techniques. If no unacceptable impacts are detected over an acceptably long period of time, the site can proceed with abandonment.

Specific Monitoring Programs for Each Mine Component

Each mine component will require monitoring for the effectiveness of treatment and control technologies and for the detection of failures of these systems at the earliest time. Therefore, the monitoring stations should include at least one advance warning station at a direct discharge point in addition to other stations located at different distances in the receiving environment from the mine component. As discussed above, the advance warning station should be monitored more frequently than other stations.

The objective of the advance warning stations is to detect any significant change from the background quality that indicates the onset of acidic drainage from the mine component. Because the stations are monitored monthly for various parameters, including pH, sulfate, alkalinity, and acidity, trends can be established for background conditions.

The sampling frequency of the advance warning stations or other stations may be revised to incorporate a special sampling frequency. For example, at some sites dry spells may be followed by heavy rains; therefore, sampling of the first flushing of the waste rock stockpile after a storm event can be incorporated into the monitoring program.

Emphasis of the receiving-environment monitoring program should be oriented directly downgradient from the mine site as defined by the direction of the surface and groundwater flow from the site. Receiving-environment monitoring defines existing conditions and identifies changes in the conditions during and after mining. Because the advance warning stations are nearest to the area of impact and are monitored at frequent intervals to detect a failure in the treatment/control techniques, the receiving-environment monitoring stations determine the longer term trends in water quality over a wider area.

A proposed monitoring program for each mine component is described in the following subsections. A summary of the surface and groundwater flows around and through each component and how they are potentially impacted during and following mining operations is outlined. Some of the geotechnical aspects of each component related to geotechnical monitoring are also described. The minimum recommended monitoring program for each component is given along with a variety of options for monitoring each component.

Environmental Monitoring of Open Pits

Potential sources of water entering a pit are precipitation, surface-water drainage, and groundwater discharge. A pit is open to the atmosphere so variations in precipitation affect the daily volume of water in the pit, particularly if precipitation represents a major total of pit flow. Surface-water drainage is frequently diverted away from the pit perimeter in order to lower pumping costs. Groundwater frequently moves toward a pit because an excavation often represents a depression in the water table. Piezometers or monitor wells may be installed around the pit perimeter to intercept groundwater before it enters the pit if water control or geotechnical stability is a concern. All water entering a pit is usually combined and directed to sumps where it is dumped from the pit. The resulting surface flow downgradient of the pit may be directed into a holding pond for monitoring and any required treatment.

From a geotechnical viewpoint, a major concern is the stability of the pit walls. Instability represents a danger to the work force and a delay in mining. Because water decreases friction through increase in pore pressures, groundwater can be a major cause of instability in a pit and may be controlled by interceptor wells or underground galleries at the pit perimeter.

During operation and at closure, the location for the advance warning station in the vicinity of the open pit should be at the discharge from the sump and/or the retention pond. The advance warning station is monitored once every month throughout operation and closure.

In addition, the seeps from the pit walls should be sampled once every 6 months to monitor the exposure of new zones over time, and to determine the contribution of each wall to the ARD from the pit. Therefore, it may be possible to separate nonacidic and acidic flows from the pit walls to reduce the amount of impacted water.

Because no groundwater flow from the pit occurs during operation, no advance warning groundwater stations are necessary. After closure, discharge of mine water to the surface is terminated. This will allow the pit to flood and the water level in the pit will recover toward the preoperational water table. As this occurs, the pit will become an integrated part of the groundwater flow system with groundwater flowing into, through, and out of the flooded pit. The surface water within the pit or flowing out of the pit can be established as the advanced warning station that will be monitored monthly at the closure phase.

Receiving-environment surface monitoring stations should be located on surface watercourses that receive direct discharge from the sump and/or retention ponds holding pit water, and at selected stream junctions downstream of the mine site. A minimum recommended sampling program would be one sampling location in each surface water flow directly receiving discharge from the sump and/or retention pond. The minimum monitoring frequency for the surface water locations should be once every 6 months, with the option for increased frequency depending on results of the advance warning station.

Groundwater monitoring at closure for the receiving environment is dependent on the hydraulic conductivity of the subsurface strata. If aquifers having high permeability are located near the pit, a groundwater monitoring network should be designed that is downgradient and hydraulically connected to the aquifers. The groundwater in an area that has a low permeability moves more slowly, and monitoring stations located nearer the pit should be able to detect any changes. A minimum recommended sampling program would be one groundwater receiving-environment sampling location in each aquifer in the area. The monitoring frequency for the groundwater locations should be once every 6 months.

During the operational phase of an open-pit mine, surface water is usually diverted away from the perimeter. As a result, any impacts on preexisting surface drainage will be primarily a consequence of diversion. Monitoring of the diversion at established stations is an option but would be considered only if a concern over the ditch bank and bed material were raised.

Precipitation may represent a major portion of pit water in wet climates. Precipitation and evaporation may be monitored at weather stations in or around the perimeter of the pit.

A pit often represents a depression in the local water table, resulting in groundwater flow toward and into the mine from the walls and bottom. This movement can be defined and monitored with piezometers and monitor wells installed around the pit perimeter. The flow rate and quality of groundwater can be monitored on pit walls, and surface water quality can be monitored at the collector ditches within the pit, at the sump, and at the retention pond outside the pit.

Environmental Monitoring of Underground Workings

The primary source of water in underground workings is groundwater because the workings often represent a depression in the local groundwater regime. A secondary source of water may be surface water directed into a shaft, adit, or decline intentionally for drilling water or unintentionally. Unlike an open pit, water flow in underground workings does not usually respond as rapidly or as extremely as in open pits because of the buffering process of infiltration to the groundwater system. Mine water flows under gravity to an adit or is pumped from the workings, resulting in a surface-water flow that may be directed to a retention pond.

Geotechnical concerns in underground workings focus on wall and roof stability to minimize dangers to the work force and delays in mining. Additionally, any collapses in the workings may lead to increases in groundwater flow due to permeability enhancements and to subsidence on upper levels and the land surface.

During operation and at closure, the location for the advance warning station in the vicinity of the underground workings should be the discharge from the sump and/or the retention pond. The advance warning station should be monitored once every month.

In addition, the seeps from the walls could be sampled once every 6 months to monitor the exposure of zones over time and to determine the contribution of each wall or zone to the ARD from the workings. Therefore, it may be possible to separate the nonacidic and acidic flows to reduce the amount of impacted water.

After closure, the discharge of mine water to the surface should be terminated. This will allow the workings to flood and the groundwater levels to recover toward preoperational levels. As this occurs, the workings will become an integrated part of the groundwater flow system with groundwater flowing into, through, and out of the mine. The advance warning station would remain the same as during operation — the discharge point from the sump and/or retention pond.

Receiving-environment surface monitoring stations should be located on surface watercourses that receive direct discharge from the underground workings and at selected stream junctions downstream of the mine site. A minimum recommended sampling program would be one sampling location in each surface-water flow directly receiving discharge from the sumps and/or retention ponds. Minimum frequency of monitoring the surface water locations should be once every 6 months, with the option for increased frequency depending on the water quality of the advance warning station.

Groundwater monitoring at closure for the receiving environment is dependent on the hydraulic conductivity of the subsurface strata. If aquifers with high permeability are located near the underground workings, a groundwater monitoring network should be designed that is downgradient and hydraulically connected to the aquifers. The groundwater in an area that has a low permeability moves slower, and monitoring closer to the mine component should be capable of detecting any changes. A minimum recommended sampling program would be one groundwater receiving-environment sampling location in each aquifer in the mine site area. The monitoring frequency for the groundwater locations should be once every 6 months.

During the operational phase of the underground workings, surface water is usually diverted away from any shafts, declines, or adits. As a result, any impacts on preexisting surface drainage will be primarily a consequence of diversion. Monitoring of the diversion at established locations will indicate the extent of the impacts.

Underground workings often represent a local depression in hydraulic head, resulting in groundwater flow toward and into the mine. This dewatering of the groundwater system can be defined and monitored with piezometers and monitor wells installed from the land surface or from the workings. Additionally, the flow rate and quality of groundwater can be monitored at stations in the workings, in the collector ditches within the workings, or at any sumps and retention ponds outside the workings.

Environmental Monitoring of Waste Rock Dumps, Ore Stockpiles, and Heap-Leach Sites

Dumps, stockpiles, or heap-leach sites are usually exposed to climatic events, so that precipitation represents a primary source of water moving through the rock. The resulting infiltration moves downward under gravity through the rock to the base of the pile. At the base, infiltration may (1) mix with upwelling groundwater and flow from the basal perimeter of the pile, (2) completely enter the underlying groundwater flow system if hydraulic conductivity is sufficiently high to accept all infiltration and hydraulic gradients have a downward component, and (3) partially enter the groundwater system if hydraulic conductivity is restricted and partially exit at the basal perimeter of the pile. Heap-leach sites may be constructed with low-permeability pads so that negligible amounts of infiltration reach the underlying groundwater system during the active life of the site.

Geotechnical monitoring of dumps, stockpiles, and heap-leach sites addresses the physical integrity of the rock after placement. Consolidation/settlement and slope stability could cause changes in the hydraulic conditions within the structure and could cause migration of rock from the site through physical movement such as slumping and toe collapse.

During operation and at closure (if the piles are not being removed), the location of the advance warning stations should be at the discharge point from a retention pond or drainage ditches located at the base of the piles. If there is groundwater drainage is found in the area to be primarily due to the hydraulic conductivity, additional advance warning stations should be established that require the installation of a groundwater well network. Both surface and groundwater advance warning stations should be monitored once a month.

Receiving-environment surface monitoring stations should be located on surface watercourses that receive direct discharge from the retention pond or ditches around the pile and at selected stream junctions downstream of the mine site, as well as one upgradient water station. A minimum recommended sampling program would be one upgradient and one downgradient sampling location in each surface water flow directly receiving discharge from the pile. The minimum monitoring frequency should be once every 6 months for the surface water locations with the option for increased frequency depending on the water quality at the advance warning station.

Groundwater monitoring at closure for the receiving environment is dependent on the hydraulic conductivity of the subsurface strata. If aquifers having high permeability are located near the pile, a groundwater monitoring network should be designed that is downgradient and hydraulically connected to the aquifers, in addition to one upgradient groundwater station. The groundwater in an area that has a low permeability moves more slowly, and monitoring locations nearer the mine component should be able to detect any changes. A minimum recommended sampling program would be one upgradient and one downgradient groundwater sampling location in each aquifer in the receiving environment. Monitoring should occur every 6 months for the groundwater locations.

Hydraulic monitoring of dumps, stockpiles, and heap-leach sites may include periodic measurements of precipitation at exposed surfaces and with periodic measurements of rates and quality of infiltration into the rock. Within a pile, monitoring would indicate the preferred water pathways and the variation in quality during flow, although monitoring of water in unsaturated, coarse-grained material is difficult and not always feasible. If a water table is positioned within the pile, a piezometer could be installed to monitor the water quality. Additionally, subsurface stations consisting of piezometers or monitor wells may be installed around the perimeter to monitor the direction, flow rate, and quality of groundwater in the area.

Monitoring of the temperature gradients may provide early indications of changes within the pile and can be accomplished using temperature probes within the piles. This method may indicate the initiation of acid generation in localized areas.

Environmental Monitoring of Tailings Impoundments

Two primary sources of water entering a tailings impoundment are the mill effluent discharge and precipitation. Water remaining above the tailings surface will flow to low-lying areas and form ponds. From a groundwater perspective, high water levels in an impoundment lead to downward movement through the tailings pile with a lateral component of flow toward the perimeter. Groundwater may leave an impoundment through the base or the retaining walls and enter the local groundwater flow system. This seepage may then enter deeper flow systems or discharge locally to the surface.

During operation and at closure, the location of the advance warning station should be at the direct discharge points from the impoundment. Direct discharge points can include seepage locations through dams and embarkments to monitor the groundwater quality, and discharge from the spillways to monitor the surface water quality. These discharge locations should be monitored every month.

When the tailings become consolidated after closure, additional advance warning stations can be located within the tailings impoundment by installing piezometers in the tailings to monitor the quality of the groundwater pore water.

Receiving-environment surface monitoring stations should be located on surface watercourses that receive direct discharge from the tailings impoundment and at selected stream junctions downstream of the impoundment. A minimum recommended sampling program would be two sampling locations in each surface watercourse which directly receive discharge from the tailings impoundment, one upgradient and one downgradient. The minimum monitoring frequency for the surface water locations should be once every 6 months, with the option to increase frequency depending on water quality at the advance warning station.

Groundwater monitoring at closure for the receiving environment is dependent on the hydraulic conductivity of the subsurface strata. If aquifers with high permeability are located near the tailings impoundment, a groundwater monitoring network should be designed that is downgradient and hydraulically connected to the aquifers. The groundwater in an area that has a low permeability moves more slowly, and monitoring nearer the impoundment should be able to detect any changes. A minimum recommended sampling program would be one upgradient groundwater station for each hydraulically connected strata impacted by the tailings impoundment, and one groundwater sampling location in each aquifer in the mine site area. The monitoring frequency for the groundwater locations should be once every 6 months.

Hydraulic monitoring within an active impoundment is not always possible due to the unconsolidated nature of fresh tailings, sometimes limiting access only to the perimeter dams and retaining walls. As a result, measurements of mill effluent rate and quality can be made at the mill. Measurements of local precipitation and evaporation can be made at the perimeter. The movement of surface water within the impoundment can be visually defined, and the movement of groundwater can be generally assessed at the perimeter with piezometers.

Geotechnical monitoring of an impoundment at established stations addresses the physical integrity of an impoundment during and after operation. Dam instability, for example, may lead to the migration of tailings solids from an impoundment to a downstream environment. Tailings settlement may affect the rate and direction of water movement, which is a primary pathway for the interaction of tailings with the surrounding environment.

Environmental Monitoring of Quarries

Like open pits, potential sources of water entering a quarry are precipitation, surface-water drainage, and groundwater discharge. A quarry is open to the atmosphere and surface-water drainage is not usually diverted around the site, therefore, variable flow rates through the quarry can be expected. Groundwater frequently moves toward a pit because an excavation often represents a depression in the water table; however, a shallow quarry may lower the water table below the base so that groundwater inflow becomes minimal. All water entering a quarry usually combines and is either pumped out or flows under gravity from the site. The resulting surface flow downgradient of the quarry may be directed into a holding pond for monitoring and any required treatment.

The prevention of acid generation from a quarry should primarily be the assessment of its acid-generation potential. The use of a potentially acid-generating quarry is unlikely; therefore, a minimum monitoring program is used only to confirm that acid generation is not occurring.

From a geotechnical viewpoint, the major concern is the stability of the quarry walls while equipment and workers are in the area. Shallow quarries may be relatively stable without any stability engineering.

The advance warning station can be located at the retention pond or at the surface discharge flow from the quarry and is monitored once a month. No groundwater advance warning stations are necessary as not all quarries will have a groundwater component. Receiving-environment surface monitoring stations are optional.

The water derived from a quarry may be monitored for flow rate and quality. If unacceptable quality is anticipated, this water may be directed to a holding pond for analysis and treatment.

The groundwater monitoring component for receiving-environment monitoring is dependent on the hydraulic conductivity of the subsurface strata. If aquifers with high permeability are near the quarry, a groundwater monitoring network should be designed that is downgradient and hydraulically connected to the aquifers. The groundwater in an area that has a low permeability moves more slowly, and monitoring nearer the quarry should be able to detect any changes.

Receiving-environment surface monitoring stations should be located on surface watercourses that receive direct discharge from the quarry and at selected stream junctions downstream of the quarry. An optional sampling program could be one sampling location in each surface water flow directly receiving discharge from the quarry.

Environmental Monitoring of Haul Roads

For road construction, stability, and maintenance, rock is often taken from mine quarries and mine waste rock. This rock is crushed to an appropriate size and is further crushed by road vehicles. As a result, these roads can be thought of as small-scale waste rock dumps, but their geochemical reactivity is higher because of the finer grain size and continual grinding. Precipitation falling onto roads passes through the rock and either enters the underlying groundwater system or moves as overland flow to low-lying areas. Because the prevention of ARD from a haul road is in the assessment of acid generation prior to use of the material, the construction of an acid-generating haul road is unlikely. However, due to the use of marginal acid-generating material or material used without knowledge of its acid-generating capability, a minimum monitoring program is proposed.

Because monitoring of acid generation from haul roads may be very difficult, the minimum recommended program would be a visual inspection of the area adjacent to the haul roads every 6 months to check for discoloration of local materials. Stations may be located in sediment traps if they have been constructed and would be monitored every month.

IMPACT OF AN ABANDONED MINE ON WATER QUALITY

The Argo Tunnel, located 30 mi (55 km) west of Denver, CO, was completed in 1904. The tunnel is 21,968 ft long and intersects 27 mines. The tunnel and the mills are no longer in use, but the tunnel continues to drain the mine workings.[6]

Table 5.11. Composition of the Argo Drainage

Parameter	Conc. (mg/L)	Amount/Day (lb)
Fe	340	2,000
Mn	160	1,000
Zn	75	500
Cu	13	85
As	0.43	3
Cd	0.32	2
Pb	0.12	1
SO_4	2,700	18,000
Dissolved solids	4,000	27,000

From Boyles, M. J. Impact of Argo tunnel acid mine drainage, Clear Creek County, Colorado, in *Water Resources Problems Related to Mining* (Baltimore, MD: American Water Resources Association, 1974), pp. 41–53. With permission.

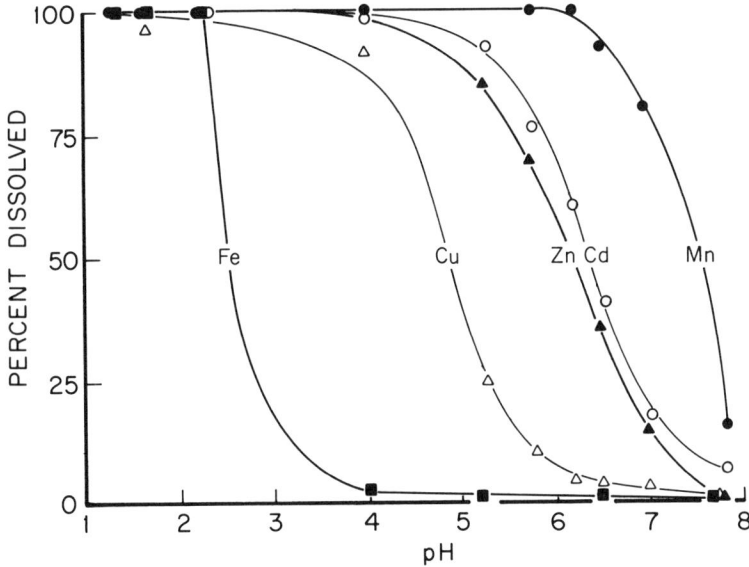

Figure 5.3. Heavy metal behavior as a function of pH in Argo water. (From Boyles, M. J. Impact of Argo tunnel acid mine drainage, Clear Creek County, Colorado, in *Water Resources Problems Related to Mining* (Baltimore, MD: American Water Resources Association, 1974), pp. 41–53. With permission.)

The flow of the drainage was 1.2 cfs. The pH of the drainage averaged 2.8 and the average metal concentrations are given in Table 5.11. The effect of pH on dissolution of metals in the mine water is given in Figure 5.3.

Water samples were taken from Clear Creek, into which the Argo Tunnel drains. The analysis of the samples is given in Table 5.12.

Table 5.12. Impact of the Argo Drainage on Clear Creek, August 16, 1973

Metal	USPHS Drinking Water Standards (µg/L)	Max. Suggested Conc. for Fish and Other Aquatic Life	State	Clear Creek Upstream from Argo	Argo Drainage	Clear Creek 1:1 Miles Downstream from Argo[a]
Fe	300 SL[b]	300	Dis	30	320,000	80
			Total	350	320,000	1,400
Cu	1,000 SL	10–20	Dis	10	13,000	40
			Total	25	13,000	80
Mn	50 SL	1,000	Dis	200	150,000	860
			Total	240	150,000	860
Zn	5,000 SL	30–70	Dis	240	71,000	560
			Total	250	71,000	610
Cd	10 CR[c]	10	Dis	<1	320	<1
			Total	<1	320	<1
Pb	50 CR	5–10	Dis	<1	110	<1
			Total	10	110	10
SO$_4$	250,000 SL	—	Dis	15,000	2,700,00	22,000

[a] Past the point of total mixing.
[b] SL = Suggested limit.
[c] CR = Cause for rejection limit.

From Boyles, M. J. Impact of Argo tunnel acid mine drainage, Clear Creek County, Colorado, in *Water Resources Problems Related to Mining* (Baltimore, MD: American Water Resources Association, 1974), pp. 41–53. With permission.

ACID ROCK DRAINAGE AND METAL MIGRATION

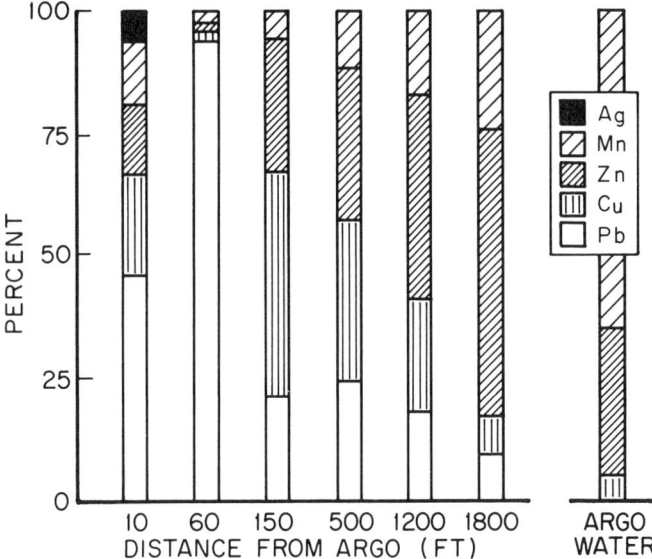

Figure 5.4. Composition of bottom sediments compared to Argo water. (From Boyles, M. J. Impact of Argo tunnel acid mine drainage, Clear Creek County, Colorado, in *Water Resources Problems Related to Mining* (Baltimore, MD: American Water Resources Association, 1974), pp. 41–53. With permission.)

In addition to water analysis, sediments from the bottom of Clear Creek were analyzed and the results are given in Figure 5.4.

HYDROLOGIC SOLUTION TO ACID MINE DRAINAGE

Hydrologic solutions may be found to solve or reduce AMD problems. The study of water movements in a lead-zinc mine in northern Idaho revealed alternatives to costly treatment of the poor-quality effluent. Delineation of areas of acid production and areas of recharge is important in this process.

The discharge from the Kellog Tunnel, which includes drainage from most of the mine, averages 2700 gal/min with an average pH of 3.3.

Mining methods used in the mine have included vertical slice, cut, and fill, with square sets and caving. Wastefill from development headings was used for backfilling stopes. Some of the wastefill used in the upper levels contains a high concentration of lead, zinc, and pyrite.

Vertical drainage of water occurs through a series of interconnected stopes between the mine levels. As the water drains down the mine openings, some of it moves through stopes containing ore and wastefill rich in pyrite. The water becomes more acidic as it passes over the waste. Movement of water in the mine provides the

228 ENVIRONMENTAL IMPACTS OF MINING

transport mechanism for the acid water. The reduction or control of water movement in the mine can be an alternative procedure to reduce the acid drainage problem. Three possible approaches include:

- Block the water movement on a selected level
- Route the water around potential acid-producing areas
- Reduce recharge to the mine

Water discharge data collected from various points in the mine indicate a close correlation with surface runoff events. Three probable areas of recharge to the mine have been identified:

- A subsidence area and related fracturing resulting from caving
- A drift and a stope intersecting the streambed
- Proximity of upper-level stopes to the land surface

Three projects for reducing recharge to the mine have been considered. One project includes building an impermeable cutoff to bedrocks at the dams built for the raises. This would provide a more effective diversion around the caving area. The groundwater and surface water recharge to the caving can be effectively eliminated with the repositioning of the raises in the actual bedrock lows and construction of more impermeable cutoffs.

The diversion of water around the areas of losses in the creeks is another project. Dams to bedrocks with collector inlets and pipes capable of carrying the spring runoff past the recharge area could be installed. Elimination of recharge in these areas could reduce most of the flow to upper levels of the mine.

WATER RESOURCE PROBLEMS IN A LEAD BELT

Southeastern Missouri contains rich lead deposits. Several major mines have been opened in that area. Several creeks flow through the lead belt. Table 5.13 summarizes the physical and chemical parameters of water quality in the creeks. Tables 5.14 and 5.15 summarize metal contents of the water. Studies of the stream sediments in the region indicated only low levels of heavy metals. Generally, the heavy metals were incorporated in the #325 mesh or finer fractions.[7]

One of the evident major environmental problems is the transport of heavy metals under conditions in which wastewater has stimulated excessive biological growth in receiving streams. In some instances, these dense, gelatinous mats of algae and their associated aquatic populations have coated streambeds, blocking photosynthetic energy input, eliminating local stream populations, and causing aesthetic problems. These biological mats act as filters that trap nutrients and sediments that are high in

Table 5.13. Summary of Physical and Chemical Parameters of Water Quality

	Station No.									
	1	2	3	4	5	6	7	8	9	10
pH	8.10	7.45	7.74	7.73	7.73	7.66	7.58	8.2	8.1	8.0
Temperature	0–25°C	0–25°C	0–25°C	0–25°C	0–25°C	0–25°C	0–25°C	0–25°C	0–25°C	0–25°C
Turbidity (JTU)	3.7	2.9	1.2	1.0	7.7	1.3	0.76	1.0	1.0	1.0
Dissolved oxygen (DO) (ppm)	5.4	4.6	4.7	5.5	5.5	5.6	5.3	5.6	5.2	5.5
Alkalinity (mg/L)	180	200	198	152	150	150	150	144	156	135
Hardness (mg/L)										
Calcium	150	140	210	100	95	100	100	75	80	75
Total	300	280	360	250	235	200	210	155	160	135
Chloride (mg/L)	0	0	30	20	0	0	0	0	0	0
Chem. Ox. Dem. (COD) (mg O_2/L)	30	20	50	40	75	95	35	15	20	30
Phosphorus (mg/L)										
Ortho-	0.1	0.1	0.1	0.1	0.1	0.1	0.1	0.1	0.1	0
Total	0.3	0.3	0.4	0.3	0.2	0.1	0.1	0.2	0.1	0.1
Nitrite (mg/L)	0	0	0	0	0	0	0	0	0	0
	0.1	0.1	0.2	0.1	0.2	0.1	0.1	0.2	0.1	0.1
Nitrogen (mg/L)										
Ammonia	0	0	0	0	0	0	0	0.1	0	0
Total organic	26.4	8.4	30.2	25.7	16.4	18.0	12.0	15.6	10.0	22.4
Specific conductance (μmho/cm)	360	200	260	450	100	160	340	300	380	280

From Hardie, M. G. and J. C. Jennett. Water resources problems and solutions associated with the new lead belt of S.E. Missouri, in *Water Resources Problems Related to Mining* (Baltimore, MD: American Water Resources Association, 1974), pp. 109–122. With permission.

Table 5.14. Metal Concentration of Streams at Sampling Sites in the New Lead Belt Study Area — Dissolved Metal Content

Element		1	2	3	4	5	6	7	8	9	10
Pb	max.	0.280	0.022	0.010	0.130	0.061	0.075	0.066	0.012	0.045	0.160
	min.	<0.005	<0.005	<0.005	<0.005	<0.005	<0.005	<0.005	<0.005	<0.005	<0.005
	mean	0.026	0.009	0.006	0.012	0.014	0.011	0.009	0.006	0.009	0.008
Mass flow (lb/yr)		581	196	135	232	86	6900	396	606	147
Zn	max.	0.460	0.084	0.027	0.250	0.097	0.034	0.036	0.023	0.128	0.088
	min.	<0.010	<0.010	<0.010	<0.010	<0.010	0.005	<0.010	<0.010	<0.010	<0.010
	mean	0.063	0.018	0.011	0.068	0.028	0.013	0.013	0.011	0.019	0.014
mass flow (lb/yr)		1391	394		771	478	103	9832	729	1251	261
Cu	max.	0.039	0.015	0.010	0.010	0.011	0.018	0.018	0.010	0.024	0.010
	min.	<0.010	<0.010	<0.010	<0.010	<0.010	<0.010	<0.010	<0.010	<0.010	<0.010
	mean	0.012	0.010	0.010	0.010	0.010	0.010	0.010	0.010	0.011	0.010
Mass flow (lb/yr)		263	226		114	172	84	7835	666	712	181
Cd	max.	0.010	0.010	0.010	0.012	0.010	0.013	0.010	0.010	0.010	0.015
	min.	0.000	0.000	0.000	0.000	<0.001	0.000	0.000	0.000	0.000	0.000
	mean	0.007	0.006	0.006	0.006	0.006	0.007	0.006	0.006	0.006	0.007
Mass flow (lb/yr)											
Mn	max.	0.550	0.054	0.011	1.82	0.680	0.012	0.010	0.018	0.840	0.020
	min.	<0.010	<0.010	<0.010	<0.010	<0.010	<0.010	<0.010	<0.010	<0.010	<0.010
	mean	0.062	0.014	0.010	0.128	0.111	0.010	0.010	0.010	0.052	0.011
Mass flow (lb/yr)		1363	312		1459	1895	82	7534	692	3455	202

From Hardie, M. G. and J. C. Jennett. Water resources problems and solutions associated with the new lead belt of S.E. Missouri, in *Water Resources Problems Related to Mining* (Baltimore, MD: American Water Resources Association, 1974), pp. 109–122. With permission.

ACID ROCK DRAINAGE AND METAL MIGRATION   231

Table 5.15. Metal Concentration of Streams at Sampling Sites in the New Lead Belt Study Area — Total Metal Content

Element		1	2	3	4	5	6	7	8	9	10
Pb	max.	0.200	0.100	0.100	0.100	0.830	0.100	0.100	0.680	0.120	0.920
	min.	<0.005	<0.005	<0.005	<0.005	<0.005	<0.005	<0.005	<0.005	<0.005	<0.005
	mean	0.029	0.011	0.010	0.013	0.081	0.011	0.014	0.035	0.019	0.009
Mass flow (lb/yr)		643	249		151	1,389	111	10,315	2,316	1,246	158
Zn	max.	0.180	0.153	0.026	0.280	0.175	0.032	0.018	0.026	0.208	0.057
	min.	<0.010	0.008	<0.010	<0.010	<0.010	0.008	0.008	<0.008	<0.010	<0.008
	mean	0.043	0.017	0.011	0.060	0.043	0.011	0.011	0.011	0.018	0.012
Mass flow (lb/yr)		966	369		686	743	88	8,107	751	1,196	221
Cu	max.	0.025	0.027	0.029	0.036	0.036	0.035	0.028	0.032	0.026	0.030
	min.	<0.010	<0.010	<0.010	<0.010	<0.010	<0.010	<0.010	<0.010	<0.010	<0.010
	mean	0.011	0.011	0.011	0.011	0.012	0.011	0.011	0.011	0.011	0.011
Mass flow (lb/yr)		246	237		128	206	88	8,000	716	702	201
Cd	max.	0.010	0.016	0.010	0.010	0.016	0.010	0.010	0.012	0.010	0.012
	min.	0.000	0.000	0.000	0.000	<0.001	0.000	0.000	0.000	0.000	0.000
	mean	0.008	0.008	0.007	0.007	0.008	0.007	0.007	0.008	0.007	0.008
Mass flow (lb/yr)		168	169		83	134	59	5,484	500	489	139
Mn	max.	0.640	0.620	0.042	0.070	0.690	0.028	0.200	0.064	0.960	0.034
	min.	0.010	0.010	0.010	0.010	0.019	0.010	0.010	0.010	0.010	0.010
	mean	0.062	0.037	0.012	0.032	0.133	0.011	0.017	0.012	0.049	0.012
Mass flow (lb/yr)		1,391	821		367	2,279	88	13,125	807	3,238	225

From Hardie, M. G. and J. C. Jennett. Water resources problems and solutions associated with the new lead belt of S.E. Missouri, in *Water Resources Problems Related to Mining* (Baltimore, MD: American Water Resources Association, 1974), pp. 109–122. With permission.

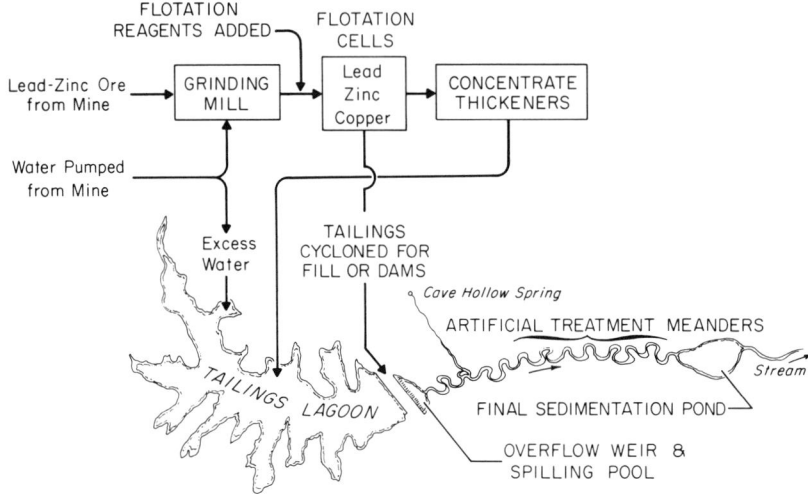

Figure 5.5. A mine and mill separation using a lagoon followed by a meander section and final sedimentation. (From Boyles, M. J. Impact of Argo tunnel acid mine drainage, Clear Creek County, Colorado, in *Water Resources Problems Related to Mining* (Baltimore, MD: American Water Resources Association, 1974), pp. 41–53. With permission.)

trace metals. They also trap finely ground rock flour, tailings, and minerals that escape from the flotation and tailings reservoirs.

A treatment system was designed to encourage the growth of algae on the mine property and to trap any suspended heavy metals on the algae. A series of broad, shallow, rapidly flowing meanders was built with a sedimentation pond placed at the end of the meanders to prevent heavy metals from escaping the system. The meanders have been successful in eliminating the algal problem downstream. Lead concentration in aquatic vegetation has significantly decreased (Figure 5.5).

ENVIRONMENTAL CONTROL MEASURES AFTER THE CLOSURE OF A LEAD ZINC MINE IN GREENLAND

The Black Angel Mine is situated at Maarmorilik, a remote area in the Bay of Uummannaq on the west coast of Greenland about 500 km north of the Arctic Circle. The main orebodies occur in an 1100-m high mountain with very steep slopes along two fjords. The mountain is partially covered by the Greenland ice cap. The orebodies are located about 700 m above sea level.[8]

The concentrator, the mine town, and all the services were located at sea level in Maarmorilik, separated from the mine by the fjord Affarlikassaa. The only access to the main orebodies is via two cableways, with a span of 1.5 km.

ACID ROCK DRAINAGE AND METAL MIGRATION

The aims of the demolition and cleanup in Maarmorilik are

- It shall be possible for persons to stay in the area without risk of life and health due to former mining activity.
- The town area shall be left neat and tidy. Equipment and visible waste shall be removed from former areas of prospecting.
- Existing and potential sources of pollution shall be eliminated or limited so that the original environmental condition before the mining activities began can be recreated as completely as possible within a few years.

The main sources of pollution from the mining activities at Maarmorilik are

- Tailings, in particular, the plume of tailings in the fjord Affarlikassaa
- Dispersed metal-rich minerals in the surroundings
- Remains of ore and concentrates in the industrial area
- Dust from various places where ore and concentrates are handled
- Waste rock dumps; in particular, the one reaching the fjord (North Face Dump)

The first source will cease when the flotation plant stops the discharge of tailings in seawater suspension. Undisturbed tailings on the bottom of the fjord will release insignificant small amounts of lead as shown by laboratory experiments.

The pollution from the second source is believed to be insignificant and relatively quickly terminating.

The pollution from the third source will be minimized by clearance and cleaning.

Pollution from the fourth source, dust, will decline significantly when the mining activities stop and the mine is abandoned.

The fifth source, the waste rock dumps, however, will be unaffected by the closure of the mill and mine, and will be the most serious pollution source in the future. Since one dump, the North Face Dump, is already an important source of pollution, a decision has been made to remove it. Taking practical possibilities into consideration, the safest place to deposit the approximately 400,000 tons of waste rock is thought to be the bottom of the fjord Affarlikassaa, where approximately 8 million tons of tailings have already been deposited. Laboratory experiments have shown that the amounts of lead released over time after depositing the waste rock on the fjord bottom will be

$$\text{kg lead} = 7.53 \times \sqrt{\text{days}} \qquad (5.10)$$

The removal of the North Face Dump from the mountain slope to the bottom of Affarlikassaa is the most expensive single operation of the cleanup after the mining activities. There are other waste rock dumps in the surroundings of Maarmorilik, but none of them reaches the sea. It has been decided to await the results of some years' monitoring before determining if it is necessary to take action against their pollution.

In order to fulfill the third aim, that existing and potential sources of pollution shall be eliminated, all places and equipment contaminated with ore or concentrates must be cleaned as described above, and waste rock dumps with an unacceptable environmental impact must be removed.

The tailings discharge at a depth of 30 m in the fjord Affarlikassaa results in a high degree of pollution of the fjord by zinc, cadmium, and lead. Typical concentrations are seen in Table 5.16. Unpolluted seawater collected and analyzed with the same equipment typically gives 0.5, 0.03, and 0.2 µg/kg for zinc, cadmium, and lead.

Obviously, the tailings release metals to the receiving seawater. In September, when the water in the fjord is stratified, the metals are kept mainly below 30 m (a sill lies between Affarlikassaa and the outer fjord Qaamarujuk at a depth of approximately 25 m); however, in March, when the density stratification is broken down, the metals are brought up to the upper part of the fjord and spread out into the neighboring fjords. The amounts of metal dissolved in Affarlikassaa and the adjacent fjord Qaamarujuk are seen in Table 5.17.

Blue mussel, *Mytilus edulis*, and seaweed, *Fucus vesiculosus* and *F. disticus*, were collected and analyzed annually; i.e., 1 year before the start of the mining activity.

It soon became obvious that a strong influence was exerted by the mining on the zinc and lead content of these intertidal organisms. The geographical distribution of lead in *Mytilus* and *Fucus* is seen in Figures 5.6 and 5.7. It is important to note that the maximum for lead (and also zinc) contamination in *Mytilus* and *Fucus* is not where the tailings are discharged in Affarlikassaa nor where the concentrations are shipped, but below the North Face Dump. This dump contains approximately 400,000 tons of waste rock from the mining in the above mountain. It is mainly composed of marble and dolomite and contains 0.8% Pb and 2.5% Zn. The dump is situated on a mountain slope from an altitude of 260 m to below sea level. It is believed that it is the action of waves and tidal movements that results in mobilization of lead and zinc from the dump. A similar effect has been observed on polluted sediments at Sørfjord, Norway.

The lead pollution of fish caught in the immediate vicinity of the mine was at its maximum in the late 1970s. Work done by the mining company to abate the pollution has probably resulted in a decrease in fish pollution; therefore, in 1989 the only lead-polluted fish parts were capelin, liver of shorthorn sculpin, and head and shells of prawn.

The spreading of lead in dust outside the mining town has been estimated by collecting and analyzing the lichen *Cetraria nivalis*. By using the correlation

$$\text{Lead deposition mg/m}^2 \text{ year} = \text{Lead concentration mg/kg}/2.7 \quad (5.11)$$

one can correlate that between 1.6 and 1.4 t/year of lead has been spread outside Maarmorilik Town and up to 23 km toward the west.

ACID ROCK DRAINAGE AND METAL MIGRATION 235

Table 5.16. Dissolved Metals in the Fjord Affarlikassaa (μg/kg) (Average of 6–8 Samples)

	Zn	Cd	Pb
Below 30 m March 1989	44.0	0.370	47.1
Above 30 m March 1989	17.4	0.191	18.8
Below 30 m September 1989	138.0	2.45	242.0
Above 30 m September 1989	4.1	0.043	2.5

From Asmund, G. Rehabilitation and demolition after the closure of the zinc and lead mine Black Angel at Maarmorilik, Greenland, in *POLARTECH, Proc. Int. Conf. Development and Commercial Utilization of Technologies in Polar Regions* (Copenhagen, Denmark: Danish Hydraulic Institute, 1990), pp.744–759. With permission.

Table 5.17. Tonnes of Dissolved Metals in the Two Fjords near Maarmorilik, Average of September and March, 1985—1988

	Zn	Cd	Pb	Vol 10^6 m^3
Affarlikassaa	6.6	0.050	4.4	70.3
Qaamarujuk	8.4	0.077	2.8	1338.0

From Asmund, G. Rehabilitation and demolition after the closure of the zinc and lead mine Black Angel at Maarmorilik, Greenland, in *POLARTECH, Proc. Int. Conf. Development and Commercial Utilization of Technologies in Polar Regions* (Copenhagen, Denmark: Danish Hydraulic Institute, 1990), pp.744–759. With permission.

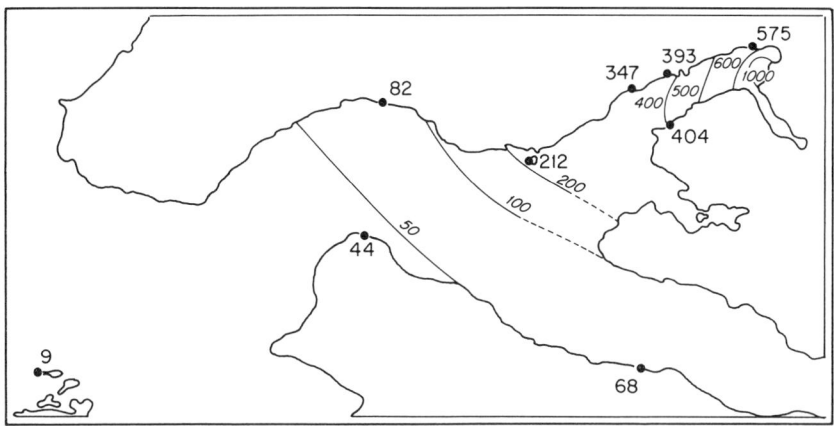

Figure 5.6. Geographic distribution of lead in the soft part of the blue mussel (mg/kg dry weight). (From Asmund, G. Rehabilitation and demolition after the closure of the zinc and lead mine Black Angel at Maarmorilik, Greenland, in *POLARTECH, Proc. Int. Conf. Development and Commercial Utilization of Technologies in Polar Regions* (Copenhagen, Denmark: Danish Hydraulic Institute, 1990), pp.744–759. With permission.)

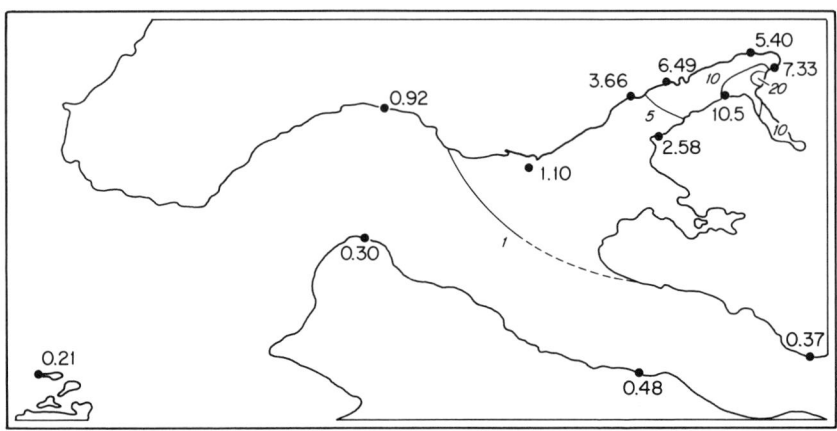

Figure 5.7. Geographic distribution of lead in new growth tips of seaweed, mg/kg dry weight. (From Asmund, G. Rehabilitation and demolition after the closure of the zinc and lead mine Black Angel at Maarmorilik, Greenland, in *POLARTECH, Proc. Int. Conf. Development and Commercial Utilization of Technologies in Polar Regions* (Copenhagen, Denmark: Danish Hydraulic Institute, 1990), pp.744–759. With permission.)

The guidelines for demolition and cleanup have been written in order to fulfill the aims mentioned above.

Accessible mine entrances and shafts will be closed or fenced in to prevent entrance to the mine by humans and possible falling accidents.

All installations and equipment will be left in the mine. Explosives and other dangerous goods must be removed.

To prevent pollution from seepage, ore plants and silos will be cleaned, and fuel, lubricants, and other toxic wastes removed. Furthermore, a concrete wall must be cast in order to hold back intrusive water in a drift under the ice cap.

Cables, machinery, and terminals in Maarmorilik will be removed. Cables will be removed by blasting and dumped into the fjord.

Ore silos, ore conveyors, concentrator, and concentrate conveyors will be cleaned either by sweeping, vacuuming, or high-pressure washing. The water used in cleaning will be filtrated and recirculated. The lead-zinc-containing sludge will be disposed of underground, as mentioned below.

Pits and other major cavities will be filled with waste from demolitions of buildings or with waste rocks without lead-zinc mineralization.

The storage for lead-zinc concentrates is located in an old marble quarry. After shipment of concentrate, the old marble surface will be cleaned by scraping and brushing with heavy equipment followed by high-pressure washing. The building covering the storage will be cleaned by vacuuming before demolishing. The waste from the cleaning will be stored underground, as mentioned below.

After cleaning, the lead-zinc concentrate storage will be used as a landfill dump.

Power, fuel, and other plants including wiring and pipes will be demolished when the fuel, lubricants, and chemicals have been removed and the plants cleaned. Pits and other major cavities will be filled with waste rocks.

Surplus explosives will be destroyed. Waste, fences, signs, and damaged crash fences will be removed from the heliport, roads, and areas used for storage. Areas polluted with lead-zinc concentrate will be cleaned by digging up the surface layer.

Equipment, chemicals, and waste harmful to the environment will be collected from the former areas of prospecting and transported to Maarmorilik for treatment as waste generated here. Prospecting camps will be burned and remains will be buried in the area together with other waste that is not harmful to the environment.

The waste generated during demolition and cleanup will be handled in different ways depending on its harmfulness to the environment.

Waste harmful to the environment consists of:

- Waste with lead-zinc concentrate or ore
- Transformers with toxic oil
- Waste oil, surplus fuel, and lubricants
- Surplus chemicals from the production

Waste with lead-zinc concentrate will be disposed of underground in drifts that are permafrozen and dry so that no possibilities exist for seepage.

Transformers with toxic oil will be shipped to Denmark for destruction. Transformers without toxic oil will be emptied of oil and disposed of in the landfill dump. The oil will be treated in the same manner as other oil-containing wastes.

Waste oil and surplus fuel and lubricants will be burned in the generators as long as the power plant is in operation. After that period, oil-containing waste will be placed into steel drums and shipped to Denmark for destruction.

Chemical waste will be shipped to Denmark for destruction.

Waste not harmful to the environment, i.e., machinery and waste generated from demolitions of buildings, will be disposed of in a landfill dump. Machinery and heavy equipment that cannot be compacted effectively on the landfill dump will be disposed of underground. After use the landfill dump will be covered with 0.5 m of waste rocks free of lead-zinc mineralization.

The North Face Dump will be removed and deposited on top of the marine tailings depot on the bottom of the fjord Affarlikassaa.

Environmental investigations during close down will measure the effect of the waste dump removal. Several seawater samples will be collected before depositing of waste rock begins and several samples will again be collected 1 week afterwards. Two months later a new collection of seawater will be made. Samples of seawater will be taken in order to estimate the release of metals from waste rock to seawater during the deposition in Affarlikassaa.

238 ENVIRONMENTAL IMPACTS OF MINING

Fish will be collected and analyzed both during the cleanup operations and 6 months later. Furthermore, liquid effluents from the concentrate storage and the North Face Dump will be sampled during work at the respective locations. The spreading of solids in the fjord system will be investigated by sediment traps. Later, studies will be made of sediment cores. Whenever activity occurs at Maarmorilik, two fixed stations in Affarlikassaa and Qaamarujuk will be sampled every month at approximately ten different depths.

When all activities, including waste rock removal, have ceased, the environment will begin its restoration process. This process will be monitored once a year for 10 to 15 years. The sampling frequencies will be incredibly longer concurrently with the rejuvenation of the investigated items to their natural states. It is expected that fish will return to the natural lead values in a few years, while blue mussels and seaweed will be polluted for many years to come. It is expected, however, that the zone in which collection and consumption of blue mussels have been restricted during the monitoring period can be gradually reduced from the 37-km distance west of Maarmorilik to a smaller area.

MINE ENVIRONMENTAL REHABILITATION

Bersbo is a mining town in the southern part of Sweden. Except for a short period in the 1930s, mining activities ceased, after 600 years or more, during the first years of the 20th century. In the mid-1800s, the mine had its heyday and was, for a decade, when the productivity of the Falun Mine was temporarily in decline, the greatest copper producer in Sweden. Compared to present conditions, the copper production was not very impressive: 600 to 900 t/year. The ore was handsorted on site and smelted in a plant some 8 km away. Therefore, the only waste left in Bersbo is waste rock, as well as a small tailings pond from the experimental activities in the 1930s to concentrate for zinc.[9]

The waste rock was disposed of in several varying-sized heaps spread over a large area (Figure 5.8). The most important are specified, together with the tailings, in Table 5.18, along with data on size and areas. Figures 5.9 to 5.11 show the discharges of copper, zinc, and cadmium from the major waste units. They have been calculated from a variety of investigations carried out in Bersbo.

More than 200,000 m³ of the waste rock masses were dumped in the waterfilled shafts; i.e., submerged disposal was applied. All the small heaps and the shallow deposits were concentrated in two large heaps that were contoured to fit a suitable covering. In this way the total deposit area was reduced from around 19 to about 9 ha.

Some of the shafts (about 1 dozen) were encompassed by the new deposits, but most could be avoided. The former had to be sealed off by a special reinforced-concrete, self-bearing roof in order not to extend the settlements of the rock masses in the shafts to the covering layers, which almost certainly would have destroyed

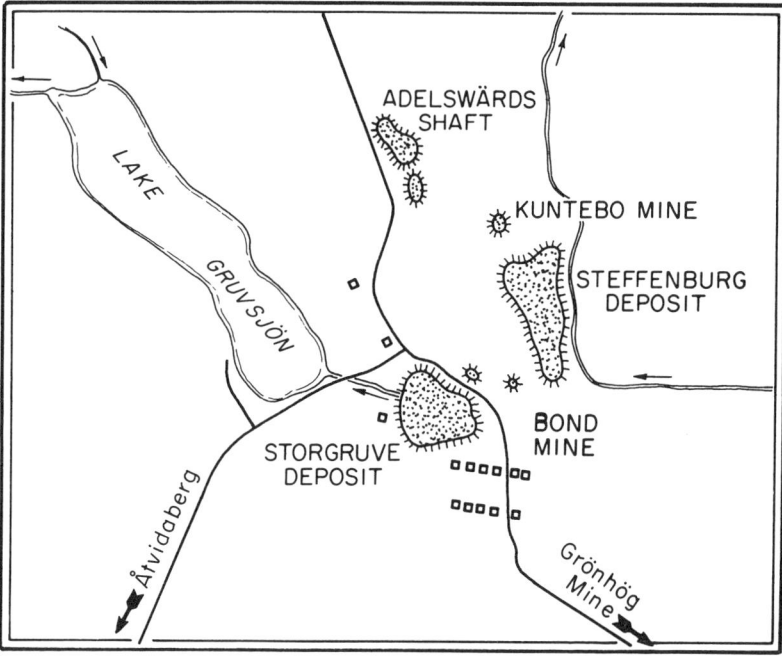

Figure 5.8. The siting of the mine waste units before remedial action. (From Lundgren, A. T. Bersbo — the first full scale project in Sweden to abate acid mine drainage from old mining activities, in *Acid Mine Drainage, Designing for Closure, GAC/MAC Annual Meeting* (Vancouver, Canada: BiTech Publ., 1990), pp. 241–253. With permission.)

their sealing properties over time. The latter shafts were sealed by a thick, unreinforced plug of concrete or Cefill (cement-stabilized fly ash).

The tailings were partly dumped into the shafts, and partly used as a filter layer separating the coarse waste rock from the fine-grained or concrete-like cover material. The top surface was smoothed and compacted before applying the sealing cover. On one of the heaps the sealing layer consisted of a crushed rock aggregate grouted with Cefill, which is a pumpable mixture of coal fly ash (about 91% by dry weight), cement (about 8% by dry weight), additives (about 1%) and water (an additional 35 to 40%) (Figures 5.12 and 5.13). On the other heap the sealing layer was constructed of a compacted, rather dry clay found about 2 km from the site. The compaction of the clay took place in three different sublayers and had to be accomplished with a heavy bulldozer that made 6 to 15 passes before the clay became plastic enough to create a continuous layer free of hollows and cracks.

The specification on the sealing layer was that it should be less permeable to water than 1×10^{-9} m/sec (saturated hydraulic conductivity). The Cefill layer met that requirement except in some minor areas in which the permeability was slightly

240   ENVIRONMENTAL IMPACTS OF MINING

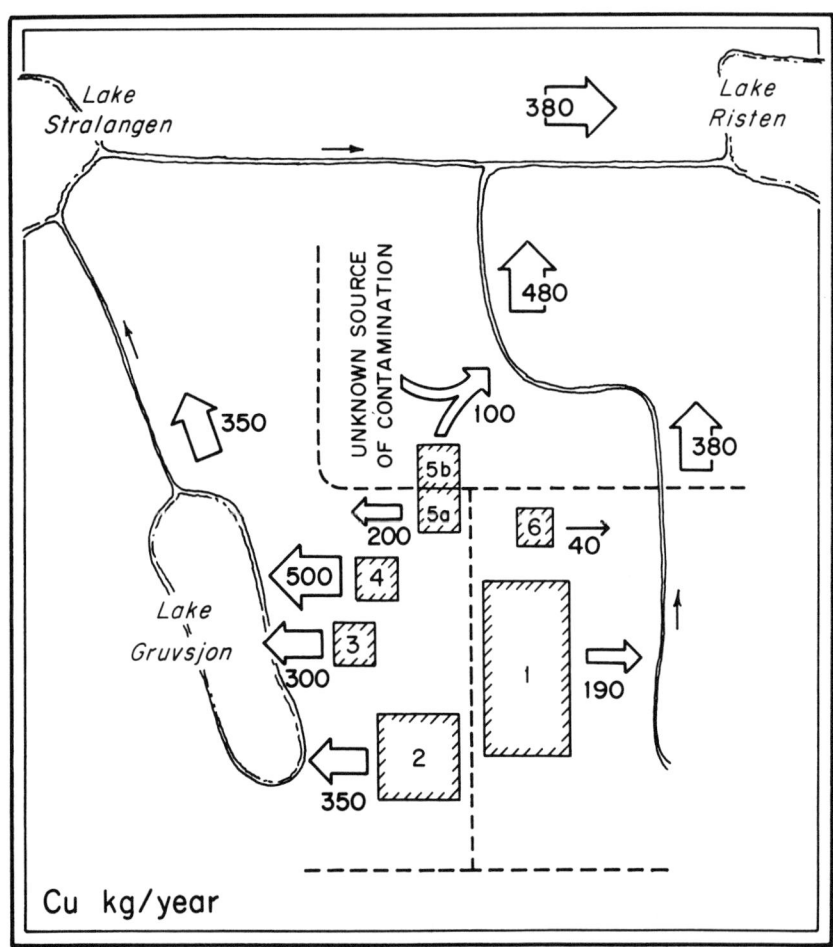

Figure 5.9. The discharge of copper from major waste units. (From Lundgren, A. T. Bersbo — the first full scale project in Sweden to abate acid mine drainage from old mining activities, in *Acid Mine Drainage, Designing for Closure, GAC/MAC Annual Meeting* (Vancouver, Canada: BiTech Publ., 1990), pp. 241–253. With permission.)

higher. Due to the puzzolanic properties it was judged that the layer would harden and become more tight over time. The clay layers became even less permeable than the specifications, almost one order of magnitude lower.

The protective layer was constructed out of a 2-m-thick layer of a glacial till found close to the native clay that was used as a sealing material. The till was applied in two sublayers, the first of which was only 0.5-m thick and with the maximum grain size of 300 mm in order not to create impacts in the sealing layer. The size specification of the upper 1.5-m layer was a maximum 600 mm. The layers were

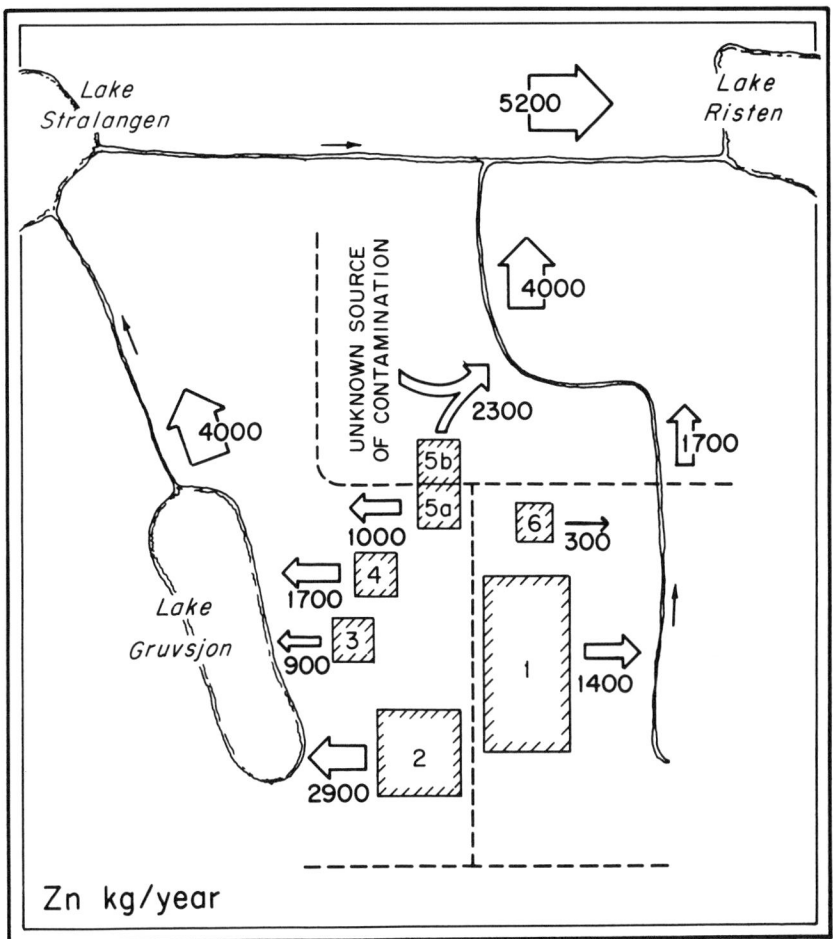

Figure 5.10. The discharge of zinc from major waste units. (From Lundgren, A. T. Bersbo — the first full scale project in Sweden to abate acid mine drainage from old mining activities, in *Acid Mine Drainage, Designing for Closure, GAC/MAC Annual Meeting* (Vancouver, Canada: BiTech Publ., 1990), pp. 241–253. With permission.)

compacted by the weight of the loaded trucks, and the final surface was smoothed and vegetated by pine plants.

All the remedial works carried out were carefully controlled. The materials used were examined continuously or on a daily basis; for example, every square meter of all three sublayers of clay was videographed, and every shipment of fly ash was controlled by laboratory tests. Several hundred permeability tests in the field and in the laboratory were performed. During the field tests many permeability cells were installed permanently in the sealing layers (15 cells in the cover of each heap).

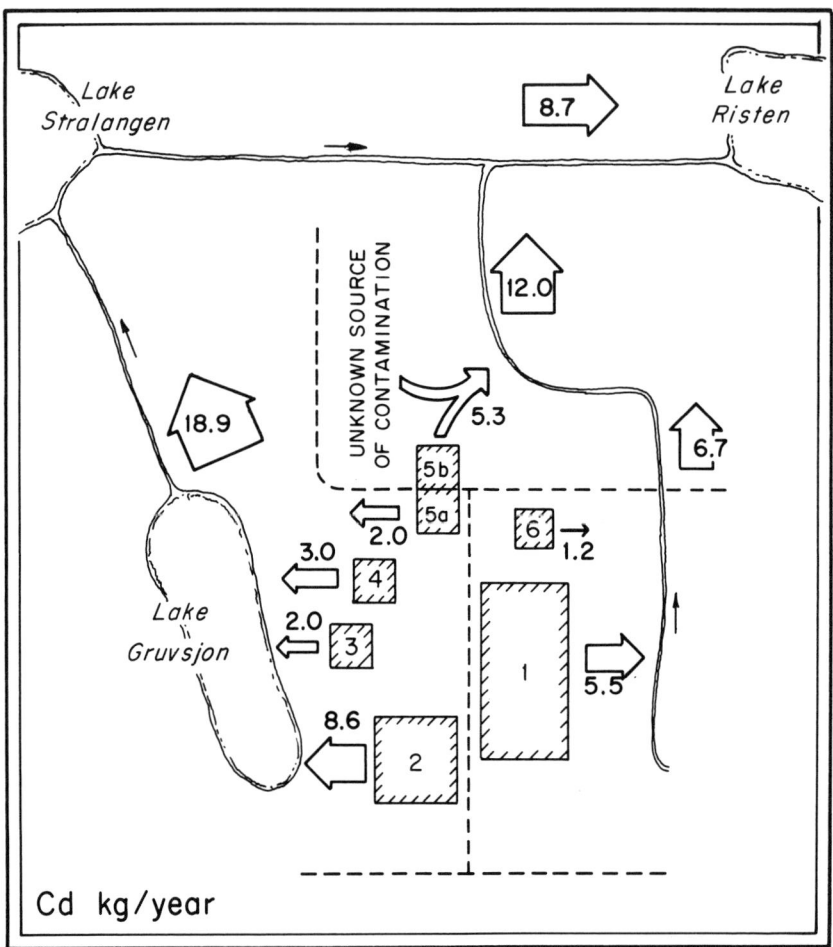

Figure 5.11. The estimated discharge of cadmium from the major waste units. (From Lundgren, A. T. Bersbo — the first full scale project in Sweden to abate acid mine drainage from old mining activities, in *Acid Mine Drainage, Designing for Closure, GAC/MAC Annual Meeting* (Vancouver, Canada: BiTech Publ., 1990), pp. 241–253. With permission.)

Some lysimeters were installed directly under the sealing layer to collect the percolate through this layer. Some 15 lysimeters were installed in direct contact with the sealing layer to act as oxygen diffusion cells. They were carefully sealed and thoroughly flushed with nitrogen before the oxygen concentration was measured. After some months the oxygen measurements were repeated.

Table 5.18. The Major Units of Solid Mine Waste Before Remedial Actions Took Place, Together with Size and Volume

Mine Waste Units	Amout (kton)	Conc. (kg/ton)		Metal Source (tonnes)	
		Copper	Zinc	Copper	Zinc
Storgruve deposit	500	2.9	7.0	1,450	3,500
Steffenburg deposit	800	2.9	7.0	2,320	5,600
Kuntebo deposit	10	2.0	6.0	20	60
Fillings (roads, etc.)	20	2.9	7.0	58	140
Grönhög Mine	30	2.5	19.6	75	588
Adelsvärd deposit	10	2.5	19.6	25	196
Tailings pond	15	2.0	8.6	30	129
Tailings delta	5	2.0	8.6	10	43
Small heaps in the forest	10	1.5	5.0	15	50
Total (tonnes)	1,400,000	—	—	4003	10,306

From Lundgren, A. T. Bersbo — the first full scale project in Sweden to abate acid mine drainage from old mining activities, in *Acid Mine Drainage, Designing for Closure, GAC/MAC Annual Meeting* (Vancouver, Canada: BiTech Publ., 1990), pp. 241–253. With permission.

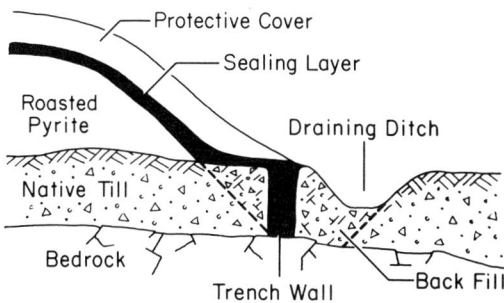

Figure 5.12. Sealing on tailings. (From Lundgren, A. T. Bersbo — the first full scale project in Sweden to abate acid mine drainage from old mining activities, in *Acid Mine Drainage, Designing for Closure, GAC/MAC Annual Meeting* (Vancouver, Canada: BiTech Publ., 1990), pp. 241–253. With permission.)

Lysimeters have also been installed under the reclaimed areas where all the waste rock was removed. The objective has been to monitor the quality of the percolating water before it reaches the groundwater.

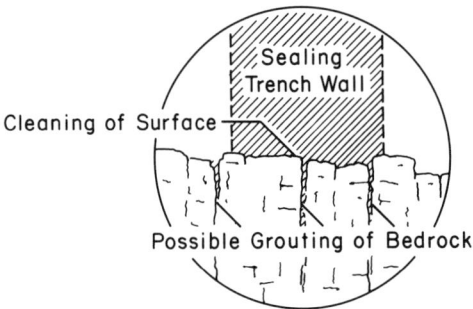

Figure 5.13. Sealing on tailings. (From Lundgren, A. T. Bersbo — the first full scale project in Sweden to abate acid mine drainage from old mining activities, in *Acid Mine Drainage, Designing for Closure, GAC/MAC Annual Meeting* (Vancouver, Canada: BiTech Publ., 1990), pp. 241–253. With permission.)

If the sealing layer is as impervious as it should be according to the specifications, i.e., if the full-scale permeability follows the small-scale permeability without interferences of defects such as cracks and hollows, then the sealing would be fully saturated with moisture and the oxygen diffusion rate would be very low. Hence, monitoring the moisture in the cover would be the perfect control of the efficiency of the sealing layer.

Tests were made with neutron/neutron probes in tubes that were drilled through the layers and 0.5 m into the waste rock. However, the measurements were not found to be reliable. Instead, calculations showed that the lateral transport capacity of the protective layer was too small to drain a normal rain percolation, and if the sealing layer acted as it was supposed to, it should result in a free groundwater table in the cover. Therefore, about 30 ordinary groundwater tubes were installed in the protective cover of the two heaps down to their respective sealing layer. The 300-mm bottom parts of tubes were perforated and dressed with fabrics.

The primary objective of covering the waste rock was to significantly reduce the transport of oxygen to the waste. Monitoring the oxygen concentration within the heaps is therefore of greatest importance because this full-scale project would act as a pilot project for the rest of the sulfidic waste deposits. Thus, a total of 11 small tubes was installed at different depths in the two heaps before the covering took place. Cocks were installed on the tubes after they had been brought through the protective cover.

Water quality monitoring has been performed in two separate programs, covering the surface water and the groundwater separately.

The covering operation was completed in March 1989. Due to the great variations in water quality over time, it is still too early to draw definite conclusions about the surface and groundwater conditions. However, it is already clear that the discharge conditions are different at the two deposits. The discharge of the Storgruve deposit (which was covered by Cefill plus till) is also the drainage of the mine, as well as that of the waste that was dumped in the mine. The water composition of this stream seems to have been changed only slightly. The major change is the increase of iron discharge and the heavy precipitation of ferrous iron in the stream.

Obviously, covering the waste has caused a reduction in the ferric weathering products, which results in an increased solubility of iron and a possibility of secondary oxidation of sulfides. The ferrous iron then oxidizes again, resulting in the iron precipitation outside the heap, and thereby causing a secondary lowering of the pH in the effluent water. Despite this secondary and temporary production of AMD, the other metal concentrations have not increased in this stream since the covering.

The water quality of the drainage of the Steffenburg deposit (covered by clay plus till) has decreased significantly around to the deposit. Farther downstream the metal concentrations are still high and the water quality improvement in the main recipient is small.

The water percolation rate through the sealing layer is measured in three lysimeters on the Steffenburg deposit (clay) to be between 0.5 and 3 mm annually, which corresponds well to the hydraulic conductivity measurements at the construction of the layers. At the Storgruve deposit no water percolation to the five lysimeters monitored has occurred thus far. This is believed to be due to the fact that unsaturated conditions still prevail in the Cefill layer. From the experiences of an excavation in a previous Cefill test cover, it is obvious that the puzzolanic reactions of the Cefill consume all the water in the matrix, and that it takes >1 year to exchange the pore gas with water. This must be especially slow if the lower part of the covering material is completely saturated. Under such a continuous layer the pathways to escape for the pore gas are scarce.

The groundwater levels in the protective layer were measured after the layer was completed. It should be noted that the summer and fall were extremely dry, which is verified by the descending groundwater table. The table did not rise significantly until November of that year. The clay layer seems to be more tight than the Cefill layer, which conforms with the permeability measurements, but not with the lysimeter results.

It is also evident that wet spots and dry spots are found on both deposits that cannot always be explained by their position on the slope. This means that parts must exist in both sealing layers that are less efficient than average to act as a water barrier. The frequency of such parts may be higher in the relatively thin (0.25 m) Cefill layer than in the thicker (0.5 m) clay layer. On the other hand, the slope gradients are higher, generally, and the rim connections were more delicate to get well sealed off on the Cefill deposit than on the clay deposit.

Oxygen measurements in the heaps were carried out at least 2 years before the remedial actions started. The concentration was always the same as in the atmosphere at all depths, which indicated that the heaps of rather coarse rock fragments were well ventilated. Also, during the works proceeding the covering, the oxygen concentration was above 20%, with the exception of some calm days when the filter layer was almost completed and the concentration temporarily dropped a few percent.

A new series of oxygen measurements in the heaps started as soon as the coverage was completed on the Steffenburg deposit (clay). At that time the Cefill cover at the Storgruve deposit was not fully completed, yet a substantial decrease in oxygen concentration took place in the deposit. Some of the small holes through the Cefill cover were not sealed off until several months later, and this probably explains the slower response in this heap. The apparatus used for the measurements has an uncertainty of 0.2%, and the measurements showed values sometimes well below zero (despite calibration with nitrogen). Therefore, all measurements below 0.5% are uncertain. The variation of the values over time is also rather great. However, the measurements have shown that the sealing layers are very tight against oxygen transport.

What does a decrease in oxygen concentration to values below 0.5% mean? According to the laboratory experiments, the metal concentrations should be reduced, as a consequence of the decreased oxidation of sulfides, by a factor of at least 5. Together with the reduction of water percolation, which seems to be on the order of at least 99%, this should lead to a total reduction of the metal transport of at least 99.8%. This is probably too optimistic, as the leaching experiments use a lot more water than that corresponding to the percolation in the covered heaps. Regardless, the result should be very good, good enough to be followed up for several years to come.

DESIGNING CLOSURE OF AN OPEN PIT MINE IN CANADA

Based on present ore reserves, mine closure at Equity Silver Mines Ltd. is currently set at year-end 1992. As waste materials are predominantly acid-generating, mine closure will not be a "walk-away" situation but, rather, will require ongoing maintenance of environmental and treatment facilities for an indefinite period.[10]

In developing the various scenarios for closure and anticipated cost liabilities, Equity Silver submitted a conceptual closure plan to MEMPR (Ministry of Energy, Mines & Petroleum Resources) in October 1988. The plan has been under intense review since submission. The review process initially went through the Equity Silver Surveillance Committee, an independent body consisting of federal and provincial government and interest groups in the area. This process took the better part of a year with recommendations from the review being used to set the conditions of a revised reclamation permit issued by MEMPR. These conditions formed the basis for future studies and refinement to the plan.

One specific condition of the permit is to establish a bond of size sufficient to meet environmental cost in perpetuity. To ensure no erosion of this fund base, annual surplus incurred is based on real interest rates; i.e., interest less inflationary increases. The ultimate bond size will be established on the annual projected maintenance cost to sustain environmental fixtures times the real interest-generating potential of the portfolio in which the fund is held.

Equity Silver actively pursues system improvement and research to optimize environmental systems. Data from research programs developed onsite are reported annually and are used to update closure concepts.

Until adequate mitigative measures can be implemented, collection and treatment of AMD is required to safeguard the surrounding environment, hence, the bond placement to sustain the operation over the long term.

Equity Silver employs conventional AMD treatment using quicklime as a base, followed by agitation, then clarification within settling ponds. Approximately 800,000 m3 of AMD is treated annually, producing approximately 80,000 m^3 of sludge consisting of 6 to 7% solids by weight. Sludge is presently pumped from the ponds and mixed with tailings. As long as the mine is in operation, this is considered to be an acceptable method for handling the material. However, with mine closure imminent this disposal will no longer be available and will require an alternate practice. High-density sludge (HDS) treatment is being considered and is at the feasibility stage. By using this process, sludge volumes can be reduced to 15 to 20% of that presently produced. Disposal options being considered include placement on filter beds for further dewatering or subaqueous disposal in one of the flooded open pits. The latter option carries some risk as pH depressions below 6.0 can redissolve and mobilize metals (especially zinc) within the sludge. Testwork does, however, exhibit that sludge has adequate buffering capacity to hold pH in the 7.5 range, with no mobilization of soluble metal.

Quicklime purchased from southern British Columbia represents a major portion of the cost to treat AMD. Equity Silver has staked local claims of limestone that has been kiln tested and confirmed to produce a good grade of quicklime. A study has been awarded to a consultant to investigate the viability of constructing a lime kiln locally to offset the high cost of transporting lime to the mine site.

Overburden is used as a cover and growth medium for sealing acid-generating wastes. This overburden, which consists primarily of glacial till, is placed on dump surfaces at an average thickness of 0.75 m. Covers are graded and seeded to complete the reclamation process. Because it is virtually impossible to assess the mitigative effect these covers have on the waste dump acid-generation process, assessments are carried out using scaled-down test plots. These plots are constructed of well-compacted till basins with drained underflows terminating in 45-gal plastic drums to capture runoff. Seven of these plots have been constructed. Four contain oxidized waste material from the Southern Tail Pit, with the remaining three consisting of unoxidized material from the Main Zone Pit.

Each set of tests for the two waste materials is comprised of control piles (uncovered) to gauge water infiltration and metal loading vs that from piles covered with glacial till. The covered piles were not compacted, but have been seeded to establish a vegetation cover.

Samples are taken from the collection vessels after each major rainfall with volumes recorded and samples analyzed for pH, sulfate, acidity, and dissolved and total copper, iron, and zinc. Total loading of metal out of the pits is calculated on the basis of volume times qualitative analysis.

After tabulating the annual loading results for the Southern Tail Pit, we find that water infiltration is reduced by about 50% by the mill covers, metal loading by about 75%, and sulfate and acidity loading around 67%. Little change was observed in the Main Zone Pit material as acid-generation processes have not fully developed.

Although strong evidence exists that clay covers reduce oxidation and metal loadings out of waste piles, concern remains as to the longevity of the process. In order to eliminate metals leaching from wastes, infiltration of precipitation must be completely eliminated. A test program is being designed to evaluate the practicality of covering a section of waste dump with a high-density PE membrane. Parallel tests will be run to assess the practicality of dump resloping, followed by till placement and compaction on steep slopes. The outcome of the test results may dictate an alternate cover technique to that presently used.

The Southern Tail Pit, which was the first zone to be mined, has been backfilled with acid-generating waste from the Main Zone Pit. Disposal of waste into this pit not only provides a readily available dump site, but also provides an area in which a portion of the waste can be disposed below water.

Upon completion of mining, water quality deteriorated in the Southern Tail Pit with pH dropping to 3.0 and copper and zinc values increasing to 13 and 11 mg/L, respectively. This deterioration was a product of oxidation of the pyrite in wall material and fractured waste in two slide areas. Prior to backfilling the pit, tests were carried out to assess the impact of waste rock disposal on final water quality. Results indicated that pH would recover to 7.5 to 8.0 and metal values would significantly decline.

Backfilling of the pit proceeded in three stages. The first step was to backfill to a horizon 1 m below the projected floodplain, followed by placement of a 2-m layer of inert non-acid-producing waste. This layer serves as a buffer zone to decrease AMD production at the water interface. The third stage of backfilling involved placement of waste above this barrier and water table. These wastes are also acid generating and are reclaimed to reduce oxidation rates. Water has been discharged to the environment from mid-1987 to the present and has not shown any deterioration in downstream water quality. An increase in zinc values has been noted over the last 2 years which is believed to be a product of AMD inflow from slide material. The pH of the receiving pool of water is not sufficient to fully precipitate zinc. Dump resloping and sealing tests were projected to be carried out during the summer of 1990. The objective of this testwork was to eliminate seepage through the wastefills, at least to a degree to which AMD input can be buffered by alkalinity within the system.

When mining of the Main Zone was completed, the waste from the adjacent Waterline Pit was placed within the Main Zone. The pit was flooded to cover wastes to eliminate acid generation. As a portion of the wall within the Main Zone Pit is acid generating, a dam was constructed at the entrance to the pit to raise the water level. Postclosure water quality in the Main Zone Pit was modeled under a program. The study concluded that pit water would turn slightly acidic during flooding, but would return to a neutral condition shortly thereafter due to the influence of alkaline groundwater. The model did not include the option of backfilling the pit with mine waste, which will further enhance alkalinity of the ponded water.

Tailings, like the greater distribution of waste rock, have the potential to produce acid given proper conditions. Fortunately, Equity Silver elected to build water-containing dams that allow tailings to be flooded through the operating stage and after closure of the mine.

Two tests were set up to kinetically evaluate the acid-generating properties of tailings submerged below water. In the first test, tailings were placed in an aquarium, then flooded with freshwater. The water was allowed to fluctuate through evaporation, with tailings exposed to the air for months at a time. After being subjected to numerous wet and dry cycles, it was found that zinc and, to a lesser extent, copper appeared in solution. The supernatant remained neutral throughout the test.

With the second test, tailings were covered with water throughout the test. To simulate wave action and oxygen entertainment, air was continuously introduced to the water cover using a bubbler system. Zinc values reached 8 mg/L in a short time frame and remained at this level throughout the remainder of the test. The pH remained neutral.

Results of this testwork indicate that acid-generating tailings must be covered with water at all times but do not necessarily preclude oxidation and mobilization of soluble metals (zinc).

Although no formal testwork has been adopted to fully evaluate passive treatment options, strong evidence has come to light that swamps in the immediate area of the mine site have the capacity to remove metal from solution. In one particular swamp, to which treated water is discharged, removal of dissolved zinc is at 70%. Reliance on this swamp to remove trace values of metal, particularly zinc, may lead to the development of cultured swamps downstream of the mine site.

METAL CONTENTS AND TREATMENT OF MINE WATER

The Berkeley Pit is a large abandoned open pit mine in Butte, MT. Since 1982 it has been filling with water; the level is currently over 213 m (700 ft) deep. At the present rate of inflow, the pit water level will rise to the low point in the rim by the year 2011. Before that time, around 1997, water rise will reach exposed alluvium in the pit walls, creating a threat to groundwater quality in the valley south of the pit. In terms of contained volume of water and quantity of metal pollutants, the Berkeley Pit is unmatched by any acid-producing mine in the United States and possibly the world.

The Berkeley Pit intersects a multitude of openings to the old underground workings to the north and west, and below the pit.

The Butte area is drained by Silver Bow Creek, which is the headwaters of the Clark Fork River, a major tributary of the Columbia. The pit is at an altitude of 1676 m (5500 ft). Because of the altitude and latitude, the climate is cool, with harsh winters.

The deposits at Butte were huge. Prior to 1983, Butte had produced over 20 billion lb of copper, 4.9 billion lb of zinc, 3.7 billion lb of manganese, 850 million lb of lead, 750 oz of silver, and 2.9 million oz of gold.

Water Types and Contents

Essentially two types of mine water are found in Butte: underground and pit. The two have the same elements in solution, but most of them are present in much higher concentrations in pit water than in underground water. The underground solutions are more reducing than the pit water, and the ionic species present in the two waters are somewhat different.

Table 5.19 shows the composition of some underground water samples and pit water samples, as well as federal monthly discharge standards. The Kelley Mine is near the Berkeley Pit and the two are intimately connected by underground workings. Water flows from the underground mines to the pit. The Travona Mine is 2440 m (8000 ft) west of the pit and is not directly connected to the Kelley/Berkeley system, as most of the underground connections between the two areas were bulkheaded before mining stopped.

Pit water is more oxidized than underground water and is highly acidic. Water at the surface has an Eh of 820 mV and a pH of 2.8, while the water at 130 m below the surface has an Eh of 468 mV and a pH of 3.14. In general, metal concentrations in near-surface waters (0 to 10 m) are lower than in deeper waters, representing the bulk of the solution in the pit.

About 40% of the water inflow to the Berkeley Pit is from underground mines. The other 60% is from surface sources, both natural flow from the adjoining perched alluvium, and leakage from the active tailings pond and leach pads through the alluvium. At present, no way has been found to precisely calculate the relative amounts from the two alluvial sources, but the leakage from the tailings pond is believed to contribute much more water than the natural alluvial flow.

Table 5.20 is an estimated average composition of the surface flows to the pit. It was calculated by subtracting the underground contribution of the various elements from the measured quantities in the pit.

Table 5.21 shows the estimated amounts and percentages of the various elements contributed by the underground flow and the surface sources to the pit. Clearly, surface sources supply most of the metal ions.

Water from the underground mines, even though it is in contact with large amounts of sulfides, remains relatively low in metals because no source of air or

Table 5.19. Compositions of Various Mine Waters of the Butte Area

(mg/L)	Pit Water[a] 1 m	Pit Water[a] 100 m	Kelley[b] 120 m	Travona[b] Surface	Federal Std.[c] Monthly
Al	103	193	3.26	<0.040	1
Total As	0.031	1.15	1.04	0.106	0.5
As (III)	0.001	0.087			
As (V)	0.005	0.768			
Cd	1.08	1.87	0.11	<0.005	0.05
Ca	433	482	407	179	
Cu	133	203	0.14	<0.004	0.15
Total Fe	202	1020	285	4.64	1
Fe (II)	60	958			
Fe (III)	142	140			
Mg	153	280	156	51.1	
Mn	73.7	162	52.4	12.84	2
K	10.4	18.7	37.4	5.4	
Na	73.1	70.8	49.8	66	
Pb	0.112	0.522	N/A	<0.04	0.3
Zn	212	497	114	0.11	0.75
Cl	9.82	22.1	20	37.7	
SO_4	4850	6760	N/A	495	
SiO_2	85.6	111	N/A	17.9	
pH	2.700	3.150		6.88	6–9
Ion storing	0.100	0.160			

[a] Davis, 1988. SiO_2 at 100 m was analyzed and provided by Dr. J. Sonderegger at the Montana Bureau of Mines.
[b] Duaime, 1990.
[c] *Federal Register*, 1988.

From Huang, Hsin-Hsiung.[11]

Table 5.20. Estimated Composition of Surface Water Flows to the Berkeley Pit

Substance mg/L	Ca 495	Mg 314	Na 81.7	K 4.2	As 1.08	Al 295
Substance mg/L	Cd 2.74	Cu 318	Fe 1238	Mn 205	Zn 662	

From Huang, Hsin-Hsiung. Characteristics and treatment problems of surface and underground waters in abandoned mines at Butte, Montana, in *Mining and Mineral Processing Wastes* (Littleton, CO: Society of Mining, Metallurgy, and Exploration, 1990), pp. 261–270. With permission.

other oxidants are found in the flooded mines. The sulfides are not being oxidized and therefore do not contribute ions to the solution. The percentages of the various metals present in the tailings are much lower, but the collective tonnage is vast, the particles in the ponds are very small in size (as they have all been through crushing and grinding circuits), and well-aerated solution is steadily supplied to the ponds.

Table 5.21. Amount and Sources of Berkeley Pit Constituents

	Pit Content		Bedrock Contribution		Surface Contribution	
Substance	Tons	%	Tons	%	Tons	%
Al	6,400	100	4.79	0.01	6,395	99.9
As	25	100	1.53	6.12	23.5	93.9
Cd	61	100	1.62	2.66	59.4	97.3
Ca	16,700	100	5,970	35.7	10,730	64.3
Cu	6,900	100	2.06	0.03	6,898	99.97
Fe	31,000	100	4,180	13.5	26,820	86.5
Mg	9,100	100	2,290	25.2	6,810	74.8
Mn	5,200	100	769	14.8	4,431	85.2
K	640	100	549	85.8	91	14.2
Na	2,500	100	731	29.2	1,769	70.8
Zn	16,000	100	1,670	10.4	14,330	89.6

From Huang, Hsin-Hsiung. Characteristics and treatment problems of surface and underground waters in abandoned mines at Butte, Montana, in *Mining and Mineral Processing Wastes* (Littleton, CO: Society of Mining, Metallurgy, and Exploration, 1990), pp. 261–270. With permission.

Under these conditions, sulfides are continuously oxidizing in the tailings pond and the waste and leach dumps, contributing metals to solutions that eventually deposit in the Berkeley Pit.

The same reactions that contribute metals to solution provide sulfate from oxidized sulfide minerals. The resulting high concentrations of sulfate (4.2 g/L at the pit water surface and 12 g/L at 130 m) promote metal dissolution and hinder precipitation. Most of the metal ions present in pit water complex strongly with sulfate. For instance, 90% of the zinc dissolved at depth is present as $ZnSO_4$ (aqueous). This greatly lessens the activity of Zn^{2+} and prevents precipitation of the zinc compounds. Although stable metal sulfate solids are seen, their stability is not high enough to offset the increased solubility of the metals.

Some minerals are near saturation in pit water and may now be precipitating from solution. The most common measure of saturation is the Saturation Index or SI, defined as the difference between the logarithm of the solubility constant of the mineral of interest and the logarithm of the product of the activities of the constituent ions in solution. This expression, however, does not correctly predict the degree of saturation. A better measure is the Individual Degree of Saturation or IDS, which is the actual amount of mineral that must be added to or removed from the solution to achieve saturation.

Table 5.22 shows both the SI and the IDS for a number of minerals suspected of being near saturation in pit water. Both surface and deep pit waters are near saturation with the ferric minerals iron hydroxide and jarosite, the aluminum minerals alunite and kaolinite, the calcium mineral gypsum, and amorphous silica. Thus, if ferric, aluminum, calcium, or silica ions are added to pit water without a correspond-

Table 5.22. Individual Degrees of Saturation and Saturation Indices of Pit Waters

Water at 1-m Depth: Initial pH 2.7

	Ferrihydrite	Geothite	Jarosite	Na-Jarosite	Hematite	Alunite
Initial metal Conc (mM)	3.51	3.51	3.51	3.51	3.51	3.74
Final pH	2.551	2.311	2.586	2.682	2.312	2.910
IDS (mM)	0.718	2.492	0.262	0.037	1.240	−0.343
SI	0.580	2.843	2.991	0.165	5.486	−1.923

	Jurbanite	Gypsum	Quartz	SiO$_2$ (am)	Kaolinite	Scorodite
Initial metal Conc (mM)	3.74	10.70	1.42	1.42	3.74 (Al)	0.005 (As)
Final pH	2.537	2.705	2.700	2.700	3.865	3.007
IDS (mM)	2.328	−1.778	1.213	−0.770	−0.748	−1.375
SI	0.571	−0.070	0.829	−0.188	−16.891	−5.361

Water at 100-m depth: initial pH 3.15

	Ferrihydrite	Geothite	Jarosite	Na-Jarosite	Hematite	Alunite
Initial metal Conc (mM)	0.251	0.251	0.251	0.251	0.251	7.26
Final pH	3.040	3.006	3.067	3.168	3.000	2.909
IDS (mM)	0.193	0.263	0.072	−0.014	0.131	0.226
SI	0.893	3.156	2.801	−0.294	6.111	1.710

	Jurbanite	Gypsum	Quartz	SiO$_2$ (am)	Kaolinite	Scorodite
Initial metal Conc (mM)	7.26	12.06	1.85	1.85	7.26 (Al)	0.001 (As)
Final pH	2.547	3.153	3.150	3.150	3.933	3.496
IDS (mM)	5.536	−1.563	1.638	−0.346	−0.175	−0.720

Table 5.22 (Continued). Individual Degrees of Saturation and Saturation Indices of Pit Waters

			Water at 100-m depth: initial pH 3.15			
	Jurbanite	Gypsum	Quartz	SiO2 (am)	Kaolinite	
Scorodite						
SI	1.248	−0.056	0.942	−0.074	−10.091	−2.915
	CuSO4:5H2O	FeSO4:7H2O		MgSO4:7H2O	ZnSO4:7H2O	MnSO4
Initial metal						
Conc (mM)	3.20	17.15		11.52	7.60	2.95
Final pH	3.518	3.587		3.801	3.787	
IDS (mM)	−546	−763		−2.431 (M)	−3.025 (M)	<−5.0 (M)
SI	−2.513	−1.962		−2.688	−2.892	−7.967

From Huang, Hsin-Hsiung. Characteristics and treatment problems of surface and underground waters in abandoned mines at Butte, Montana, in *Mining and Mineral Processing Wastes* (Littleton, CO: Society of Mining, Metallurgy, and Exploration, 1990), pp. 261–270. With permission.

ing addition of water, the result should be mineral precipitation rather than increased concentration in solution.

The concentrations of the metals in the surface water are much lower than in deep water (as shown in Table 5.22), even though the surface water is more acidic. This is due, at least in part, this is due to the oxidative behavior of iron. Near the water surface, atmospheric oxygen, aided by sunlight, causes iron to be oxidized from the ferrous or 2+ state to the ferric or 3+, which promptly forms a solid hydroxide, such as FeOOH or $Fe(OH)_3$. As the solid particles form, they tend to adsorb or incorporate other metal ions into the ferric hydroxide structure. Eventually, the particles sink, carrying any incorporated ions down with them. This would account for the surface iron concentration of 202 mg/L, as compared to 1020 mg/L at depth. Adsorption can explain the lowering of the deep arsenic concentration of 1.21 mg/L to 0.031 at the surface more completely than formation of the most likely arsenic mineral, scorodite ($FeAsO_4$), which would require an arsenic concentration of around 100 mg/L. Manganese may behave in a manner similar to that of iron, being oxidized near the water surface from the 2+ state to 3+ or 4+, forming a solid oxide, and coprecipitating with the iron.

In the first few years of pit flooding, winter freezing may have played a role in lowering surface metal concentrations. When part of a solution freezes, the ice tends to have a lower concentration of soluble ions than the original solution. The resulting unfrozen portion of the solution is enriched in soluble ions, is denser than the original solution, and sinks, while the relatively clean ice, of course, floats. This process has been invoked to explain certain natural brines derived from seawater. It is certainly not a factor at present in the Berkeley Pit, as no ice cover formed in the winters of 1988 to1989 and 1989 to 1990. The pit did freeze in previous years.

A practical process is needed for treating pit water to produce water meeting applicable drinking water/discharge standards and to recover the valuable metals. Because the volume to be treated is large and a plant must run for many years, simple processes are preferred over those more advanced requiring careful control and highly skilled operators.

A major concern is the disposal of unstable waste sludges produced by pit water treatment. The ideal disposal site for sludge is the area in which the sludge constituents originated: abandoned mine workings. The inactive underground mine workings in Butte would be very difficult to maintain access to and do not have sufficient volume to be a viable long-term disposal site, but the pit itself could be a very satisfactory permanent sludge pond. Any sludges placed there should be well understood in terms of composition and chemical interactions with pit water. Unstable ions in the sludges will certainly concentrate in the pit water, and ions introduced during treatment may interact with the native solution.

Metal concentrations in the pit solution can be lowered significantly by adding lime to raise the pH and sparging the solution with air to oxidize the metals in solution. First, the lime dissolves

$$CaO + H_2O = Ca^{2+} + 2OH^- \quad (5.12)$$

and raises the pH. With the rise in pH, aluminum precipitates as the hydroxide

$$Al^{3+} + 3OH^- = Al(OH)_3 \quad (5.13)$$

and the increase in calcium concentration precipitates gypsum

$$Ca^{2+} + SO_4^{2-} + CaSO_4 \quad (5.14)$$

Other metal-sulfate complexes dissociate as free sulfate is removed from solution

$$MSO_4 = M^{2+} + SO_4^{2-} \quad (5.15)$$

so that still more sulfate is available for gypsum formation.

Exposure to air and light oxidizes iron in solution

$$2Fe^{2+} + 1/2\, O_2 + H_2O = 2Fe^{3+} + 2OH^- \quad (5.16)$$

but at low pH the rate is slow. The rate is strongly dependent on the pH and a major concern is the pH level needed to have a reasonable rate of iron oxidation.

Oxidation is slower when sulfate, as Na_2SO_4, is increased. However, even though the sulfate concentration in pit water is high, iron oxidation is faster even than in the simple solution. This is due to the presence of other metal ions in the solution. The catalytic effects of at least some of them are enough to more than offset the extra sulfate.

Raising the pH does not significantly change the iron oxidation rate. Tests at a pH of 8.5 have shown a rate similar to that observed at 5.25. Other influences on iron oxidation might include photooxidation and iron bacteria.

As gypsum and ferric hydroxide coprecipitate, they tend to adsorb other ions from solution. Zinc, for instance, should theoretically remain in solution until the pH rises well above 5, but about 10% of the zinc is removed from solution at a pH of 5. Other metals behave similarly. In particular, arsenic is removed from solution by ferric hydroxide precipitation.

As the pH rises above 7, metals other than iron start to precipitate as hydroxides:

$$M^{2+} + 2OH^- = M(OH)_2 \quad (5.17)$$

This process should be effective at removing most of the objectionable metals from solution at a reasonable pH. An exception is manganese. The stable oxide of manganese is MnO_2, which should form readily if the manganese is present in solution in the 4+ state. Because manganese in the pit solution is primarily present in the 2+ stage, an oxidation step for manganese is needed. Manganese can be oxidized by atmospheric oxygen, but the rate is slow below pH 9. That pH turns out to be the upper limit of water discharge standards, so it would be desirable to remove manganese from solution at a lower pH. Other possible approaches include formation of $MnCO_3$, which would not require manganese oxidation, solvent extraction, or use of an alternative oxidant.

The most practical approach to Berkeley Pit water treatment appears to be the use of a multistage process. The first step would raise the pH of raw pit water, while simultaneously aerating the solution. The objective would be to remove as much aluminum and iron as possible from solution as hydroxides. The large quantity of iron oxide produced would adsorb almost all of the arsenic present, while leaving most of the zinc and manganese in solution to be dealt with in later steps. The first-stage sludge would carry a large portion of the initial sulfate, as well as arsenic, iron, and aluminum and would be suitable for disposal.

The second stage of treatment would have the objective of precipitating zinc and possibly manganese, along with the remaining sulfate. The sludge from the second stage could then be treated to recover these metals, with residual sludge being recycled to stage 1. The primary means of treatment would be to raise the pH.

The reagents being considered to raise the pH in the first stage are lime (CaO), calcium carbonate ($CaCO_3$), and soda ash (Na_2CO_3). These are all commonly available and reasonably safe and cheap. Both lime and calcium carbonate are effective at raising the pH to over 5 and with both iron is air oxidized and removed as ferric hydroxide. The major difference is the behavior of zinc. The use of lime resulted in larger zinc losses from solution than with the use of calcium carbonate. When powdered lime was used to raise the pH to 5.38, 67% of the iron in solution precipitated, while 50% of the zinc was lost to the sludge. When calcium carbonate was used to raise the pH to 5.34, 84% of the iron precipitated and only 7% of the zinc.

This may occur because although calcium carbonate raises solution pH, it does so in a gentler manner than lime. As calcium carbonate reacts in acid solution

$$2CaCO_3 + 2MSO_4 = 2CaSO_4 + 2M^{2+} + 2H_2CO_3 \quad (5.18)$$

it does not produce local pockets of high pH. Lime, added as the solid or slurried, dissolves faster than gypsum can form and creates local areas with much higher pH than the average solution pH. Reactions can occur in these regions that are not in general equilibrium with the system, such as the formation of zinc hydroxide. Thus, zinc and other metals coprecipitate to a lesser degree in a calcium carbonate system than in a lime or soda ash system. Calcium carbonate has other advantages over lime in that it is cheaper and has been reported to produce better settling and less voluminous sludges than lime.

The primary means of treatment would be to raise the pH. Calcium carbonate cannot raise the pH to a level that is high enough to be useful in the second stage. Unless the calcium carbonate particles are scrubbed, surface films inactivate them at a pH of about 6.5. Both lime and soda ash can raise the pH high enough to work. Lime would be a calcium source to precipitate the remaining sulfate, but soda ash would be a carbonate source for manganese and possibly zinc precipitation. It may be that a mixture of the two will turn out to be most effective.

It would be desirable to keep the final pH below 9, as discharge would have to be acidified if it is necessary to go higher for thorough metals precipitation. Methods that might be used to remove residual metals after second-stage treatment include zeolite and ion exchange and solvent extraction.

Other metals of concern in Berkeley Pit water include copper, lead, and cadmium. High levels of copper are present in pit water, and the recovery of copper is of economic interest. Cementation with iron

$$CuSO_4 \text{ (aq)} + Fe(s) = Cu(s) + FeSO_4 \qquad (5.19)$$

as a pretreatment before stage 1 is an obvious method to be tried, and preliminary tests indicate that copper in the pit solution will respond well to this approach.

Lead and cadmium are present in pit water at rather low levels, although their concentrations exceed discharge standards. Lead and possibly cadmium will probably precipitate in the stage 1 and 2 sludges.

As mentioned previously, when water freezes, ions in solution tend to stay in the unfrozen solution, resulting in a relatively clean ice. This principle works, to a degree, on Berkeley Pit water. Concentrations of metals in solution were lowered by about 50% in melted ice produced by freezing 10% of the solution. Freezing thus shows some promise as a treatment step. Freezing could also play a role in sludge disposal. It has been used to lower the water content and volume of sewage sludges and might be useful in lowering the volume of treatment sludges.

The climate in Butte is semiarid and the air is quite dry much of the time. Evaporation has been suggested as a means of releasing clean water to the atmosphere rather than to a stream. It turns out that an evaporation pond would be prohibitively large. Studies of evaporation rates in the vicinity indicate that 414 acres of pond area would be required to evaporate an average of 1 million gal/day, taking into account that a pond in Butte would freeze over in the winter and be successful only in the warm months. Treatment of 7.5 million gal/day would require a pond, presumably lined, of over 7200 acres.

REFERENCES

1. *Draft Acid Rock Drainage Technical Guide,* Vol. 1, (Vancouver, Canada: Bi Tech Publ., 1989).
2. Errington, J. C. and K. D. Ferguson. Acid mine drainage in British Columbia, in *Proc. Acid Mine Drainage Seminar/Workshop, Halifax, Nova Scotia* (Ottawa: Environment Canada, 1987), pp. 67–87.
3. Morin, K. A. and J. A. Cherry. Migration of acidic groundwater seepage from uranium tailings impoundments, *J. Contam. Hydrol.* 2:323–342 (1988).
4. Lapakko, K. Prediction of AMD from Duluth complex mining waste in North Eastern Minnesota, in *Proc. Acid Mine Drainage Seminar/Workshop, Halifax, Nova Scotia* (Ottawa: Environment Canada, 1987), pp. 187–221.
5. Asmund, G., M. Hansen, and P. Johansen. Environmental Impact of Marine Lake Tailings Disposal at the Lead-Zinc Mine at Maarmorilik, West Greenland, presented at the International Conference of Environmental Problems from Metal Mines, Roros, Norway (1988).
6. Boyles, M. J. Impact of Argo Tunnel acid mine drainage, Clear Creek County, Colorado, in *Water Resources Problems Related to Mining* (Baltimore, MD: American Water Resources Association, 1974), pp. 41–53.
7. Hardie, M. G. and J. C. Jennett. Water resources problems and solutions associated with the new lead belt of S.E. Missouri, in *Water Resources Problems Related to Mining* (Baltimore, MD: American Water Resources Association, 1974), pp. 109–122.
8. Asmund, G. Rehabilitation and demolition after the closure of the zinc and lead mine Black Angel at Maarmorilik, Greenland, in *POLARTECH, Proc., Int. Conf. Development and Commercial Utilization of Technologies in Polar Regions* (Copenhagen, Denmark: Danish Hydraulic Institute, 1990), pp. 744–759.
9. Lundgren, A. T. Bersbo—the first full scale project in Sweden to abate acid mine drainage from old mining activities, in *Acid Mine Drainage, Designing for Closure, GAC/MAC Annual Meeting* (Vancouver, Canada: Bi Tech Publ., 1990), pp. 241–253.
10. Patterson, R. J. Mine Closure Planning at Equity Silver Mines, in *Acid Mine Drainage, Designing for Closure, GAC/MAC Annual Meeting* (Vancouver, Canada: Bi Tech Publishers, Ltd., 1990), pp. 441–445.
11. Huang, Hsin-Hsiung. Characteristics and treatment problems of surface and underground waters in abandoned mines at Butte, Montana, in *Mining and Mineral Processing Wastes* (Littleton, CO: Society of Mining, Metallurgy, and Exploration, 1990), pp. 261–270.

CHAPTER 6

Hydrologic Impact

INTRODUCTION

Surface mining affects surface streams through increased runoff and subsequent channel erosion as a result of reduced infiltration rates. The streams may also be affected by decreased surface runoff, which may be good or bad, depending on the specific regional, climatic, and geologic setting. Decreased runoff results from diversions and increased infiltration rates where more permeable rock strata become exposed by surface mining.

Modifications of local or regional recharge zones by surface mining involve the alteration of infiltration rates by removal of vegetative covers, the alteration of soil profiles, and compaction. Reduced infiltration rates decrease groundwater storage and reduce water availability.

Disruptions of groundwater systems and flow patterns by surface coal mining are of particular concern in the semiarid western region of the United States due to the importance of groundwater to livestock and irrigating agriculture in the region. Shallow and coal seam aquifers can be drained by mining activity, causing the temporary or permanent loss of existing wells near mined areas. The impact may not be restricted to the mined area and could extend several miles away from the site. Direct distribution of aquifers may be a very definite hazard where coal extraction is planned on a regional basis.

In southeastern Montana the Decker Mine, located in an area of local discharge, was opened in the summer of 1972.[1] The D1 coal aquifer is about 15 m thick and

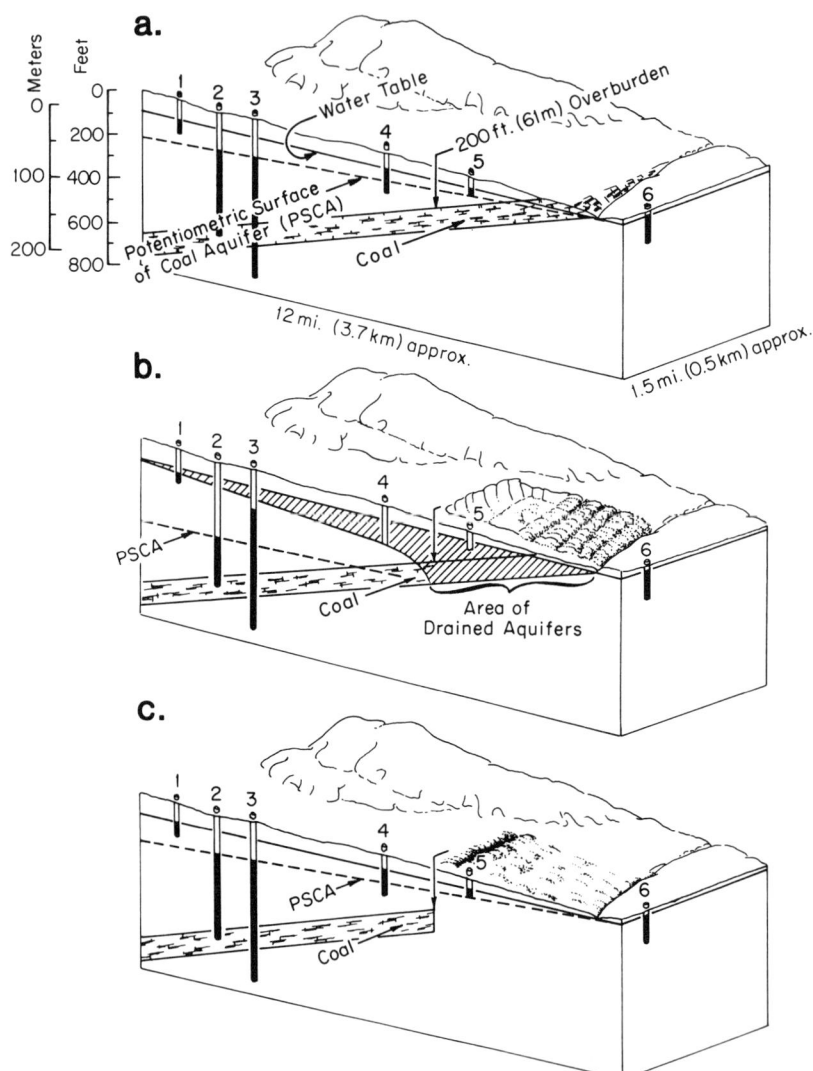

Figure 6.1. Impact of surface coal mining on water table. (From El-Ashry, T. M. *Impacts of Coal Mining on Water Resources in the U.S.* (Denver, CO: Environmental Defense Fund, 1976).)

provides ample water to stock and domestic wells. A study 1 year later reported that water levels in wells declined up to 8 m. The greatest water level declines occurred close to the mine in wells that were cut off from recharge areas. The rates of decline did not begin to recede after the first year, reflecting the constantly increasing stress on the groundwater flow system (Figure 6.1).

With underground mining, subsidence and fracturing of the overlying strata may cause surface runoff to be diverted underground and may disrupt aquifers, causing local water level declines and change in the direction of groundwater flow near the mine. An aquifer currently discharging in a single spring used by livestock and wildlife may, as a result of subsidence, discharge over a larger area by diffuse seepage.

In hydrologic systems in which multiple level aquifers exist, breach of the confining layer separating the aquifers will cause an increase in vertical flow and hydraulic connection between the affected aquifers. The effects will vary according to location of the breach relative to regional and local flow patterns.

Contamination of shallow aquifers may also take place through downward seepage of poor quality mine water. The removal of considerable amounts of overburden could have an effect on the artesian pressure in the underlying aquifers. Reduced overburden pressure could cause upward movement of water and increase discharge into the mine pit.

Disruption of groundwater aquifers during mining results in a spoil that is generally less permeable than the original overburden. Reduced permeability inhibits lateral movement of groundwater through a previously undisturbed aquifer. Any water passing through such spoil will contain increased mineralization.

HYDROLOGIC IMPACT OF PHOSPHATE MINING

Each year, Florida mines approximately 40 million tons of phosphorite, most of which is ultimately converted to various commercial fertilizers. The phosphorite is mined from within the bottom portion of the 40- to 60-ft thick surficial aquifer, using open pit mining techniques. The ore zone or "matrix" is generally overlain by relatively permeable fine sand to silty and slightly clayey fine sand. The ore itself is highly variable in nature but, in central Florida, typically consists of less permeable clayey phosphatic sands and sandy clays with stringers of permeable sand. The ore is much more sandy in the north Florida mining area.[2]

Prior to and during mining, the surficial aquifer is dewatered to at least the top of the ore zone to improve mine-cut stability and to improve ore recovery. The overburden sands are excavated and cast into the previous mine cut, after which the ore is excavated, slurried, and pumped to a beneficiation plant where the clay, sand, and phosphate "concentrate" are separated. Until the mined areas are reclaimed, generally 2 to 3 years postmining, the surrounding unmined areas continue to dewater.

Mined land reclamation in the Florida phosphate industry generally takes one of the following three forms: waste clay area, sand tailings fill area, and land and lakes area. In the first reclamation option, the mine pit is filled with waste clay, which is pumped directly from the beneficiation plant at approximately 3% solids. Embankments, up to 40 ft high, are constructed around the mine pit to contain the clay slurry. The clay is stored above ground to allow for long-term consolidation of the clay and

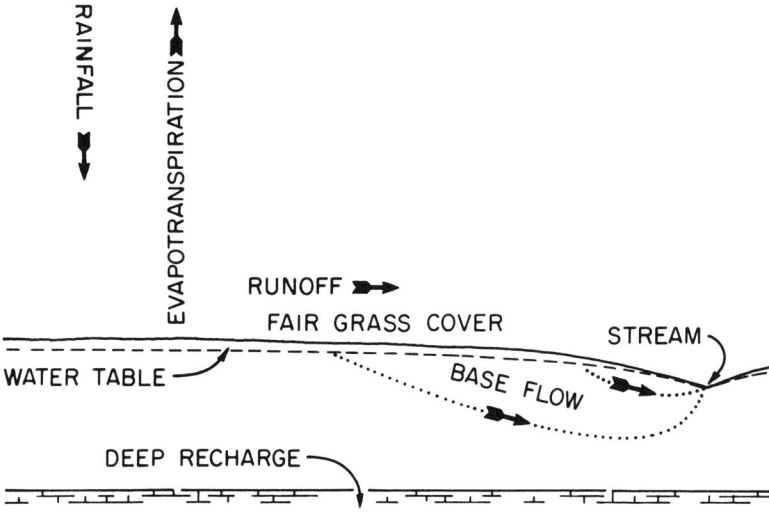

Figure 6.2. Natural conditions. (From Garlanger, J. E. Ground water restoration in mined areas, in *SME Annual Meeting Preprint 90-76* (Denver, CO: Society of Mining Engineers, 1990).)

to make certain that the surface of the clay does not ultimately settle below the surrounding topography. After the clay has consolidated and the upper 2 to 3 ft of clay becomes desiccated, the surrounding embankments are pushed down, sometimes onto the surface of the clay and sometimes onto the surrounding land, and the area is vegetated to complete reclamation.

The second reclamation option consists of pumping sand tailings from the beneficiation plant into the mine pit up to the level of the surrounding ground. Because the sand tailings dewater rapidly as they are deposited, no surrounding embankments are necessary. The surface of the sand tailings is ultimately vegetated.

The third reclamation option consists of flattening the spoil piles and allowing the mine pit to fill up with rain- and groundwater, creating a land and lakes area. Wetland-type vegetation is usually planted around the shore of the lake before the water reaches its ultimate level.

One of the primary objectives of reclamation is to return the mined areas, to the greatest practical extent, to their natural premining condition, particularly with respect to surface water and groundwater resources.

To determine the impact of mining and subsequent restoration of the surficial aquifer of the ground and surface water resources of the affected watershed, it is first necessary to model the premining condition. Figure 6.2 illustrates a typical phosphate mining area with a surface stream on one side and an unmined area on the other. The average hydraulic conductivity and thickness of the surficial aquifer in the study area were selected as 0.9 m/day and 20.7 m as determined from several aquifer

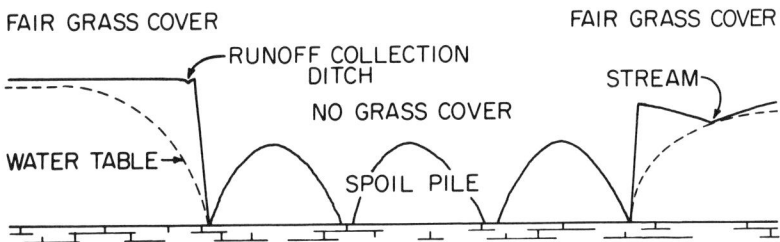

Figure 6.3. Mined scenario, 2 years after mining. (From Garlanger, J. E. Ground water restoration in mined areas, in *SME Annual Meeting Preprint 90-76* (Denver, CO: Society of Mining Engineers, 1990).)

pump tests and numerous test borings. A deep recharge rate of 25 mm/year under a hydraulic head difference of about 25 m (from published data) was judged representative for the study area. The surficial soils were considered to have a fair grass cover, typical of natural pasture land, and a minimum infiltration rate of about 10 mm/hr. An SCS curve number of 66 was selected based on the minimum infiltration rate and the vegetative cover.

The first step in modeling the premining condition was to estimate the net recharge to the surficial aquifer, i.e., the base flow to surface streams. This was done both analytically, using a simple seepage model, and empirically, using measured stream flows. Once the base flow component was estimated, daily water balance calculations were performed using the computer model to determine the daily, monthly, and average annual runoff and evapotranspiration. These calculations were performed for each of the three sections of the affected watershed, i.e., the unmined area, the future mining area, and the stream setback area.

Note that as the distance to the stream decreases, the average annual surface runoff and evapotranspiration values decrease while the base flow increases. The base flow component is negligible in the unmined area of the watershed because as shown by the flow lines, most of the base flow is derived from the land closest to the stream. The weighted average annual surface runoff and evapotranspiration values agree quite well with published data for the modeled region.

Figure 6.3 illustrates a typical mined area prior to reclamation. The overburden soils are redistributed within the mined area as rows of spoil piles. The surrounding areas remain dewatered (as shown) until the mined area is reclaimed, generally 2 to 3 years after mining. Seepage and other water collected in the mine area are consumed within the beneficiation process.

In this scenario, the net recharge to the surficial aquifer, i.e., the amount of annual rainfall that reports as seepage into the mine cut and as deep recharge, was estimated using the simple analytical seepage model. The seepage model was also used to calculate the amount of base flow that is diverted from the stream to the mine cut. The water balance model was then used to determine the other hydrologic parameters, i.e., surface runoff and evapotranspiration.

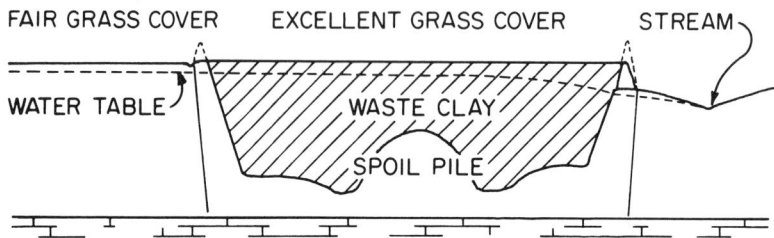

Figure 6.4. Final reclaimed scenario, waste clay area. (From Garlanger, J. E. Ground water restoration in mined areas, in *SME Annual Meeting Preprint 90-76* (Denver, CO: Society of Mining Engineers, 1990).)

As expected, the mined area experiences the greatest impact with respect to all of the hydrologic quantities. Surface runoff is virtually zero because of containment in the mine pit. Evapotranspiration decreases substantially because of the lack of vegetation. Base flow becomes negative as groundwater flows toward the mine both from the unmined area and the stream setback area. Furthermore, because the water table is drawn down to pit bottom, deep recharge is reduced significantly. As can be seen, the impact of mining on the adjacent unmined area and the stream setback area is also significant.

The waste clay generated during the beneficiation process is pumped to a mine pit, which is surrounded by earthen embankments constructed of the overburden soils. A typical waste clay pond is schematically shown in Figure 6.4. The waste clay usually contains a high level of nutrients essential for good vegetative cover. An excellent grass cover was considered in the modeling, as per normal practice. The consolidated waste clay has a relatively low coefficient of permeability, on the order of 10^{-0} to 10^{-7} cm/sec, and consequently, has a low infiltration rate. Based on a minimum infiltration rate of 1 mm/hr and considering an excellent grass cover, an SCS curve number of 80 was assigned to the reclaimed ground surface of the waste clay area.

The computations for this scenario were fairly straightforward. Base flow in the stream setback area was estimated using the seepage model in combination with the water balance model. The surface runoff and evapotranspiration were then estimated using the water balance model.

Note that the waste clay, being a relatively impervious soil, does not contribute to the base flow of the stream. Most of the base flow to the stream is derived from rain that falls on the embankment and within the stream setback area. Because groundwater inflow from the clay pond area is decreased, the water table is lowered, net recharge is increased, and evapotranspiration is reduced significantly in the stream setback area. In the reclaimed clay pond and unmined area, the evapotranspiration and surface runoff are close to premining values. Interestingly, the weighted average hydrologic parameter values for this reclaimed scenario are also close to the premining values.

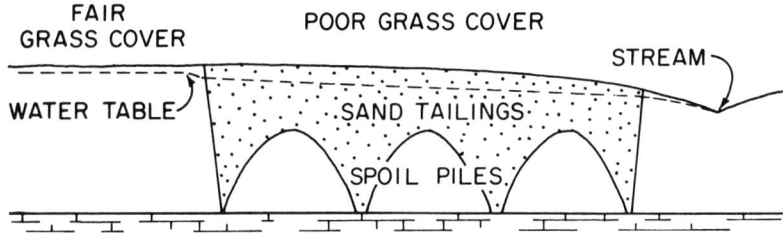

Figure 6.5. Final reclaimed scenario, tailings area. (From Garlanger, J. E. Ground water restoration in mined areas, in *SME Annual Meeting Preprint 90-76* (Denver, CO: Society of Mining Engineers, 1990).)

Sand tailings generated during the beneficiation process are also deposited in some of the mined areas, generally up to the surrounding ground level. A typical sand tailings area is illustrated in Figure 6.5. The sand tailings generated in Florida are uniformly graded quartz particles with a fines content (percent passing the U.S. No. 200 sieve size) usually <8%. Consequently, the sand tailings have a high coefficient of permeability, on the order of 10 cm^2/sec, and hence, a high infiltration rate. A minimum infiltration rate of 11.4 mm/hr was selected for this study. The sand tailings also contain some nutrients essential for vegetative growth. However, because of a low water retention capacity (i.e., available water for vegetative demand), sand tailings cannot sustain a good vegetative growth. Therefore, poor grass cover was selected for the modeling of the sand tailings areas. An SCS curve number of 73, based on the high minimum infiltration rate and the poor grass cover, was assigned to the reclaimed ground surface in the sand tailings area.

The computation procedures for this scenario are similar to those used for the waste clay scenario. The seepage model was first used to estimate baseflows. The water balance model was then used to determine surface runoff and evapotranspiration.

The results of the modeling indicate that evapotranspiration from the reclaimed tailings pond is reduced significantly from the premining condition. Even though the sand tailings are permeable enough for most rainwater to infiltrate, the available storage in the sand tailings is used up early in the rainy season, subsequently causing relatively high surface runoff. Because of the high transmissivity, base flow to the stream under this scenario is also relatively high. Note that the recharge from the tailings pond area keeps the water table in the stream setback area at a higher level. Because the water table has not changed significantly from the premining level, deep recharge remains at the premining rate. On average, base flow is higher but evapotranspiration is lower than for the premining condition.

After reclaiming most of the mined areas as either waste clay ponds or sand tailings areas, some of the mined areas are reclaimed as land and lakes. In land and lakes reclamation, the perimeter of the mine pit is graded to a slope of 4 horizontal to 1 vertical or flatter and is vegetated. The mine pit is then allowed to fill up with

268   ENVIRONMENTAL IMPACTS OF MINING

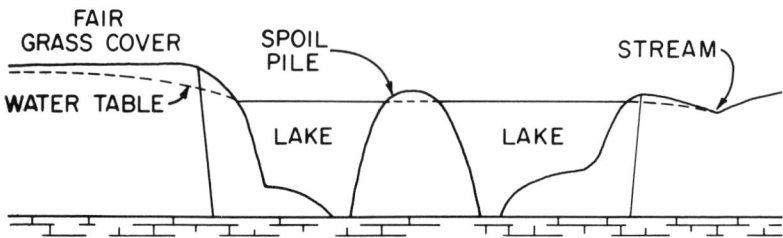

Figure 6.6.   Final reclaimed scenario, land and lakes. (From Garlanger, J. E. Ground water restoration in mined areas, in *SME Annual Meeting Preprint 90-76* (Denver, CO: Society of Mining Engineers, 1990).)

rain- and groundwater. A few tall spoil piles form islands within the lake. In some cases, artificial wetlands may also be formed along the perimeter of the land and lakes. A typical land and lakes reclamation scheme is illustrated in Figure 6.6.

As in the other scenarios, seepage into the lake and base flow to the stream were estimated using the analytical seepage model and runoff and evapotranspiration were estimated using the water balance model. Lake evaporation in Florida is typically on the order of 1270 mm/year. Evapotranspiration from the land portion of the land and lakes area was considered to be insignificant relative to the lake evaporation. The surface runoff was calculated by simply subtracting evaporation, base flow, and deep recharge from the rainfall.

As shown, the surface runoff from the lake (i.e., the overflow to the stream) is small compared to the surface runoff from the unmined areas. The evapotranspiration and surface runoff from the unmined areas are smaller than, but fairly close to, the natural background levels. On average, however, the evapotranspiration and surface runoff from the affected watershed are significantly different that the premining levels. The base flow component is also relatively high. The deep recharge rate remains at the premining level.

Table 6.1 summarizes the average annual hydrological quantities for the affected watershed under premining conditions, during mining, and for the various reclaimed conditions. After reclamation, however, the waste clay scenario provides hydrological values closest to premining conditions, indicating virtually no impact on the ground and surface water resources of the watershed. The reclaimed sand tailings scenario results in higher average base flow and lower evapotranspiration than premining levels. The reclamation scenario with the largest difference between premining and postreclamation scenarios is the land and lakes scenario. This scenario has a higher base flow to the stream and higher evapotranspiration/evaporation than the premining watershed. Consequently, surface runoff is relatively small as compared to the premining level.

Because all three reclamation options are generally used within a particular watershed, the overall impact of reclamation on surface and groundwater resources is intermediate between the range of impacts presented above for each option.

Table 6.1. Weighted Average Hydrological Quantities (mm) for the Affected Watershed

	Natural Conditions	During Mining	Postreclamation		
			Tailings Area	Waste Clay Area	Land and Lakes
Rainfall	1445	1445	1445	1445	1445
Surface runoff	371	51	391	363	160
Evapotranspiration/ evaporation	993	726	902	1011	1151
Base flow to stream	56	<173>	127	46	109
Deep recharge	25	10	25	25	25
To beneficiation plant	—	775	—	—	—

From Garlanger, J. E., Ground water restoration in mined areas, in SME Annual Meeting Preprint 90-76 (Denver, CO: Society of Mining Engineers, 1990).

Simple analytical seepage models in conjunction with a computerized water balance model and some evaluation have made it possible to rationally measure the impact of mining and subsequent restoration of the surficial aquifer on the ground and surface water resources of the watershed affected by mining. The water balance program used for the analyses presented in this chapter was developed for evaluation cover requirements for sanitary landfills. Considerable experience and judgment were required in applying it to the mining and reclamation scenarios in this chapter. A more comprehensive model capable of integrating groundwater seepage calculations with water balance calculations would be very helpful for studies of the type presented herein. It is to be emphasized, however, that a computer model can never replace actual field observations in evaluating the impacts of mining and reclamation on the surface and groundwater resources of an area.

HYDROLOGIC IMPACT OF PHOSPHATE GYPSUM DISPOSAL AREAS IN CENTRAL FLORIDA

Large quantities of phosphate rock are mined and processed in central Florida. The phosphate rock from the mines is further processed in "chemical plants" to produce phosphoric acid. A by-product of the processing of the phosphate rock to produce fertilizer chemicals is an impure form of gypsum referred to as phosphogypsum. For each ton of phosphate rock processed, approximately 1.5 tons of phosphogypsum is produced. The typical method to dispose of this by-product gypsum is to stack it in large piles, locally referred to as gypsum stacks or gypsum fields.[3]

The gypsum stacks have been the focus of many studies in recent years in an attempt to identify their potential for creating groundwater and/or air pollution.

As expected, most gypsum disposal stacks are located as close as possible or practical to the chemical plant in order to keep pumping costs to a minimum. They are often located next to the mining area. A typical gypsum stack is 400 to 600 acres and has an associated cooling water pond of approximately 250 acres. The gypsum slurry is transported from the chemical plant to the top of the stack using acidic process water. The gypsum slurry is deposited on the top of the stack, the gypsum settles out, and the process water is reused. The process water used to transport the gypsum to the top of the stack is recirculated to the plant generally via the cooling water pond. The process/cooling water is acidic, containing sulfuric and phosphoric acid from the digestion of phosphate rock with sulfuric acid.

In most cases runoff from the side slopes of the stacks is collected in ditches surrounding the perimeter of the stacks. The process water is returned to the chemical plant for reuse in unlined ditches or pipelines.

Typically, once the stack reaches a height of 100 to 150 ft, another stack is started in a new location and/or the existing stack is expanded. However, due to difficulties in obtaining permits, recently some stacks have been proposed for heights of up to 200 ft.

In the past these plant facilities were generally located in areas away from population centers. However, in recent years, Florida has experienced unprecedented growth and areas which were once remote and removed from population centers are now being surrounded as suburbs grow.

The upper surficial aquifer and the Floridian aquifer are the principal groundwater sources in central Florida. In most instances these two aquifers are separated by a confining bed which may have an intermediate aquifer system. Underlying the lower Floridian aquifer is another confining bed. The upper surficial or water table aquifer is principally composed of sand, clayey sands, and, in some areas, shell and gravel beds.

The Floridian aquifer consists principally of porous limestones. The confining units are generally sandy or silty clays, clays and marls, and/or dense limestones and dolomites or dolosilts.

The surficial aquifer is unconfined and rises or falls in response to rainfall and discharges to streams and underlying aquifers. The water level of the surficial aquifer lies below the land surface generally from about 4 to 10 ft in the area of most of the gypsum stacks.

The water in the Floridian aquifer is generally confined. Recharge to the Floridian aquifer is principally by lateral flow, leakage through confining beds, and recharge in Karst regions of Florida. Fortunately, most gypsum stacks are located in an area of low recharge to the Floridian aquifer. The general natural flow of groundwater in the central Florida phosphate district is southwestward toward the Gulf of Mexico. Since about 1975 the U.S. Geological Survey (USGS) has monitored and mapped the wet and dry season potentiometric level of the Floridian aquifer. During the winter months agricultural pumpage in southcentral Florida can reverse the discharge flow.

The sandy surficial sediments which comprise the water table (surficial) aquifer are typically 5 to 50 ft thick. These surficial sediments are underlain by 20 to 80 ft of innerbedded phosphatic, sandy, shelly, clayey, marley sediments that comprise the

Pliocene Bone Valley formation. The Miocene Age Hawthorn Formation underlies the Bone Valley Formation. The Hawthorn is an impure marine dolomitic limestone which contains varying concentrations of phosphate and quartz sands, clay, marl, and dolomite and reaches thicknesses of 100 ft. In many areas the lower portion is an intermediate aquifer-producing zone. Underlying the Hawthorn is the Miocene Tampa Formation. The Tampa is similar to the Hawthorn, but contains less dolomite and has more clay beds. The Tampa ranges from a few feet thick to 100 ft thick. Portions of the upper Tampa are a thick sequence of Oligocene to Eocene aged limestones. These limestones are hundreds of feet thick and comprise the principal Floridian aquifer. Granular evaporites generally underlie the Floridian aquifer.

The processed water from the chemical plants that is used to slurry the gypsum to the disposal area is highly acidic (pH of 1.4 to 1.8) and has a high dissolved-solids concentration at about 28,000 ppm. The predominant contaminants are sodium, phosphate, fluosilicates, hydrogen, and sulfate.

Native groundwater has a dissolved-solids concentration of approximately 500 ppm with a pH of generally <7.0.

Migration of radionuclides, fluosilicates, phosphates, and trace metals is easily precipitated as the acid is neutralized by the carbonate in aquifer fabrics.

Recent monitoring data for some operating plants indicate the chemical front is slowly creeping out from the field as the "carrying or absorptive properties" of the aquifer fabric are reached. As a result of the increasing chemical fronts, regulatory agency personnel are putting increasing pressure on the operators to contain/prevent the leaking of process water from the gypsum stacks.

In the past, the gypsum disposal fields were constructed either on natural unmined land or in many cases directly in the mined lands associated with the phosphate mining process. This meant that the gypsum was deposited directly upon the existing land surface or on the top of the Hawthorn Formation (Figure 6.7). However, during the past 10 to 15 years, several approaches have been taken to locate the stacks in areas which would alleviate the potential for groundwater contamination. Initially, to protect the surface water, ditches were dug around the stacks to collect the runoff and seepage from the slopes of the stacks. This worked to collect the surface water runoff from the gypsum disposal areas.

In the early 1980s attempts were made to site stacks in areas in which naturally occurring thick clays could be used as a natural liner. In some areas of the central Florida district, the Hawthorn Formation is very impermeable and is quite thick. In the early 1980s USS Agri-Chemicals used a modification of this approach in an area in which the Hawthorn was very impermeable and waste clays existed. In addition, a ditch was dug around the stack to prevent lateral migration of leachate. However, the water level in the ditch had to be carefully controlled to prevent migration of contaminated groundwater from the stack into the surrounding surficial aquifer.

In the mid-1980s Gardinier Chemical proposed an extensive system consisting of a compacted clay liner and underdrains overlying a thick sequence (15 to 20 ft) of naturally occurring Hawthorn clays in their permit application for a new gypsum

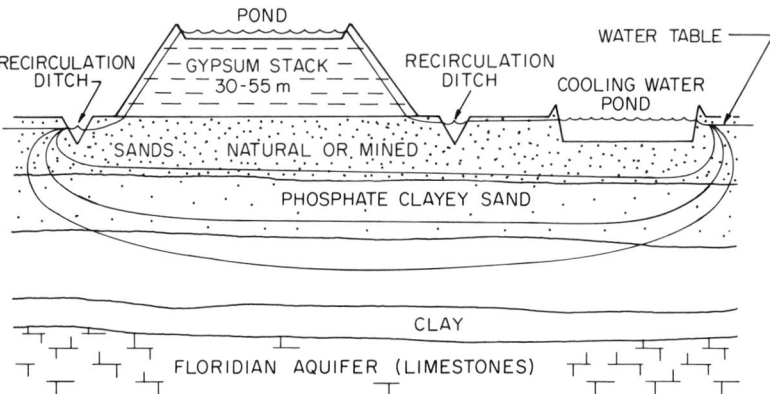

Figure 6.7. Gypsum stack over existing surface. (From Garlanger, J. E. Ground water restoration in mined areas, in *SME Annual Meeting Preprint 90-76* (Denver, CO: Society of Mining Engineers, 1990).)

stack. This was a very elaborate system of underdrains, liners, slurry walls, etc. Due to the Gardinier Chemical Plant's location on Tampa Bay and proximity to nearby population centers, these measures were required to ensure that the stack would be permitted and that the groundwater would be protected. More recently, IMC Fertilizer (IMCF) has proposed the construction of a new gypsum disposal stack. Initially, IMCF proposed more conventional stack construction techniques in which the stack would be built directly upon the Hawthorn Formation in a mined-out area. Recently, due to increasing pressure from the regulatory agencies, it has revised its plans and proposed the installation of a synthetic liner beneath the stack and a slurry cutoff wall along portions of the cooling water pond (Figure 6.8). The State of Florida is presently considering a measure requiring all new gypsum stacks constructed in Florida to be built with a liner to protect the groundwater.

In summary, recent studies have indicated that some potential groundwater impacts are associated with phosphogypsum disposal areas in Florida. Most of these studies have indicated that the lateral groundwater impact to the surficial aquifer system extend beyond the existing nonlined gypsum disposal stacks for approximately 1500 ft. In some cases contamination has been reported in the intermediate aquifer system. The various regulatory agencies including the Environmental Protection Agency (EPA), FDER, and other state and local governments have continued to increase the pressure for permit applicants to design gypsum stacks which will protect the groundwaters of the state. In the past 10 years gypsum stacks have been designed and sited in order to use the natural confining layers and buffering sedi-

HYDROLOGIC IMPACT 273

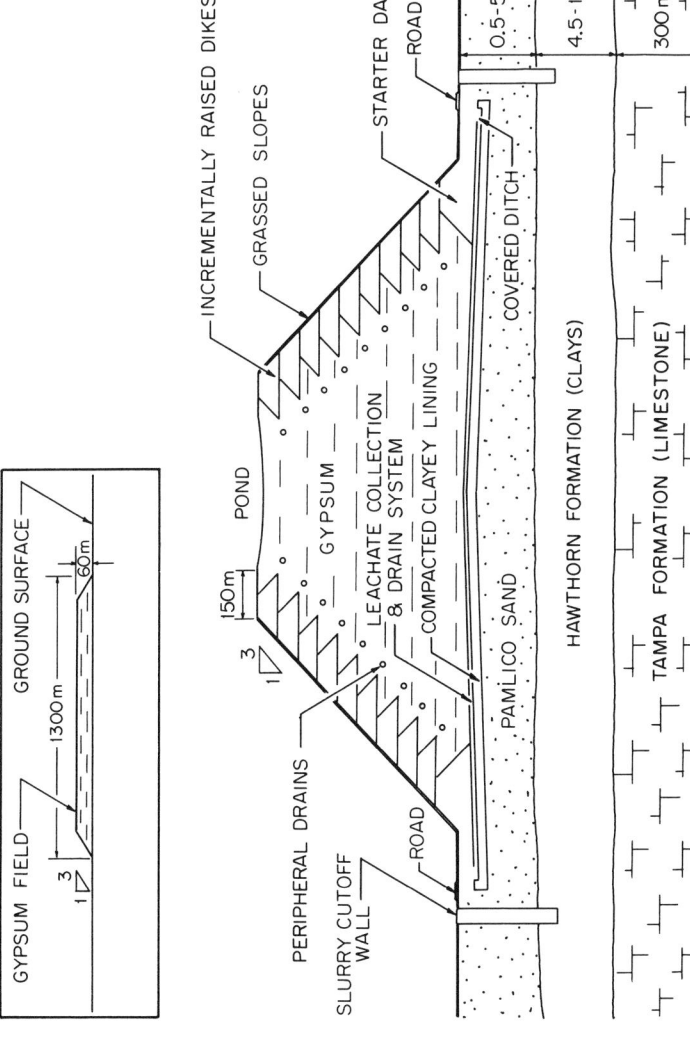

Figure 6.8. Modified gypsum stack to protect groundwater. (From Garlanger, J. E. Ground water restoration in mined areas, in *SME Annual Meeting Preprint 90-76* (Denver, CO: Society of Mining Engineers, 1990).)

274 ENVIRONMENTAL IMPACTS OF MINING

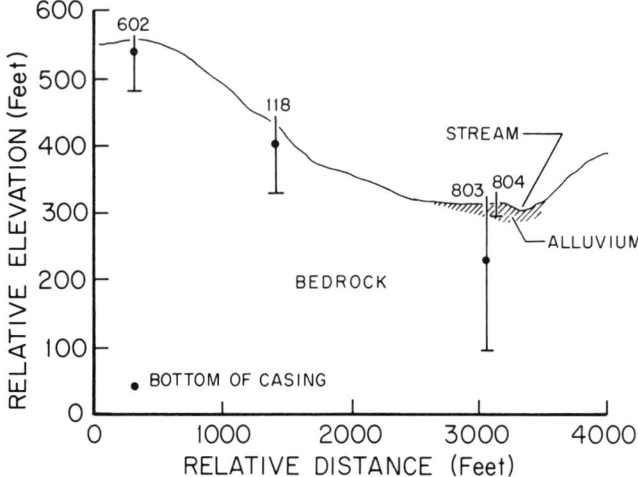

Figure 6.9. Water-level hydrographs. (From Stoner, J. D. Probable hydrologic effects of subsurface mining, *Ground Water Monit. Rev.* Winter:128–136 (1983).)

ments that occur in Florida. Artificial compacted clay liners and slurry walls, and, more recently, synthetic membranes overlying the natural confining carbonate sediments to mitigate and control the leachate from gypsum disposal systems have been designed. The results of groundwater monitoring for these newly proposed stacks, once constructed, will be used to determine the next generation of controls and constraints to be applied to gypsum stack permit conditions.

HYDROLOGIC EFFECT OF SUBSURFACE COAL MINING IN THE APPALACHIAN REGION

The bedrock in the region consists of interbedded sandstone, siltstone, shale, limestone, and coal. Sandstones are usually <6 m thick and limestones are usually <1 m thick. Coal beds typically are <3 m thick.4

The principal economic seams that have been developed include Pittsburgh (PA), Sewickley (PA), and Waynesburg (WV). The eastern part of the Pittsburgh coal reserve has been removed largely by subsurface mining.

Fractures and joints occur in most of the rock types. Surface joints that are often exhibited tend to follow a regional structure.

Precipitation varies from 1 to 1.3 m annually. A part of the precipitation evaporates and transpires, another part flows overland into streams, and the remaining part seeps downward through soils and rock to the zone of saturation. Within the zone of saturation, movement continues downward and laterally toward discharge locations.

HYDROLOGIC IMPACT 275

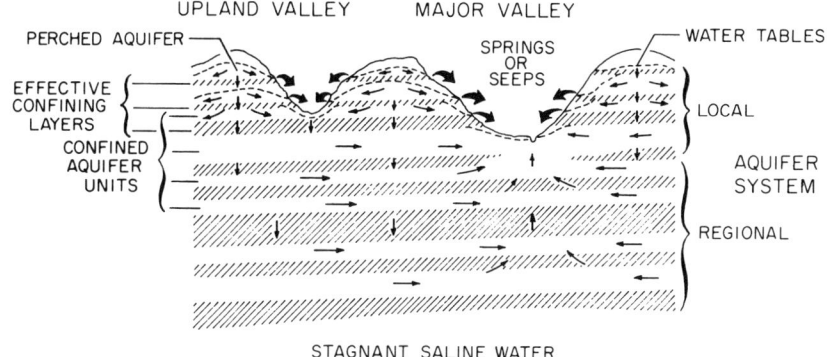

Figure 6.10. Conceptual model of groundwater flow. (From Stoner, J. D. Probable hydrologic effects of subsurface mining, Ground Water Monit. Rev. Winter 128–136 (1983). With permission.)

Water-level hydrographs in the four wells shown in Figure 6.9 showed the greatest fluctuations in the hilltop wells and the least change in the valley wells. The differences were primarily attributed to variations in potential infiltration aquifer storage capacity, permeability, and location with respect to the groundwater flow system.

A conceptual model of groundwater flow based on the data analysis from the region is shown in Figure 6.10. Most groundwater circulation occurs within 45 m of land surface where the more fractured shallow bedrock aquifers occur. Downward movement beneath hills is retarded by low permeability confining layers. At shallow depths below the water table such confining layers can cause perching of groundwater and lateral flow to hillsides where the water discharges at springs or seeps. At greater depths, the hydraulic head is commonly large enough to force vertical leakage through confining layers into the underlying confined aquifer unit.

Water loss due to the mining of the Pittsburgh coal seam has been reported in a number of cases. Groundwater levels have been recorded in wells located over three active underground mines. The type of mining and the distance between the well bottom and the mined coal seam differs at each site. Generally, the magnitude of water level decline is expected to be inversely proportional to the thickness of bedrock between the mine and the well bottom. In addition, retreat (pillar removed) and longwell methods of mining may induce fracturing above the coal seam and hydraulically connect the shallow aquifers to the deep mine. Such problems may explain why no decline is seen in well GR-118 which is 73 m (240 ft) above a room and pillar type mining operation and a 3-m (10 ft) decline is seen in well GR-543 which is 185 (608 ft) above a longwall face.

A simulation of probable hydrologic effects of underground mining was conducted using a two-dimensional cross-section and a digital finite difference program.

276   ENVIRONMENTAL IMPACTS OF MINING

The simulation results indicated the following:

- Water levels could decline as much as 4.6 m (15 ft) in 46-m (150 ft) deep wells located along undermined villages. The maximum effects of water level decline would occur within 1 year of mining.
- Springs and shallow wells above drainage probably will not be affected.
- Streamflow may be reduced by 6.6 L/sec/km^2 (0.6 ft^3/sec/mi^2) in 1 year after completion of mining. Larger reductions could occur with higher permeability vertical fracture zones.
- The presence of vertical fracture zones could magnify and accelerate the drawdown effects and mine inflow.

EFFECTS OF LONGWALL MINING ON HYDROLOGY

Longwall mining relies upon overburden caving to restabilize the large rock mass that is disturbed during coal extraction. Caving characteristics are known to play a very important role in the surface subsidence and the overall pressure redistribution underground. Since the advent of high-extraction mining techniques, concern for local, domestic, and agricultural water supplies has increased. Detailed assessments of the impact of high-extraction mining techniques to local domestic water supplies are very limited. Prior investigations have addressed changes in well yield and water quality, but few have included a complete hydrological study addressing results both before and after undermining occurs.[5]

The U.S. Bureau of Mines conducted a complete hydrologic study at a mine site to investigate the effects of longwall mining on shallow local water supplies. A total of eight water wells were drilled at the mine site and hydrologic parameters such as specific capacity, transmissivity, and well yield were determined both before and after mining. Electronic water level indicators were installed on each water well to provide continuous observations of water level fluctuations. In addition, periodic measurements were made at select well locations of mining-induced ground movements to characterize changes in the ground surface. Although this information provides a portion of the necessary parameters for the development of a better understanding of the changes to the hydrologic regime associated with undermining, numerous site-specific studies must be conducted before all of the effects are understood. Such studies are inherently important to the development of predictive models.

The study site is located in southwestern Pennsylvania (Greene County). The topography of the site is typical of the Northern Appalachian Coal Region, consisting of hilly terrain, with steep to moderate slopes. The land use in the vicinity of the study site is primarily agricultural. Other areas are relatively inaccessible due to dense vegetation.

The study area overlies two adjacent longwall panels. Panel 1 is approximately 720 ft wide and 7600 ft long and panel 2 is approximately 750 ft wide and 7700 ft long. The overburden or depth of cover over the study site varies from approximately 975 to 585 ft.

According to the USGS, the Greene County groundwater flow system is complex and strongly controlled by secondary permeability in the form of fracture zones and bedding plane fractures. The annual precipitation for Greene County ranges from 38 to 41 in., where 22 to 25% of the mean annual precipitation circulates through the aquifer systems via these fracture zones and bedding plane fractures. Stoner reports that fracture zones or openings tend to diminish as the overburden depth increases and aquifers beneath valleys exhibit larger average hydraulic conductivity or permeability than aquifers beneath hilltops. Also, shallow groundwater systems have been documented to be dependent on seasonal climatological data in relation to flow quantity.

A total of eight (6-in. diameter) wells were drilled and placed strategically above the two longwall panels. The wells were not completed in a unique, individual aquifer; rather they were drilled in a manner similar to that of the local area and were thus completed in the relatively shallow saturate zone. To ensure an open wellbore for the life of the study, Schedule 80 (4-in. diameter) PVC casing was installed to total depth. The wells were positioned roughly perpendicular to the trend of the longwall panels. This alignment permitted observations of mining effects along a profile line extending across both longwall panels. Well Nos. 2, 4, 6, and 8 were located at quarter-panel width. Well Nos. 1 and 5 were placed at the center of each respective panel. Well Nos. 3 and 7 were located above the gateroads between the two panels. The drilling depths of all wells are shown in Table 6.2.

Data were collected from all wells to monitor the hydrologic parameters associated with shallow aquifers before, during, and after mining. In addition, various hydrologic parameters were determined and included well yield, transmissivity, specific capacity, and the water level fluctuations. Initial data collection began 8 months prior to undermining of Well No. 1 to determine baseline data on water levels and hydrologic parameters.

Drawdown and recovery tests were performed before and after undermining on all wells to measure significant hydrologic parameters.

Electronic recorders that monitor continuous water level fluctuations were installed on all wells. The wells were configured to permit continuous recording of water level fluctuations. The set-up consisted of a pulse generator, encoder, and a 12-V battery. Fluctuations of water levels are monitored via the pulse generator and are transmitted to the encoder. The encoder is a battery-operated instrument for the automatic real-time monitoring of water level data and is capable of computing a data point every 10 min up to once every 24 hr. For this study, the encoder was programmed to monitor water levels every 4 hr. Remote communication of the encoder is provided via a modem. The instrument can be accessed by a telephone. On-site communication with the encoder was performed utilizing a portable computer.

Table 6.2 displays the initial or static water levels, well yields, specific capacities, and transmissivity values for the respective wells. Well yield is simply the volume of water per unit of time discharged from a well, either by pumping or by free flow. It is measured commonly as the pumping rate in gallons per minute (gpm). For this

Table 6.2. Premining Data

	Well No.							
	1	2	3	4	5	6	7	8
Well depth (ft)	268.0	227.0	167.0	167.0	167.0	167.0	167.0	167.0
Initial water level (ft)	140.50	202.25	103.00	111.00	71.75	68.00	28.25	14.00
Well yield[a] (gpm)	4.85	NA	5.53	12.07	6.22	4.94	7.34	9.46
Specific capacity[a] (gpm/ft drawdown)	0.071	NA	0.128	1.069	0.115	0.085	1.471	0.076
Transmissivity[a] (gpd/ft)	10.67	NA	17.57	127.79	39.80	29.64	712.23	32.43

NA = not applicable due to limitations of equipment used for testing.

[a] All values were obtained utilizing a pump test.

From Owili-Eger, A. S. C., Geohydrologic and hydrogeochemical impacts of longwall coal mining on local aquifers, in SME Annual Meeting Preprint 83-876 (Denver, CO: Society of Mining Engineers, 1983).

study, all well yields prior to undermining were determined by tests. Well yield was calculated as the volume of water that was released or pumped from storage in a given length of time. The specific capacity of a well is its yield per unit of drawdown, usually expressed as gallons per foot of drawdown. Dividing the yield by the drawdown, each measured at the same time, gives the value of the specific capacity. Specific capacities for all wells prior to mining are shown in Table 6.2.

Transmissivity is the rate at which water will flow through a vertical strip of water-bearing formation 1 ft wide and extending through the full saturated thickness, under a hydraulic gradient of 1.00 or 100%. The transmissivity of a formation is especially important because it indicates how much water will move through the formation. It was originally decided to collect transmissivity information via standard pumping tests and an observation well. However, the low yield rates observed precluded the use of the standard pumping test. As an alternative, a procedure was used for testing. The time-recovery data from the pumped well were used for determining the transmissivity of the formation. The time-recovery data for the pumped well are more accurate than the time-drawdown curve because during the recovery period water-level measurements can be made without interference from pump vibration and without the effects of momentary variations in the pumping rate. The premining transmissivity values are shown in Table 6.2.

Postmining well yield data for Well Nos. 5, 6, 7, and 8 were determined through pump tests. Well Nos. 1, 2, 3, and 4 experienced effects or casing restrictions due to the mining of panel 1. Well yield values for these wells could not be determined using a pump test due to these conditions. Table 6.3 displays postmining well yield data for the four remaining wells.

Table 6.3. Postmining Data

	Well No.							
	1	2	3	4	5	6	7	8
Final water level (ft)	248.07	NA	109.41	107.00	67.50	69.00	22.50	26.00
Well yield (gpm)	NA	NA	NA	NA	4.34	5.10	6.91	5.57
Specific capacity (gpm/ft drawdown)	NA	NA	NA	NA	0.094	0.097	1.256	0.056
Transmissivity[a] (gpd/ft)	17.08	NA	47.92	207.16	60.82	96.55	1462.00	45.16
Transmissivity[b] (gpd/ft)	NA	NA	NA	NA	20.64	56.24	776.35	10.95

NA = not applicable due to offset in well casing limiting installation of downhole instruments.

[a] Values obtained utilizing a "slug" test procedure.
[b] Values obtained by utilizing a pump test.

From Owili-Eger, A. S. C., Geohydrologic and hydrogeochemical impacts of longwall coal mining on local aquifers, in SME Annual Meeting Preprint 83-876 (Denver, CO: Society of Mining Engineers, 1983).

The specific capacity of Well Nos. 1, 2, 3, and 4 could not be determined (well yield information was needed and such postmining information was unavailable). The postmining specific capacities of Well Nos. 5, 6, 7, and 8 are shown in Table 6.3.

Postmining transmissivity parameters for all wells were determined utilizing the pump test procedure and a "slug" test procedure. This "slug" test method of determining transmissivity values was used on Well Nos. 1, 2, 3, and 4. In this method a known volume or "slug" of water is suddenly injected into or removed from a well and the decline or recovery of water level is measured at closely repeated intervals. Table 6.3 displays postmining transmissivity data for all wells utilizing both the pump test and "slug" test procedures.

Water level fluctuations were observed to varying degrees during the life of the study. A portion of the variation in fluid level can be attributed to the regional drought that occurred during the study. Some of the effects, however, were caused by mining. It was believed that the effects of mining on the remainder of the well array would be minimal due to the proximity of the wells to the mine workings. Furthermore, it was impossible to isolate the effects of mining in the remainder of the wells due to the masking effects of the regional drought.

As mentioned earlier, postmining values for well yield on Well Nos. 1, 2, 3, and 4 were not performed due to downhole restrictions. Well yield tests on Well Nos. 5, 6, 7, and 8 were performed and showed no considerable change that can be attributed to the undermining of panel 1. A minor change in well yield occurred in Well No. 8. This observation, however, was considered to be insignificant.

A comparison of the pre- and postmining specific capacity measurements for Well Nos. 1, 2, 3, and 4 is limited due to downhole restrictions. Measurements for Well Nos. 5, 6, 7, and 8 were obtained and showed no major increases or decreases due to undermining of panel 1.

Tables 6.2 and 6.3 provide transmissivity values for all wells pre- and postmining, respectively. As mentioned earlier, postmining transmissivity values for Well Nos. 5, 6, 7, and 8 were performed utilizing both a pumping test and a "slug" test procedure. Therefore, a direct comparison utilizing one method of determining transmissivity was not obtainable for these wells. Given that two different testing methods were utilized, Well No. 1, located at the center of panel 1, experienced minimal change in transmissivity. Perhaps subsurface fracturing, which may have occurred as a result of mining, had little effect on Well No. 1. Well Nos. 3 and 4 experienced small increases in transmissivity. Well Nos. 5, 6, and 7 showed slight increases in transmissivity after the undermining of panel 1, whereas a small decrease in transmissivity was observed at Well No. 8 (corresponding to the decrease in well yield and specific capacity previously mentioned).

A dramatic fluctuation in water level occurred in Well No. 1, located over the centerline of the panel, when the longwall was in the vicinity of the well array. In fact, the fluid level began to change when the longwall was about 100 ft from the well. The fluid level began to rise, then fell abruptly when the longwall passed directly below the well. This change is exactly the same as that observed at another test site. Subsequent measurements showed the fluid level beginning to rebound. However, a downhole problem (perhaps an offset or constriction of the casing) occurred when the longwall face was 425 ft beyond the well. This condition limited further data collection. A measurement made 659 days after undermining showed the fluid level to have recovered 20 ft.

Well No. 2, positioned approximately one half the distance between the centerline and the gateroad area, experienced a decline through a large portion of the study. The decline was perhaps related to the regional drought. In fact, the rate of fluid level decline did not substantially change as the longwall passed below. Approximately 27 days after undermining occurred (or when the longwall face was 535 ft beyond the well), the water level dramatically declined below the bottom of the well. Shortly thereafter, the casing offset and further measurements were impossible. It is unknown whether the fluid level recovered because the casing offset occurred at a position some 120 ft above the highest recorded water level for this well.

Well No. 3, located over the gateroad area between longwall panels 1 and 2, showed a decline in fluid level; however, the effects of mining and the regional drought could not be separated.

The following conclusions were drawn:

- Well yield and specific capacity for wells located beyond the current panel (i.e., the panel being mined) appear to be unaffected by mining.
- Transmissivity did not change for the well located in the center of the current panel.

Table 6.4. Initial and Final Water Levels

Well No.	Initial Water Level (ft)	Final Water Level (ft)[a]	Relative Change (ft)
1	140.50	248.07	−107.57
2	202.25	NA	NA
3	103.00	109.41	−6.41
4	111.00	107.00	+4.00
5	71.75	67.50	+4.25
6	68.00	69.00	−1.00
7	28.75	22.50	+6.25
8	14.00	26.00	−12.00

NA = Not applicable due to offset in casing restricting measurement of water level.

[a] Final values are those water levels observed (where possible) on day 400.

From Owili-Eger, A. S. C., Geohydrologic and hydrogeochemical impacts of longwall coal mining on local aquifers, in SME Annual Meeting Preprint 83-876 (Denver, CO: Society of Mining Engineers, 1983).

Changes in transmissivity values for the remaining wells in the array did not appear to correlate with mining.
- The largest water level fluctuations occur in wells that are directly undermined. Changes in water level position appear to be minimal for wells located in adjacent longwall panels.
- Water level fluctuations for wells located in the current panel begin to occur when the longwall face is less than one overburden thickness from the well.
- Water level fluctuations and ultimate water loss occur before the process of subsidence is completed. Fluid level recovery also appears to begin prior to completion of the subsidence process.
- A loss of water may only be a temporary condition; fluid level recovery was observed in all wells affected by mining (with the exception of the well, where measurements were impossible). Furthermore, a loss of water does not imply that all of the overburden rock mass has been lost to the gob zone. Water may indeed be at some level below the bottom of the well (Table 6.4).

REFERENCES

1. El-Ashry, T. M. *Impacts of Coal Mining on Water Resources in the U.S.* (Denver, CO: Environmental Defense Fund, 1976).
2. Garlanger, J. E. Ground water restoration in mined areas, in *SME Annual Meeting Preprint 90-76* (Denver, CO: Society of Mining Engineers, 1990).
3. Gurr, T. M. Hydrologic impacts of phosphate gypsum disposal areas in central Florida, in *SME Annual Meeting Preprint 90-193* (Denver, CO: Society of Mining Engineers, 1990).

4. Stoner, J. D. Probable hydrologic effects of subsurface mining, *Ground Water Monit. Rev.* Winter:128-136 (1983).
5. Owili-Eger, A. S. C. Geohydrologic and hydrogeochemical impacts of longwall coal mining on local aquifers, in *SME Annual Meeting Preprint 83-376* (Denver, CO: Society of Mining Engineers, 1983).

CHAPTER 7

Erosion Sediment Control

Sediment is one of America's greatest pollutants. More than 1 billion tons of sediment reach the major streams of the United States annually.[1] Damages are reflected in the reduced carrying capacity of streams, clogged reservoirs, destroyed habitat for fish and other aquatic life, filled navigation channels, increased flood crests, degraded facilities for water-based recreation, increased industrial and domestic water treatment costs, and premature aging of lakes by enrichment of the water with silt-carried fertilizer that promotes algae growth, destroys crops, and reduces productivity of floodplain soils.

Erosion and sedimentation are natural processes that are usually gentle actions releasing a controlled amount of silt from watersheds to receiving streams. Surface mining activities accelerate these natural processes and short-duration, high-intensity storms can become a violent force moving thousands of tons of soil in a brief period of time. Cover is a very important factor. With the removal of ground cover, water moves across the denuded area on its own terms, picking up soil particles as it flows and leaving gullies behind. The susceptibility of strip-mined land to erosion depends on:

- Physical characteristics of the overburden
- Degree of slope
- Length of slope
- Climate
- Amount and rate of rainfall
- Type and percentage of vegetative ground cover

284   ENVIRONMENTAL IMPACTS OF MINING

Via development of erosion and sedimentation control plans before disturbance of the area, many of the detrimental effects of strip mining can be prevented.

PRELIMINARY SITE EVALUATION

The preliminary site evaluation includes a thorough surface reconnaissance of the potential development site and familiarization with local geology, hydrogeology, and soil characteristics. Wherever extensive grading is anticipated, a preliminary subsurface investigation is desirable to provide information on the geologic, soil, and groundwater conditions.[2]

For a good preliminary site evaluation, topographic, geologic, soils, and zoning maps and aerial photography should be utilized. These maps provide information on physical features relating to erosion and sediment control, and are often used as base maps on which locations of critical physical features and preliminary layouts of the potential development can be indicated.

In addition to maps, available publications on soil and geologic conditions should be utilized. Engineering characteristics generally considered in sediment and erosion control include:

- Depth to bedrock
- Depth to water table
- Soil classifications
- Grain size gradations
- Permeability
- Available water capacity
- Reactions (pH)
- Shrink-swell potential
- Moisture-density relationship

In the soil survey reports, the following aspects are generally covered:

- Suitability as source for topsoil
- Pipelines (construction and maintenance)
- Road and highway locations
- Pond and reservoir sites
- Dikes, levees, and urban embankments
- Drainage systems
- Irrigation
- Terraces and diversions

Assessment of prime physical features critical to erosion and sediment control is required in preliminary site evaluation. These features should be studied and delineated on a site map.

Land Type

The proposed mine development site should be categorized into three basic land types: barren area, agricultural area, and woodland area.

Barren areas are nearly or totally void of any vegetation. These areas often require considerable grading and elaborate vegetative and structural measures to control erosion and sedimentation.

Agricultural areas are open areas under cultivation or are potentially cultivable. Unless these areas are planted in grain crops, agricultural areas generally support a stand of grasses, legumes, or herbaceous plants.

Woodlands are described as mature stands, pole stands, and mixed stands. Mature stands generally contain trees with trunk diameters of 10 in. (25 cm) or greater. Pole stands are thick stands of tall, small, round trees with trunk diameters between 6 to 10 in. (15 to 25 cm). Mixed stands contain both mature- and pole-sized trees. Mixed stands are generally found in woodlands that have been selectively lumbered, whereas pole stands generally occur on tree plantations or in woodland areas that have been previously nearly cut or burned over.

Mature trees with full crowns have the highest aesthetic value. However, they are less likely to recover from injury than smaller and younger trees. Mature trees, due to their crown size, bark structure, trunk strength, and broad root structure are more capable of resisting changes in wind and sun exposure resulting from extensive clearing. Crowded pole-sized trees often have constricted crowns and root systems in proportion to their height. This condition reduces their ability to withstand exposure to intense sun and wind.

Soil and Rock

The presence of highly erodible soils is a very important physical feature, especially if these soils occur on moderate to steep slopes. It is not always possible to easily recognize a highly erodible soil horizon, as a stand of vegetative cover can mask the soil layer or the erodible soil can occur beneath a surface soil of different character.

Highly erodible soils are usually characterized by a deficiency of soil particles that have cohesive strength. This cohesive strength is usually a function of the clay-sized (colloidal) fraction of the soil horizon. The presence of highly erodible soils should be confirmed at an early stage of the site survey.

Occurrence of rock outcrops in the proposed development site should be noted. When rock is excavated, the rock can be stockpiled for use in erosion and sediment control.

Streams

Streams are very often the recipient of sediment from the development site as well as the means of transporting it to private and public properties downstream from the mine development site. Streams themselves can contribute to the sediment load through channel degradation and bank erosion.

Three major factors contribute to increased stream erosion:

- Restriction of the stream channel due to sedimentation
- Increased runoff due to decreased infiltration in the runoff area
- Destruction of the natural vegetation along streambanks

For small streams flowing through major development areas, the erosional effect of increased runoff is certainly a major consideration in sediment and erosion control.

The stream gradient will, to a large extent, affect its sediment transport capabilities. Wide floodplains, meandering courses, and sediment buildup in the channel indicate shallow gradient. Increased sediment load will cause additional sedimentation of the channel.

Streams are dynamic entities in nature. Room for normal channel migration and adjustment to newly imposed runoff stress must be maintained.

Where the streambanks are high and steep, additional runoff from the watershed causes serious streambank erosion problems. Every attempt should be made to preserve or enhance the vegetative cover of streambanks, especially grasses, sedges, and woody shrubs with dense fibrous root systems. In poorly vegetated areas, it may be necessary to flatten the slopes and establish a good vegetative stand in order to control streambank erosion.

Floodplains

During periods of intense runoff, floodplains become inundated by flood flow and act as an extension of the stream itself. Integrity of the floodplain must be preserved.

Impoundments

Impoundment structures can be utilized for storm water retention and sediment collection. They should have sufficient area and capacity. Natural impoundments such as lakes are aesthetically valuable physical features and should be protected against sediment damage. During the preliminary site evaluation, the existing conditions of the lake and shoreline should be evaluated in order to determine possible effects of sedimentation on the ecological and physical features of the impoundment.

In designing manmade ponds, the pond capacity should be sufficient to handle the anticipated water flow. When the impoundment capacity is impaired by sediment buildup, the sediment should be removed and disposed in a manner that will not reintroduce it into the system.

Groundwater Conditions

Groundwater conditions affect erosion and sedimentation. Groundwater seepage prevents the development of a vegetative cover and causes soil to slough into the ditches where it is directly introduced into the drainage system. The presence of springs and mottling in the soil indicates a high water table.

Vegetative Cover

A dense vegetative cover of grass, weeds, shrubs, vines, or trees is very effective in preventing erosion on steep slopes, swales, and along drainageways and impoundment waters. Vegetation should be evaluated in terms of its benefit to erosion and sediment control.

Vegetative cover along waterways and impoundments must be protected, as it is both a soil stabilizer and a filter for sediment-laden water flowing into watercourses.

PLANNING

In the planning stage, a course of development is formulated. The planning stage is divided into four progressive steps:

- Preliminary site investigation
- Preliminary design
- Subsurface investigations
- Final design

Preliminary Site Investigation

As discussed earlier, during preliminary site investigations the critical physical features must be evaluated in terms of their relationship to erosion and sediment control. These features should be delineated on a base map.

The complete drainage system should be shown on the topographic map. Important woodland tracts must be delineated.

A complete detailed topographic map should be utilized to record the critical physical features.

Preliminary Design

During the preliminary design phase, the mine development plan should be made in a manner that will minimize damage to physical features critical to erosion and sediment control. Grading damage should be minimized as avoiding steep slopes, which result in high cuts and fills, and by following natural ground controls as closely as possible.

In locating drainageways, care must be taken to ensure that the resulting channel gradient and related discharge velocity will not cause erosion of the vegetative drainageway liner.

Subsurface Investigations

The subsurface investigation on geological features and soil characteristics should be utilized to determine the erodibility of the soils and their capability of sustaining a long-term vegetative cover.

As a general rule of thumb, a 50% (2:1) slope is assumed to be the maximum slope upon which vegetation can be satisfactorily established and maintained. However, a maximum vegetative stability cannot be attained on slopes steeper than 33% (3:1). Optimum vegetative stability requires slopes of 25% (4:1) or less. The maximum slope should be applied only to ideal soil conditions in which the soil is not highly erodible and has an adequate moisture-holding capacity.

For droughty soils (those that exhibit a poor moisture-holding capacity due to excessively high permeability and a low percentage of fines) and for highly erodible soils, the maximum permissible slope should be considerably <50%.

Droughty soils generally have <30% fines. When these soils are encountered in cut areas and where reconditioning by the addition of fines or suitable topsoil is not planned, the cut slopes should not exceed 35% (3:1) so that a suitable vegetative cover is established. Furthermore, these soils must be planted with warm season, deep-rooted, drought-resistant grasses and legumes suited to that particular region. For more drought-resistant soils with >30% fines, conventional cool season grasses and legumes of the region can be utilized.

Where fills are to be constructed with droughty soils and where some finer grained, drought-resistant soils are available, a portion of the drought-resistant soil should be segregated for later use in topdressing the fill.

Soils containing excessive amounts of fines, especially clay-sized particles such as clays and clayey silts, can also be difficult to stabilize with vegetation. The dense structure of many of these soils inhibits root development and moisture penetration. Cut slopes in these types of soils should be kept as flat as possible in order to enhance infiltration. On flatter slopes, where sloughing will not occur, the slopes should be dressed with topsoil or other suitable soil, or the existing soil can be upgraded by the addition of organic material and fertilizer. On steep cut slopes, the existing soil should be reworked as the cut progresses and while the slope is accessible to scarification, spreading, and compaction equipment.

The vegetation growth on any soil slope can be enhanced by roughening the soil surface. This practice helps germination by reducing sheet erosion and increasing water infiltration. The soil surface should be roughened along the contour in order to reduce the chance of rilling.

Since a long-lived vegetative erosion control cover on critical slopes can be achieved via the establishment of locally adaptable ground covers and shrubs, it is desirable to include seeds of these plants along with the conventional hydroseeding of grasses and legumes. Overplanting grassed slopes with ground covers or shrubs before vegetative deterioration occurs results in slope erosion. Planting directly to ground covers any shrubs using erosion control mattings or marshes to prevent erosion during the period is also a good practice.

Soil erodibility depends upon several physical features. The relative proportion of sand, silt, and clay in the soil, the organic content of the soil, the soil structure, and the permeability of the soil are major factors. After delineating these parameters, the grass erosion measure of soil erodibility is expressed as tons of removed sediments per acre of land surface area.

Well-graded soils generally exhibit a relatively high resistance to erosion because they have both cohesive and intergranular strength. Loose granular, fine-grained soils are highly erodible when exposed on steep slopes. Some types of clay soils are less erodible because they have greater cohesive strength. However, many of the indurated clays and silty clays that contain expansive clay minerals are susceptible to excessive erosion by slaking and alternate wet and dry cycles.

The soil testing program should include the determination of pit and nutrient levels of soils brought to the surface by mining activities, as these soils will support vegetation. In these areas in which toxic soil compounds are commonly encountered, testing must be performed to determine these compounds so that corrective measures can be taken.

Problems involving pH values in soil are common and must always be investigated. Excessively acidic soils will require regular applications of crushed or pulverized limestone so as to support a vegetative cover. In some instances, the use of vegetation with acid-tolerant characteristics is possible. The nutrient level of the soil is influenced by its nitrogen, potassium, and phosphorous content; soils deficient in these nutrients will require regular applications of proper fertilizers selected on the basis of soil tests.

Groundwater seepage is caused by the exposure of the groundwater table and can cause serious erosion and sediment control problems. Where subsurface investigations reveal severe high water table conditions, every effort must be made to minimize disturbance of these potential problem areas. It may be better to avoid disturbance where these conditions exist.

This is especially true with fluid clay formations. On steep cut slopes, seepage can cause sloughing in erodible soils. Excessive seepage also prevents the maintenance of a satisfactory long-term vegetative cover.

When seepage is encountered in cuts, costly structural measures may be required to control erosional problems. Where the seepage is confined to a small localized area, the water generally can be trapped below the surface by using perforated drain pipe, graded stone, and sand filters; it should lead to a disposal area. Where the seepage extends over a considerable distance along a slope, a longitudinal pipe and stone under drain might be necessary.

For use as topsoil, it must contain at least 30% fines (−#200 mesh) and should meet the state standards for organic content, seed content, and noxious weed content. If the topsoil quality does not meet the standard, additional nutrients and chemicals to the soil surface will be necessary. Nutrient and chemical additive quantities must be established on the basis of soil tests.

Final Design

Stabilization of major waterways is an important task in the final design process. Major waterways include all natural or constructed waterways that can be described as either permanent or intermittent streams.

Increased runoff, channel constriction caused by siltation or construction, and destruction of natural vegetation can accelerate waterway erosion.

In large development work the corrosive effect of increased runoff can be controlled through the construction of storage ponds that will collect and store runoff water during periods of heavy precipitation. The ponds should be constructed to allow the gradual release of the stored runoff during low flow periods. Storage ponds also affect infiltration and evaporation, both of which reduce the total runoff. They also collect sediment which would otherwise be deposited in the waterways.

Filling of floodplains must be avoided except at roadway crossings. Properly sized conduits should be placed at crossings. All natural vegetation, whether it be grass, brush, or tree, adjacent to natural waterways must be protected from excavation activities and preserved in its natural conditions.

Vegetation along waterways helps erosion and sediment control in three important ways: the dense rootmat helps hold the soil in place; the foliage (grasses, legumes, and other low-growing plants) and dead litter such as leaves filter out the sediment from the overland flow; and vegetation dissipates the erosive energy from falling raindrops.

In the event the natural conditions cannot induce satisfactory erosion control, induced vegetative and structural practices will be needed.

Vegetative measures are adopted when one or more of the following conditions exist:

- Poor quality vegetative cover
- Relatively flat terrain
- Low stream bed
- Tangential flow
- Low flow velocity
- Fertile soil

Prior to planting the vegetation, the banks should be graded to a fairly flat slope, preferably 25 to 33% grade or flatter. Such excavation will destroy any existing vegetation in the bank, but the bank excavation will increase the chemical capacity. The grading should be performed during periods of low precipitation, and the soil exposure time should be minimized.

Vegetation for planting in the streambank stabilization process should be selected after considering the following factors:

- Erosive forces
- Soil moisture

- Sedimentation
- Soil conditions

In most cases strip planting techniques should be used. The technique involves planting a strip of wet soil-tolerant, highly erosion-resistant vegetation in the critical area of the waterline, and conventional robust-rooted grasses and legumes above the critical zone. For added protection in selected locations, wet soil-tolerant bushes and trees can be planted near the waterline. For the protection of graded and planted areas until a strong stand of vegetation is established, an erosion control melting or blanket can be utilized in addition to normal mulching practices. The local soil and water conservation district, the state forest service, and fish and wildlife agencies can help in selecting local types of vegetation for use in waterway stabilization.

Obstructions such as logs and boulders should not be removed indiscriminately, as they are required by fish and other aquatic life. These obstacles act as natural energy dissipaters. Their removal may increase velocity of flow and thereby can intensify erosion at critical areas. Straightening may be undesirable because the stream gradient can be steepened. Steepened gradient can increase the rate of downcutting in the channel, or it may rejuvenate the downcutting cycle in a stable channel. Headwater gully erosion can increase. The removal of recently deposited sediments from the stream channel is beneficial as it returns the stream to a more stable and more material alignment and channel configuration.

Structural measures for protection of natural waterways against erosion are divided into two types: grade control structures and bank protection structures. Grade control structures are used to control the gradient of the waterway channel so that the velocity of flow is reduced, thereby reducing both channel and bank erosion.

The most common type of grade control structure is the check dam. Check dams are short dams constructed of a variety of materials such as logs, treated lumber, stone, concrete, and synthetic materials. The check dam flattens the slope of the stream and dissipates energy. Stone or concrete is placed in the high energy area at the downstream of the check dam so as to prevent undercutting of the structure. In streams susceptible to flooding, check dams should be used with caution as they reduce flow rates and increase the chance of flooding.

Two types of bank protection structures are used: revetments and deflectors. Revetments comprise a wide variety of both rigid and flexible structures which are used as erosion-resistant facing on streambanks and lakeshores. The flexible type of structure is preferred and is generally more economical for streambank protection. Flexible revetments such as riprap, fabriform mats, gabions, etc., have advantage over rigid revetments such as asphalt paving or monolithic concrete, because they are capable of adjusting to minor changes in foundation conditions without losing their integrity. The most common type of flexible revetment used for streambank protection is randomly placed stone riprap, composed of loose stone placed on sand/gravel filter and/or filter cloth. Other types of flexible revetments, although not nearly as flexible as stone riprap, are gabions, fabriform mats, interlocking concrete blocks, and steel or concrete tetrahedrons. Selection of the revetment type for particular bank condition will depend upon the strength requirements, length of required service, and aesthetic factors. In areas in which the extreme durability of randomly placed stone riprap is unnecessary or in areas in which rock is not readily available, other types of revetments may be necessary, considering also fish and wildlife habitat or aesthetic reasons.

Common types of rigid to flexible revetments include concrete or asphalt paving, grouted stone riprap, and sacked concrete. To be effective, a rigid-type revetment requires a firm, stable foundation and careful construction. Where fills are being protected a high degree of compaction is required beneath the revetment to prevent excessive settlement. To prevent undercutting at the toe, all revetments should be placed below the existing ground surface. In addition, adequately sized loose stones should be placed at the toe for additional safety.

All flexible and most rigid revetments are placed on a grade of approximately 50% (2:1). Some types of rigid revetments, including some varieties of gabions, wood sheet piling, and metal sheet piling, are built with a vertical face. These types of revetments are commonly used where water access, e.g., boat traffic, is essential or are used as retaining walls involving the filling of floodplains.

The other type of bank protection structure, the deflection structure, usually consists of a stone, concrete, or wooden groin that angles outward from the shore in a downstream direction and deflects the current away from a critical area of the streambanks. This type of structure should be used in wide streams in which the deflected current will not damage the opposite streambanks.

Many material types covering a wide range of costs are available for construction of bank and shore protection structures. The following factors are considered for selection of materials:

- Ability of the material to withstand the stress conditions at the site
- Initial cost and availability of construction material
- Maintenance and replacement costs
- Aesthetic considerations

Stabilization of minor waterways is an important requirement in sediment and erosion control. Minor waterways include all natural and constructed waterways such as roadway draining ditches, drainage swales, or diversion ditches, which are not classified as either permanent or intermittent streams.

The location and design of minor waterways are of considerable importance. The waterways that collect and transport the surface runoff to the streams in the watershed can be major sources of sediment pollution if they have been poorly constructed or inadequately maintained.

Whenever possible the natural drainage system should be preserved. When natural waterways are utilized, the natural vegetation system should be preserved. Traffic should not be allowed in the waterways. The natural vegetation may not, by itself, be able to resist the additional erosive force from increased runoff caused by mine excavation and development activities. Additional reinforcement with structural measures and additional planting may be necessary.

The vegetated waterways facilitates the loss of surface runoff through infiltration. Whenever vegetation cover is not adequate, a series of short check dams or a stone lining can be used. The use of vegetative waterways can be limited by factors such as high soil erodibility, steep slope, high water table, excessively droughty soils, and high soil toxicity. Such problems can be alleviated by using a series of check dams to flatten the gradient of the waterway and dissipate the flow energy.

Wet soil conditions due to a high water table can often be resolved by using pipe underdrains. In droughty or toxic soil conditions, undercutting and backfilling or topdressing with nontoxic drought-resistant soil can be utilized. Care should be taken to establish a good bond between the native soil and the placed soil and to compact the placed soil.

Robust-rooted grasses that germinate rapidly and grow fast are most suitable for stabilizing waterways. With adequate maintenance they form a dense rootmat and a dense uniform surface cover which does not restrict the movement of water and benefits both surface water infiltration and transpiration loss of near-surface groundwater.

For protection against erosion in grassed waterways, the following factors should be considered:

- The erodibility of the soil for the proposed slope
- The flow velocity limitation for the vegetation
- The flow resistance of the selected vegetation
- The method of vegetation establishment required to accommodate the volume and velocity of the design flow

In general, seeding is practiced in waterways in which the designed water velocity is <1.3 m/sec. Sodding is used in waterways when the flow velocity is expected to be between 1.3 to 2.6 m/sec. Seasonal considerations may sometimes rule out seeding. If the soil is erodible at the above-mentioned velocities, additional structural measures will be required.

Temporary stabilization measures are required in seeded as well as sodded waterways to protect against erosion until the vegetation is firmly established.

Jute netting is commonly used for temporary stabilization of waterways. The jute netting is placed directly over the prepared seedbed and with proper anchoring it minimizes soil erosion. Due to its thick fibrous composition the jute also functions as a mulch.

Plastic, paper, and fiberglass nettings are also available. They have a longer service life than that of jute netting. However, due to the dense structure of the individual material strands used in forming the nettings, they do not function as a mulch. Therefore, the plastic and fiberglass nettings are used over a long fiber mulch such as straw or hay.

294   ENVIRONMENTAL IMPACTS OF MINING

Subsurface drainage is a common problem in the establishment of vegetation in waterways when using nettings and mulch. Permeable, granular soils often cause piping and subsequent loss of soil from beneath the mulch and netting. Periodic erosion checks must be established across the waterways and beneath the netting.

Soil slopes include natural soil slopes and all denuded cut and fill. Manmade cut and fill slopes are usually constructed at a grade of 0 to 50%. Under some exceptionally good soil and hydrological conditions the grade can be steepened to 67%. A slope of 33% is considered to be the maximum for the safe operation of excavation equipment. Factors usually considered in designing a slope include slope stability, soil erodibility, and the ability of the soil to support vegetation.

Slope stability is determined by soil mechanics. The stability of slopes is analyzed to predict the possibility of landslides.

Soil erodibility is a function of:

- The quality of vegetative and cover
- The soil gradation and permeability
- The degree of soil compaction
- Clay mineralogy
- The slope grade
- The slope length
- The quantity of water collected by the slope

As the length of the slope increases the erodibility of the slope and the water quantity collected increase. Their effect on erosion can be controlled with different types of diversions such as terraces, benches, top of cut ditches, temporary diversion dikes, and interrupter dikes. Benches and terraces break the length of cut and fill slopes and collect runoff water, and direct it to a disposal point.

Diversion dikes are temporary berms of soil placed along the top of cuts and fills or at intervals along graded natural slopes for diverting runoff water away from the denuded slope. The runoff is diverted to a stabilized disposal point such as a level spreader, temporary flexible downdrain, temporary sectional downdrain, or flumes. Diversion dikes are utilized during the development phase, and in the case of fills, they should be maintained until an adequate vegetative cover is established. In cuts, diversion ditches are established at the top of the cut.

Compaction of fills is a critical factor in erosion control. Poor compaction can cause serious problems. As a minimum criteria for erosion and sediment control the upper 1 ft of all fills should be compacted to at least 85% of optimum. The surface of cut and fill should be roughened at a perpendicular direction to the flow to reduce the flow velocity and enhance water infiltration. Discing and light scarification will accomplish this effect. Where the slope is too steep to allow parallel movement of vehicular traffic, cleated dozers traversing the slope can achieve a satisfactory texture on newly placed soil.

The gradation of the soil on the surface of the slope affects both the erodibility of the soil and its ability to support vegetation. For example, many types of well-drained silty sands are highly erodible and may be droughty. When the soil exhibits either or both of these conditions, the configuration of the slope should be adjusted to accommodate this condition, or the slope should be topdressed with an erosion- and drought-resistant soil. As an alternative to topdressing, a suitable soil can be mixed with the existing soil. When a topdressing is applied, the dressing soil must be firmly bonded to the existing soil surface in order to prevent slippage downslope. This bonding can be increased by scarification of the slope.

The quality of the vegetative soil is determined both by the physical characteristics of the soil (density, permeability, moisture-holding capacity) and by its chemical composition (the pH of the soil, the presence of nutrients in the soil, and toxic elements). All those factors affect the quality of the vegetative cover.

When the pH and nutrient levels of the soil are known, it may be possible to adjust these values to a desired level by addition of lime, fertilizers, etc. When the vegetative condition of the soil cannot be modified, topsoil should be placed.

Seasonal factors also influence vegetation quality. Cool, moist periods of the year are favorable for seed germination and plant growth. Different types of vegetation exhibit a widely differing ability to tolerate certain climatic conditions. In a given region, the most suitable vegetation can be determined with the guidance of the local soil and water experts.

In some cases fast-growing annuals are planted as a temporary ground cover until climatic conditions are favorable for the germination and growth of more desirable perennial grasses and legumes.

Soil stabilization measures include both short- and long-term vegetative measures. They are utilized to control water and wind erosion of soil during and after grading operations.

Interim stabilization measures are used to retard erosion over the short-term, i.e., hot summer and winter months, until conditions are more favorable for long-term vegetative stabilization. Such measures include mulching and use of nettings, blankets, etc., along with the seeding of annual grasses. Fiber mulches such as straw, hay, and woodchips, as well as chemical soil binders, are commonly used to stabilize graded areas prior to seeding for permanent vegetation. The chemical soil binders can penetrate and bind the near-surface soil or bind the surface of the soil. Chemical soil binders are used primarily to protect denuded soil from wind and water erosion during delays in grading operations or until permanent seeding is possible.

Some chemical soil binders also function as mulches (e.g., hay) to aid in germination and the growth of seeded vegetation as they conserve moisture in the soil and provide temporary protection against erosion.

The use of woodchips for short-term soil stabilization and mulching has become common. Woodchips are long lasting, due to their weight and shape, and they require little or no tacking to keep them in place.

Fiberglass mulch is also available. All organic and inorganic fiber mulches require some form of attachment in order to prevent them from being blown or washed away. Three methods are commonly used to secure fiber mulches. The first method, called crimping, is used on straw and hay mulches and is carried out by a crimping machine which partially punches the mulches into the soil. The second method of securing mulches, known as tacking, applies an asphalt or chemical binder to the mulch. The third method includes the use of various types of nettings made of jute, plastic, paper, and fiberglass.

Sediment retention structures are designed to remove sediment from the runoff water. The basic function of these structures is to hold the runoff water for sufficient lengths of time so that the sediment settles out of suspension.

The most common type of retention structure is the sediment pond which is excavated along a natural or manmade waterway. Sediment basins are usually considered short-term structures. The capacity of the sediment basin is determined upon the area of its watershed, the topography of the watershed, the infiltration rate of the soils in the watershed, and the regional hydrological factors.

Most states have regulations covering the design and construction of all sizes of sediment basins. These regulations must be reviewed before designing the impoundment structures.

FORMULATION OF AN EROSION AND SEDIMENT CONTROL PLAN

A detailed erosion and sediment control plan should be developed along with mine plans. The plan must be presented for review, approval, and certification to all cognizant agencies empowered by sediment and erosion control legislation.

During mine development the amount of land to be exposed at any time should be minimized. The site should be developed in stages. All clearing, grading, and stabilization operations in an area should be completed before moving into another area. The sequence and scheduling in which development will occur must be clearly established in maps.

The same map that delineates the sequence of development must show the location of all sediment retention structures. The plan documents must indicate sequence of the construction of these structures. All sediment control structures for natural waterways should be installed before any clearing or excavation work is initiated.

Traffic control is important in woodland developments where uncontrolled traffic can cause severe tree damage. Vegetative filter strips along waterways and all undisturbed open spaces should be delineated on a site map and designated as "off limits" areas for all vehicular traffic. For woodland areas, all vehicular traffic will stay within the roadways, access corridors, or utility rights-of-way. All traffic should be restricted from crossings streams or stabilized drainageways except at approved stabilized crossing locations.

All critical areas along streams must be marked on the site maps and the recommended method of stabilization indicated. Stream stabilization work should be scheduled for periods of low precipitation during the growing season and should be performed prior to the initiation of clearing and grading operations in the watershed.

The erosion and sediment control plan must clearly define vegetative practices, both temporary and permanent. The plan must state and show where and when sod, temporary seeding, and permanent seeding are to be used. Specifications should be provided regarding ground preparation, sod quality, seed type and quantity, fertilization, and mulching.

The construction specifications must clearly define the maximum length of time that a graded area can be left uncovered after the completion of grading and also after grading shut downs as is common practice during the winter months. Short-term stabilization practices during any lengthy grading delays should also be specified.

Figures 7.1 through 7.4 illustrate sedimentation control practices.

Operation

The implementation of a sediment and erosion control plan is included in the operation procedures.

Clearing operations should be planned in detail. In wooded sites detailed procedures for removal and disposition of trees should be established after considering any legal aspect of the clearing operations. Salvaged wood products should be used in the implementation of sediment and erosion control practices.

Special attention should be given to the completion of additional clearing required for equipment travel corridors. Corridors for equipment travel should be established across the areas that will not be immediately denuded, especially in denuded areas. The natural filter strip areas should be given careful attention. Disturbances to vegetated floodplains should be avoided.

Stockpile areas should be selected and designated on a grading plan. These areas should be allocated after considering sediment and erosion control requirements such as direct delivery of sediment to waterways, damage to vegetation, and unnecessary destruction of trees. Temporary or interim stabilization of soil stockpiles should be instituted. Critical slopes on stockpiles should be avoided. Stockpiling immediately adjacent to watercourses should not be allowed because the stockpiled material will provide a direct and a high volume source of sediment to storm runoff.

Temporary vegetative measures planned for implementation on major stockpile areas should be established immediately after the stockpile operation is completed. Proper mulching and soil stabilization in conjunction with these seeding operations should be carried out.

As the rough grading operations near completion, the installation of structure and vegetative practices must be promptly accomplished. Diversion dikes, interceptor dikes, filter berms, etc. should be constructed according to plan. The use of woodchips on a cut and fill plan should begin as soon as slopes are available to receive them. Once chemical soil stabilization has been planned, it must be applied as soon as the slopes are completed. The establishment of temporary vegetative cover should be initiated as soon as slopes are available.

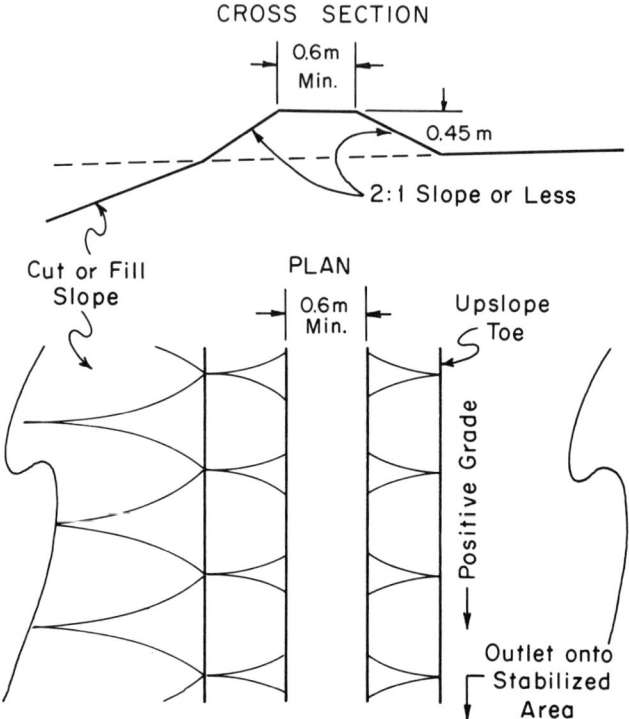

Figure 7.1. Diversion dike. (From Becker, B. C. and T. R. Mills. "Guidelines for Erosion and Sediment Control," U.S. EPA Report R2-72-015 (1972).)

Timely implementation of sediment and erosion control practices is essential. Each day a potential sediment source remains unstabilized, it exists as a source of pollution.

Drainageway protection must be an integral part of the grading operations. Traffic should not be allowed to cross waterways except at specified locations.

As soon as the sediment and erosion control measures begin to function, their maintenance process must be initiated. Sediment removal from structures designed to trap and filter must begin. Prompt replacement of items such as straw bales, woodchips, and seedbeds is vital.

Maintenance

For timely clean-out and stable disposition of trapped sediment from sediment retention basins, specific schedules should be established. A rule of thumb is to clean out a basin when it has lost 50% of its storage capacity due to sediment deposition. The removed sediment should not be indiscriminately piled or dumped because the sediment can move back into the storm drainage system by successive storms.

EROSION AND SEDIMENT CONTROL 299

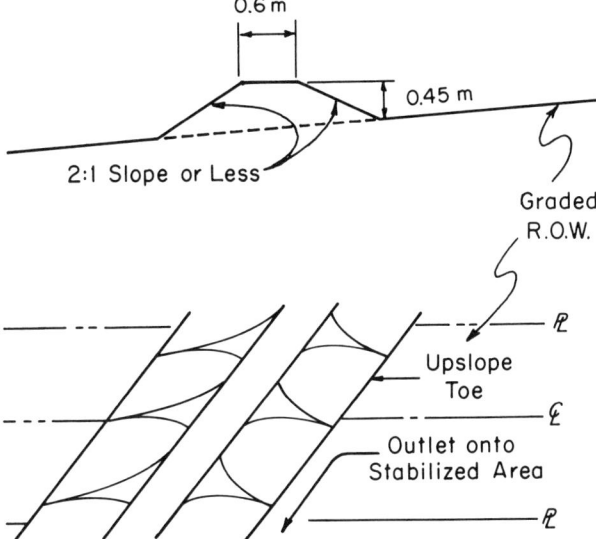

Figure 7.2. Interceptor dike. (From Becker, B. C. and T. R. Mills. "Guidelines for Erosion and Sediment Control," U.S. EPA Report R2-72-015 (1972).)

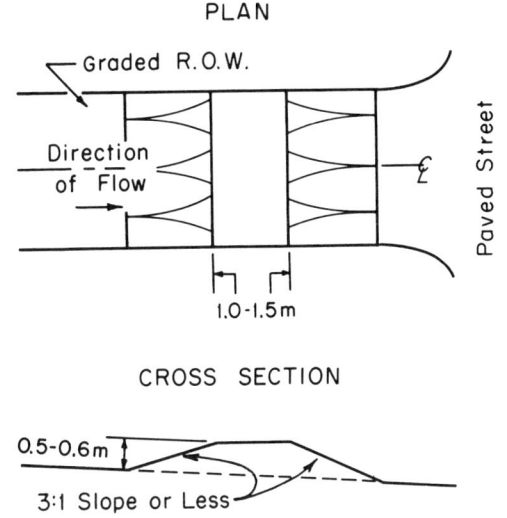

Figure 7.3. Filter berm. (From Becker, B. C. and T. R. Mills. "Guidelines for Erosion and Sediment Control," U.S. EPA Report R2-72-015 (1972).)

300 ENVIRONMENTAL IMPACTS OF MINING

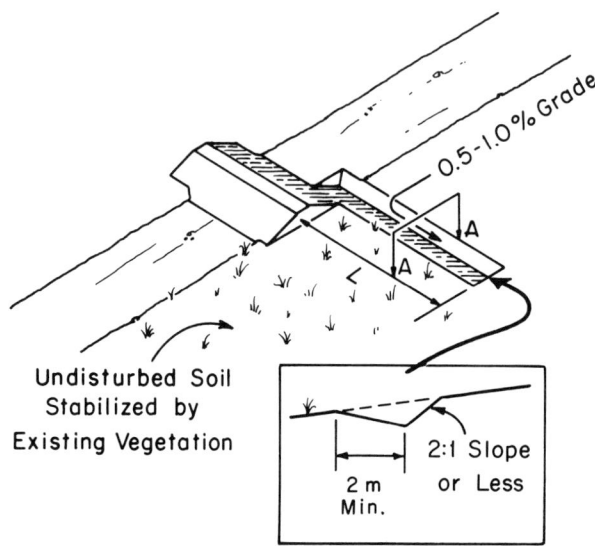

Figure 7.4. Level spreader. (From Becker, B. C. and T. R. Mills. "Guidelines for Erosion and Sediment Control," U.S. EPA Report R2-72-015 (1972).)

Disposition behind a protective berm or filter strip can often suffice if large quantities of sediments are not involved. With large quantities, hauling to a disposal area may be required.

Vegetative practices require maintenance in two general areas. The first is periodic refertilization. In areas in which failures have been experienced in establishing vegetative protection, prompt attention must be given.

SEDIMENTATION CONTROL IN A SURFACE COAL MINE

As operator of the mine, the company is responsible for safeguarding the environment from the impact of surface mining operations. It was recognized from the beginning that siltation of the streams in the vicinity of the mine had to be prevented as part of the mining operations.

The plan adopted was to divert the runoff water into settling ponds where natural gravitational settling would clear the water. It soon became apparent that natural gravitation alone would not do the job. The high clay content of the soil, as carried by the runoff, would become colloidal. These particles carry a negative electrical charge, therefore repelling each other and causing turbid water conditions.

Bench testing indicated that the addition of an organic polyelectrolyte, with a positive electrical charge, effectively caused flocculation of the suspended particles into an agglomeration of particles. Gravitational settling out then occurred. The next phase of the program was to devise a system to accomplish this under field conditions.

The questions that needed answers were

1. What was the rainfall pattern?
2. What was the watershed capacity and probable runoff?
3. What was the correct design level for dams and outlet structures, 10-year, 25-year, and 100-year flood frequency?
4. What was the anticipated silt loading?
5. What was the range of effectiveness of the polyelectrolyte?
6. What hardware was available to continually meter the correct volume of chemical to cause flocculation and water clarification?

A unique system was developed in a coal-mining district in Washington State that effectively answers these questions on an operational basis. The system continually monitors the water flow and automatically meters the correct parts per million of chemical into the turbid water. Rapid mixing follows, then flocculation. Settling out and clarification take place in the quiet water of the second pond. Clear water was decanted into the receiving water.

The coal mine is within the Centralia Coal District from which coal has been mined since the late 1800s, the heaviest period of activity being the days of the coal-burning railroad locomotives. Earlier mining involved different methods. The actual coal field dedicated to this project is contained within a 21,000-acre property. During the projected 35-year life of the operation, approximately 7000 acres was projected to be disturbed.

The land is typical southwestern Washington State rural low-elevation woodland. The topography is a rolling landscape with rounded hills that rise somewhat steeply from flat, poorly drained valley bottoms. Second-growth conifer and hardwood stands cover the hills. Douglas fir is the primary commercial tree species. Red alder is the most plentiful hardwood species, found either mixed with the conifer or in nearly pure stands. Alder is of minor commercial importance. From a forest classification standpoint, the area is classed as a Douglas fir area, with the usual plant species commonly associated in this type.

The valley bottoms consist of small farm units that pasture beef cattle and raise hay. Farming is marginal in these valleys, with the farmer usually "working out" and working the land on a part-time basis. Many of the farmers who have sold their land to the mine remain on their farms as lessees.

As stated, the mine is in southwestern Washington State, west of the Cascade Mountain range. Mild temperatures and a lengthy winter rainy season is the normal climate of the region. The average annual rainfall is considered to be 45 to 48 in./year, with autumn rains beginning in mid-October and continuing almost uninterrupted until mid-April. Snowfall occurs rather infrequently; some winters pass without any snowfall at the lower elevations. Snow depth seldom exceeds 4 to 8 in. and remains on the ground for no more than a few days.

302   ENVIRONMENTAL IMPACTS OF MINING

The rainfall pattern follows a reasonably predictable pattern, i.e., long periods of steady rainfall between early October and late April. During these months, a high probability exists for some daily precipitation. This means that the system must be designed to accommodate extended periods of high runoff as well as sudden storms having peak periods of precipitation.

The main watercourse in the mine locale is Hanaford Creek, which rises to the east of the mine, approximately 15 mi away, in the low-elevation timbered mountains. It flows due west, draining approximately 56 mi^2 before emptying into the Skookumchuck River at Centralia. During periods of high precipitation and runoff, when the Skookumchuck's banks are full, the Hanaford overflows its channel and floods the adjacent pasture land. Winter flows may exceed 500 cfs while summer flows can be as low as 2 cfs.

Modest as it may appear, Hanaford Creek is important to the area and a significant migratory fishery is located there. Near its headwaters and in its mountain locale it becomes a sparkling stream with spawning areas for salmon and steelhead. It is for this reason, and also because it has a very limited assimilative capacity, that such an effort to guard its quality is essential.

The mining operations do not take place in Hanaford Creek proper. However, the mine shop-office, coal preparation plant, power plant, and coal inventory area are located close together on its south bank. Coal stripping takes place along the ridges to the south and eventually to the north.

The minor streams draining these adjacent valleys, with their small watersheds, are tributaries to Hanaford Creek, and these are directly affected by mining activity. Because these tributaries receive the runoff from the mine area, they have been incorporated into a system to prevent siltation of the main stream, Hanaford Creek.

The mining process at the Centralia mine exposes the parent material (overburden) to the weather in a "bottoms up" operation. This parent material is called the Skookumchuck Formation and is made up of marine and nonmarine deposits. It is a fine-grained, readily erodible material.

From the beginning it was recognized that sedimentation is the primary water quality problem in the mine. Because of the low sulfur content of the coal (0.4 to 0.7%), the problem of acid mine water did not present itself. Before mining progressed, drainage settling ponds and ditches to intercept any silt-carrying runoff were proposed to be built. As proposed, these settling ponds would cause natural gravitational settling out of the silt, and only clean water would be discharged into Hanaford Creek. In addition, a program of as early as possible grass seeding was needed to hold the soil in place and also to prevent siltation. The proposal included the possibility that the water would have to be treated "chemically" to correct any condition that may cause degradation of the receiving waters ecosystem. At this juncture the state's newly organized Department of Ecology (reorganized/renamed Water Pollution Control Commission) insisted that the runoff passing through the mine area may become contaminated and thus constitute industrial waste. A Waste Discharge Permit was negotiated along the lines of intended course of action and policy. The Waste Discharge Permit listed the standards with which the mine was to comply.

Fortunately, the management of both the mine and the power plant had the foresight to initiate a water quality study of Hanaford Creek and its tributaries around the mine before mining began. For nearly 3 years, consistently in each season, monthly water samples were taken from many "stations" in the creek area. The information obtained from this sampling program was extremely beneficial as the program progressed from water monitoring to actual water quality control. Cooperating and participating in this premining water monitoring program were the Department of Ecology and the State Department of Fisheries.

Approximately 600 acres was cleared of all vegetative cover, exposing the mineral soil to oncoming rains. Three settling ponds were constructed to receive the runoff from three small watersheds, the largest of these was 1100 acres. The decant from these ponds was to come through a half-round culvert riser with removable boards on the flat side; control of the discharge level was regulated by installing or removing these 4-in. boards. Thus, the decant would always be at the surface, allowing for maximum resident time and maximum gravitational settling of the waterborne silt.

A bench test was conducted to determine the physical and chemical properties of the soil when subjected to runoff conditions. The test strongly indicated that gravitational settling alone would be insufficient to clarify the muddy water and showed that a significant fraction of the soil (clay) that would be subjected to runoff would become near-colloidal and remain suspended in the water for more than 1 week. The actual empirical time for complete natural settling to occur was not determined because the time required extended far beyond the resident time in any of the settling ponds. Residence time in the ponds was a matter of 8 to 23 hr, depending upon the obvious factors of rainfall intensity and storm duration.

Because of the bench test results, it would be necessary to go beyond mere diversion of the runoff into settling ponds. To prevent the mining activity from adversely affecting the watercourses, a method of removing the silt suspended in the runoff and thus decanting only clean water was essential. The problem was to find the "something" that would clarify the muddy water quickly, and to do so effectively under winter weather field conditions. Bench tests indicated that the soil-water suspensions were quite stable; the particles possessed negative charges and would not settle out unless this negative polarity was altered.

To effect clarification of the runoff water, a chemical, Nalco 634, was introduced by Nalco Chemical Co. Nalco 634 is a high molecular weight organic polyelectrolyte with a high charge density, which is effective as a primary coagulant, providing the positive charge that neutralizes negative colloidal particles and also "bridges" these particles to form a floc. It is designed to flocculate silt or other finely divided matter in aqueous systems; it neutralizes the charge on siliceous material and clays, producing a rapid settling floc; it adheres to the substrate and remains in the settled sludge; and it does not add to the biochemical oxygen demand (BOD) of the wastewater because it adheres to the settled solids. The company was able to demonstrate via bench tests that when introduced into turbid water at the rate of 10 to 20 ppm, Nalco 634 neutralized the negative charges of the colloidal clay particles, causing them to flocculate and settle out, thereby rapidly clarifying the water.

The chemical was bioassayed by a recognized fisheries biologist and found not to be harmful to the aquatic life found in the streams at the dilution rates anticipated. The Department of Fisheries had authorized use of this compound.

The first step in clarification of the runoff water was to determine the anticipated silt loading at the point at which the chemical was to be introduced into the runoff water. Settling ponds were arranged in sequence, each sequentially downstream. The heavy silt-laden runoff entered the upstream end of the stream pond. As soon as stream flow velocity diminished, the heavy silt load settled out. Turbidity at this point was over 1000 JTU (Jackson turbidity units), 10,000 to 15,000 mg/L, and 1.5 to 2% solid. By the time the runoff passed through the stream pond the heavier material settled out. At the discharge weir, turbidity was measured to be 85 to 120 JTU, 120 to 130 mg/L, 0.4 to 0.7% solids by volume, and 0.012 to 0.013% oven dry solids.

The Waste Discharge Permit required that the discharge be no more than 5 JTU above normal background. Records indicated that the range in turbidity of Hanaford Creek during the rainy season was between 20 and 55 JTU.

Bench tests indicated that Nalco 634 would effect clarification to the 25 to 50 JTU range if the dosage levels were as follows:

Turbidity (JTU)	Dosage (ppm)
40–80	5
80–120	10
120+	20

These were to be initial guidelines. Experience and continual bench testing were to be the watchword.

It was next necessary to estimate the expected flow rate of runoff at the discharge weir. This was necessary so that the chemical dosage to the gallons per minute flow, parts per million were converted to cubic centimeters per minute. The weir resembled, for the most part, a sharp-crested weir. Stream flow in inches over the weir equal to gallons per minute was calculated:

Estimated Flow Across Weir		Dosage Nalco 634 (cm^3/min)
cfs	gpm	10 ppm
2.2	1000	40
8.9	4000	160
15.6	7000	270

Estimated Flow Across Weir		Dosage Nalco 634 (cm³/min)
cfs	gpm	
26.8	12,000	450
40.0	18,000	680

The average flow rate was estimated to be in the 6000 to 10,000 gpm range and chemical dosage to be at 10 ppm or 230 to 340 cm³/min range.

The next step in going from bench test to field operation was to devise a method of introducing the chemical at the discharge weir of the upstream pond. Recommendations were to (1) dilute the chemical with fresh clean water, (2) cause rapid mixing to provide for maximum contact between the chemical and the suspended particles, (3) follow with a period of slow, gentle mixing to allow for the floc building and particulate stripping, and (4) create a period of relatively quiet water to allow for the floc to settle out. These recommendations pertain to an industrial plant complex in which mixing facilities, freshwater, and electrical power are readily available and taken for granted. In the initial stages no mixing facilities existed except the velocity of the runoff passing through the culvert. The freshwater source was the domestic water supply of a farm tenant. No electrical power supply existed.

The first "chemical treatment station" was set up at a point at which the farmer's water supply could be utilized. A simple open-sided structure resembling a school bus stop was constructed at the site above the stream channel to shelter the feed pump from the weather. In order to effect adequate mixing of the chemical and the turbid runoff, a 50-ft half-round culvert with baffles every 5 ft was placed in the stream channel. All the water from this drainage had to pass through this mixing area.

At the "chemical treatment station" four barrels of Nalco 634 were set in place. The waterline was tapped and a freshwater supply of approximately 15 gpm was made available. A Precision feed pump model 9101-21 was chosen to pump the chemical from the drums. A dilution-tee was installed in the waterline and the mixture was carried to the mixing culvert in a $1/2$-in. plastic line. Flow of the mixture was forced more by gravity than by pump pressure. A portable gasoline generator was set up to supply the power to the feed pump.

Facilities were soon proved inadequate. The concept of settling ponds in sequence was a viable concept. The chemical did cause flocculation and clarification, but at a higher dosage than expected. The method of introducing the chemical into the water was underdesigned, to say the least.

The chemical worked in the field. The ponds effected settling. What was needed was the hardware to meter the chemical to match the flow on a continual basis for 7 to 8 months each year. The parameters of the problem were

- As the runoff increased or decreased due to precipitation, the volume across the

discharge weir increased and decreased accordingly. This change would occur continually. Part of the system had to sense this change and automatically initiate action.
- The testing of water three times daily showed that the turbidity of the water would remain within the range of 85 to 120 JTU. Because these turbidities were well within the effective range of the chemical, only changes in flow (inches of water) over the weir and not changes in turbidity be registered. However, because flow is not proportional to level, a device was necessary to proportion the polyelectrolyte to the flow.
- The feed pump would have to respond automatically and accurately to the continually changing demand.
- The system should be maintenance-free and rugged enough to withstand field conditions.
- 220-110 V electrical power was an absolute necessity.

This system consisted of four components:

- Liquid level capacitive probe
- Liquid level monitoring and control instrument with flow proportional output circuit
- Automatic pump control
- Chemical metering pump

Necessary support equipment included a chemical holding tank, various plastic plumbing lines, tees, PVC pipe, etc.

Located at the upstream pond spillway, the liquid level capacitive probe sensed the changes in the level of the outflow. It had no external electronic moving parts, but continually emitted a signal that indicated the exact water level in the pond. The liquid level monitoring and control instrument accepted the signal from the capacitive probe and converted the signal to a 10 to 50 mA DC output for a 0 to 100% pond height variation. This variation is equal to 0 to 24 in.H_2O over the weir. This, in turn, is equal to 0 to 25,000 gpm. The 10 to 50 mA analog output is flow proportional. The automatic pump control accepts the 10 to 50 mA DC output from the control instrument and amplifies it to produce a signal which is used to control the output of the chemical feed pump. This component is an interrupted stroke metering pump with an output range of from 1 to 10 gal/hr.

As stated before, daily testing demonstrated that the turbidity level would remain in the range of 85 to 120 JTU and would not be appreciably affected by an increase in flow. The bench testing proved that the polyelectrolyte was quite effective through and beyond this range at 5 ppm. It is then simply a matter of calibrating the system to meter in this amount. After calibration the correct amount is automatically metered according to the flow. The exact amount of the chemical is pumped from the holding tank through plastic lines into a spreader tube hung across the spillway. The chemical drips into the water that passes below.

Strong, rapid mixing follows immediately; diluting the chemical with nonturbid water is unnecessary. This mixing is the key to good chemical dispersion and must be provided to produce an effective flocculation result. This mixing phase has been accomplished by designing iron obstructions that cause turbulence and mixing into the spillway fins and angle.

The remaining step in clarification is the period of quiet water provided by the second pond. Here, settling out and clarification take place. Clean water is decanted from this pond.

While field testing of the above system was going on, bench testing of other flocculating formulations was being carried out. Bench tests indicated that American Cyanamid's Super Floc 330 would be effective over a wider range of turbidity at half the dosage rate of Nalco 634. The Cyanamid chemical was placed in actual field operation. At the same time two drainages with identical systems were operating, and the Super Floc 330 was used in one of the systems. After some initial problems with viscosity, it was found that Super Floc 330 was a more effective, efficient chemical.

The proof of the system is that it works. Under the regulations of the Waste Discharge Permit, the water above and below the mine is analyzed three times a day. On an operational basis, during periods of high rainfall and resulting runoff, the two-pond chemical flocculating system prevents the siltation of Hanaford Creek. Pertinent numbers for comparison are as follows:

Entering silt load	1.5 to 2.0% solids by vol	1000+ JTU
Silt load of chemical station	0.4 to 0.7% solids by vol	85–120 JTU
Decant from second pond	Clear water	4–15 JTU

The flocculated silt accumulates quite quickly in the downstream ponds. It is necessary to periodically, depending upon the size of the pond, remove this silt or in some way allow for its accumulation.

SURFACE MINE SEDIMENTATION CONTROL

Surface mining and reclamation has the potential to at least temporarily increase surface runoff and the resultant erosion and gully development. This increase in erosion is due in part to the compaction of the soil surface with heavy equipment, the creation of large relatively unvegetated watersheds, and the elimination of natural dendritic ephemeral drainage patterns. Methods for stabilizing gully channels which may develop prior to the establishment of sufficient vegetation to control erosion should be adopted. Utilization of various sizes of riprap check dams, reinforcement of riprap check dams, and the relationship of the type of control structure to the area of watershed and resultant flow velocities will be necessary.

Drainage patterns on reclaimed areas depend on the remedial landscaping that is applied to an area and the natural erosional processes that subsequently modify the landscape. Remedial landscaping includes determining the topographic layout of an area, the size of its watershed, the type of vegetation to be used, the need for terraces and/or other drainage features, and the shape and layout of the total drainage system needed. In this area, three basic types of drainage patterns were found:

- Predesigned
- Naturally formed on the reclaimed surface
- Predesigned and subsequently modified by natural erosional processes

Predesigned drainage patterns were characterized by single, linear terraces and/or waterways. Natural drainage patterns have not been established on a regional scale, but on a local scale, dendritic, parallel, braided, and yazoo drainage patterns exist. Where modification of predesigned drainage features had taken place, it generally occurred by means of yazoo drainage.

A quantitative measure of a basin's drainage characteristics can be determined from the peak rates of runoff that occur in a particular drainage system. The riprap size capable of maintaining a stable waterway is a function of a waterway's corresponding peak rate of runoff.

The peak rates of runoff analysis for the gullies investigated are shown in Table 7.1. The peak rate of runoff ranged between 21.7 and 386.2 cfs. The average peak rate of runoff was 109.6 cfs, but only six gullies were determined to have peak rates of runoff >100 cfs.

Most of the material used to construct the riprap structures consists of crystalline limestone and associated chert, most likely from the Mississippian-age Burlington Formation. The chert appeared to contain a type of silica which is susceptible to water absorption, expansion, and subsequent fracturing.

The characteristics of the riprap structures used for gully control essentially consist of three components:

- Shape and layout
- Particle size distribution
- Usage of check dams

In layout, most of the gullies had a linear channel, while some had single bends and others were sinuous.

The lengths of the waterways varied, but the width of the channels was generally found to be about 14 ft. Most likely this width is a function of the dozer blade used to grade the waterway. A plan view of the waterways showed channel sides were generally parallel. In some cases, the width of the channel was irregular because placement of the riprap in the channel was highly variable. As a result, some of the waterways had jagged, elliptical, or cone-shaped channel sides in outline. Riprap-lined channels generally were between 1 and 2 ft in depth. In cases in which a series of check dams were used, the channel depth was between 2 and 4 ft.

Table 7.1. Time Concentration Calculations for Gullies Investigated at the Prairie Hill Mine, MO, Central Missouri Method

Gully No.	Peak Rate Runoff (cfs)	Watershed Location Factor (L)	Soil Infiltration Factor (I)	Topographic Factor (T) (ft)	Watershed Shape Factor (S)	Vegetative Cover Factor (V)	Contour Farming Factor (C)	Surface Storage Factor (P)	Runoff Frequency Factor (F)	Peak Rate Runoff (Q) = qLITSSVCPF (cfs)
1	29.8	1.01	1.2	0.79	1.05	2	1	1	1	59.9
2	16.3	1.01	1.2	0.79	1.03	2	1	1	1	32.2
3	166.5	1.01	1.2	0.79	0.99	2	1	1	1	315.7
4	17.1	1.01	1.2	0.86	1.03	2	1	1	1	36.7
5	51.8	1.01	1.2	0.86	1.07	2	1	1	1	115.5
6	22.1	1.01	1.2	0.86	1	2	1	1	1	46.1
7	201.7	1.01	1.2	0.79	1	2	1	1	1	386.2
8	80.2	1.01	1.2	0.79	1.1	2	1	1	1	168.9
9	131.5	1.01	1.2	0.79	0.98	2	1	1	1	246.8
10	15.9	1.01	1.2	0.79	1.05	2	1	1	1	32
12	48.9	1.01	1.2	0.86	0.97	2	1	1	1	98.9
13	47.5	1.01	1.2	0.79	0.96	2	1	1	1	87.3
14	28.2	1.01	1.2	0.79	0.86	2	1	1	1	46.4
15	33.6	1.01	1.2	0.79	0.95	2	1	1	1	61.1
16	38	1.01	1.2	0.79	0.96	2	1	1	1	69.9
17	50.9	1.01	1.2	0.79	0.94	2	1	1	1	91.6
18	51.2	1.01	1.2	0.79	0.85	2	1	1	1	83.3
19	155.7	1.01	1.2	0.79	1.05	2	1	1	1	313.1
20	39.5	1.01	1.2	0.79	1.01	2	1	1	1	76.4
21	11.2	1.01	1.2	0.79	1.05	2	1	1	1	22.5
22	10.8	1.01	1.2	0.79	1.05	2	1	1	1	21.7

From Vories, K. C. and C. D. Elifrits. Controlling large gullies on a midwest surface mine, in *SME-AIME Preprint 86-348* (Denver, CO: Society of Mining Engineers, 1986).

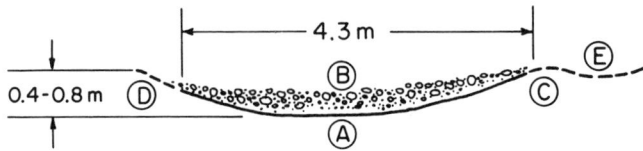

A) Gravel filter bed (5 cm)
B) Protective rip-rap channel cover (7.5-20 cm)
C) Ridge formed by construction or differential erosion
D) Ideal gradation between channel sides and channel
E) Yazoo gully formed from Ⓒ

Figure 7.5. Waterway cross-section. (From Vories, K. C. and C. D. Elifrits. Controlling large gullies on a midwest surface mine, in *SME-AIME Preprint 86-348* (Denver, CO: Society of Mining Engineers, 1986).)

In general, waterways constructed of uniformly distributed riprap (riprap generally with a range of 2 to 10 in.) placed with a pan scraper had streamlined channels, whereas gullies dump-filled with well-graded or large riprap had irregular channels. Figure 7.5 illustrates the various components of a channel in cross-section. In some channels, lack of uniformity of size in riprap utilized and differential weathering between the channel riprap and surrounding soil accounted for the formation of ridges along the sides of the channel and the subsequent development of yazoo gullies. The use of a gravel filter bed below the appropriate size channel riprap and a uniformly sized riprap should minimize this problem. Channel riprap, where placed using a pan scraper, consisted of a thin veneer of channel riprap designed to secure the gravel filter bed beneath. In places in which the size distribution was uneven or riprap was not graded using a pan scraper, the thickness of the channel cover was variable, increasing the probability of channel erosion.

Wings on check dams generally were flared upstream. Check dams were usually placed where terraces fed into channels or where special conditions dictated (Figure 7.6).

The particle size distribution of the riprap depended on:

1. Time of emplacement — In general, riprap structures earlier built had wide particle distributions. Also, later structures were placed with a tractor-pulled pan scraper.
2. Method of emplacement — Riprap placed using a pan scraper tended to have uniform particle distributions, while the riprap carried to gullies by dump trucks tended to have nonuniform particle distributions.
3. Structure under consideration — In general the particle size distribution of the channel beds tended to be more uniform than that of check dams.
4. Amount of weathering — Fracturing of the chert increases the percentage of the smaller-sized fraction at the expense of the larger-sized fraction.

EROSION AND SEDIMENT CONTROL

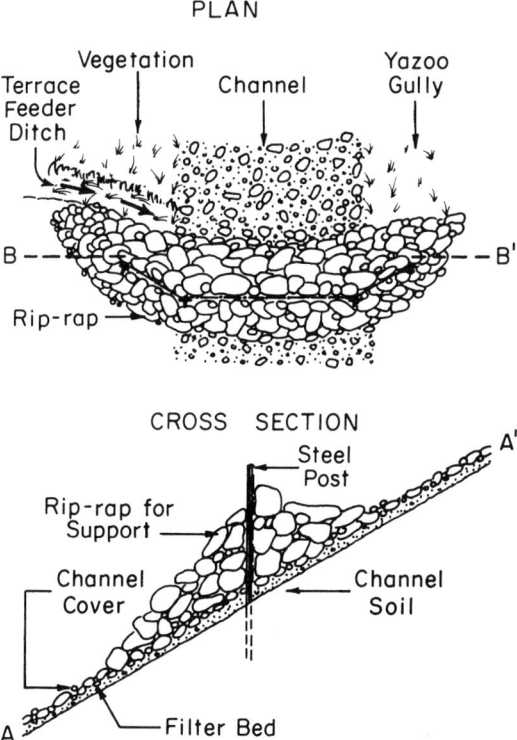

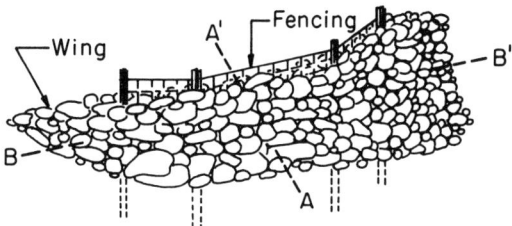

Figure 7.6. Components of a check dam. (From Vories, K. C. and C. D. Elifrits. Controlling large gullies on a midwest surface mine, in *SME-AIME Preprint 86-348* (Denver, CO: Society of Mining Engineers, 1986).)

Table 7.2 includes the range and nominal diameter of riprap used in the gullies at the site, but does not include sizing characteristics of the filter bed material. Any riprap with a maximum size of <10 in., although uniform in relation to other riprap

312 ENVIRONMENTAL IMPACTS OF MINING

Table 7.2. Waterway Analysis for Gullies Investigated at the Prairie Hill Mine, MO

Gully No.	Watershed Area (Acres)	Slope Gradient (%)	Gully Length (ft)[a]	No. of Check Dams	Particle Range (in.)	Nominal Diameter (in.)	Peak Runoff Rate (cfs)	Triangular Sectional Channel Area (ft²)	Channel Water Velocity (ft/sec)	Recommended NCSA Riprap Range/Av. (in.)	Waterway Field Status[b]
1	7.4	9	400	2F[c]	3 to 9	6	59.9	7	8.6	3 to 12/6	S
2	3.5	9	400	3F-[d]	2 to 16+	5	32.2	7	4.6	1 to 3/1.5	S
3	62.7	2	520	7[e]	2 to 20+[f]	8	315.7	17.5	18	15 to 48/24	E
		8	160	0							
		12	120	0							
4	3.7	10	280	2F	3 to 8	5	36.7	6	6.1	12 to 6/3	S
		14	160	0							
		10	440	1							
5	14.7	5	560	6e	3 to 18[f]	8	115.5	10.5	11	5 to 18/9	S
6	5.1	19	280	2F	2 to 9	5	46.1	5	9.2	3 to 12/6	E
7	79.6	4	360	3F	2 to 24+	12	386.2	12	32	15 to 48/24	S
8	25.3	25	160	2F	2 to 10	6	168.9	7	24.1	15 to 48/24	E
9	46.8	17	120	2F	2 to 21+	4	246.8	11.3	22	15 to 48/24	S
10	3.4	5	60	2	3 to 8	5	32	5	6.4	2 to 6/3	
		13	140	0							
12	13.7	8	140	0	2 to 8	5	98.9	8.5	11.6	5 to 18/9	S
		19	120	2							
13	13.2	2	120	1	12 to 24+[f]	20+	87.3	8	10.9	15 to 18/9	S
		4	100	3							
14	6.9	6	120	1	2 to 9	5	46.4	5	9.3	3 to 12/6	S
		8	50	0							
15	8.6	10	80	1	2 to 24+	9+	61.1	10	6.1	12 to 6/3	S
		12	40	1							
16	10	3	320	0	2 to 30+	12+	69.9	7	10	3 to 12/6	S
		9	100	0							
		12	40	0							
17	14.4	11	140	0	2 to 20+	8	91.6	10	9.2	3 to 12/9	S
18	14.5	11	120	0	2 to 16	9	83.3	7	11.9	5 to 18/9	S
		7	160	0							

EROSION AND SEDIMENT CONTROL

19	57.7	3	1600	12[e]	3 to 20+	11+	313.1	14	22.4	15 to 48/24	E
		9	200	0							
		4	200	1							
		6	280	1							
20	10.5	20	120	3	2 to 7	5	76.4	10.5	7.3	2 to 6/3	S
		7	200	0							
		3	240	0							
21	2.2	4	300	2	2 to 9	4	22.5	6	3.8	No. 8 to 1.5/.75	S
22	2.1	34	120	0	2 to 9+	5	21.7	7	3.1	No. 8 to 1.5/.75	S

[a] Length of riprap channel and/or check dam series.
[b] S = stable; E = erosion.
[c] F = Check dams fenced (not fenced if blank).
[d] F- = 2 check dams fenced, 1 not fenced.
[e] Series of check dams without riprap channel.
[f] Distribution applies to riprap in check dams; majority of channel was stable, but portions were scoured.

From Vories, K. C. and C. D. Elifrits. Controlling large gullies on a midwest surface mine, in *SME-AIME Preprint 86-348* (Denver, CO: Society of Mining Engineers, 1986).

distributions found, was in itself generally well-graded. This type of material was generally found in the newer, pan scraper waterways.

Three basic types of check dams were found in the gullies investigated:

1. Rock check dams reinforced with fence panels and wing structures
2. Rock check dams without fence panels, but with wing structures
3. Straw bales supported with steel rebar

Fenced check dams were used on 7 of the 21 riprap waterways; 4 waterways had no check dams; and 10 waterways had nonfenced check dams.

Straw bales were found to be ineffective in controlling erosion in large gullies because they were easily washed out or bypassed. Intact straw bales were found in their original orientation, yet the gully water had either flowed around or undermined the bales. All bales showed significant signs of physical deterioration resulting from the effects of long-term weathering.

In general the watershed size, the gully length, the gradient, the need for terrace ditches, the extent of yazoo gully formation, and the peak rate of runoff are proportional to the number of check dams needed for a particular gully.

A plot of frequency vs the number of dams per gully indicates that two dams were used per gully most often at the Prairie Hill Mine. The number of dams used per gully at the mine is a function of gully length and gradient. Although a few exceptions exist, points plotted below the line indicate conditions under which one dam or less was used; points plotted above the line indicate conditions under which two dams or more were used. The gully length refers to the length of channel in the plan view and is limited to the length between gradient changes. For example, a particular gully may be 400 ft in length, but 100 ft may correspond to a gradient of 4%. This plot is useful for determining the number of dams needed on a section of channel once an estimate of gradient and corresponding channel length is known for that section of channel.

Often the spacing of the dams used on a gully was evenly spaced over the length. For example, if a gully were 100 ft long and two dams were to be used, then a spacing of 33 ft would be used between the dams. Check dams were often used in cases in which terraces intersected the gully; however, the design of terraces was such that they were also evenly spaced along the slope. In some of the longer gullies, a series of check dams was used along portions of the gully, while other portions of the gully contained fewer check dams.

Fencing was generally limited to the central portions of the check dams, but in some cases was extended along inner portions on the wings. Fencing consisted of standard, heavy, steel-welded livestock panels cut length-wise in halves. The fencing was reinforced with between two and four steel T-posts driven into the channel floor. Only a few of the dams used riprap on the downstream side of the fence to help support the structure.

Mined land that is ready to be reclaimed is first graded with the construction of any terraces, levees, ditches, and drainageways, and then is subsequently covered with topsoil and seeded. Terraces generally are designed to have gradients of 1 to 3%, while ditches can have gradients up to 5%.

Several types of vegetative cover are utilized at the mine. Often the stands are a mixture of vegetative types. One mix includes a warm season grass: Blackwell switchgrass, birdsfoot trefoil, Korean lespedeza, and crownvetch. Another mix emphasizes cool season species for erosion control and wildlife habitat, including timothy, orchard grass, smooth bromegrass, alfalfa, and red clover.

Two basic techniques were used in the construction of riprap-lined gullies at the Prairie Hill Mine: (1) grading of the gully using a dozer blade followed by placement of the riprap in the channel by dump trucks, and (2) the same method used as in (1), and the riprap is then spread in the channel using a pan scraper. Gullies that were constructed using the second method had a better shaped channel than those constructed using the first method, especially when the riprap had a more uniform particle size distribution.

During the course of the gully investigation, it became apparent that a combination of techniques was needed to successfully control erosion and that failure of one of the components of the riprap structure will most likely lower the degree of erosion protection and may even initiate the formation of new gullies. In general, the use of appropriate, uniform-sized riprap for a particular flow velocity is considered the most important parameter in channel design.

The channel shape shown in Figure 7.5 is recommended. V-shaped channels, in contrast, promote scouring. The use of both filter beds and channel riprap were the most helpful in reducing the effects of scouring. Care should be taken to prevent the development of ridges along the channel, as the effectiveness of the waterway in collecting water and helping to create yazoo gullies is reduced. Banking of the channel in a manner similar to that of banked roads may be needed if the channel on the downslope side could be breached.

An estimate of the peak rate of runoff should be determined for the waterway of interest. If a known size of riprap is to be used, then the corresponding velocity can be determined from Table 7.3. Dividing the peak rate of runoff by the velocity determines the needed channel area. From this the appropriate channel depth and width can be determined.

The layout of the waterway can be beneficial, detrimental, or indecisive. The layout of the waterway should be designed to reduce the gradient of the channel. Bends can be designed into the channel layout, but it appears that linear waterways are sufficient in most cases. Waterway lengths are generally small in comparison to the size of the watershed, and, therefore, the effects of gradient within the channel itself should not be an important factor in determining the peak rate of the runoff. However, if the waterway length is relatively long, then the effects of the channel should be included in the peak rate of runoff analysis.

Table 7.3. NCSA Graded Riprap Stone

Flow Velocity (ft/sec)	NCSA No.	Size in. (sq. openings)			Corresponding Size of Filter Stone NCSA No.
		Max.	Av.	Min.	
2.5	R-1	1 1/2	3/4	No. 8	FS-1
4.5	R-2	3	1 1/2	1	FS-1
6.5	R-3	6	3	2	FS-1
9.0	R-4	12	6	3	FS-2
11.5	R-5	18	9	5	FS-2
13.0	R-6	24	12	7	FS-3
14.5	R-7	30	15	12	FS-3

From Vories, K. C. and C. D. Elifrits. Controlling large gullies on a midwest surface mine, in *SME-AIME Preprint 86-348* (Denver, CO: Society of Mining Engineers, 1986).

The peak rate of runoff is mainly affected by the size of the watershed. The size of the watershed may be reduced during the initial design process or by adding more waterways to a particular drainage system. Other factors also affect the peak rate of runoff, and the contribution of any of these factors will help reduce the effects of erosion in a waterway.

The gullies that appeared to be the most effective in controlling erosion utilized gravel filter beds covered by well-graded channel riprap of the appropriate size. Table 7.3 can be used to compare the actual particle size used and the recommended National Crushed Stone Association (NCSA) particle size. In cases in which the waterway experienced erosion, the problem resulted from either a lack of a properly sized riprap for a particular velocity, a sufficient range but gap-graded particle size, or a combination of both. In most cases in which the channel was stable, the actual range of particle sizes was consistent with the guidelines set by the NCSA (Table 7.3). Scouring in gully no. 6 was not caused by inappropriate riprap, but resulted from an unfinished check dam that was supposed to capture a feeder ditch.

Check dams worked best with well-graded distributions in which the dam sufficiently prevented the passage of smaller riprap, provided enough permeability to prevent the check dam from impounding water to the point at which water topped or sidetracked the dam, and did not overstress the fencing to the point of failure.

The uniformity of the channel riprap was observed to be an important factor in the construction and the stabilization of the channel. Channels having wide, well-graded particle size distributions or containing larger diameter materials tended to have irregular channel surfaces that allowed the scouring of the channel and filter beds.

No particular pattern is applied to the placement of check dams at this site, but it is apparent that the use of a check dam can be beneficial in reducing scouring in channels. The use of wings on check dams helped channel waters brought in by ditches and reduced the effects of yazoo gullies along the main channel.

When fenced structures are used, it is important to provide proper support on the downslope portion of the check dam to prevent its failure.

SURFACE MINE DRAINAGE CONTROL

Storm water runoff from mine facilities such as haul roads, waste disposal sites, and orebodies have become recognized as the primary facilities associated with nonpoint source drainage. Depending on the orebody, facility design, and geographical setting, nonpoint source drainage can pollute and have effects therefrom on local surface and groundwater resources.

Design considerations for plan development must incorporate parameters including contributing drainage area size, slope, flow velocity, vegetation characteristics, and soil scour velocities.

The success of nonpoint source drainage and erosion stabilization planning is based on implementing control measures that dissect drainage patterns, reduce flow velocities, and disperse runoff toward sediment control sites. Interim and long-range drainage control measures should begin during exploration activities.

Nonpoint source drainage differs from point sources in that the storm water cannot be traced to a specific, identifiable point of entry into the waterway or aquifer.

Impact on water resources from nonpoint sources is considered to occur more frequently or be more significant than point sources. Nonpoint source pollutants include sedimentation, changes in background pH level, alteration to the biota community, reduction to beneficial uses, and introduction of heavy metals or toxic substances. Sedimentation is frequently considered to be the most common form of nonpoint source pollution in surface water resources.

The erosion-transport-sedimentation process affects surface waters by covering stream bottoms, smothering aquatic insects and spawning gravels, clogging intergravel spaces used by young fish, and altering the actual water quality. Additional items having an impact on water quality that are related to sedimentation include potential increases in nutrient levels, such as nitrogen and phosphorous compounds, temperature, pH, and a reduction in dissolved oxygen.

The primary source of erosion-related factors associated with mining operations is haul roads. These are typically controlled through the implementation of "best management practices" (BMPs). The development of BMP programs for control of nonpoint source pollution should include interim and longitudinal water management planning. An effective surface-drainage water management plan incorporates interim drainage stabilization for erosion control with reclamation activities.

The Thunder Mountain Mine is located in central Idaho adjacent to the Frank Church River of No Return Wilderness Area. The project is situated at an elevation of about 2438 m (8000 ft) at the headwaters of two primary tributaries to the Middle Fork Salmon River. The location of the mine increased concern and interest within the regulatory and public sectors. Primary concerns associated with project development and operations are related to the impact of sediment on local anadromous fisheries resources and the use of cyanide in the recovery process, as well as the potential impact on surrounding water resources.

318 ENVIRONMENTAL IMPACTS OF MINING

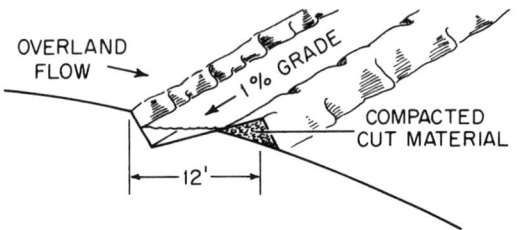

Figure 7.7. Dispersion terrace system. (From Mohr, R. Thunder Mountain Mine, *Min. Eng.* February:210–212 (1991). With permission.)

Thunder Mountain is operated as an on-off heap-leach gold mine. Leached ore is treated with alkaline chlorination to destroy residual cyanide in the heaps. Following the cyanide destruction process, spent ore is disposed of by off-loading individual pads and placing the material in approved waste disposal facilities.

Water quality monitoring for selected parameters is conducted on a weekly or monthly basis at monitoring wells, springs, and adjacent streams. The water quality parameters currently of most interest to state and federal agencies include pH, cyanide (weak acid dissociable form), turbidity-chloride, and key heavy metals.

To address issues and concerns identified during the permit process, an innovative surface water management plan was developed. The plan was initially designed based on evaluating potential flood events associated with rainfall and snowmelt characteristics representative of the project region.

Following a review of several hydrologic model results, a combination of rain-on-snow was selected for drainage control facility design. This approach provided maximum protection from typical snowmelt runoff events, plus an additional conservative safety factor for the occasional rain-on-snow runoff event.

Sediment control facilities were implemented during the exploration phase of project development. Sediment control facilities were constructed during autumn of the first exploration season. Early storm water runoff management structures were designed to prevent concentrations of water from building up on areas cleared of vegetation for exploration purposes. Dispersion terraces were constructed based on design parameters, including total drainage area, slope, soil scour velocities, and runoff potential per acre of drainage area (Figure 7.7).

Location of dispersion terraces was initially conducted via the use of site contour maps generated from aerial mapping of the project area. Watershed drainage areas contributing runoff to exploration sites were divided into subdrainage areas to control the amount of runoff contributing to any one dispersion terrace.

After the preliminary design of drainage control facilities was completed, dispersion terraces were located in the field by surveying their alignments as laid out in site maps. All terraces were surveyed at a constant slope of 1%. Establishment of this design criteria prevents runoff collected in the dispersion terraces from exceeding a flow velocity of 0.6 to 0.8 m/sec (2 to 2.5 fts).

Grade control of terraces was accomplished by placing a stake every 15 m (50 ft) along the length of the facility during the surveying process.

Construction of dispersion terraces was accomplished via a bulldozer. The bulldozer developed an inward-sloping cut, placing soil material from the cut on the outslope side of the terrace, compacting the material as it follows the grade stakes along the length of the facility.

At the end of each dispersion terrace, a level spreader was constructed to enhance the dispersion of runoff onto undisturbed vegetation cover. Through this technique of combining dispersion terraces and level spreaders, runoff was prevented from concentrating on sites cleared for exploration, minimizing the erosion process. Additionally, dispersion terraces act to collect settleable solids which are the result of the 1% grade. By aligning terraces at this gradient, scour of the terrace surface is prevented and deposition of settleable solids is enhanced.

Dispersion terraces and level spreaders were used in combination throughout the exploration phase of Thunder Mountain Mine.

Project construction began by clearing and grubbing all vegetation from project facility siting areas. As a result of creating large, cleared areas, soil was exposed to the erosion-transport-sediment deposition process over about 61 ha^2 (150 acres). This exposure of soil to the erosion process on steep slopes created the need for an intensive drainage stabilization and erosion control plan to protect local water resources from sediment impacts.

The basis of the plan incorporated dividing the disturbed areas into small subdrainages with a maximum size of about 2 ha^2 (5 acres). Additionally, all runoff from upland watershed areas was diverted around project facility sites to minimize the contributing area of drainage to project facility sites under construction.

Aerial mapping of the entire project area identified the small subdrainage basins. At the base of each subdrainage area, a dispersion terrace was developed at a 1% slope. Dispersion terraces collected all surface storm water runoff and routed it off and away from the project's area of disturbance.

Actual alignment of dispersion terraces was modified in the field to incorporate existing exploration road cuts to the extent that they were practical. Use of existing exploration roads minimized the amount of additional disturbances required.

Dispersion terraces were developed such that the upper terrace extended farther into the forested area adjacent to the cleared zone than the next downslope terrace. By aligning terraces in this manner, runoff from the uppermost terrace would disperse over forest ground cover vegetation while concentrating runoff to the next downslope terrace. This technique was applied to all terraces as they were constructed down the slope of a cleared area. This enhanced the dispersion of runoff by preventing it from reconcentrating into gully flow conditions.

Due to the large areas of disturbance associated with project construction, additional sediment trap facilities were incorporated at the end of all dispersion terraces and level spreaders (Figure 7.8). Two similar sediment trap techniques were applied based on the anticipated flows expected from each terrace.

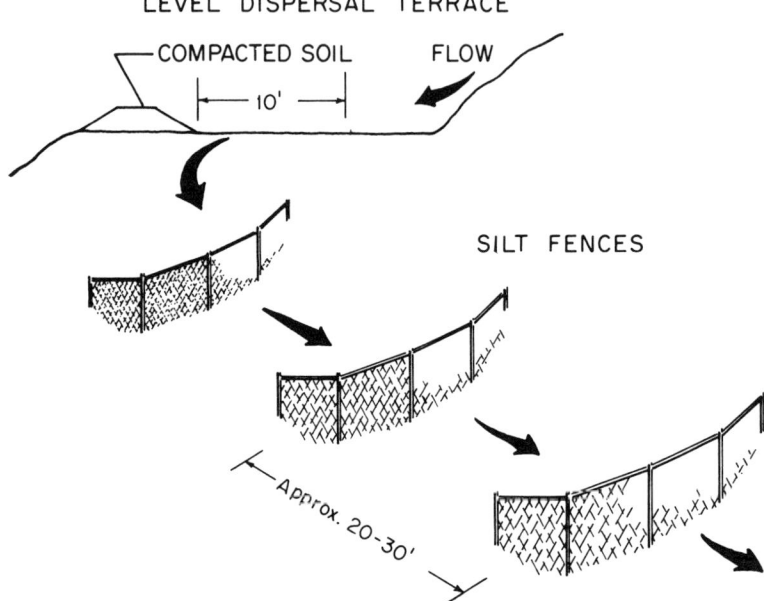

Figure 7.8. Level spreader/silt fence sediment trap. (From Mohr, R. Thunder Mountain Mine, *Min. Eng.* February:210–212 (1991).)

The primary sediment trap technique used consisted of placing silt fence structures at the outfall point of dispersion terraces and level spreaders. Silt fences were incorporated due to the minimal costs associated with the material and the life expectancy of the product, thought to be longer than the life of the project.

Placement of the silt fence structures was based on forming the fence material in an arch pattern such that runoff entering the fence would not spill around the uphill edge. Forming an arch out of the silt fence structure in this manner forced the runoff to pond before spilling over the top of the material at the central point of the arch. Rocks were placed on the downslope side of the silt fence arch at the point at which runoff spilled over the fence. The use of rocks in this location prevents additional scour of soil, minimizing sediment impacts.

A combination of silt fence and log barriers was incorporated as the alternative sediment trap technique for the construction phase of the project. The use of log barriers, in conjunction with silt fences, was applied to the drainage stabilization and erosion control plan in order to evaluate the efficiency of the two techniques and to determine if cost savings could be realized by increased applications of log barrier sediment traps.

After evaluating the efficiency of the silt fences vs the log barriers, a decision was made to increase the use of silt fences. This was because of the long life span of the silt fence material in field applications at elevations similar to that of Thunder

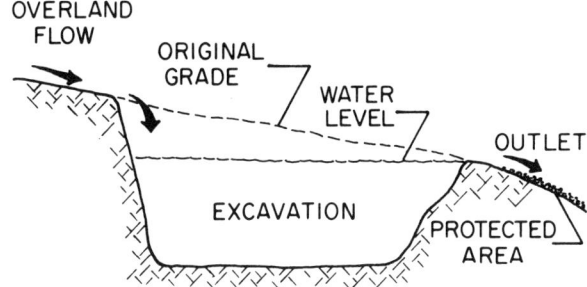

Figure 7.9. Excavated sediment sump. (From Mohr, R. Thunder Mountain Mine, *Min. Eng.* February:210–212 (1991).)

Mountain. Also, the material is flexible enough to reuse at different sites once an area becomes stabilized or project development precludes the need for future drainage control.

During the operating life of the Thunder Mountain Mine, water management planning and reclamation progressed on an interim basis. This allowed the operations to expand annually while surface runoff control measures were incorporated as needed. The reclamation of disturbed sites was conducted when practical based on short- and long-term stabilization needs.

Interim water management control measures employ dispersion terraces, level spreaders, silt fences, and several new techniques that were developed in response to drainage conditions generated from major mine facilities.

The primary source of runoff from mine facilities developed for project operations was generated from mine haul roads. Haul road compaction and road gradients combined to create large flow rates during major snowmelt and rain-on-snow events.

To control the sediment load resulting from project facilities, sediment sumps, rock filter sediment traps, roadside culvert sediment traps, and protection of culvert outlets were added to the water management plan. These additional structures were combined with existing and new water management facilities (Figures 7.9 and 7.10).

Sediment sumps were located in two general areas. The main location was at the end of dispersion terraces before runoff entering silt fence or log barrier sites. Sufficiently large amounts of sediment were trapped in these sediment sumps to require an annual cleaning with a backhoe. The other location was along the length of dispersion terraces. Sediment sump placement about every 15 m (50 ft) along the terraces acted to extend the functional life of silt fences before maintenance of the structures was required.

Water management facilities, similar to those described, have been constructed throughout the project area. This extensive use of nonpoint source water management techniques has effectively allowed the Thunder Mountain Mine to operate in a wilderness setting without having a major impact on the local water resources. The

322 ENVIRONMENTAL IMPACTS OF MINING

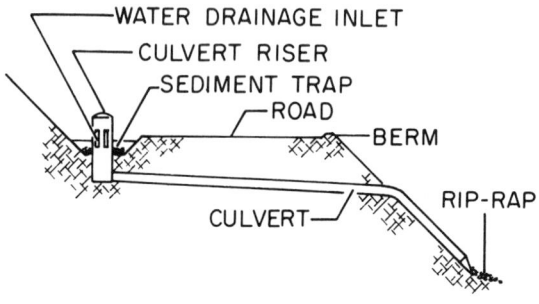

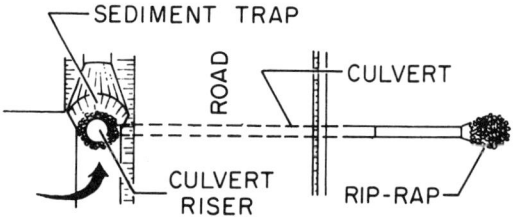

Figure 7.10. Roadside culvert sediment trap. (From Mohr, R. Thunder Mountain Mine, *Min. Eng.* February:210–212 (1991).)

key aspect of the overall water management plan has been to maintain an interim program on an annual basis that is designed around the project's yearly progression.

Practical applications to sediment control can be applied to surface mining operations in a cost-effective manner. The key to establishing a functional water management and reclamation program is to begin with a thorough understanding of the hydrologic regime of the project area.

Once the hydrologic design parameters are set, the next step is to divide the drainage areas within the project area of disturbance into individual subdrainages. The intent here is to minimize the total flow that can be generated from any one area to a level that will not cause erosion of native soils. By dispersing runoff into small components, nonpoint source pollution from surface runoff can be minimized.

Once an effective water management plan has been established, the program can be modified annually over the life of the mine. Use of cost-effective projects such as silt fences that can be used throughout the life of the project help control costs.

Coordinating water management with reclamation planning helps expedite overall postmining closure requirements. Additionally, interim reclamation speeds the revegetation process such that areas reclaimed early in the project life can reduce the total number of acres under disturbance. This minimizes the total area contributing to the erosion-transport-sediment process.

This approach can also provide an effective program for meeting anticipated state and federal nonpoint source requirements.

REFERENCES

1. Grim, C. E. and R. D. Hill. "Environmental Protection in Surface Coal Mining," National Environmental Research Center, U.S. National Technical Information Service, PB-238 538 (1974), pp. 101-115.
2. Becker, B. C. and T. R. Mills. "Guidelines for Erosion and Sediment Control," U.S. EPA Report R2-72-015 (1972).
3. Vories, K. C. and C. D. Elifrits. Controlling large gullies on a midwest surface mine, in *SME-AIME Preprint 86-348* (Denver, CO: Society of Mining Engineers, 1986).
4. Mohr, R. Thunder Mountain Mine, *Min. Eng.*, February:210-212 (1991).

CHAPTER 8

WETLANDS

INTRODUCTION

A multitude of definitions exists for the word "wetland." Several definition and classification systems have been developed for differing needs and purposes. Wetlands are identified in terms of soil characteristics, the types of plants supported, and the degree of wetness. A wetland[1] is a transition between terrestrial and aquatic systems, in which water is the dominant factor determining the development of soils and associated biological communities. It is also where, periodically, the water table is at or near the surface, or the land is covered by shallow water. The wetlands meet one or more of three conditions:

- Areas predominantly supporting hydrophytes
- Areas with predominantly undrained hydric soil, which are wet long enough to produce anaerobic conditions that limit the types of plants that grow there
- Areas with nonsoil substrates such as rock or gravel that are saturated or covered by shallow water at some point during the growing season

In general terms, wetlands can be described as areas flooded or saturated by surface water or groundwater often or long enough to support those types of vegetation and aquatic life that are specially adapted for saturated soil conditions.

In popular terms, shallow water or saturated areas dominated by water-tolerant woody plants and trees are called swamps, those dominated by soft-stemmed plants are considered marshes, and those with mosses are called bogs.

In the United States, the principal saltwater swamps are mangrove wetlands found along the southern coast of Florida. Mangroves are among the few woody plants that tolerate saltwater environments.

Freshwater swamps can contain various woody plants and water-tolerant trees. Southern swamps commonly contain bald cypress; tupelo gum; water, willow, and swamp white oak; and river birch.

Saltwater marshes are dominated by salt-tolerant herbaceous plants, e.g., cordgrass and black rush.

Freshwater marshes are dominated by herbaceous plants. Submerged and floating plants may exist in abundance. Common emergents include cattails, bulrush, reed, grasses, and sedges. A wet meadow may be intermittently flooded or saturated with shallow water, but marsh species, especially sedges and wet grasses, are supported.

Bogs are characterized by stable water levels and acidic, low-nutrient water- and acid-tolerant mosses. Typical bog plants like cranberry, tamarack, black spruce, etc., may have roots in deep, spongy accumulations of dead sphagnum moss and other plant materials, partially decomposed under bog conditions.

In geological terms, natural wetlands are ephemeral components of the landscape. They are dependent upon disturbances — long-term, large-scale tectonic forces or localized seasonal events such as daily or annual flooding and drying. In the absence of tectonic or hydrologic disturbance, wetlands gradually progress to relatively dry upland-type lands.

Large-scale natural wetlands developed from recent glaciation, mountain-building processes, and changes in sea level. After streams and rivers were reestablished by erosional forces, most wetlands drained.

Wetlands developed in North America by continental glaciation, valley glaciers, and changing sea levels along the coastal plains in the east and southeast. Bottomland hardwood swamps along the Mississippi and other large river systems were formed by changing sea levels, silt deposition from upland erosion, and natural channel alterations.

The survival of all natural wetlands depends upon natural disturbance or cyclic fluctuations in hydrology. Without disturbance, a bog or a marsh will gradually fill in, becoming a forest or prairie.

Natural wetlands undergo continuous internal changes seasonally and annually. Periodic flooding and drying support some plants and inhibit others, creating changes in microbial, invertebrate, and larger animal populations. Over time, major changes in microbial, invertebrate, and larger animal populations take place. The dependency of wetland communities on hydrologic patterns is clearly seen through changes in species composition resulting from changes in water levels and flow patterns. Alterations in depths and flow can help maintain a desirable composition of plant species.

The productivity of natural wetlands can sometimes exceed that of the most fertile farm fields. Wetlands receive, hold, and recycle nutrients continually washed from upland regions. These nutrients support the growth of an abundant macro- and microscopic vegetation, which converts inorganic chemicals into organic materials. Wetlands join with worms, insects, crustaceans, reptiles, amphibians, fish, birds, and mammals who feed on plant materials. Other animals from nearby aquatic and terrestrial environments are drawn to feed on plants and animals at the highly productive "edge" environment of wetlands, thereby extending the productive influence of wetlands far beyond their borders.

A direct relationship between wetland destruction and the declining populations of valuable species of fish, shellfish, birds, reptiles, and fur-bearing animals that rely on certain types of wetland habitats has been observed. Many studies have related the destruction of summer breeding wetlands and winter feeding wetlands to declines in the populations of migratory birds. The effects of wetland destruction can be most significant in regions where such habitat is the least common.

Natural wetlands along coasts, lakeshores, and riverbanks have drawn increased attention for their role in stabilizing shorelands. The inland wetlands offer protection from natural floods by slowing the flow of floodwaters, desynchronizing the peak contributions of tributary streams, and reducing peak flows on main rivers.

Some wetlands function as recharge areas for groundwater, allowing water to seep slowly into underlying aquifers. Wetlands may serve as discharge areas for surfacing groundwaters.

The most important function of wetlands is water quality improvement.[2] Wetlands provide effective, cost-free treatment for many types of water pollution. Wetlands can effectively remove or convert large quantities of pollutants from point sources (industrial and municipal effluents) and nonpoint sources (mine, agricultural, and urban runoff), including organic matter, suspended solids, metals, and excess nutrients. Sedimentation, natural filtration, and other processes participate in cleaning the water of many pollutants. Some pollutants are physically or chemically immobilized and are permanently retained until disturbed. Chemical reactions and biological decomposition break down complex substances into simpler ones.[3] Wetland plants remove nutrients through absorption and assimilation, and produce biomass. Oxygen is produced in abundance, increasing the dissolved oxygen content of the water and the soil in the immediate vicinity of plant roots, increasing the capacity of the system for aerobic bacterial decomposition of pollutants and for supporting a wide variety of oxygen-consuming aquatic organisms, some of which may utilize additional pollutants.

Many of the nutrients are retained in the wetland system and recycled through successive seasons of plant growth, decay, and death. If water seeps to groundwater, filtering through soils, peat, or other substrates removes excess pollutants and nutrients. When water leaves the surface, nutrients trapped in substrate and plant tissues do not contribute to the growth of noxious algae blooms or excessive aquatic weed growth in downstream lakes and rivers.

Natural wetlands can remove iron (Fe), manganese (Mn), and other metals from acid drainage. Accumulations of limonite, known as bog iron, were mined long ago as a source of Fe. Similarly, mixed oxides of Mn, called wad or bog manganese, are

Table 8.1. Comparison of Physical and Chemical Attributes of Organic and Mineral Soils

Parameter	Mineral Soil	Organic Soil
Organic content (%)	<12–20	>12–20
pH	6.0–7.0	<6.0
Bulk density	High	Low
Porosity	Low (45–55%)	High (80%)
Hydraulic conductivity	High (except clays)	High (fibric) Low (sapric)
Water holding capacity	Low	High
Nutrient availability	Generally high	Often low
Cation exchange	Low, dominated by major cations	High, dominated by hydrogen ion

From Donald, A. H. and R. K. Bastian. Wetland ecosystems: natural water purifier, in *Constructed Wetlands for Wastewater Treatment* (Chelsea, MI: Lewis Publishers, Inc., 1990), pp. 5–19.

the product of less acidic wetland removal processes. These wad deposits contain mixed oxides of Fe, copper (Cu), and other metals.

CONSTRUCTED WETLANDS

Constructed wetlands are manmade and designed systems consisting of saturated substrates, emergent and submergent vegetation, animal life, and water, that simulates natural wetlands. Bogs and swamps have been constructed or used for wastewater treatment, but most constructed wetlands for wastewater treatment resemble marshes. Best for wastewater treatment are marshes with herbaceous emergent and submergent plants. Water-tolerant woody plants in swamps may require 5 to 20 years for development and full operational capability. Bogs dominated by mosses are difficult to establish. They have less retention capacity and limited adaptability to fluctuating water levels. If water level and nutrients are increased, the bogs tend to become marshes. Marshes with cattail, bulrush, rush, or giant reed adapt to varying water and nutrient levels and are tolerant of high pollutant concentrations.

Natural and constructed wetlands commonly consist of five principal components:

- Substrates with different rates of hydraulic conductivity (Tables 8.1 and 8.2)
- Plants adapted to water-saturated anaerobic substrates
- A water column (Table 8.3)
- Invertebrates and vertebrates (Tables 8.4 and 8.5)
- An aerobic and anaerobic microbial population

Microbes — bacteria, fungi, algae, and protozoa — modify containment substances to obtain nutrients or energy and carry out their life cycles. The effectiveness

Table 8.2. Range of Physical Characteristics for Peat and Mineral Soils

Soil Type	Total Porosity (%)	Hydraulic Conductivity (m/day)	Bulk Density (g/cm)
Peat			
Fibric	>90	>1.3	<0.09
Hemic	84–90	0.01–1.3	0.09–0.20
Sapric	<84	<0.01	>0.20
Mineral			
Gravel	20	100–1000	⁻2.1[a]
Sand	35–50	1–100	1.2–1.8
Clay	40–60	<0.01	1.0–1.6

[a] Calculated based on porosity.

of wetlands designed for wastewater treatment is dependent on the success in developing and maintaining optimal environments for desirable microbial populations. Microbes are usually ubiquitous, naturally occurring in most waters, and have large populations in wetlands and contaminated waters with nutrient or energy sources. With unusual pollutants, inoculation of a specific group or strains of microbes may be required. Some naturally occurring microbial groups are predatory and will forage for pathogenic organisms.

Wetland plants have two important functions. Within the water column, stems and leaves significantly increase the surface area for the attachment of microbial populations. Wetland plants can transport atmospheric gases, including oxygen (O), down into the roots, thus enabling their roots to survive in an anaerobic environment. Some leakage can occur, producing a thin-filmed, aerobic region called the rhizosphere which surrounds each roothair. Some oxidation takes place in this microscopic region, but more importantly, the rhizosphere supports large microbial populations that conduct desirable modifications of nutrients, metallic ions, and other compounds. The configuration on a microscopic scale of an aerobic region surrounded by an anaerobic region, multiplied by the large area of the rhizosome boundary, is critical to nitrification-dentrification and many other pollutant transformations.

Many desirable wetland plants are widely available.[4] Giant reed and various species of cattail occur on every continent. Naturally occurring plant species are adapted to local climate and soil conditions and, hence, are likely to succeed and provide treatment in a constructed wetland. Exotic plant species may not survive or may perform poorly.

Some plant-substrate combinations are more efficient in wetlands constructed for treating wastewater, and some may be more tolerant to high pollutant concentrations. Many wetland projects have included a single plant-substrate combination. Maintaining a monoculture may incur high costs, and an insect pest outbreak could seriously damage a monoculture system. A mixed species system may offer more resistance to pest attacks and fluctuating loading rates.

Table 8.3. Plant Species Tested for Use in Constructed Wetlands for Wastewater Treatment

Emergent
  Scirpus robustus
  Scirpus lacustris
  Schoenoplectus lacustris
  Phragmites australis
  Phalaris arundinacea
  Typha domingensis
  Typha latifolia
  Canna flaccida
  Iris pseudacorus
  Scirpus validus
  Scirpus pungens
  Glyceria maxima
  Eleocharis dulcis
  Eleocharis sphacelata
  Typha orientalis
  Zantedeschia aethiopica
  Colocasia esculenta
Submerged
  Egeria densa
  Ceratophyllum demersum
  Elodea nuttallii
  Myriophyllum aquaticum
Floating
  Lagorosiphon major
  Salvinia rotundifolia
  Spirodela polyrhiza
  Pistia stratiotes
  Lemna minor
  Eichhornia crassipes
  Wolffia arrhiza
  Azolla caroliniana
  Hydrocotyle umbellata
  Lemna gibba
  Lemna spp.

From Guntenspergen, G. R., F. Stearns and J. A. Kadlec. Wetland vegetation, in *Constructed Wetlands for Wastewater Treatment* (Chelsea, MI: Lewis Publishers, Inc., 1990), pp. 73–88.

Fostering the growth of a single species may require significant changes in water levels, which deactivates part or all of the wetland system. Three plant varieties — cattail, bulrush, and giant reed — commonly used in wetlands treatment systems tend to create and maintain single species by inhibiting the growth of other plants.

Substrates — soils, sand, and gravel — provide:

- Physical support for plants
- Considerable reactive surface area for complexing ions, anions, and other compounds
- Attachment surfaces for microbial populations

Table 8.4. The Ten Major Bioelements, Their Sources, and Some of Their Functions in Microorganisms

Element	Source	Function in Metabolism
C	Organic compounds, CO_2	Main constituents of cell material
O	O_2, H_2O, organic compounds, CO_2	
H	H_2, H_2O, organic compounds	
N	NH_4^+, HO_3^-, N_2, organic compounds	
S	SO_4^{2-}, HS^-, S^0, $S_2O_3^{2-}$	Constituent of cysteine, methionine thiamin pyrophosphate, coenzyme A, biotin, and lipoic acid
P	HPO_4^{2-}	Constituent of nucleic acids, phospholipids, and nucleotides
K	K^+	Principal inorganic cation in the cell; cofactor of some enzymes
Mg	Mg^{2+}	Cofactor of many enzymes (e.g., kinases); present in cell walls, membranes, and phosphate esters
Ca	Ca^{2+}	Cofactor of enzymes; present in exoenzymes (amylases, proteases); Ca-dipicolinate is important component of endospores
Fe	Fe^{2+}, Fe^{3+}	Present in cytochromes, ferredoxins, and iron-sulfur proteins; cofactor of enzymes

Table 8.5. Minor Bioelements, Their Sources, and Some of Their Functions in Microorganisms[5]

Element	Source	Function in Metabolism
Zn	Zn^{2+}	Present in alcohol dehydrogenase, alkaline phosphatase, aldolase, RNA and DNA polymerase
Mn	Mn^{2+}	Present in bacterial superoxide dismutase; cofactor of some enzymes (PEP carboxykinase, recitrate synthase)
Na	Na^+	Required by halophilic bacteria
Cl	Cl^-	
Mo	MoO_4^{2-}	Present in nitrate reductase, nitrogenase, and formate dehydrogenase
Se	SeO_3^{2-}	Present in glycine reductase and formate dehydrogenase
Co	Co^{2+}	Present in coenzyme B_{12}-containing enzymes (glutamate mutase, methylmalonyl-CoA mutase)
Cu	Cu^{2+}	Present in cytochrome oxidase and oxygenases
W	WO_4^{2-}	Present in some formate dehydrogenases
N	Ni^{2+}	Present in urease; required for autotrophic growth of hydrogen-oxidizing bacteria

From Portier, R. J. and J. P. Stephen. Wetland microbiology: form, function, processes, in *Constructed Wetlands for Wastewater Treatment*, Hammer, D. A., Ed. (Chelsea, MI: Lewis Publishers, Inc., 1990).

Surface and subsurface water transports substances and gases to microbial populations, removes by-products, and provides the environment and water for the biochemical processes of plants and microbes.

Constructed wetlands offer broad applicability as wastewater treatment systems. This is more likely to be true for naturally occurring organic and inorganic substances, as well as for some anthropogenic compounds. Natural wetlands occur at low topographic areas receiving runoff waters from various sources. Wetlands have adapted to substances carried by runoff water, adapting them to support high productivity rates. Many wetlands have higher rates of carbon fixation and biomass production than even the best-managed agricultural farms.

The microbial populations of the wetlands conducting critical processing of pollutants have short generation times, high reproductive rates, and considerable genetic plasticity, which allows these organisms to rapidly adapt to and exploit new nutrient and energy sources.[5]

Summarizing, the wetland treatment processes include:

- Adsorption — ion exchange
- Consumption — plant uptake
- Filtration
- Biological transformation
    Fe, Mn oxidizing bacteria
    Sulfate reduction
    Dentrification
- Geochemical removal
    Pyrite formation
    Reduction/oxidation

It must be noted that many mechanisms that modify and immobilize pollutants are not clearly understood. Some wetland systems remove or modify many complex toxics, but plant harvesting, incineration, and ultimate disposal can present difficulties. Long-term accumulation of heavy metals and unaltered toxic compounds in vegetation and sediments may reduce widespread distribution in the environment, but the concentrated deposits may create the detrimental effects of bioaccumulation and biotransport. Periodic recovery/recycling procedures may be necessary. Other uses of these areas may be restricted.

A process schematic is shown in Figure 8.1. The processes involved in wetlands include the following:

- Conditions and processes vary over depth
- Mosaic of oxidizing and reducing conditions in root zone, requiring further research
- Removal by vegetation, organic soil, microorganisms
- Anaerobic conditions for acid, metal removal

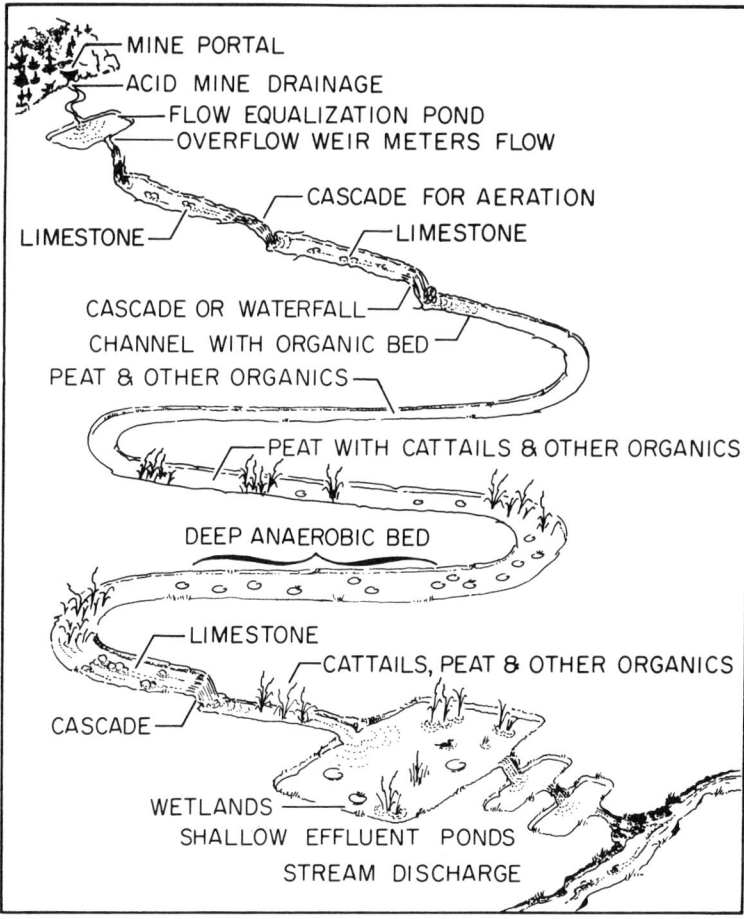

Figure 8.1. Process schematic of a wetland.

$$SO_4^{2-} + 8e^- + 8H^+ \rightarrow S^{2-} + 4H_2O \qquad (8.1)$$

- Generation of S^{2-} → metal removal
- Consumption of H^+ → increase in pH

Advantages of wetland treatment include:

- Adaptability to acid drainage (pH as low as 2.7) and elevated metals content
- Capital costs of wetland system are relatively low
- Operational and maintenance costs of constructed wetland systems are relatively low

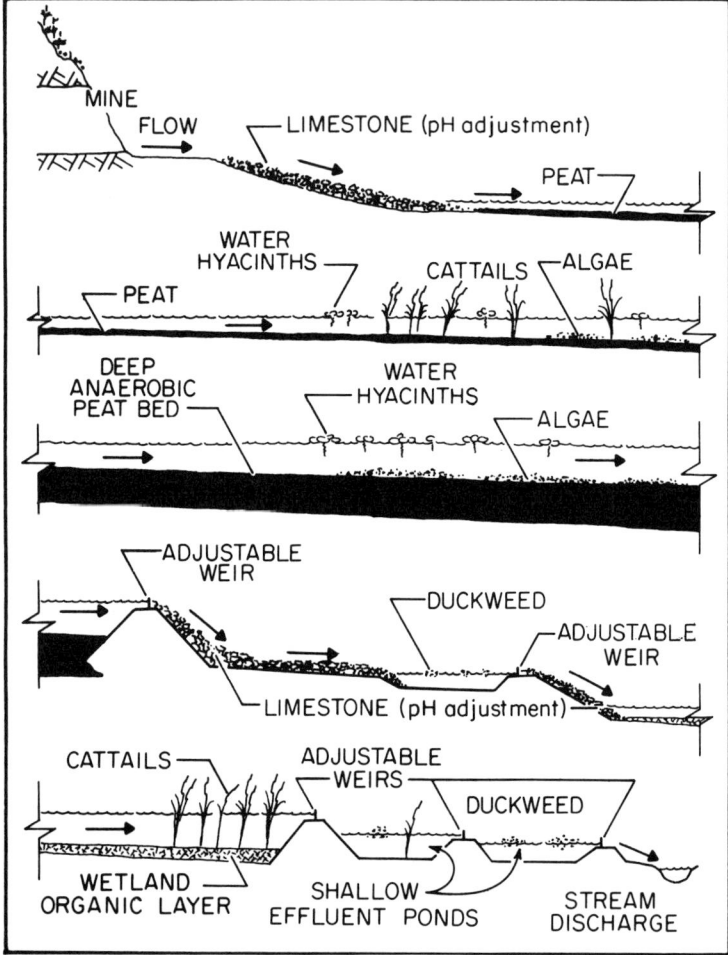

Figure 8.2. A schematic of an engineered ecosystem treatment process.

- Generally self-maintaining and require little or no operator supervision
- Provide wildlife habitat and flood control

Disadvantages include:

- Availability of sufficient flat or gently sloping land
- Land area requirement per unit flow of water to be treated can be large (typically around 50 m^2/m^3 day^{-1})
- Treatment during winter is reduced
- Impact on wildlife is still unknown

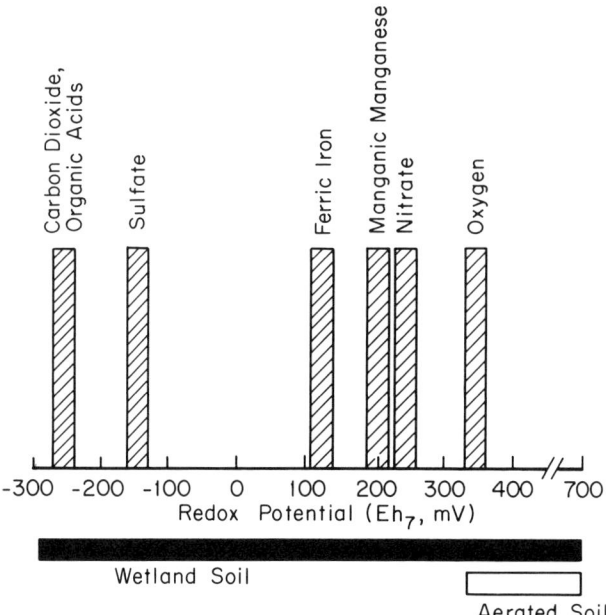

Figure 8.3. Redox stability for the major soil redox systems. (From Faulkner, S. P. and J. L. Richardson. Physical and chemical characteristics of freshwater wetlands soils, in *Constructed Wetlands for Wastewater Treatment* (Chelsea, MI: Lewis Publishers, Inc., 1990), pp. 41–72.)

A schematic of an engineered ecosystem treatment process is shown in Figure 8.2. The sequential reduction of wetland soil is shown in Figure 8.3.

CONSTRUCTED WETLANDS FOR MINE DRAINAGE TREATMENT

Contamination of streams and rivers by acidic mine water is one of the most persistent industrial pollution problems in the United States. The acid lowers the pH of the water, making it corrosive and unable to support many forms of aquatic life. In the Appalachian region, over 8000 km of streams and rivers are adversely affected by drainage from abandoned coal mines. This pollution will likely continue unabated for decades.[6]

Wetlands are a potential natural treatment system for small flows of acid mine water. Two independent studies of *Sphagnum* moss bogs undertaken to determine what adverse effects acid mine water was having on the wetland vegetation indicated a lack of adverse effects and, in fact, natural treatments of the mine water. First, a group from Wright State University studied a site in the Powelson Wildlife area in Ohio where *Sphagnum recurvum* was found growing in pH 2.5 water. Fe, magne-

Table 8.6. Construction and Maintenance at Wetlands Surveyed to Date[a]

Parameter	Type	Frequency (%)	No. of Sites Reporting
Inflow structure	None	32	19
	Seeps into collection basin	26	
	Pipe	26	
	Ditch	26	
	Seeps into French drain	10	
	Level spreader	5	
	Weir	5	
Outflow structure	PVC pipe (10–20 cm or 4–8 in.)	76	20
	Spillway onto pond	14	
	Level spreader/overland flow	5	
	Weir/ditch	5	
Vegetation type	Cattail (*Typha*)	100	20
	Sphagnum	45	
	Horsetail (*Equisetum*)	20	
	None planted	10	
Substrate	Hay	58	12
	Clay	33	
	Spent mushroom compost	25	
	Limestone	17	
	Spoil	8	
	Topsoil	8	
	Manure	8	
	Forest floor litter layer	8	
Survival — cattail	100%	NA[b]	9
Survival — *Sphagnum*	Marginal	NA	3
Changes noted over time	System performance (pH decrease in winter)	NA	7
	Algal population growth	NA	
	Volunteer vegetation	NA	
	New seeps	NA	
	None	NA	
Maintenance	None	NA	9
	Grading to limit surface runoff into wetland	NA	
	Haybale dikes to limit channelization	NA	
	Cleaning of debris behind weir	NA	
Fertilizer/lime application	Agricultural lime	NA	7
	Super phosphate	NA	
	10-10-10	NA	
	10-10-20	NA	
	15-15-15	NA	
	Unspecified	NA	
Problems	Banks breached	NA	8
	Regulation of water depth	NA	
	Sedimentation at outflow	NA	
	Sedimentation at inflow	NA	
	Muskrats	NA	

Table 8.6 (continued). Construction and Maintenance at Wetlands Surveyed to Date[a]

Parameter	Type	Frequency (%)	No. of Sites Reporting
	Flow measurement	NA	
	Washout	NA	
	Channelization within wetland	NA	

[a] These wetlands were all constructed for AMD treatment prior to 1986. Frequencies may total more than 100% for a given parameter due to presence of multiple parameter types.
[b] NA: Frequency data are not available for some parameters because of the limited response to related questions.

From Girts, M. A. and R. I. P. Kleinmann. Constructed Wetlands for Treatment of Acid Mine Drainage, National Symposium on Mining, Hydrology, Sedimentology and Reclamation, Lexington, KY (1986), 165.

sium (Mg), sulfate, calcium (Ca), and Mn all decreased while pH increased from 2.5 to 4.6 as the water flowed through the bog. A natural outcrop of limestone located at the downstream end provided sufficient neutralization to raise the effluent pH to between 6 and 7.

A similar study was conducted by a West Virginia University group at Tub Run Bog in northern West Virginia. They found that acid drainage flowing into the wetland area rapidly improved in quality. Within 20 to 50 m of the wetland border, pH rose from 3.05 to 3.55 to 5.45 to 6.05, while within 10 to 20 m, sulfate and Fe concentrations decreased from 210 to 275 mg/L to 5 to 15 mg/L, and from 26 to 73 mg/L to <2 mg/L, respectively. Overall, they found that the water quality of the bog effluent was equal to that of nearby streams unaffected by mine drainage.

The use of natural wetlands for acid mine water treatment is seldom feasible. While examples of long-established natural wetlands receiving acid water from nearby mining operations have been found, it is rare that a natural wetland system is available on a permitted mine area. Also, there are cases in which wetland vegetation has been killed as a result of the inflow of both treated and untreated mine water.[25] To use natural wetlands as mine water treatment systems, numerous legal barriers must be overcome; so it would appear that in general established wetlands are unlikely to be used for acid water treatment.

A logical alternative is to construct wetlands where needed to substitute for chemical treatment facilities. The feasibility of this concept was first demonstrated in pilot-scale tests, and is now being used on a full-scale basis at over 25 mine sites in Pennsylvania, West Virginia, Ohio, Maryland, and Alabama. These wetlands have been constructed by mining companies specifically for water treatment.

The U.S. Bureau of Mines conducted a survey of existing wetlands in the eastern U.S. for mine drainage treatment in 1986. Surveys have been received for 20 wetlands constructed prior to 1986. The average wetland age (as of July 1986) was 17 months, and the median age was 12 months. A majority of these wetlands are located in a region approximately 75 mi north and northeast of Pittsburgh; five are found in the Sligo Quadrangle of Pennsylvania.[6] Most of the wetlands are situated at the base of a reclaimed hillside, collecting seepage from the hillside; some are at the toe of spoil or at the base of an unreclaimed highwall.

Wetland construction and maintenance data are summarized in Table 8.6. A majority of the survey respondents provided information on inflow and outflow structure, vegetation type, and planting substrate. Fully one third of the wetlands use no inflow structure whatsoever. At approximately 25% of the wetlands, pipes or ditches are used to channel waterflow in the wetland. Other structures, such as level spreaders and French drains, are also used, and at some sites several structures are used in conjunction (ditch and level spreader, for example). Only the collection basin, level spreader, and weir provide any kind of flow regulation, and this occurs through water storage rather than diversion of high flow. For outflow structures, PVC pipe is commonly used, although spillways, level spreaders, and weirs provide regulation of water level at some sites.

Cattail, *Sphagnum* moss, and horsetail were planted, sometimes in separate sections of multisection designs, and occasionally side by side. Cattails were planted at all surveyed sites, and *Sphagnum* at approximately half the sites. Substrate used for planting was related to vegetation type. Hay without underlying limestone was used with *Sphagnum* moss, and hay or the soil present at the planting site was used for cattail and horsetail plots. Other mixtures, including topsoil, were only used in cattail and horsetail plots.

Survey questions related to vegetation survival, changes observed in the wetland over time, maintenance, and general problems which developed were answered by less than half of the respondents. Because of the sketchiness of this portion of the data set, Table 8.6 reports the answer categories without associated frequencies. Even with the limited information, however, some patterns are apparent. The survival of the initial planting of cattail plants or rhizomes was generally judged to be good. In contrast, although data were obtained for only one third of the moss sites on this question, all three sites showed poor survival, and replanting was attempted at least once. Because all of the *Sphagnum* sections reported in this survey were used concurrently with cattail sections, the death of the moss did not result in the failure of the entire treatment system.

Several of the respondents commented that they had observed seasonal changes in system performance. In particular, it was noted that pH dropped during the winter months, although metal removal continued. While pH showed a decrease during the winter months in the surface water of a natural Appalachian *Sphagnum* bog, decreases in pH were also found in nonwetland watersheds during the same period. A study of an established cattail wetland that has been receiving acid mine water since August 1985 did not reveal lower pH readings of either influent or effluent water during the winter of 1986. Because no mention was made in survey responses of seasonal pH fluctuations of influent water, it is possible that a decrease in the pH of

Table 8.7. Wetland Area, Hydrologic Regime, and Detention Time

Parameter	Mean	Median	Range
Size (m^2)	1,550	929	93–6,070
No. of sections	3	3	1–7
Size of section (m^2)	795	276	19–6,070
Flow rate (L/sec)	1.3	0.5	0.06–13
Area: flow rate (m^2/L/sec)	2,390	928	61.1–10,700
Water depth (m)	0.3	0.4	0–2
Treatment capacity (m^3)	565	132	21.2–2,470
Detention time (day)	10.2	4.79	0.17–75.4

From Girts, M. A. and R. I. P. Kleinmann. Constructed Wetlands for Treatment of Acid Mine Drainage, National Symposium on Mining, Hydrology, Sedimentology and Reclamation, Lexington, KY (1986), 165.

water entering the wetland system was simply being passed through in decreased effluent pH. At any rate, the question of seasonal fluctuations in treatment efficiency needs to be investigated over a long period of monitoring inflow and outflow wetland water chemistry.

Maintenance of the constructed wetland systems surveyed was generally limited to the application of fertilizer or agricultural limestone, if any maintenance occurred at all. Agricultural limestone was added periodically at 75% of the sites reporting, when low pH measurements at the discharge point indicated it was necessary.

A clear majority of the difficulties encountered with the constructed wetlands was related to waterflow and depth control. Bank breaching, regulation of water depth, washout/scouring, channelization, and the inability to measure flow rates were listed as problems. We have observed that many of the wetland "failures" (not included in this study) to date have been related to high flow damage. The survey results confirm that this is not an isolated problem.

The most frequently mentioned consequence of high flows was the breaching of banks. As most of the wetlands are sited in backfill or spoil material, banks are composed of only moderately consolidated material. With high water inflow rates and resulting high water levels, banks tend to give way or to be overtopped and eroded at lower spots. Muskrats and other burrowing animals have been implicated in bank damage as well.

The slowing of water as it enters the wetland caused the problem of sedimentation at inflow and outflow structures, more often at the latter. This process may be related to the high frequency of pipe outflow structures, as opposed to their frequency of use as inflow structures. Pipes, in fact, were most often found in conjunction with outflow sedimentation problems.

Design parameters related to wetland area and volume of water treated are given in Table 8.7. Wetlands constructed prior to 1986 usually consist of three compartments of approximately 300 m2 (2970 ft^2, median value) each. Some were much larger with fewer sections; this did not appear to be related to high flow rates or to

Table 8.8. Water Chemistry[a] and Treatment Efficiency for 11 Constructed Wetlands

Parameter	Inflow			Outflow			Treatment Efficiency	
	Mean	Median	Range	Mean	Median	Range	Mean	Mean
pH (units)	4.9	5.8	3.1–6.3	6.0	6.5	3.5–7.7	0.92[b]	0.80
Acidity[c] (mg/L)	170	50	ND[d]–600	40	16	ND–140	0.76	0.68
Fe (mg/L)	33	9.6	0.4–220	1.2	0.5	0.05–7.3	0.96	0.95
Mn (mg/L)	26	20	8.7–54	15	15	0.3–52	0.42	0.25
SO_4 (mg/L)	950	1000	270–1600	740	500	160–1500	0.22	0.50

[a] Reported values rounded to two significant figures due to unknown precision of data.
[b] As [H+].
[c] Acidity is reported as calcium carbonate equivalent. Samples were titrated to an endpoint of pH 8.3.
[d] ND: below detection limits.

From Emerick, J. C. and E. A. Howard. Using Wetland for the Control of Western Acid Mine Drainage, in *SME Annual Meeting Preprint 88-173* (Denver, CO: Society of Mining Engineers, 1988).

water chemistry, and may have been dictated by site or economic constraints. Median flow into these wetlands is 0.5 L/sec (8 gal/min or gpm); again, the distribution is skewed by one extremely high inflow rate and is bimodal, with peak frequencies at 0.4 L/sec (6 gpm) and 1.6 L/sec (25 gpm). The median area-to-flow rate ratio of 928 m^2/L/sec (630 ft^2/gpm) is more conservative than the rule-of-thumb (200 ft^2/gpm) which has been suggested.[26] Evidently, an allowance for high flow rates has been incorporated into the wetland designs.

Reported water depths range from 0 to 2 m (0 to 6 ft), with a median of 0.4 m (1.3 ft). As water depth and vegetation type planted in the wetland are related, with cattails and horsetails planted in water deeper than 0.2 m (0.5 ft), these data reflect the preponderance of cattail wetlands found in this survey. However, it may also point to a causal factor in the observed mortality rate of the *Sphagnum* moss planted; most species of *Sphagnum* have difficulty surviving in areas where the water level above the plant is >2 cm (0.07 ft) due to limited diffusion of carbon dioxide. Deep water also adversely affects water treatment efficiency, which has been observed to decrease at water depths above 0.1 m (4 in).

The wetland capacities (excluding freeboard) range from 21.2 to 2470 m^3, with three large, deep wetlands skewing the distribution. Detention times range from 0.17 days (4 hr) to 75.4 days, with a median detention time of 4.79 days. In comparison, the detention times in a review of wetlands used for wastewater treatment ranged from 2.16 to 145 days. The wetlands for acid mine drainage (AMD) treatment thus provide low to moderate detention times. As studies of optimal detention time for metal removal and acid neutralization in these systems have not yet been undertaken, the significance of these detention times is unknown.

Treatment is ongoing in the wetlands surveyed. Results of water chemistry information supplied on the survey sheets (Table 8.8) suggest that hydrogen and Fe ion removal efficiencies are high, from 80 to 96%. Acidity (titrated to pH 8.3) effects are less, showing a 68 to 76% decrease from inflow to outflow. Mn and sulfate are reduced 22 to 50%. Median removal efficiencies are lower than those calculated from mean values, except for sulfate, for which the concentration values are normally distributed at the inflow but abnormally distributed at the outflow. A qualifier must be added here, however, as it is possible that the outflow samples reported by the mining company may be collected at the absolute outflow point, i.e., after additional chemical treatment to raise pH. In such cases, the results do not represent only the effect of the wetland, but of the entire treatment system.

In comparison, maximum removal efficiencies found in a 70-m transect from the edge of a surface mine toward the center of a natural *Sphagnum*-dominated wetland were 99, 98, and 97%, respectively. A field test of the effects of *Sphagnum* moss in a shallow portable bog unit found little reduction in pH or acidity within the moss section, but a 43 to 90% reduction in both after the water flowed from the moss through a section filled with coarsely crushed limestone. Sulfate concentrations were similarly unaffected by flow through the moss bed. Fe concentrations, on the other hand, were decreased by 50 to 70% at the outflow point. In a bench scale laboratory experiment, Fe removal ranged between 65 and 95% in troughs with differing pH and Mn concentrations; Mn removal ranged from <5% to 76% in the same experiments.

In a study of natural cattail wetlands receiving mine drainage, pH and Mn in effluent waters in July were found to both decrease and increase, while total Fe concentrations decreased by 26 to 88% and sulfate concentrations decreased by 8 to 87% relative to inflow chemistry. At a small cattail wetland in southwestern Pennsylvania, which has been receiving acid mine water since August 1985, removal efficiencies (as determined from median values of weekly samples) are 37, 58, 58, 14, and 47% for hydrogen ions, acidity, Fe, Mn, and sulfate, respectively.

Thus, while removal efficiencies vary between natural wetlands, laboratory models, and constructed wetlands, and between *Sphagnum*- and *Typha*-dominated wetlands, several common trends are noted between these wetlands: Regulation of pH is variable; with the presence of limestone the hydrogen ion concentration can be lowered by as much as 90%. The variation in acid neutralization may be linked to exposure to limestone removal of metals, while the variation in sulfate removal may be due to the presence or absence of an anaerobic environment in which sulfate reduction can take place. Fe is consistently removed to a greater extent than Mn, possibly because of the preferential binding of Fe to cation exchange sites and to the presence of metal oxidizing bacteria.

Metal Removal in Constructed Wetlands

One of the major environmental problems in western mining districts is the control of mine discharge, which is often characterized by high metal concentrations and low pH. In Colorado, over 1400 mi of streams and rivers exceed basic water quality standards for aquatic life, agriculture, or for the domestic water supply, because of high metal loading. While elevated metal concentrations are an expected natural consequence of waters flowing over and through mineralized rock, the problem is greatly exacerbated by mining activity, particularly from abandoned or inactive mines in which no present treatment is in place.

A wetland treatment can remove metals from influent waters in a variety of ways. As chemical oxidation takes place, oxides and hydroxides of Fe can be expected to precipitate from surface flow, resulting in the characteristic yellow sediment in mine drainage known as "yellowboy." These precipitates coat the surfaces of the bottom or settle as a flocculate that is subsequently filtered out as the water passes through the soils and around vegetation. Because precipitation of metals is strongly pH-dependent, and because the pH of mine drainage often is low (usually from 2.5 to 5.5), ferric hydroxide will be the main precipitate, with most other metals remaining soluble in such acid conditions. Where the pH of the water is buffered at higher levels, such as might occur in the presence of carbonate orebodies, precipitation of other metals will take place, often as coprecipitates with the Fe oxyhydroxides.

Under conditions of neutral or alkaline pH, precipitation is the dominant form of metal removal — one reason that conventional treatment techniques often specify the addition of lime. Under low-pH regimes in wetlands, other processes become more important in removing metal ions.

Organic soils, especially peat, typically have a high cation exchange capacity (CEC), generally attributed to humic and related organic acids. The CEC is a measure of the ability of a substance to adsorb positively charged ions (such as metal ions) by exchanging them for other ions bound to the molecular structure of the substance. Values of several commercially available peats analyzed at the Colorado School of Mines were in the range of 0.13 to 1.05 meq/g dry peat. Actual CEC values of wetland soils vary considerably, depending on the conditions of soil formation, the nature of the ions already occupying molecular exchange sites, and the percentage of the inorganic soil fraction. For example, the soil CEC of seven high-elevation wetland sites (3280 to 3770 m) in central Colorado ranged from 0.07 to 0.65 meq/g dry soil.[7] With time, metal adsorption by cation exchange will diminish as exchange sites in the organic soils become saturated with metals. Thus, the main limiting factor for subsequent adsorption by cation exchange would be the annual addition of fresh organic matter produced by the wetland plants.

Bacteria and other microorganisms are important biological components of wetland water and soils. Bacteria, algae, and fungi can affect dissolved metal concentrations by direct accumulation of metals within their body structure or by producing sufficient changes in the aqueous environment to facilitate the precipitation of metals from solution. Bacteria, in particular, have been shown to possess the ability to scavenge large amounts of metal from their environment. Sulfate-reducing bacteria are capable of producing metal and hydrogen sulfides that can be accumulated in the oxygen-poor reducing environment of the sediments. A beneficial side effect of this microbial process is the removal of excess hydrogen ions by the release of hydrogen sulfide gas, resulting in less acidic conditions.

Vegetation growing in the wetland is also an important component of the metal removal process. The annual cycle of biomass production provides potential exchange sites as old leaves, stems, and roots die. Decomposing plants are also a source of organic carbon that supplies energy for microbial processes. Wetland plants can absorb certain metal species directly through their root systems, and some metals may be translocated to other plant parts. At the end of the growing season, some accumulated metals are released as the aboveground portions die. It is possible that metal-rich root systems in oxygen-poor zones may retain high concentrations, even after death, if redox potentials are sufficiently low to permit conversion of the metals to an insoluble sulfide form. The retention of accumulated metals by plant parts has not been well studied for wetland situations, so the relative importance of annual net removal of metals by vegetation is not known. In the Rocky Mountains, the composition of wetland plant species changes significantly from lower to higher elevations. For example, foothill sites are dominated by cattails (*Typha*), bulrushes (*Scirpus*), and large sedge species (e.g., *Carex nebraskensis* and *C. utriculata*). High-elevation

Table 8.9. Surface Water Quality in a Wet Meadow Below the Shoe Basin Mine, Central Colorado

Distance Below Adit (m)	pH	Conductivity (µmhos/cm)	Conc. (ppm)			
			Fe	Mn	Zn	Cu
0	3.35	700	13.3	55.3	18.4	0.69
50	3.15	600	6.9	36.6	13.2	0.57
80	3.45	350	2.0	28.1	6.7	0.15
100	3.65	280	0.5	20.5	6.0	0.13
150	3.85	220	0.4	14.5	4.5	0.09
180	4.40	100	<0.1	1.0	0.9	0.02

From Emerick, J. C. and E. A. Howard. Using Wetland for the Control of Western Acid Mine Drainage, in *SME Annual Meeting Preprint 88-173* (Denver, CO: Society of Mining Engineers, 1988).

Table 8.10. Surface Water Quality in a Wet Meadow Below an Inactive Mine at Chattanooga, Southwestern Colorado

Distance Below Adit (m)	pH	Conductivity (µmhos/cm)	Conc. (ppm)			
			Fe	Mn	Zn	Cu
0	6.2	280	7.7	1.9	0.3	0.01
15	5.0	280	6.9	2.0	0.3	0.01
150	5.1	280	0.3	2.2	0.4	0.02
200	5.0	240	0.1	1.5	0.5	0.02

From Emerick, J. C. and E. A. Howard. Using Wetland for the Control of Western Acid Mine Drainage, in *SME Annual Meeting Preprint 88-173* (Denver, CO: Society of Mining Engineers, 1988).

sites above 3280 m are often dominated by a near-monoculture of aquatic sedge (*C. aquatilis*). Differences in plant composition and climatic conditions will likely affect rates of metal uptake by the vegetation as well as the rate of peat formation, although data are lacking to provide a clear estimate of the magnitude of this effect. We generally see deeper peat accumulations at higher elevations, presumably because of slower decomposition rates in colder wetland soils.

While wetland systems offer potential as a treatment medium, no long-term scientific studies have been conducted of their ability to maintain a given level of metal removal, nor has much information been published regarding their cost-effectiveness. Investigators from the U.S. Bureau of Mines have summarized the results of studies on a large number of systems installed in the eastern coal mining regions. While some successes are evident, most of the systems are only a few years old. Few systems have been constructed in the Rocky Mountain region, and data on their metal removal ability are incomplete. Available land area will more than likely be one of the most important considerations, assuming that ongoing studies substan-

Table 8.11. Containment Removal Mechanisms in Aquatic Systems Employing Plants and Animals

Mechanism	Contaminant Affected[a]	Description
Physical		
Sedimentation	P — Settleable solids S — Colloidal solids I — BOD, nitrogen, phosphorus, heavy metals, refractory organics, bacteria, and virus	Gravity settling solids (and constituent contaminants) in pond/marsh settings
Filtration	S — Settleable solids, colloidal solids	Particulates filtered mechanically as water passes through substrate, root masses, or fish
Adsorption	S — Colloidal solids	Interparticle attractive force (van der Waals' force)
Chemical		
Precipitation	P — Phosphorus, heavy metals	Formation of or coprecipitation with insoluble compounds
Adsorption	P — Phosphorus, heavy metals	Adsorption on substrate and plant surface
Decomposition	S — Refractory organics P — Refractory organics	Decomposition or alteration of less stable compounds by phenomena such as UV irradiation, oxidation, and reduction
Biological		
Microbial metabolism[b]	P — Colloidal solids, BOD, nitrogen, refractory organics, heavy metals	Removal of colloidal solids and soluble organics by suspended, benthic, and plant-supported bacteria; bacterial nitrification/denitrification; microbially mediated oxidation of metals
Plant metabolism[b]	S — Refractory organics, bacteria, and virus	Uptake and metabolism of organics by plants; root excretions may be toxic to organisms of enteric origin
Plant absorption	S — Nitrogen, phosphorus, heavy metals, refractory organics	Under proper conditions, significant quantities of these contaminants will be taken up by plants
Natural dieoff	P — Bacteria and virus	Natural decay or organisms in an unfavorable environment

a P = primary effect; S = secondary effect; I = incidental effect.
b Metabolism includes both biosynthesis and catabolic reactions.

tiate the efficacy of wetlands treatment. The Bureau of Mines has been recommending 200 ft^2 (21.5 m^2) of wetland surface area for each gallon per minute (3.79 L/min) of discharge, which would amount to nearly 0.5 acre (2023.4 m^2) for a discharge of 100 gpm (378.5 L/min). Even that amount of area may not be enough if the site is at a relatively high elevation where the cooler climate would slow geochemical and biological processes.

The composition and pH of the mine discharge will also affect the potential degree of treatment that might be expected from a wetland. Wetlands appear to remove Fe relatively efficiently, but the same may not necessarily hold true for other metals. Dissolved metal concentrations from two natural high-elevation wetlands receiving mine discharge in the Rocky Mountains of Colorado show different trends. At the Shoe Basin Mine, located at 3638 m in the Front Range of central Colorado, approximately 56 L/min of discharge flows into a small wetland. Table 8.9 shows pH, conductivity, and concentrations of selected dissolved metals in the surface water as it flows through the wetland. The wetland in this case is a wet meadow (technically a fen) dominated by aquatic sedge and underlain by several feet of peat soils. Each metal decreased in concentration with greater distance from the mine adit, and pH increased, with the exception of the first 50-m stretch downstream from the mine adit. We believe the decrease in pH in this initial segment was due mainly to the oxidation of ferrous Fe. The total metal concentration in a soil sample from the meadow was 128 mg/g dried soil, which included 95.5 mg/g of Fe, 29.2 mg/g of Al, and 1.4 mg/g of Zn. The CEC of the soil at this site was relatively low (0.13 meq/g soil) and represented only perhaps 2% of the soil metals. Total metal concentrations in the leaves of aquatic sedge growing in the meadow exceeded 2000 mg/kg, and included 151 mg/kg of Zn, 882 mg/kg of Fe, 967 mg/kg of Mg, 96 mg/kg of Al, and 26 mg/kg of Pb.

Surface water quality data from a wetland at the Chattanooga Townsite show a similar decrease in Fe, but relatively little change in other metal concentrations. This site is located in the San Juan Mountains, approximately 5 mi northwest of Silverton, CO, at an elevation of 3346 m. Here, the initial pH was relatively high (6.2) where the discharge exits the adit at a flow of approximately 75 L/min, which dropped to 5.0 as it flowed through the wet meadow (Table 8.10). Both this and the Shoe Basin site have been abandoned for several decades. It is unknown to what extent the wetland processes are in equilibrium with the metal load from the mines, but it is apparent that the vegetation, which was probably disturbed to some degree during the mining operations, its acid and metal tolerance is sufficient to recolonize both of these drainage-affected areas. Improvement of water quality as mine discharge flows through wetland areas such as that above, which have become naturally established, may be taken as an indication that wetlands may retain a long-term capability for removing some metal species. Contaminant removal mechanisms are given in Table 8.11.

Given the incomplete knowledge regarding process rates and long-term effectiveness, we are not yet in a position to design a wetland system to meet the needs of a specific mine drainage situation. Expectations of using such systems for complete treatment, particularly for discharge with excessively high metal concentrations, are premature or at best should be viewed with caution. However, using the systems as a pretreatment or as a finishing or polishing treatment in conjunction with conventional treatment methods should be considered.

Site Selection

The site of a constructed wetland is often determined by the location of the wastewater source. The wastewater source can seldom be relocated. Therefore, siting a wetland system is usually limited to the immediate area, which is often unsuitable. However, siting may be optimized through a systematic investigation process. The site investigation process includes site selection, temporary and permanent engineering design, environmental effects analysis, construction evaluation, remedial works design, and construction and operational checks. Geological, geotechnical, hydrological, and other environmental factors are considered in the site selection process. Site selection is influenced by the availability of a suitable site and geotechnical merits; for example, well-developed soils, good access, or low flood potential.[8]

Site investigation and selection include:

- Preliminary fact-finding survey
- Aerial photography interpretation
- Initial field survey
- Limited subsurface exploration, site soils classification, and environmental data
- Potential environmental effects and regulatory requirements

Site selection considerations include the following factors:

- Land use considerations
- Hydrology
- Geology
- Environmental considerations

In siting a wetland the most important considerations may be land use and access. The wastewater to be treated must be accessible to the site. For mine drainage wetland systems, wastewater access to numerous sites may be limited if more natural, self-sustaining gravity-flow systems are desired. The site must be accessible to construction equipment, operational personnel, and chemical delivery vehicles.

Land availability is an important issue. Some states may require surface control of the site by the operator. Required land area is determined by the wetland size, flow control structures, building and equipment storage areas, access roads, utility rights-of-way, potential chlorine contact basin, a flow equalization basin, and a final chemical treatment cell.

Hydrologic considerations include characterizing surface and groundwater flow patterns, use, quantity, and chemistry. Hydrology, if not properly evaluated during the siting process, may significantly impair the operation and effects of a wetland.

Drainage basin characteristics for candidate sites should be evaluated along with flooding and scouring possibilities. Minimum, maximum, and average seasonal water levels influencing the wetland hydroperiod should be determined from site date. Wetlands should be sited away from major streams or springs because of potential flooding, scouring, sedimentation, erosion, high groundwater tables, saturated ground, and variable water quantity and quality.

The wastewater and receiving surface waterbodies should be chemically characterized for existing and potential use downstream. Downstream users should be identified and if possible, sites that would adversely affect downstream users should be avoided.

For AMD treatment, the minimum water quality analyses for baseline and wastewater characterization should include:

- pH
- TSS (total suspended solids)
- DO (dissolved oxygen)
- Fe
- Mn
- SO_4

Groundwater hydrology characterization should include:

- General flow pattern
- Depth
- Quality
- High water tables
- Existing and potential use
- Nearby wells

Depending on the wastewater type, wetlands may be located in either groundwater recharge or discharge areas. AMD system substrates and dikes, if constructed of inferior material, may allow infiltration or flow-through seepage. Therefore, siting investigations should evaluate the sensitivity of the hydrologic environment to alterations. The wetland area should include monitoring points downstream of the final wetland cell. Sites overlying karst geology, parched water tables, arid streambeds, fissured igneous or metamorphic rock, or known groundwater recharge areas should be avoided if the wastewater has the potential for recharging groundwater with contaminants. In certain soils, low-permeability or impermeable liners or compaction can help. Ideally, a constructed wetland should be sited in a groundwater discharge area to minimize the potential groundwater pollution.

Geologic considerations in the site selection process include:

- Surface materials and soil characterization
- Bedrock depth
- Topography
- Other geotechnical aspects

In mined areas, land subsidence or sinkholes may influence wetland siting. Soil surveys, geologic and geologic hazard maps, and aerial photographs should be evaluated for potential land subsidence, mine shafts, sinkholes, faults, or other factors that could affect engineering works or cause groundwater contamination from a constructed wetland.

Environmental and regulatory considerations, both state and federal, should be considered.

Performance Expectations

Wetland systems reduce many contaminants, including biochemical oxygen demand (BOD), suspended solids (SS), nitrogen, phosphorus, trace metals, trace organics, and pathogens. This reduction is achieved by different mechanisms: sedimentation, filtration, chemical precipitation, adsorption, microbial interactions, and uptake vegetation.

Settleable organics are rapidly removed in wetland systems by quiescent conditions, deposition, and filtration. Attached and suspended microbial growth is responsible for the removal of soluble BOD. SS are filtered and settled within the first few meters beyond the inlet. Cell length can be important in Fe removal. The relation can be attributed to oxidation rates.

Metal removal processes include sedimentation, filtration, adsorption, complexation, precipitation, plant uptake, and microbially mediated reactions, especially oxidation. Early mine drainage systems attempted to construct bogs predominantly with *Sphagnum* or other mosses because Fe and Mn removal was observed in natural bogs receiving mine drainage. Systems with humic substrates can be effective because of their ion exchange capacity. A natural peatland (bog) in Minnesota removed 80% of the nickel and nearly 100% of the copper from tailings drainage. Uptake by vegetation accounted for <1% of total metal removal. However, the ion exchange capacity of humic materials or mosses has only limited functional longevity. A *Sphagnum* bed with an area of 1600 m^2 and a depth of 30 cm treating a mine drainage flow of 40 L/min and an Fe concentration of 200 mg/L may be saturated in only 2.2 years.

Due to comparatively high process reactive surfaces in marshes and available knowledge on marsh construction and marsh wastewater systems, most recent AMD treatment systems include marshes dominated by cattail (*Typha*) or other emergents with in situ substrate materials. Metallic ion concentrations occurring on stems, leaves, and roots and knowledge of bacterial oxidation of Fe and Mn indicated the important role of microbial processes.

In recent research, extracellular microscopic encrustations were analyzed on filamentous algal cells, and intracellular crystals were observed in suspended algal cells. Algal death, precipitation, and burial immobilized these metals for long periods.

Constructed wetlands (marshes) have high removal efficiencies for Fe and lower efficiencies for Mn. Fe and Mn removal efficiencies ranged from 0 to 99% and –8 to 96%, respectively, in research involving ten constructed wetlands.

Most systems treating AMD are surface flow systems. Subsurface flow systems can provide more contact surface for absorption, but substantial deposits of insoluble metal compounds may clog the system. For example, a 0.6-ha surface flow constructed wetland reduced dissolved Fe of 80+ mg/L to <1 mg/L in 110 L/min average flow, immobilizing 4636 kg/year of Fe plus other metal oxides and hydroxides.

Modification of pH is often needed to treat AMD. Water quality standards for receiving streams commonly require the pH of AMD to be as low as 2. For AMD systems, pH changes in *Sphagnum* wetlands range from a decrease to no change. The reduction of pH was postulated from the exchange of hydrogen cations on moss for Fe cations. Cattail and mixed-species systems may tend to increase from inlet to outlet. Mean influent pH of 20 systems was observed as 4.9 and mean effluent pH was 6.0.

Changes in pH within a wetland can be caused by several mechanisms. The production of humic substances, much of which consist of organic acids, can increase acidity. In wetlands treating AMD, the oxidation of ferrous Fe to ferric Fe and subsequent hydrolyzation to ferric hydroxide may be an important source of acidity. Formation of CO_2, carbonates, and bicarbonates will cause an increase in acidity.

Wetlands are essentially reducing ecosystems. Reduction of nitrate and sulfate causes the production of alkalinity. In algal-dominated systems, removing carbon dioxide increases pH and alters forms of alkalinity.

Design loading rates are based on these considerations:

- Treatment objectives — stringent water quality standards require lower loading rates
- System use — basic treatment following pretreatment, secondary/advanced treatment, or polishing treatment
- System type — surface flow or subsurface flow
- System configuration — multiple cells in series, in parallel, or in combination
- Safety factors

For AMD treatment systems, the median loading rate has been found to be 928 m^2/L/sec, ranging from 61.1 to 10,700 m^2/L/sec. A suggested rule-of-thumb value is 294 m^2/L/sec. According to Tennessee Valley Authority (TVA) -developed guidelines, loading rates of 2.0 and 7.0 cm^2/mg/min for Fe and Mn, respectively, are recommended for waters with pH <5.5. For pH >5.5, the loading rate is increased to 0.75 $m2$/mg/min for Fe and 2.0 m^2/mg/min for Mn.

Hydraulic Design and Control Structures

Constructed wetlands are attached-growth reactors. Performance is given by the first-order, plug-flow kinetics:

$$\frac{C_c}{C_o} = \exp\left[-K_t t\right] \qquad (8.2)$$

where C_c = effluent concentration (mg/L)
C_o = influent concentration (mg/L)
K_t = temperature-dependent first-order reaction rate constant (days^{-1})
t = hydraulic residence time (days)

According to the above equation, as the hydraulic residence increases, effluent concentrations of biodegradable contaminants decrease. Consequently, hydraulic residence time is a key design and operational parameter.[11]

Hydraulic residence time is defined as:

$$t = \frac{LWnd}{Q} \qquad (8.3)$$

where L = length of system (m)
W = width of system (m)
n = porosity of the bed (%)
d = depth of submergence (m)
Q = average flow through the system (m^3/day)

In AMD treatment systems, the basic design is surface flow type. A surface flow system consists of a cell(s) with wastewater routed at shallow depths over a substrate supporting emergent vegetation. Flow is controlled by the shallow depth, low flow velocity, and the plant stems and litter.

The configuration of a wetland system affects important hydrologic factors; for example, water velocity, water depth, fluctuation, detention time, circulation and distribution patterns, turbulence, and wave action. Configuration should enhance wastewater distribution to maximize contact between wastewater, substrate, and vegetation by minimizing short-circuiting. Configuration choice should consider these factors:

- Degree of pretreatment
- Required treatment area
- Available land shape
- Slope
- Length-to-width ratio

- Desired bed slope
- Needed excavation and grading to obtain cell depth and slope
- Substrate type
- Internal dikes
- Operational and maintenance flexibility

A wetland cell can be designed to use one or more of the three types of flow patterns:

- Plug flow
- Step feed
- Recirculation

Plug flow is now used for most acid drainage systems. It requires minimal piping, energy use, operation, and maintenance.

Alternative configurations include a single cell, parallel cells, and serial cells (longitudinal or serpentine), and a combination of wetland cells and pond.

A single cell is the simplest design and the least expensive to build, but operational flexibility is limited. A single cell is recommended only for small flows (<4 m³/day).

At least two parallel cells are desirable for operational and maintenance flexibility. Flow is distributed equally or proportionally between the cells based on loading rates or other reasons.

Serial cells can be longitudinal or serpentine so as to obtain a greater variety of treatment mechanisms.[9]

For surface flow systems, the length:width ratio is typically 10:1 or greater, to ensure plug-flow conditions.

The porosity of the system is defined as:

$$n = \frac{V_v}{V} \tag{8.4}$$

where V_v is the volume of voids, and V is the total volume.

In surface flow systems, V_v is practically the volume not occupied by vegetation and varies with the type and density of live and dead vegetation. Preliminary data indicate the potential volumes of several common species are cattails (*Typha*), 10%; bulrush, 14%; reeds, 2%; woolgrass, 6%; and rushes, 5%. Corresponding porosity values range between 86 and 98%.

Organic loading in AMD wetlands is light, and water depth is not critical in maintaining aerobic conditions. The system capacity should allow for substantial deposition of metal compounds during operational lifetimes. Fe removal efficiencies of 98% of 135 mg/L Fe in influents represents a deposition of 1.5 kg/m2/year of Fe plus associated substances. Water depths should vary from 2 to 3 cm at the inlet to

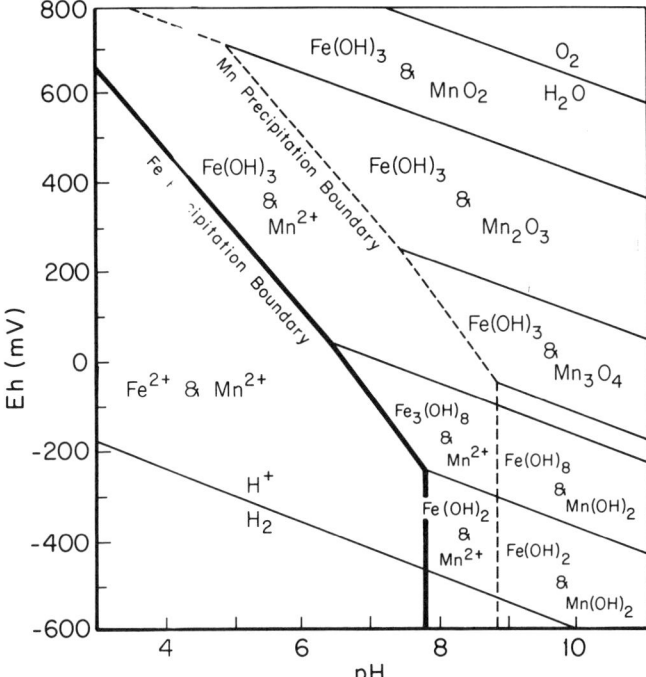

Figure 8.4. Composition Eh-pH iron and manganese stability diagram. (From Faulkner, S. P. and J. L. Richardson. Physical and chemical characteristics of freshwater wetlands soils, in *Constructed Wetlands for Wastewater Treatment* (Chelsea, MI: Lewis Publishers, Inc., 1990), pp. 41–72.)

provide optimal habitats for emergent plants to 70 to 100 cm at the outlet to accommodate metal deposits.

To minimize short-circuiting, the lateral bed slope should be almost 0, and a uniform longitudinal slope should range from 0 to 1%.

The cross-sectional area needed for a given flow is determined by the bed hydraulic gradient and hydraulic conductivity. Hydraulic conductivity changes as vegetation and microbiological units mature. In mature systems, hydraulic conductivity may vary from inlet to outlet.

Inlet and outlet structures for surface flow systems include an open-end pipe or channel/spillway flowing to and from the wetlands. They are sized to handle maximum-design storm flows and are normally sited to minimize short-circuiting, but vegetation, internal dikes, water depth, solids deposition, bottom uniformity, and length:width ratios affect short-circuiting. An example of an inlet and outlet structure is shown in Figure 8.4.

Dikes or other carriers are used in wetland streams to control flow paths and minimize short-circuiting. Finger dikes are commonly used to create serpentine

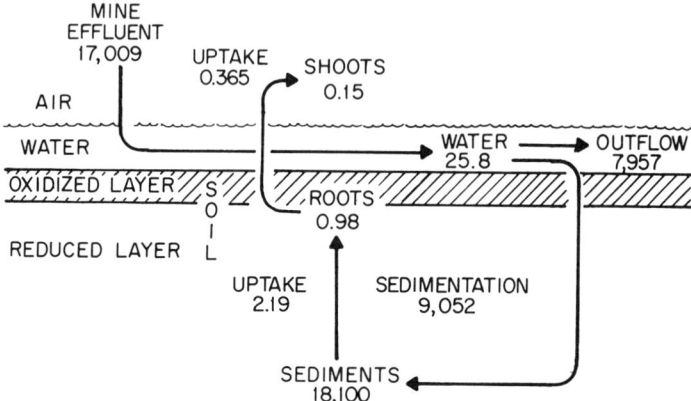

Figure 8.5. Iron fluxes and pools in a constructed wetland. (From Faulkner, S. P. and J. L. Richardson. Physical and chemical characteristics of freshwater wetlands soils, in *Constructed Wetlands for Wastewater Treatment* (Chelsea, MI: Lewis Publishers, Inc., 1990), pp. 41–72.)

configurations (Figure 8.5). Divider dikes separate cells and attain desired length:width ratios within site-specific constraints. Dikes are usually constructed of native soils, but finger dikes are also constructed with sandbags and treated timber.

SUBSTRATE

Substrate-water and substrate-root interfaces are critical in the development of aerobic-anaerobic treatment mechanisms. The substrate supports vegetation, provides surface area for microorganism attachment, and is associated with the physical and chemical treatment mechanisms. Substrates influence treatment capability by affecting detention time, contact surfaces for organisms with the wastewater, and oxygen availability.[10]

Substrate selection is based on cost and treatment requirements. Substrates include natural soils, soil mixtures, coal-fired power plant ash, and combinations.

Substrate type has little influence on SS, organic removal, and biological degradation of organics. Substrates influence the removal of some contaminants (e.g., metals) through ion exchange and adsorption onto humic and fulvic substances and clay particles. Organic soils and clay minerals have higher exchange capacities than coarse mineral substrates such as gravel. Organic soils with a high humic content readily remove metallic ions through ion exchange.

Substrate Evaluation for Acid Mine Drainage Systems

As shown in Figure 8.6, each of the 20 cells consisted of a buried, half-round, fiberglass culvert 6.3 m (20 ft) long and 1.1 m (42 in.) in diameter. An unlined pond

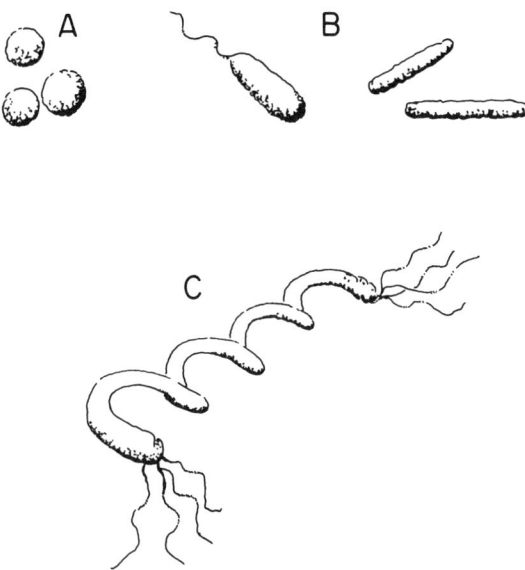

Figure 8.6. Bacterial morphology: (A) cocci form, (B) bacilli, (C) spiral form. (From Portier, R. J. and J. P. Stephen. Wetland microbiology: form, function, processes, in *Constructed Wetlands for Wastewater Treatment* (Chelsea, MI: Lewis Publishers, Inc., 1990), pp. 89–105.)

in spoil material was planted with marsh/wet meadow vegetation. Pond and pea gravel cells served as controls.[12]

Water from a mine seep was pumped into a head box and gravity discharged into each cell. A PVC pipe collection system discharged into a field-scale wetlands treatment system and then to a stream. Appropriate valving provided flow control to each cell. Seep water had a pH of 5.9 s.u., with 37 mg/L Fe and 16 mg/L Mn.

Clay was obtained from the B soil horizon at an undisturbed site near the facility. Mine spoil was obtained on site, and pea gravel was purchased from a river gravel operation. Topsoil consisted of soil from the A horizon harvested from nearby agricultural fields. The acid wetland consisted of substrate from a natural wetland below an acid seep nearby. The natural wetland consisted of substrate from a nearby natural wetland without acid drainage.

Samples of each substrate type (except topsoil) were analyzed for standard soil composition and chemistry parameters (Table 8.12). After planting, ten times the desired level of 0.32 kg of 6-12-12 fertilizer was inadvertently placed in each cell, resulting in an equivalent rate of 4638 kg/ha (4097 lb/acre).

Initial evaluations were designed to test cattail (*Typha latifolia*) with bulrush (*Scirpus cyperinus*) planted in three cells. Planting material was obtained from nearby wetlands unimpacted by mine drainage. Vegetative material was washed to remove soil, cut to the standard stem length of 30 cm, and weighed before planting.

Table 8.12. Composition of Five Substrate Types Tested

Soil Parameters	Substrate Types				
	Natural Wetland	Acid Wetland	Clay	Mine Spoil	Pea Gravel
Organic matter (%)	0.6	0.9	0.5	0.1	1.3
Phosphorus (mg/L)	5.0	6.0	2.0	3.0	1.0
Potassium (mg/L)	94.0	98.0	72.0	67.0	1.0
Magnesium (mg/L)	114.0	80.0	29.0	37.0	1.0
Calcium (mg/L)	650.0	290.0	40.0	20.0	4.0
Sodium (mg/L)	11.0	17.0	9.0	10.0	4.0
pH (mg/L)	7.0	4.4	5.1	4.8	5.7
Nitrate (mg/L)	4.0	4.0	4.0	6.0	1.0
Sulfur (mg/L)	30.0	384.0	41.0	131.0	26.0
Zinc (mg/L)	2.4	7.0	0.9	2.1	0.3
Manganese (mg/L)	122.0	45.0	9.0	6.0	1.0
Iron (mg/L)	94.0	138.0	20.0	8.0	1.0

From Steiner, G. R. and R. J. Freeman. Configuration and substrate design considerations, in *Constructed Wetlands for Wastewater Treatment* (Chelsea, MI: Lewis Publishers, Inc., 1990), pp. 363–377.

Fifty-seven cattails, with an average total weight of 6132 g, were planted on 30-cm centers in 3 rows of 19 plants in 16 cells. Three cells were similarly planted with bulrush, having an average total weight of 10,639 g. Stem densities and heights were periodically measured throughout the study.

Relatively high flow rates to each cell (1.0 L/min) were established, then reduced to 0.5 L/min to accelerate the development of differential treatment results. Individual cell flow rates were measured, recorded, and adjusted daily.

Weekly or biweekly, inflow water samples obtained from the manifolds, and outflow samples from the upper level of each cell, were analyzed for pH, temperature, DO, conductivity, TSS, oxidation-reduction potential, total and dissolved Fe, total and dissolved Mn, ferrous Fe, Al, and sulfate. A two-way analysis of variance (ANOVA) with a Ryan-Einot-Gabriel-Welsch multiple F-test and paired t-tests were used in data analysis.

Because these results represented only the first year of operation, the statistical analyses were limited to those parameters most important to operational permit discharge limits, i.e., pH, dissolved Fe and Mn, and TSS.

All substrate types significantly reduced dissolved Fe and TSS and increased pH levels (Table 8.13). Effluent concentrations of dissolved Fe ranged from 4.8 mg/L (acid wetland/cattail) to 7.2 mg/L (topsoil/cattail) (Table 8.14). TSS in cell effluents ranged from 13.9 mg/L (acid wetland/cattail) to 14.5 mg/L (natural wetland/cattail). Variation in effluent pH values was lower, ranging from 6.2 s.u. (natural acid wetland/cattail) to 6.4 s.u. (topsoil/cattail). Effluent values for dissolved Mn, ranging from 13.9 mg/L (acid wetland/cattail) to 14.5 mg/L (natural wetland/cattail), tested significantly different (four of six substrates) from influent values (14.8 mg/L), but

Table 8.13. Comparisons of Influent-Effluent Concentrations of Fe, Mn, TSS, and pH between Six Replicated Substrates

Substrate/Vegetation	N	Mean Effluent Conc. (mg/L)	Prob > t
	Influent 32.3 mg/L	Dissolved Iron	
Topsoil/cattail	57	7.2	<0.0001
Natural wetland/cattail	57	6.1	<0.0001
Clay/cattail	57	5.6	<0.0001
Acid wetland/bulrush	57	5.4	<0.0001
Mine spoil/cattail	57	5.3	<0.0001
Acid wetland/cattail	57	4.8	<0.0001
	Influent 14.8 mg/L	Dissolved Manganese	
Natural wetland/cattail	57	14.5	0.254
Clay/cattail	57	14.3	0.007
Mine spoil/cattail	57	14.3	0.006
Topsoil/cattail	57	14.2	0.006
Acid wetland/bulrush	57	14.1	0.038
Acid wetland/cattail	57	13.9	0.076
	Influent 32.1 mg/L	Total Suspended Solids[a]	
Topsoil/cattail	55	16.6	0.008
Clay/cattail	55	13.6	0.004
Acid wetland/cattail	55	12.2	0.001
Mine spoil/cattail	55	12.1	0.0006
Natural wetland/cattail	55	12.1	0.002
Acid wetland/bulrush	55	11.8	0.001
	Influent 5.9 s.u.	pH[b]	
Topsoil/cattail	56	6.4	0.0001
Acid wetland/bulrush	52	6.3	0.0001
Clay/cattail	54	6.3	0.0001
Mine spoil/cattail	53	6.3	0.0001
Natural wetland/cattail	54	6.2	0.0001
Acid wetland/cattail	53	6.2	0.0001

[a] Two influent (and corresponding effluent) values over 400 mg/L were excluded from the analysis.
[b] pH values over 7.5 were excluded from the analysis.

From Brodie, A. G. et al. An Evaluation of Substrate Types in Constructed Wetland, U.S. Bureau of Mines, IC 9183 (1988), pp. 389–398.

absolute removal values were very low and statistical significance was inconsequential.

Seasonal comparisons of effluent values revealed that acid wetland/cattail significantly reduced dissolved Fe during the winter (January–February) but other seasonal differences (March–September) in Fe or in other parameters were unclear.

Table 8.14. Effect of Substrate Type on Effluent Concentrations of Dissolved Fe and Mn, TSS, and pH

	Mean Effluent Conc. (mg/L)	N	Wetland Type
Dissolved Fe (mg/L)	5.5	52	Topsoil/cattail
	5.2	54	Clay/cattail
	5.1	53	Natural wetland/clay
	4.8	54	Mine spoil/cattail
	4.7	54	Acid wetland/bulrush
	4.6	53	Acid wetland/cattail
Dissolved Mn (mg/L)	14.3	53	Natural wetland/cattail
	14.2	54	Mine spoil/cattail
	14.1	54	Clay/cattail
	14.0	52	Topsoil/cattail
	14.0	53	Acid wetland/cattail
	13.9	54	Acid wetland/bulrush
TSS (mg/L)	13.7	52	Topsoil/cattail
	11.8	54	Clay/cattail
	11.2	53	Acid wetland/cattail
	10.7	54	Mine spoil/cattail
	10.2	53	Natural wetland/cattail
	10.0	54	Acid wetland/bulrush
pH (standard units)	6.4	51	Topsoil/cattail
	6.3	51	Clay/cattail
	6.3	49	Acid wetland/bulrush
	6.3	50	Mine spoil/cattail
	6.3	50	Natural wetland/cattail
	6.2	49	Acid wetland/cattail

From Brodie, A. G. et al. An Evaluation of Substrate Types in Constructed Wetland, U.S. Bureau of Mines, IC 9183 (1988), pp. 389–398.

Examination of the results for these parameters over time exposes a strong temporal pattern in removal efficiencies for all substrate/plant treatment types that were obscured by overall influent/effluent means. Declines in influent Fe and Mn concentrations, a concomitant increase in pH, and extremely variable TSS values during the study period are unexplained, though possibly related to regional drought conditions.

Effluent values for TSS and dissolved Fe and Mn suggested that low removal efficiencies during the first winter may have been due to (1) system initiation in late autumn and (2) overloading during the first 10 weeks of operation. At an application rate of 0.5 L/min, efficiency improved but leveled off until the warmer weather of early February. Removal efficiency gradually improved until mid-April and rapidly thereafter. The pronounced pattern, common to all substrate types, suggested that microbial populations increased dramatically with the warmer temperatures from mid-February onward, and renewed plant growth in April developed system capability to remove much of the Fe from the acid drainage influent. Results from the pea gravel cell showed similar but more variable declines in Fe content.

Because the effluent results are not statistically significant between substrates, the removal efficiency improvement pattern common to all experimental cells suggests a major biological component in the removal process. The plant-substrate-microbe complex that is important to water quality improvement will develop, without inocula, in the six substrates tested to date, indicating the ubiquitous nature of desirable microbial organisms in wetlands treatment systems.

Relatively poor removal of Mn, even at the end of the growing season, compared to much greater removal in many operational systems suggested that the plant-soil-microbial complex does not substantially alter dissolved Mn compounds until after dissolved Fe is relatively low or unavailable for microbial metabolism. A lower influent application rate resulting in better dissolved Fe removal is likely to result in the improved treatment of dissolved Mn.

Variations in influent and effluent pH seemed closely related, although a similar gradual improvement over time was apparent. In contrast, a strong pattern of improved removal of TSS over time was evident, despite widely varying influent values.

Though trends for pH and dissolved Mn are not readily apparent, substantial and consistent reductions in dissolved Fe and TSS, although influent values showed considerable variation, suggest that the plant-soil-microbial complex in wetlands treatment systems is amenable to considerable fluctuation in loading rates.

Interpretation of absolute effluent values in terms of wetland treatment system efficiencies must incorporate the application rates in use during these studies. A high application rate (0.5 L/min, ca. 500 ft^2/gal/min, 0.4 m^2/mg Fe/min, Brodie et al. 1988) was deliberately selected to accentuate differences, if any, among different substrate types. Valve imprecision caused daily average flows to vary from 0.34 to 0.43 L/min as compared to desired rates of 0.5 L/min. Acid and normal wetland substrates inadvertently had higher flow (application) rates. Because these rates are substantially greater than the recommended rates employed at 11 operating treatment systems, the results must not be extrapolated to field-scale operating systems.

In addition, cattail growing in all substrate types exhibited a typical root growth above the substrate and within the water column around each stem that was not present in operating wetland treatment systems. Visual inspection also revealed a consistent pattern of more vigorous stem and leaf growth in the upper portion of each cell as compared to the middle or lower portions.

A comparison of the average number of cattail stems from new shoots present four times during the study revealed normal increases throughout the growing season, but it did not show significant differences between substrate types. Similarly, little difference was apparent in the height of cattail grown in different substrate types, though topsoil was a significantly better growth medium than pea gravel in July and significantly better than spoil and pea gravel in October. Analysis of

Table 8.15. Nutrients in Water and Sediment and Cattails in Three Experimental Wetland Cells

Wetland Cell	Location in Cell	Water (mg/L)	Sediment (mg/kg)	Cattail (mg/kg dry weight)
Total Phosphorus				
A5	Upper	0.01	170	420
A5	Middle	0.02	280	320
A5	Lower	0.03	180	260
B6	Upper	0.01	240	520
B6	Middle	0.05	400	320
B6	Lower	0.03	120	200
C5	Upper	<0.01	350	960
C5	Middle	0.01	240	400
C5	Lower	0.02	170	320
Total Kjeldahl Nitrogen				
A5	Upper	0.20	330	4900
A5	Middle	<0.02	310	3740
A5	Lower	<0.02	310	2720
B6	Upper	0.20	430	5810
B6	Middle	<0.02	430	1980
B6	Lower	<0.02	310	2780
C5	Upper	0.16	340	7880
C5	Middle	<0.02	360	3380
C5	Lower	<0.02	250	4870
Ammonia Nitrogen/Nitrate Nitrogen				
A5	Upper	0.51/0.02	—	—
A5	Middle	<0.01/0.01	—	—
A5	Lower	<0.01/<0.01	—	—
B6	Upper	0.51/0.02	—	—
B6	Middle	0.01/0.02	—	—
B6	Lower	0.01/0.01	—	—
C5	Upper	0.50/0.03	—	—
C5	Middle	<0.01/0.02	—	—
C5	Lower	<0.01/0.01	—	—

From Brodie, A. G. et al. An Evaluation of Substrate Types in Constructed Wetland, U.S. Bureau of Mines, IC 9183 (1988), pp. 389–398.

phosphorus, total nitrogen, and NH_3/NO_3 nitrogen showed considerably more nitrogen content in the water in the upper portion of each cell and a similar difference in phosphorus and total nitrogen in cattail stems and leaves (Table 8.15).

Within-cell differences in cattail vigor appeared to be related to available nitrogen and phosphate in seep water and relatively low concentrations in substrate types. Excess fertilizer applied at initiation may have flushed out, leaving influent seep water as the principal source for important plant nutrients, resulting in atypical root growth in the water column. Since all cells exhibited atypical root growth and differences in cattail vigor were greater within each cell than between cells, these variables were unlikely to have differentially influenced the comparison of removal efficiencies in substrate types.

In summary, the results suggested (1) that substrate type is relatively unimportant to removal treatment efficiency because the desired plant-substrate-microbe complex became established in each type; (2) that microbial inocula were unnecessary; and (3) that vegetation may substantially improve treatment efficiency. From a practical standpoint, substantial differences must exist between substrate removal efficiencies to justify the considerable construction costs entailed in deliberately installing a specific substrate in field-scale operating systems.

The functional values of wetlands vegetation previously identified from operating system results were supported by analysis of these data; i.e., substrates with emergent vegetation had higher removal efficiencies than substrates lacking vegetation.

Vegetation

In wetland system development, the vegetative component is a major factor. The plants selected should:

- Have active vegetative colonizers with spreading rhizome systems
- Have considerable biomass or stem densities to achieve maximum translocation of water and nutrient assimilation
- Offer maximum surface area for microbial populations
- Have efficient oxygen transport into the anaerobic root zone to facilitate the oxidation of reduced toxic metals and support a large rhizosphere
- Be a species combination that will provide coverage over the broadest spread of water depth in the terrain conditions

It is important to determine the reaction of particular species for use in a wetland environment. Water depth dictates zonation of wetland plants. Species should be selected primarily using this criterion.[13]

Wetland plants can be purchased from nurseries, collected in the wild, or grown for a specific project.

Water and Soil Parameters Affecting Growth of Cattails

Cattail-dominated wetlands have been shown to reduce Mn and Fe. Their high survival rate is another reason for utilizing cattails in manmade wetlands; cattail survival after transplanting by various procedures was higher (78%) than 11 other species.[14]

Little, if anything, is known about the various water quality or soil chemical properties that may limit the growth of wetland plants, including cattails. This information is needed so that these limitations to growth can be overcome when building wetlands to treat AMD. Three pilot studies were completed to (1) survey water quality where cattails grow naturally; (2) compare water and soil parameters from sites where cattails were found to immediately adjacent spots where no cattails were found; and (3) determine the effects of wetland vegetation on total Fe and Mn concentrations and removal in water as compared to an identical site with no vegetation.

Water samples were taken at 56 sites that had cattails growing. If an adjacent site appeared to have the same flow of water and the same substrate, but no cattails growing, it was sampled. In addition, other isolated sites with no cattails that appeared to have good waterflow and a mud substrate were sampled. Water temperature was measured to the nearest 0.5°C. Readings were made within 1 sec of removing the thermometer from the water. Water temperature was used in standardizing pH and conductivity meters.

Electrical conductivity was measured with an 18-V Fisher conductivity meter, model no. 152. A Fisher mini pH meter, model no. 640, was used to measure water pH. Prior to each pH measurement the meter was standardized to pH 4 and 7 buffers that had been allowed to equilibrate at water temperature. The combination electrode was cleaned with distilled water between each buffer and each water measurement. Waterflow velocity and depth were measured for flow rate calculations to be made at a later time.

Two water samples were collected in 250-mL nalgene bottles at each sampling point. One of these samples was filtered through preweighted metrical membrane filters, pore size 0.45 µm. Both samples were treated with 1 mL of 50% nitric acid to fix metals. A third unfiltered, unacidified water sample was collected in 1-L plastic cubitainers.

In the laboratory, unfiltered, acidified water samples were used to determine the concentrations of total Fe and Mn cations, and of Ca^{2+}, Mg^{2+}, Na^{1+}, and K^{1+}. Filtered, acidified samples were used to determine dissolved Fe and Mn levels. Cation concentrations were determined by atomic absorption spectrophotometry (AAS) by American Society for Testing and Materials (ASTM) Method D2576.

In the course of collecting information for the first objective of this study, four sites of interest were noted. These sites had dense stands of cattails growing naturally, but there were small spots, either adjacent to or within the wetlands, that had no cattails. Soil and water samples were taken at 17 locations within these four sites. Methods of water analysis were identical to those outlined above. Sediment samples were taken from the same points as the water samples, usually in the middle of the cattails wetland or in the nonvegetated areas. Sediment samples were collected with an auger to a depth of 10 cm from the soil-water interface and transferred to plastic bags. In the laboratory, cations were extracted by the double-acid method and determined by AAS. Phosphorous was extracted and measured on a UV-VIS spectrophotometer.

To study the effects of wetland vegetation on Fe and Mn removal, two plywood boxes were constructed in the middle of a cattail marsh. The marsh is approximately

4 acres in size and had an average water pH of 2.85, total Fe of 25 ppm, and total Mn of 16 ppm (based on samples taken, three locations in three seasons).

The boxes were 5 × 1.2 × 0.3 m in size and were embedded 0.3 m deep in the marsh, adjacent to each other. They were placed so that there was a slight flow of water from one end to the other. Inflow pipes allowed flow of about 5 L/min. Sediment from the marsh was placed into the boxes as substrate. In one box adult cattail plants were carefully transplanted at the same density found in the marsh, making a total of 228 cattails. The other box remained devoid of all vegetation. Water samples were taken at the inflow and outflow of each box approximately every month for 1 year. Counts of cattail density were made periodically as well. Sediment samples were taken.

Where Cattails Grow

Water at cattail sites had a wider range of pH than water at non-cattail sites. Cattail sites had a mean pH of 4.85, while non-cattail sites had a mean pH of 3.26. A wetland consultant indicated that only rarely were cattails found in water with a pH of below 2.5 or an Fe concentration above 100 ppm. Of the cattail sites in this study, 42% (11/26) had a water pH below the 3.26 mean, with 6 sites below pH 2.5 and 1 below 2.0. This suggests that cattails will tolerate lower pH, and pH alone may not be a limiting factor for the establishment of this plant.

A relationship apparently exists, however, between Fe concentrations and the presence or absence of cattails. More often sites with lower concentrations of total Fe in water had cattails; 57% (15/26) of the cattail sites had Fe levels below 25 ppm, as compared to 13% (4/30) of the non-cattail sites. In contrast 36.7% (11/30) of the non-cattail sites had Fe concentrations over 101 ppm (Table 8.16).

There were only three cattail sites with Fe concentrations of over 100 ppm, the highest being 144 ppm (Table 8.17). On that site only scattered cattails were found. A second site with 113 ppm Fe had only four scattered cattail plants, while another site with 113 ppm Fe had a dense stand of cattails.

There may well be a combination of factors that affects cattail growth. When the sites without cattails were ranked by increasing water pH (Table 8.18), higher total Fe was found at the lower pH levels. There were nine sites with pH <2.3. The average total Fe for the eight non-cattail sites with pH below 2.30 was 233.7 ppm while average total Fe for eight non-cattail sites with pH above 3.51 was 73.5 ppm (Table 8.17). Fe levels of sites with pH between 2.30 and 3.51 were highly variable (Table 8.17). When the low pH sites with no cattails were compared to the low pH sites with cattails, it was noted that Fe levels were lower in the cattail sites (Table 8.17). Of 26 sites with cattails, 17 had pH above 3.00, and only 3 of those had Fe levels of 73 ppm or above (Table 8.17). Average total Fe for the 12 cattail sites with pH above 4.02 was only 27.9 ppm (Table 8.17).

Water and soil samples were taken on a paired basis when possible (i.e., one sample from a cattail site and another sample from an adjacent spot without cattails)

Table 8.16. Sites With and Without Cattails Ranked by Fe Concentration in Water[a]

Without Cattails			With Cattails		
Fe	pH	Mn	Fe	pH	Mn
747.00	2.07	34.00	144.00	5.70	7.30
537.00	1.84	11.30	113.00	2.85	3.56
211.30	2.91	9.30	113.00	2.85	3.56
169.00	2.36	2.85	91.77	2.05	10.08
152.00	3.51	3.78	91.00	2.30	15.10
153.70	2.15	9.12	90.00	5.70	6.90
133.00	2.14	3.03	73.00	6.78	10.30
123.00	5.90	5.50	70.00	2.30	3.55
122.00	2.98	1.11	59.10	2.36	6.60
116.60	3.57	13.09	32.00	4.02	14.50
103.00	2.45	2.28	28.16	2.42	16.94
93.00	1.70	7.40	14.90	3.67	4.10
89.67	2.13	14.63	12.06	2.54	18.92
88.62	2.20	9.90	10.60	3.12	51.80
86.00	5.87	5.56	9.80	6.10	2.30
85.00	4.11	15.10	8.10	6.41	1.46
78.00	3.54	4.22	7.06	2.55	
63.50	2.79	7.70	4.98	3.56	15.80
63.00	6.32	1.76	4.43	8.25	0.42
60.00	2.33	6.30	3.45	3.03	5.40
47.63	2.66	6.36	2.14	8.10	0.24
39.00	5.99	7.80	1.44	6.10	0.34
38.60	2.54	11.10	1.05	7.40	0.12
30.80	2.43	4.19	0.39	6.63	2.41
27.94	2.19	17.05	0.14	8.25	0.01
25.20	2.47	7.40			
24.40	2.66	15.20			
23.00	2.81	21.90	0.02	7.78	0.04
23.00	2.65	24.90			
0.38	7.06	2.41			

[a] All values are in ppm units except for pH.

From Brodie, A. G. et al. An Evaluation of Substrate Types in Constructed Wetland, U.S. Bureau of Mines, IC 9183 (1988), pp. 389–398.

to test for additional differences associated with cattails. Only total Fe in water samples was significantly different (Wilcoxon's sign rank $p < 0.055$) between sites with cattails (53.8 ppm mean Fe) and adjacent sites without cattails (98.2 ppm mean Fe) (Table 8.18). For soil samples, significant differences occurred for Ca and Mn, with higher levels of both elements present in sites with cattails (Table 8.18). These differences were affected by the small sample sizes and the great variation in samples. For example, the high soil Mn level found in cattail sites was due to one site's having 184 ppm and a second having 57.6 ppm of Mn. No other cattail site soil samples contained >12.2 ppm Mn. One cattail site had 228 ppm potassium, but no other sites were >21.2 ppm. Ca levels in soil samples from cattail sites seemed inflated as well, due to two samples of 1074 and 752.

Table 8.17. Sites With and Without Cattails Ranked by pH Value[a]

Without Cattails			With Cattails		
pH	Fe	Mn	pH	Fe	Mn
1.70	93.00	7.40	2.05	91.77	10.08
1.84	537.00	11.30	2.30	91.00	15.10
2.07	747.00	34.00	2.30	70.00	3.55
2.13	89.67	14.63	2.36	59.10	6.60
2.14	133.00	3.03	2.42	28.16	16.94
2.15	153.70	9.12	2.54	12.06	18.92
2.19	27.94	17.05	2.55	7.06	
2.20	88.62	9.90	2.85	113.00	3.56
2.33	60.00	6.30	2.93	113.00	4.92
2.36	169.00	2.86	3.03	3.45	5.40
2.43	30.80	4.19	3.12	110.60	51.80
2.45	103.00	2.28	3.56	4.98	15.80
2.47	25.20	7.40	3.67	14.90	4.10
2.54	38.60	11.10	4.02	32.00	14.50
2.65	23.00	24.90	5.70	144.00	7.30
2.66	47.63	6.36	5.70	90.00	6.90
2.66	24.40	15.20	6.10	9.80	2.30
2.79	63.50	7.70	6.10	1.44	0.34
2.81	23.00	21.90	6.41	8.10	1.46
2.91	211.30	9.30	6.63	0.39	2.41
2.98	122.00	1.11	6.78	73.00	10.30
3.51	152.00	3.78	7.40	1.05	0.12
3.54	78.00	4.22	7.78	0.02	0.04
3.57	116.60	13.09	8.10	2.14	0.24
4.11	85.00	15.10	8.25	4.43	0.42
5.87	68.00	5.56	8.25	0.14	0.01
5.90	123.00	5.50			
5.99	39.00	7.80			
6.32	63.00	1.76			
7.06	0.38	2.41			

[a] All values are in ppm units except for pH.

From Samuel, D. E., J. C. Sencindiver and H. W. Rauch. Water and Soil Parameters Affecting Growth of Cattails, U.S. Bureau of Mines, IC 9183 (1988), pp. 369–374.

Al was not measured in water samples, but it was in soil samples. The ratio of the average pH:Al for ten soil samples from non-cattail sites was 0.143, while the same ratio from seven cattail sites was 0.240. Thus, it appears that lower pH and the higher Al content of the soil may present problems for cattail germination or survival.

It is obvious from Table 8.18 that no one soil or water parameter exhibits low or high levels when cattails are present compared to when they are absent in the adjacent sites. At several adjacent sites (one with and one without cattails), it was noted that both had high pH and low Fe concentrations in both soil and water samples. Other parameters such as soil Al were similar. One pair of samples showed low water pH (2.8 to 3.0), and high Fe (120 to 150 ppm) in the water, and similarly had low soil pH and high soil Fe levels. Yet, cattails thrived in one place and were totally absent, as was all vegetation, immediately adjacent to that site. In fact, a

Table 8.18. Mean Water and Soil Quality Parameters (ppm) for 10 Sites Without Cattails and 7 Adjacent Sites with Cattails

	Water Analysis		Soil Analysis	
	Cattail Sites	Non-Cattail Sites	Cattail Sites	Non-Cattail Sites
pH	3.94	3.87	4.54	3.87
Fe	53.8	98.2	133.6	93.0
Mn	6.3	6.2	43.6	4.4
K	8.7	7.7	49.0	8.9
Na	18.4	15.8	NS[a]	NS
Ca	201.5	190.1	439.1	123.0
Mg	76.2	68.1	17.5	10.0
Al	NS	NS	17.9	25.7
Po	NS	NS	5.9	3.8
Zn	NS	NS	19.8	1.6
Cu	NS	NS	0.5	0.4

[a] NS: not sampled.

From Samuel, D. E., J. C. Sencindiver and H. W. Rauch. Water and Soil Parameters Affecting Growth of Cattails, U.S. Bureau of Mines, IC 9183 (1988), pp. 369–374.

distinct demarcation existed between the area with cattails and that with none, within the same seep. This was found at several locations, even though no individual paired water or soil parameters were significantly different.

Researchers have speculated, both in academic settings and in the literature, about the role of wetland plants in the complex process of removing Fe and Mn from AMD. Cattail plants remove some Fe from the environment, and it is known that complex bacterial reactions also lead to the removal of certain metals from water. Some believe that the species of wetland plants is of little consequence in improving water quality and that the entire wetland ecosystem provides the ecological basis for the removal of Fe and Mn. Thus, the plants and the soil provide the environment for algae and bacteria to remove metals through a complex oxidation and reduction reaction.

Initially planted in the 64-ft^2 (20-m^2) box were 228 cattails. Three months later, many of the adult plants had died, but new sucker growth resulted in 261 cattail plants. Counts were also made yielding 194, 293, and 295 cattails, respectively. An increase occurred in plant density from 3.56 cattails per 10 m^2 to 4.61 cattails per 10 m^2 in 1 year.

Monthly water samples were compared for two periods. The first sample was taken 2 days after construction of the boxes and the placement of substrate and cattails. Other samples in this first period were taken. An abandoned surface mine located above the wetland was opened and water was treated, causing an increase in pH and a decrease in Fe and Mn entering the inflow of the cattail and control box. A second period was established for data analysis. No significant change took place

Table 8.19. Average Inflow and Outflow Levels (ppm) of Iron and Manganese in a 16 × 4 ft (5 × 1.3 m) System with a Flow of 1 gpm

	Iron		Manganese	
	Inflow	Outflow	Inflow	Outflow
First period				
Cattails	19.53	17.70	19.15	20.23
No cattails	20.67	18.90	18.57	20.37
Second period				
Cattails	7.77	4.63	12.98	11.08
No cattails	5.42	5.38	12.08	12.32

Note: One system had cattails and one had no cattails. In analysis there were two periods (in late September a mine became active above this wetland, and water was treated, which raised the pH and lowered the iron and manganese; thus, data are separated): first, June 20 to September 16, 1986 (four samples; the first sample in this period was taken 2 days after the wetland was constructed.); second, October 6, 1986 to May 21, 1987 (six samples).

either in Fe or Mn during the first period for either the cattail box or the control box (Table 8.18).

For the second period, there was no significant average decrease in the difference ($p > 0.05$) between the inflow and outflow for Fe in the cattail box (decrease of 3.14 ppm Fe) or the control box (decrease of 0.04 ppm Fe).

Excluding the first sample, which was taken 2 days after the wetland was constructed, the cattail site showed a reduction in Fe from inflow to outflow in eight of nine samples taken (average monthly decrease of 3.46 ppm). Excluding the first sample, the control site showed a reduction in Fe from inflow to outflow in four of nine samples (average decrease of 1.43 ppm).

During the first period, Mn increased from inflow to outflow an average of 1.08 ppm for the cattail wetland and 1.80 ppm for the control. During the second period (six samples taken), Mn increased at the outflow (average of 0.24 ppm) in the control but decreased (insignificant average of 1.90 ppm at 0.05 level) at the outflow of the cattail wetland. The pH range at the inflow was 2.85 to 3.26 for the first period and 3.65 to 6.87 for the second, with no significant difference in the average monthly change for the cattail box or the control at the outfall.

Though the findings are preliminary, it appears that cattails aid in the removal of Fe. After the cattails had been planted and were growing for 3 months, there was an average monthly decrease in Fe from inflow to outflow of 3.14 ppm. Since only 0.04 ppm was removed during this same period in the control, which contained sediment only, the plants must play some role in providing an ecological system that removes Fe.

Soil samples were taken at the inflow and the outflow of the cattail and control boxes in August 1987 (Table 8.19). The highest levels of soil Fe were found at the outflow of the cattail box. Levels of soil Fe were considerably higher in soils found in the box with cattails than in the box with sediment and no cattails.

EFFECTS OF CATTAILS (*TYPHA*) ON METAL REMOVAL

Using cattail (*Typha*) wetlands is a cost-effective way to treat AMD because these wetland systems are capable of removing Fe and Mn from the drainage. Due to saturated soil conditions and the presence of decomposable organic matter, the biochemical oxygen demand of most wetlands far exceeds the rate of oxygen diffusion into the sediments. Without oxygen, the sediments become highly reduced as microbial respiration proceeds, using electron acceptors such as Fe, Mn, and SO_4^{2-}.

Cattail plants have morphological features such as arenchyema tissues which aid in transporting oxygen from aboveground parts to the roots, thereby enabling the plant to grow in highly reduced environments. The direct ameliorative effect of cattails is the metal removal by plant uptake from AMD. Metals taken up by the plants are subsequently retained in the detritus after senescence of the plants. The indirect effect of the cattails on water quality is that they modify the chemical and biological environments of the sediments.[15]

Natural cattail wetlands (major species *Typha latifolia*) that received drainage from reclaimed surface coal mines were selected for study. The wetlands ranged in size from 40 to 1500 m^2 (Table 8.20).

Plant populations in each wetland were determined by counting the number of plants per square meter. A 1 × 1-m quadrat was randomly placed at five points in each wetland. All cattails within the quadrat were counted. The total aboveground portions (leaves and stems) and rhizomes of cattails were sampled near the water inflow, near the middle, and near the water outflow of each wetland. The plant samples were cleaned with distilled water and dried in a forced air oven at 65°C. Dried samples were ground in a Wiley mill having stainless steel blades. These samples were analyzed for Fe and Mn on an inductivity coupled plasmaspectrograph.

Sediment samples were collected from the same three points of cattail sampling. The sediment samples were collected with an auger to a depth of 10 cm from the soil-water interface and transferred to plastic bags. Soil pH and Eh for redox measurements were made in the fresh, wet sediments. After pH and Eh were measured the sediments were air dried and passed through a 2-mm sieve. Soil texture was determined by the pipette method. Extractable Fe and Mn were determined by extracting the sediments with 0.1 N HCl, using a 1:10 ratio of sediment to extracting solution. The extracted elements were determined by AAS.

Additional samples for studying the effects of vegetation on pe + pH of the sediments were collected from one point outside and on the water entry side of each wetland. The pe + pH data were calculated as described above.

Monthly water quality samples were collected. In general the mine water was near neutral pH. The lowest pH at any sampling time was 6.3 and the highest was 8.2. Only one site consistently had pH values above 6.0; water pH at all other sites was generally below 6.0 and commonly between pH 3.0 and 4.0. Total Fe and Mn varied.

Table 8.20. Plant Density, Plant Vigor, and Biomass Production of Cattails in West Virginia Wetlands Receiving Mine Drainage

Property	Season	Monongalia County	Preston County
Plant density (plants/m^2)	Spring	12.9[a] 7.9–15.9[b]	7.9[a] 0.7–19.2[b]
Plant vigor (g dry matter/plant)	Spring	39.0 20.1–81.7	16.5 11.0–24.2
	Autumn	45.5 25.3–91.5	20.6 13.2–31.4
Biomass (g dry matter/m^2)	Spring	411.9 211.9–663.0	146.2 5.8–345.0
	Autumn	554.3 264.5–1180.2	191.8 7.8–446.7

[a] Mean of seven sites in Monongalia County and five sites in Preston County.
[b] Range.

From Sencindiver, J. G. and D. K. Khumbla. Effects of Cattails (*Typha*) on Metal Removal from Mine Drainage, U.S. Bureau of Mines, IC 9183 (1988), pp. 359–368.

The direct and indirect ameliorative effect of cattail wetlands on AMD is related to primary biomass production within the wetland ecosystem. Biomass data for wetlands in this study are presented in Table 18.20. Dry matter yield was used as an index for biomass production and computed from plant density and dry matter yield per plant (plant vigor). Plant density in Monongalia County, WV, wetlands was significantly greater than in Preston County, WV, wetlands. Cattail growth in the wetlands of Monongalia County was significantly more vigorous than the cattail growth in Preston County wetlands at both sampling times. During the spring the mean per plant dry matter production in Monongalia County wetlands was 2.5 times that of Preston County. At the fall sampling, the plant vigor in Monongalia County was 2.2 times greater than in Preston County.

Biomass production was three times higher in Monongalia County than in Preston County at the spring sampling and almost three times higher at the autumn sampling. Biomass production depended on plant density and plant vigor. Low levels of biomass production were caused by low plant density. Both plant vigor and biomass were significantly higher in autumn-harvested cattails than in spring-harvested cattails.

When the dry matter production of cattails in this study was compared to the dry matter production in fertilized cattail stands managed for biomass production in Minnesota, Wisconsin, and Texas, the amount of dry matter produced by natural wetlands in the West Virginia study was <50% of that produced in the managed wetlands. Thus, with proper management, a great potential exists for increasing biomass production in cattail wetlands. The productivity of cattail wetlands is related to the nutritional status and the biochemical environment of the substrate in which

Table 8.21. Properties of Sediment in Cattail Wetlands Receiving Mine Drainage

Property	Season	Monongalia County	Preston County
pH	Spring	7.93	4.67
		7.64,–8.11	2.34–6.07
Fe	Spring	5,343	3,503
		1,992–10,154	1,306–7,616
	Autumn	6,418	5,625
		2,436–13,550	2,464–10,850
Mn	Spring	1,536	280
		430–2,216	50–713
	Autumn	1,412	357
		422–2,617	53–697

From Sencindiver, J. G. and D. K. Khumbla. Effects of Cattails (*Typha*) on Metal Removal from Mine Drainage, U.S. Bureau of Mines, IC 9183 (1988), pp. 359–368.

the cattails are growing. In general, lower biomass production in Preston County was related to the pH of the sediment (Table 8.21). A significant positive correlation coefficient ($r = 0.73$) was observed between sediment pH and biomass. Other factors such as low sediment concentration of phosphorus (P) could also have affected biomass production.

Wetland sediments have a natural tendency to convert to pH 7.0, irrespective of the initial soil pH. In Preston County wetlands, sediments with pH values <7.0 indicated the presence of nonequilibrium conditions, i.e., those sediments represented the wetlands that were constantly receiving an influx of highly acidic water.

Significant differences were found in the extractable Mn concentrations between the sediments of Monongalia and Preston Counties. The extractable Mn concentrations in the Preston County sediments were significantly lower than Mn concentrations in Monongalia County sediments. The correlation coefficient between the aboveground plant tissue Mn concentrations and concentration of extractable metals in the sediment was an insignificant 0.52.

No marked differences in the amounts of extractable Fe in sediments were found between counties or between seasons. No significant correlation was found between the amount of metals extracted from the sediment and the concentrations of the metals in the plant tissue. This lack of correlation was probably due to the nature of the substrate for plant growth. In wetlands receiving AMD, the extractable metals are indicators of the capacity factor for plant nutrition, i.e., the amount of metal that can be potentially taken up by the growing plants. The actual amount of metal removal by the plant is dependent on the intensity factor, i.e., the concentration of metal in the soil solution. In normal plant growth media the intensity factor is governed by the capacity factor, but in wetlands receiving AMD, differences in water chemistry of the drainage and sediment and the continuous influx of metals into the growth media make the intensity factor largely independent of the capacity factor. This

Table 8.22. Concentration and Uptake of Manganese in Cattails Sampled from West Virginia Wetlands Receiving Mine Drainage

Plant Parts	Season	Monongalia County	Preston County
Rhizomes (μg/g)	Spring	665 625–705	885 498–1278
	Autumn	795 567–972	684 394–1119
Leaves and stems (μg/g)	Spring	1202 702–1824	3109 1879–4071
	Autumn	2202 586–2912	3189 2653–3970
Leaves and stems (kg/ha)	Spring	5.37 2.64–7.92	5.20 0.11–11.60
	Autumn	11.47 3.72–23.22	6.19 0.25–15.95

From Sencindiver, J. G. and D. K. Khumbla. Effects of Cattails (*Typha*) on Metal Removal from Mine Drainage, U.S. Bureau of Mines, IC 9183 (1988), pp. 359–368.

continuous flow of mine drainage into the system is responsible for the lack of a significant correlation between extractable metals and concentrations in plant tissue.

The texture of the sediments in the Monongalia County wetlands was sandy loam, silt loam, or clay. The texture of wetland sediments in Preston County was either clay loam or silty clay loam. These textures probably represent the textures of the surrounding minesoils.

The plant tissue analyses for Mn (Table 8.22) suggest another reason for the yield differential between cattails growing in Preston and Monongalia Counties. The lower cattail yields in Preston County could be due to the high levels of Mn in the plant tissue. A significant increase took place in the plant tissue concentrations of Mn from spring to autumn samples for Monongalia County, but the seasonal concentration differences were not significant for Preston County. This lack of increase in Mn concentration with the growing season in Preston County could be due to the fact that the concentrations during the spring season had already reached the maximum value for bioconcentration in the cattail. In this case no further concentration increases could have occurred.

During the spring season the mean Mn concentration in cattails from Preston County was 2.5 times greater than the mean Mn concentration in cattails from Monongalia County. Even with these large differences in concentrations of Mn, the differences in uptake of Mn at the spring sampling between counties were statistically insignificant. This lack of a sizable difference in uptake is related to low biomass production in the cattail wetlands of Preston County.

Table 8.23. Concentration and Uptake of Iron in Cattails Sampled from West Virginia Wetlands Receiving Mine Drainage

Plant Parts	Season	Monongalia County	Preston County
Rhizomes (μg/g)	Spring	–	15,228[a]
		–	9,648–19,506[b]
	Autumn	5,392[a]	11,170
		1,000–10,813[b]	8,036–14,650
Leaves and stems (μg/g)	Spring	595	2,043
		109–1,539	213–5,369
	Autumn	1,919	208
		409–7,352	88–344
Leaves and stems (kg/ha)[c]	Spring	2.67	1.07
		0.23–5.47	0.04–2.80
	Autumn	3.74	0.76
		1.56–5.31	0.06–1.31

[a] Rhizome data are means of four sites in each county. All other data are means of seven sites in Monongalia County and five sites in Preston County.
[b] Range.
[c] Uptake in kilograms per hectare was calculated by multiplying concentration (micrograms per gram) by biomass.

From Sencindiver, J. G. and D. K. Khumbla. Effects of Cattails (*Typha*) on Metal Removal from Mine Drainage, U.S. Bureau of Mines, IC 9183 (1988), pp. 359–368.

Fe concentrations in cattail tissues in Preston County were significantly different at both sampling times from those in Monongalia County (Table 8.23). Fe concentration in Preston County cattails was 3.4 times greater than Fe concentration in Monongalia County cattails in the spring, but this difference was reversed at the autumn sampling when the Fe concentration in Monongalia County cattails was 9.2 times greater than the Fe concentration in Preston County cattails. These seasonal differences could be due to the formation of Fe oxide coatings that were observed on the roots of cattails growing in Preston County. Under reducing conditions Fe2+ is thought to be oxidized to the less soluble Fe^{3+} by oxidative agents released from the roots, thus creating a coating or plaque of an insoluble Fe^{3+} compound on the root surface. Fe oxide coatings are a survival mechanism for cattails growing in a contaminated environment because the coatings do not allow further uptake of Fe. Because no additional Fe was taken up by the plant, the Fe concentration decreased as the plant continued to grow. Plants growing in Monongalia County may have continued taking up Fe during the growing season. The high concentrations of Fe found in Preston County rhizomes are probably due to the coatings.

Seasonal differences in Fe uptake were significant for the cattails growing in Monongalia County, while they were not significant for those in Preston County. In absolute terms, the amounts of Fe removed by plants were extremely low: 2.67 and 3.74 kg/ha for the spring and autumn sampling periods in Monongalia County. The amounts taken up in Preston County were smaller: 1.07 and 0.76 kg/ha for the spring and autumn sampling, respectively.

Table 8.24. Redox Environment in Sediment of Cattail West Virginia Wetlands Receiving Mine Drainage

Property	Monongalia County	Preston County
pH	7.95[a]	4.63[a]
	7.64–8.11[b]	2.34–6.07[b]
pe + pH	4.57	4.31
(with cattails)	3.93–5.36	2.62–6.20
pe + pH	10.63	9.82
(without cattails)	8.79–13.25	4.97–13.27

[a] Mean of seven sites in Monongalia County and five sites in Preston County.
[b] Range.

From Sencindiver, J. G. and D. K. Khumbla. Effects of Cattails (*Typha*) on Metal Removal from Mine Drainage, U.S. Bureau of Mines, IC 9183 (1988), pp. 359–368.

In order to calculate total removal of Fe by cattails, biomass of the above and below ground parts of the plants must be known. Biomass of rhizomes was not determined in this study, so it was estimated. Biomass of *Typha latifolia* varies, but it is commonly 50% of the total biomass.

To calculate an example of Fe removal by cattails in this study, a site in Preston County was chosen because it had the highest concentration of Fe (19,506 µg/g) in the rhizomes. Above and below ground biomass each equaled 287.2 g/m².

To estimate uptake by the rhizomes, biomass must be multiplied by the Fe concentration. Thus, 287.2 g/m² × 19,506 µg/g = 5,602,123.2 µg/m², or approximately 56 kg/ha of Fe were removed by the rhizomes during the growing season.

This site had a median flow of 0.5 L/sec, therefore, 15,768,000 L/year will flow into this Preston County wetland. The median concentration of Fe entering this wetland was 10 mg/L. Therefore, 157,680,000 mg/year or 157.7 kg/year of Fe will be added to this wetland. Because the area of this wetland is 61.27 m², additional calculations show that approximately 25,700 kg/ha of Fe will enter it during 1 year. The cattail plants are removing only 57.31 kg/ha (56 kg/ha for the rhizomes and 1.31 kg/ha for the tops), therefore, the plants are removing only about 0.2% of the total Fe entering the wetland.

Metals can be removed from water by various mechanisms; data on the redox environment within a cattail wetland may provide some information on possible mechanisms. Sediment on one site in Preston County had a positive Eh of 230 mV, but the Eh of sediments at all other sites ranged from –247 to –18 mV. These Eh values suggest that reduction may be a major mechanism for the removal of Fe. The pe + pH values of the sediments with cattails were close to 4.24 where SO_4^{2-} and Fe^{3+} can be reduced to form insoluble FeS_2 at pH 7.0.

The role of cattails in enhancing the reduced environment in wetlands was confirmed when redox potentials were determined in wetland sediments with and without cattails (Table 8.24). The data indicate that the presence of cattails has a significant effect on the redox environment, making it more reducing. Cattails can sufficiently lower the redox potential of the sediments to precipitate Fe in its sulfide form.

Thus, it can be concluded that amounts of metals removed from mine drainage by cattail wetlands depend upon cattail biomass and the concentration of metals in plants. In this study biomass production appeared to be related to the quality of the sediment in the wetland. Low pH and high Mn levels were related to reduced cattail biomass in some wetlands. To increase metal removal from mine drainage by plant accumulation, emphasis should be placed on increasing dry matter production rather than on increased bioconcentration.

Data from this study also indicate that plant uptake is not a major mechanism for the attenuation of metals in mine drainage. Eh values and pe + pH calculations show that cattail wetlands have reducing environments; the presence of cattails in a wetland actually promotes a more reducing environment. Chemical reduction within the wetland may be a mechanism by which metals such as Fe are removed. Additional research will be required to determine other possible mechanisms of metal removal from mine drainage flowing through cattail wetlands.

METAL RETENTION CAPACITY OF WETLANDS FOR TREATMENT OF ACID MINE DRAINAGE

Freshwater wetlands may act as sinks for metals. Several field and laboratory studies indicate that metals in mine drainage can be retained in *Sphagnum* wetlands.[6] Because metals typically do not have a gaseous phase, it is reasonable to assume that, like P, the capacity for metal retention in wetland systems is finite. However, relatively little effort has been focused on developing quantitative estimates of the ultimate long-term capacity for metal retention in manmade wetland systems constructed specifically for mine drainage treatment.[16]

Five processes are primarily involved in Fe retention in *Sphagnum* wetlands: uptake and incorporation of Fe by growing *Sphagnum* mosses, the removal of Fe ions from solution by cation exchange, the specific adsorption of Fe onto organic matter, the formation of insoluble Fe oxides, and the formation of insoluble Fe sulfides. Based on our understanding of the biology and chemistry of each process, it is possible to estimate an upper limit for Fe retention in *Sphagnum* wetlands.

As a plant micronutrient, Fe is taken up by growing *Sphagnum*. The highest reported rate of net primary production of *Sphagnum* is 610 g/m^2/year. The highest reported Fe concentration in field-collected *Sphagnum* plant tissue is 5.8 mg/g. Multiplying these two values gives an estimate of Fe uptake by growing *Sphagnum* of 3.5 g/m^2/year. This estimate is generous not only because we used the highest reported values for both growth and tissue Fe concentration, but also because *Sphagnum* growth may be inhibited by the high Fe concentrations typical of AMD. For example, in a 33-day laboratory study, growth in length of *S. fallax* in solutions containing 100 mg/L Fe was reduced by 32.5% relative to control plants growing in solutions containing 0 mg/L Fe.

The binding of Fe^{2+} to negatively charged sites on the peat matrix provides another mechanism for Fe retention. Fe retention through cation exchange was measured by placing 5 g of dried peat into replicate flasks and adding 100 mL of a solution containing either 0, 2, 4, 6, 8, 12, or 16 meq/L of Fe (added as $FeSO_4$; pH of each solution adjusted to 4.0). After 16 hr, the mixtures were filtered and binding of Fe^{2+} with the concomitant release of other cations from exchange sites was calculated. Using the Langmuir equation, a maximum retention of Fe^{2+} via cation exchange of 204 µeq/g (5.7 mg/g) was estimated.

As with mineral soils, peat can retain Fe by a chelation-like specific binding of the metal cations to sites on the organic matter matrix. Using 0.1 M $Na_4P_2O_7$ extractions, the highest organically bound Fe concentration measured in a *Sphagnum* peat sample was 89.3 mg/g. Although little is known about the chemical nature of the binding sites on *Sphagnum* peat, it appears that the specific binding of Fe to organic matter may be a much more important mechanism for Fe retention than cation exchange in peat exposed to mine drainage.

Fe can also be retained in peat by the formation of insoluble Fe oxides. The highest concentrations of amorphous Fe oxide (oxalate extractable Fe) and crystalline Fe oxide (bicarbonate-citrate-dithionite extractable Fe) that we have ever determined in a peat sample are 62.8 and 39.0 mg/g, respectively. These peat samples were collected from Tub Run Bog, WV, a naturally occurring *Sphagnum*-dominated wetland receiving inputs of acid coal mine drainage from an adjacent abandoned coal surface mine. To what extent the formation of such Fe oxides is abiotic vs biotic is presently unknown. However, indigenous populations of both Fe- and Mn-oxidizing bacteria have been found in field-collected samples of *Sphagnum* peat.

Some studies have suggested that bacterial dissimilatory sulfate reduction and the formation of Fe sulfides could play an important role in the removal of Fe from mine drainage. Although sulfate reduction and the concomitant formation of Fe sulfides does occur in freshwater *Sphagnum* peat, there is little evidence that Fe sulfides accumulate to any significant extent in peat exposed to mine drainage. Presumably, the accumulation of Fe sulfides is precluded by their reoxidation, although at present little is known about the process of sulfide oxidation in freshwater wetland peat. Using the Cr^{2+}-reduction technique, the highest concentration of Fe sulfides (actually $H_2S + S^0 + FeS_2 + FeS$) that we have ever measured in a single peat sample is 0.7 mg/g Fe.

Assuming it is possible to obtain maximum Fe retention by each of the above processes, an upper limit on Fe retention in a hypothetical manmade *Sphagnum* wetland, 40 × 40 m, 30 cm deep was calculated (Table 8.25). In making these calculations, a bulk density for *Sphagnum* peat of 0.1 g/cm was assumed. If such a manmade wetland were exposed to mine drainage with a flow of 4 L/min and an Fe concentration of 100 mg/L (Fe loading rate of 576 g/day), it would take an estimated 44 years until Fe retention by all processes combined would become saturated.

376 ENVIRONMENTAL IMPACTS OF MINING

Table 8.25. Maximum Fe Retention in a Hypothetical Manmade *Sphagnum* Wetland

Process	Max. Retention (g/m)	Days to Reach Saturation
Growth of *Sphagnum*	3.5	9.7
Cation exchange	171	475
Adsorption onto organic matter	2,508	6,967
Formation of amorphous Fe oxides	1,884	5,233
Formation of crystalline Fe oxides	1,170	3,250
Formation of Fe sulfides	21	59
Total	5757	15,994 = 44 yr

From Wieder, R. K. Determining Capacity of Metal Retention in Man-Made Wetlands Constructed for Treatment of Coal Mine Drainage, U.S. Bureau of Mines, IC 9183 (1988), pp. 375–381.

The analysis in Table 8.25 reveals that the incorporation of Fe by growth of *Sphagnum* and retention by cation exchange and sulfide formation make relatively minor contributions to overall potential Fe retention, as compared to the specific binding of Fe to organic matter and the formation of insoluble Fe oxides. It should be noted, however, that 4 L/min represents a very low flow. If the flow of mine drainage were 40 L/min (still only a moderate flow) and the Fe concentration were 200 mg/L, Fe saturation of the wetland would occur in only 2.2 years. The time estimate obtained in Table 8.25 also assumes that it is possible to obtain the maximum Fe retention by each of the processes involved. In peat samples that have been subjected to mine drainage, either in the field or in the laboratory, the maximum saturation of Fe retention by all of the processes in Table 8.25 was rarely observed. To place the estimate of the maximum potential Fe accumulation in *Sphagnum* peat in some perspective, an Fe concentration of 5757 g/m^2 (Table 8.25) is equivalent to an Fe concentration of 19% of the dry mass.

The approach used in Table 8.25 has provided an estimate of the ultimate capacity for Fe retention in a manmade *Sphagnum* wetland exposed to mine drainage. However, data from field situations in which wetlands have been constructed specifically for mine drainage treatment are also needed.

ROLE OF *SPHAGNUM* PLANTS IN IRON UPTAKE

Before the wetland approach to AMD treatment can become feasible, however, the effects of AMD on wetland vegetation, including *Sphagnum*, must be evaluated. The vegetation in manmade wetland systems serves not only as a mechanism of metal retention by the process of plant uptake, but also as a stabilizing mechanism for the organic substrate within the wetland, minimizing the potential for erosion.

Under natural conditions, *Sphagnum* growth and species distributions are affected by environmental factors such as pH, moisture, and nutrient availability. *Sphagnum* species can accumulate Fe and other metals with no apparent inhibition of growth when subjected to chronic inputs of low concentrations of metals via atmospheric deposition. Accumulation under these conditions is species specific and not necessarily correlated with plant growth.[17]

In contrast to the considerable amount of information available on *Sphagnum* growth in relatively undisturbed environments, comparatively little is known about the response of *Sphagnum* to the low pH and high metal concentrations typical of AMD. One study has demonstrated that the discharge of chemically treated mine drainage (pH values from 6 to 9) into what was once a naturally acidic *Sphagnum* wetland (surface water pH near 4.0) resulted in both the death of the vegetation and erosion of the organic peat down to the underlying mineral soil.

In light of these findings, the objectives of this study were to assess the growth responses of two commonly occurring *Sphagnum* species in solutions with Fe concentrations typically found in AMD, and to assess the apparent uptake of Fe from solution by these two *Sphagnum* species. Also, the potential contribution that uptake of Fe by growing *Sphagnum* could make to Fe retention within a wetland is compared to the potential contribution that other chemical and biological processes could make to Fe retention.

Living plants of two *Sphagnum* species, *S. fallax* and *S. henryense*, were collected at Big Run Bog, WV, a *Sphagnum*-dominated wetland not affected by AMD. In the laboratory, individual plants of each species were cut to an initial length of 5 cm and placed in groups of seven or eight into 2.5-cm diameter, 3.5-cm tall PVC cylinders. Three of these cylinders were placed in an 8-cm diameter fingerbowl to which a particular treatment solution was added. Treatment solutions contained either 0, 1, 10, 100, 1000, or 10,000 mg/L Fe, thereby spanning the range of Fe concentrations typical of AMD. $FeCl_2$ rather than $FeSO_4$ was used in preparing the treatment solutions. A previous study of the growth of *S. fallax* and *S. henryense* in AMD waters demonstrated that growth was negatively correlated with SO_4^{2-} but not with Cl^- concentrations in AMD. The treatment solutions were prepared by adding $FeCl_2$ to synthetic bog water containing, in milligrams per liter: 0.88 Ca^{2+}, 0.19 Mg^{2+}, 0.47 K^+, 0.23 Na^+, 0.16 NH_4^+, 0.07 NO_3^-, 3.44 SO_4^{2-}, and 0.62 Cl^-. The synthetic bog water approximates the ion concentrations found in Big Run Bog surface waters. Treatment solutions were added to each bowl so that the water table was located at the tops of the PVC cylinders. Every 2 to 3 days, additional treatment solution was added to maintain the water at the tops of the cylinders, and at weekly intervals the entire volume of water in each bowl was replaced. Plants were grown for 33 days.

Plant growth was assessed by removing individual plants from their cylinders and measuring their length with a metric ruler at 11-day intervals. Because the growth of *Sphagnum* is indeterminate and occurs along the long axis of the plant, an increase in length is an appropriate measure of growth over time.

After 33 days, determinations of chlorophyll concentration were made. Using approximately half of the plants from each bowl, the uppermost 1 cm of tissue (capitulum) was excised and weighed. Chlorophyll was extracted from the fresh plant tissue with 80% aqueous acetone. Absorbance of the extract at 649 and 665 nm was measured using a double-beam spectrophotometer, and chlorophyll concentration was calculated using equations given in Dolphin.

The plants remaining in each bowl were weighed and then dried for 24 hr at 55°C for fresh mass:dry mass ratio determinations. The dried plant material was ashed in a muffle furnace at 300°C for 1 hr and at 800°C for 3 hr. The ash was extracted with 6 M HCl and brought to final volume with distilled water. Fe concentrations in the extract solutions were determined by AAS.

For each species, the effects of Fe concentration in the treatment solution on growth (increase in length over the 33-day period) and chlorophyll concentration were assessed using Friedman's tests conducted on the data from day 33. Correlations between growth, final chlorophyll concentration, final tissue Fe concentration, and Fe concentration in the treatment solutions were determined using Spearman's rank correlation tests. Differences between the species with regard to growth and final tissue Fe concentration were evaluated with Mann-Whitney tests. A significance level of $p = 0.05$ was used in all tests.

The two *Sphagnum* species exhibited both similarities and differences in growth responses to increasing Fe in the treatment solutions. For *S. fallax*, after 33 days, growth in the 1 mg/L Fe solution was significantly greater than that in the control solution (0 mg/L), and growth in the 10 mg/L Fe solution did not differ significantly from that in the control solution. However, as solution Fe concentrations increased from 100 to 10,000 mg/L, plant growth progressively decreased. In contrast, for *S. henryense*, growth at all Fe concentrations was significantly less than growth in the control solution. For each species, the increase in length achieved over 33 days was significantly negatively correlated with treatment solution Fe concentration (Spearman's ρ values of −0.82 and −0.93 for *S. fallax* and *S. henryense*, respectively). Also for both species, at solution Fe concentrations ≥100 mg/L, not only was growth reduced relative to controls, but no significant increase in growth was observed after the first 11 days. Regardless of treatment solution Fe concentration, the inhibition of growth relative to control plants was proportionately greater for *S. henryense* than for *S. fallax*.

Final chlorophyll concentration in *S. fallax* generally decreased with increasing treatment solution Fe concentration. For *S. henryense*, final chlorophyll concentration in treatment solution Fe concentrations ≤10 mg/L were not significantly different from those plants grown in 0 mg/L Fe, but as Fe concentrations in treatment solutions increased to 100 mg/L and greater, chlorophyll concentration decreased. In each species, chlorophyll concentration in plant tissues after 33 days was positively correlated with increase in length (Spearman's ρ values of 0.68 and 0.82 for *S. fallax* and *S. henryense*, respectively). Moreover, at Fe concentrations ≥100 mg/L, final chlorophyll concentration was <1.2 mg/g dry mass and no additional growth was achieved after the first 11 days.

For both species, Fe concentration in the plant tissues after the 33-day period increased with increasing treatment solution Fe concentration, yet there was no significant difference between the two species in final tissue Fe concentration across all treatment solution Fe concentrations ($p = 0.34$). For each species, growth achieved after 33 days was negatively correlated with final tissue Fe concentration (Spearman's ρ values of -0.83 and -0.86 for S. fallax and *S. henryense*, respectively).

The observed reduction in *Sphagnum* growth with increasing Fe concentration in the treatment solutions may be related to a decreased ability of the plants to take up nutrient cations. Cation exchange on the cell walls of *Sphagnum* plants appears to be a major mechanism of cation uptake by the living cells of the plants. The maximum reported values for CEC in *Sphagnum* species are close to 1 meq/g dry mass, and base saturation of *Sphagnum* (percent of total CEC accounted for by the base cations Ca^{2+}, Mg^{2+}, K^+, and Na^+) has been measured as 8.7%. Cation exchange is a dynamic process in which equilibria between adsorbed cation and soluble cation concentrations are rapidly established. Therefore, as Fe^{2+} concentration in solution increases, increasing quantities of base cations and protons will be desorbed from exchange sites because of competition from the Fe^{2+} ions. For both *Sphagnum* species, when grown in solutions with Fe concentrations ≥ 100 mg/L, not only did final Fe concentration in the plant tissues exceed the CEC of *Sphagnum* (1 meq/g dry mass), but a dramatic reduction in both growth and chlorophyll concentration relative to controls was observed after 11 days. Thus, the observed reduction in plant growth may have been caused by a deficiency of nutrient base cations induced by competition from Fe^{2+} for exchange sites.

The concentration of Fe in the tissues of both species exceeded the total CEC of the plants by as much as 650% in Fe treatment solution concentrations ≥ 100 mg/L. This excess Fe must have been incorporated into the plants as nonexchangeable Fe via specific binding to organic matter or the formation of Fe oxides or oxyhydroxides. No obvious oxide deposits were discernible during examination of the tissue using light or scanning electron microscopy (SEM). This suggests that organic binding of Fe by the plants may have been a more important process in Fe accumulation than either cation exchange or oxide formation.

The potential contribution of growing *Sphagnum* plants to Fe retention in a wetland constructed for AMD treatment can be estimated based on the growth of the plants relative to controls and measured Fe uptake. To estimate potential Fe uptake on an areal basis, field estimates of primary productivity of *Sphagnum* must be incorporated into the computation. The average annual net primary productivity of the two dominant *Sphagnum* species from Big Run Bog (*S. fallax* and *S. magellanicum*) is 5.75 g dry mass/dm²/year. Multiplying this value by the percentage of growth relative to controls and by final tissue Fe concentration for each species at each solution Fe concentration yields estimates of potential Fe accumulation by growing *Sphagnum*. The projected areal accumulation of Fe by *S. fallax* was 30 to 230% greater than that by *S. henryense* for any particular solution Fe concentration. This

Table 8.26. Estimation of the Potential Fe Uptake, on an Areal Basis, by Growing *Sphagnum* Plants Subjected to Different Fe Concentrations in AMD

Treatment Solution Fe Conc. (mg/L)	Estimated Fe Uptake (g/m²/yr)	
	S. fallax	*S. henryense*
1	2.2	1.8
10	6.5	4.1
100	16.8	5.4
1,000	29.0	8.8
10,000	17.7	13.8

From Spratt, A. K. and R. K. Wieder. Growth Responses and Iron Uptake in *Sphagnum* Plants and Their Relation to Acid Mine Drainage Treatment, U.S. Bureau of Mines, IC 9183 (1988), pp. 279–285.

difference is accounted for by the difference between the species in growth relative to controls; in all solution Fe concentrations excluding 10,000 mg/L, reduction *S. fallax* growth was significantly less than the reduction in *S. henryense* growth. The maximum Fe accumulation projected for the two species (29 g/m²/year) was more than eight times greater than a previous estimate (3.5 g/m²/year) based on a lower Fe concentration in *Sphagnum* tissue. However, in achieving an Fe uptake value of 29 g/m²/year, the plants would effectively die. At Fe treatment solution concentrations where plants were still growing and green, maximum Fe uptake was only 6.5 g/m²/year. Despite these new estimates of the potential maximum for Fe uptake by *Sphagnum* plants, the contribution that plant uptake would make to total Fe retention in a wetland system is still small compared to the contributions made by other chemical and biological processes within the peat. Especially noted are the binding of Fe to organic matter and the formation of Fe oxides, which together may account for as much as 97% of total Fe retention (Table 8.26).

These results show that *Sphagnum* plants can survive and grow in solutions with Fe concentrations up to 10 mg/L, but at 100 mg/L or greater they turn brown and stop growing. In a wetland designed to treat low flows of AMD containing relatively low Fe concentrations, the plants could remain viable. In addition, although *Sphagnum* plants do accumulate more Fe when exposed to higher Fe concentrations, the potential contribution of *Sphagnum* plants to overall Fe retention in comparison to other processes contributing to Fe retention in a wetland constructed for AMD treatment appears to be minor. Nonetheless, the need to maintain a vigorous cover of vegetation in constructed wetland treatment systems is critical to minimize the potential for erosion of an organic substrate. We have demonstrated clear differences between species in their growth and chlorophyll responses to AMD; *S. fallax* exhibits greater increase in length than *S. henryense* in Fe-rich water. Further study of *Sphagnum* species' responses to other AMD constituents, particularly to metals, is warranted to determine which species of *Sphagnum* are most likely to survive when exposed to different types of AMD.

IRON AND MANGANESE REMOVAL IN A *TYPHA*-DOMINATED WETLAND

In this study, an instrumented wetland was constructed to quantify Fe and Mn removal over a 10-month period following wetland construction. The wetland received flow from an AMD seep emerging from spoil reclaimed following mining of the Middle Kittanning (#6) and Lower Kittanning (#5) coal mines of the Allegheny Formation in Tuscarawas County, Ohio.

Based upon an average flow estimate of 10 gpm, a 110×20 ft (33×66 m) basin was constructed (see Figure 8.10) following the recommended size of 200 ft^2 (60 m^2) of wetland per 1 gpm of flow. The basin was leveled and filled with 6 in. of agricultural-grade limestone which, in turn, was covered with a 10-in. humic layer consisting of an equally proportioned mixture of peat, compost, and sandy soil. To avoid coating the limestone with a ferric hydroxide precipitate, the limestone was placed as a distinct layer and not mixed with the humic soil. Cattails (*Typha*) were collected from the surrounding area and planted in the basin at 1-ft centers. Cores taken 1 month later showed that these layers had compacted to approximately 5.5 in. of limestone and 7 in. of humic material.

The water budget is recognized as a key factor affecting water quality and wetland functions. Therefore, 40° V-notch weirs with continuous water level recorders were installed at the inlet and outlet to measure surface waterflow into and from the wetland. Flow volumes were calculated from the following equation for sharp-edged V-shaped weirs:

$$Q = C \times 8/15 \times \sqrt{2g} \times \tan(\theta/2) \times H^{2.5} \qquad (8.5)$$

where Q = discharge (ft^3/sec)
         C = weir coefficient (0.582 for a 40° V-notch weir)
         θ = total angle of notch (degrees)
         H = head above the lowest point of the notch (ft)
         g = gravitational acceleration (32.17 ft^2/sec)

A weighing-bucket continuous-recording rain gauge was placed at the site to measure rainfall; potential evapotranspiration was estimated from evaporation pan data collected during the spring and summer months. Additionally, six two-level piezometer nests were placed within the basin: the lower level sampled subsurface water within the bottom 3 in. of the humic substrate, and the upper level sampled the top 3 in. of the substrate.

Nineteen separate wetland water samples were collected on a biweekly basis from the piezometers, the inlet and outlet weir pools, and from 33 surface sites defined by an 11-column × 3-row sample grid (Figure 8.7). Frozen wetland surface conditions and site inaccessibility due to excessive rainfall precluded the collection of some samples. Chemical analysis included Ca, Fe, K, Mg, Na, Mn, SO$_4^{2-}$, and

382 ENVIRONMENTAL IMPACTS OF MINING

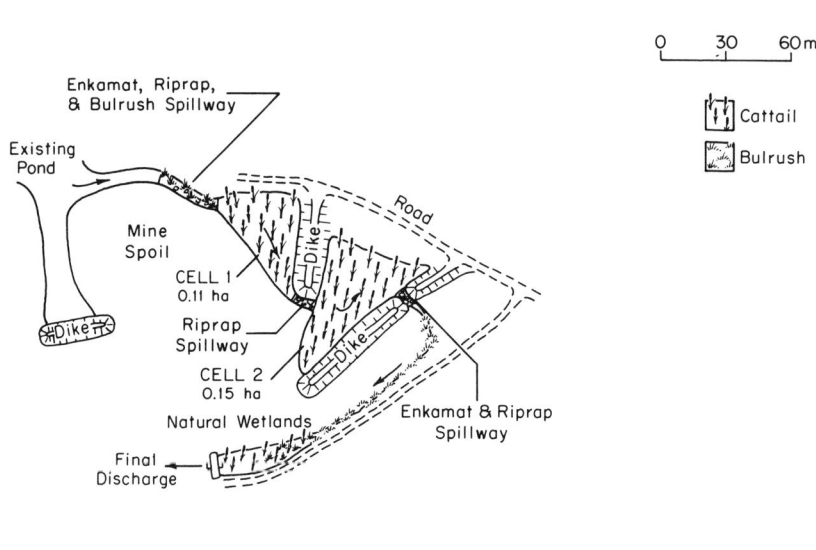

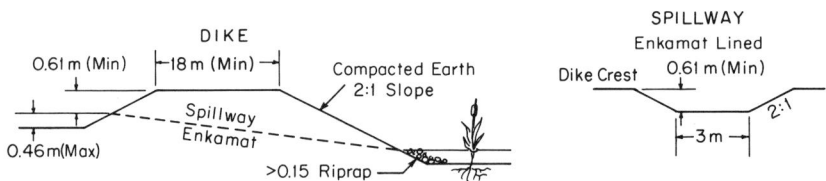

Figure 8.7. Inlet and outlet structures in a constructed wetland to treat AMD. (From Watson, J. T. and J. A. Hobson. Hydraulic design considerations and control structures, in *Constructed Wetlands for Wastewater Treatment* (Chelsea, MI: Lewis Publishers, Inc., 1990), pp. 379–391.)

orthophosphorus. Inductively Coupled Plasma spectrometry was used for all analyses except for SO_4^{2-} and orthophosphorus. These were determined by colorimetric procedures, with one modification: rather than reading the sample concentration directly from the spectrophotometer, a series of standards was analyzed with each sample batch to establish a regression relation of concentration vs absorbance. Sample concentrations were then determined using this equation.

Water budget volumes were calculated by dividing the study period into intervals whose midpoints corresponded to the date of water sampling. Budget components include inlet flow (If), outlet flow (Of), rainfall (Ppt), potential evapotranspiration (Pet), and change in volume of surface water stored (ΔS) within the basin (volume determined by multiplying wetland area × average surface water depth). All of the above components were calculated for each interval, and the following equation was used to determine the residual (Res) component:

$$If + Ppt - Of - Pet - \Delta S = Res.$$

A positive residual value denotes a net water gain or an ungauged loss of water (such as vertical seepage down through the basin), while a negative residual value denotes a net water loss or an ungauged source of water (such as vertical seepage up into the basin). Residual values might also reflect operator and/or instrument error.

Following water budget analysis, a total mass flux for each interval was calculated by multiplying inlet and outlet flow volumes by their respective inlet and outlet concentrations. Assuming that negative residual volumes resulted from an ungauged source, these were multiplied by the inlet concentrations and added to the amount of mass entering the wetland. Positive residuals were attributed to ungauged outflows, which were multiplied by the outlet mass concentrations and added to the amount leaving the wetland. Total influx was calculated as the sum of the influx values for each interval. Total outflux was determined in a similar manner. Inflow and outflow were the only water budget components used in mass budget calculations because Fe and Mn concentrations in rainfall were assumed to be negligible and the change in surface storage accounted for <1% of the total water budget.

Concentrations of Fe, Mn, and SO_4^{2-} at the seep varied throughout the study period, and while their maximum concentrations were 44, 50, and 2720 ppm, respectively, they averaged 31 ppm Fe, 34 ppm Mn, and 1130 ppm SO_4^{2-}. Mass concentrations at the inlet to the wetland averaged 13 ppm Fe, 30 ppm Mn, and 990 ppm SO_4^{2-}; evidently Fe and SO_4^{2-} were removed from solution while the water flowed through a ditch leading from the seep to the wetland (insert, Figure 8.7). The following discussion considers only the mass concentrations that actually entered the wetland.

Overall, 74% of the 186.4 kg of Fe delivered to the wetland during the study period remained within the basin. Each sample interval within this period consistently displayed a smaller mass outflux than the influx, indicating that Fe removal continued through time. The smallest percentage of Fe removal, 43%, occurred on two separate occasions: first, during the December 8 interval, when 2.4 kg of Fe were removed from the flow; and second, during the April 15 period, when 6.2 kg were removed. The largest reduction occurred with the removal of 26.4 kg of Fe (94% of the inlet mass flux).

Mn was not as successfully treated by the wetland: only 8.3% of the 368.5 kg of mass that entered the basin during the entire study period was removed from the flow. During most of this period, inlet and outlet Mn flux values barely differed. Not until mid-May did the system begin to consistently remove Mn from solution, and, as with Fe, the June 1 period displayed the greatest Mn removal, retaining 14.2 kg (26% of the inlet mass concentration) within the wetland. Later, during the July 3 sample period, inlet Mn concentrations were reduced by 40%, a percentage corresponding to the removal of 13.8 kg.

Flow does appear to influence metal removal in a wetland system. The peak inlet flow volume, 20.1 gpm, occurred over the period of April 15. Peak inlet mass concentrations followed this springtime hydrograph peak: the maximum Fe influx of 23.6 ppm occurred on May 1, and the maximum Mn influx of 53.9 occurred on June 1. This sequence — peak mass concentrations following peak flow volumes — could be caused by the springtime "flushing" of metals from the mine spoil.

The wetland corresponded to the increased flow volumes by decreasing the percentage of Fe removed from its inlet water. During the springtime increase of the inlet hydrograph (March 18 to April 15), the amount of mass removed dropped from approximately 10 kg (80 to 85% of the inlet mass) for each of the March 18 and April 7 periods, to 6.2 kg (43% of the inlet mass) for the April 15 period. The amount of Fe removed during May 1 increased to 18 kg, but because inlet Fe concentrations increased greatly during this period, this removal corresponded to only 55% of the total mass.

The outflow hydrograph exhibited a lower and broader springtime peak, suggesting that the increased inlet volume did not immediately flow through the basin, but that a portion of the water was stored within the wetland and then released at a more constant rate. Fe concentrations at the outlet increased with increasing outflow volume.

During the April 15 period, when the inlet flow was at its highest and percentage of Fe removal at its lowest, Mn concentrations were slightly higher at the outlet than at the inlet. Mass budget calculations show an extra 0.9 kg of Mn leaving the wetland during the April 15 period, and an extra 1.3 kg leaving over the May 1 period. From May 16 through the end of the study period, Mn was removed from the wetland flow. In fact, the greatest amount of removal (14.2 kg) coincided with the largest Mn influx (54.3 kg) during the June 1 period.

Outlet Mn concentrations were greatest during the intervals of June 1 and June 16, reflecting the high inlet concentrations of the same periods. The peak outlet concentration of 43.6 ppm did not occur during the same interval as the peak inlet concentration, however; instead, the outlet peak appeared during the following, June 16, period.

The effects of length and width on the removal of Fe and Mn from the surface water were examined with an analysis of variance. Removal was calculated by subtracting the concentration at each grid point from the inlet concentration. Time effects were not examined with these tests; instead, all 19 samples obtained from a particular location were considered to "replicate" the average chemical conditions at that area. While width was not found to be significant in the removal of either metal, length was significant in both cases, with Fe being the more affected. Duncan's multiple range test was used to examine the equality of the column means averaged over the three rows. The test indicated that while removal did increase with length, the means for columns 7 to 11 were not markedly different from one another; similarly, the means for columns 5 and 6 were also not significantly different.

A second-order regression analysis of Fe removal vs length produced a predicted curve with a 0.962 r^2 value. While a first-order relationship would also describe the data points, the quadratic curve provides a better fit and also suggests that the rate of Fe removal slows during the time it takes water to flow from inlet to outlet. This is a likely possibility as reaction rates can slow as the reactants become more dilute.

The importance of length in the removal of Mn cannot be as clearly defined as with Fe. Length was found to be a significant factor, but Duncan's test revealed that the greatest removal occurred in columns 7 to 9, and that columns 10 to 11 averaged the same amount of removal as columns 5 to 6. Other factors in addition to length must be influencing Mn removal. The first-order regression of Mn removal vs length demonstrates this with an r^2 of 0.53; length cannot explain all of the variability in Mn removal. A second-order fit did not significantly improve the r^2 value and, therefore, is not presented here.

The two-way interaction of length × width was important as well, and a plot of the row means per column indicated that three surface locations contained water with less Fe and Mn, 37 and 20%, respectively, than water at the outlet. Samples from these areas were also distinguished by an enrichment in K, P, sodium (Na), Mg, and Ca. Similar water chemistry, i.e., high nutrient and low dissolved metal concentrations, was also found in the interstitial waters, especially from samples in the lower piezometers located immediately above the limestone layer. Based upon this information, two hypotheses can be made: (1) the high Ca and Mg values at these three locations were due to the dissolution of limestone, which could have been mixed with the humic layer in these areas during wetland construction, and/or (2) interstitial water, rather than surface water, might have been collected at these areas. The second possibility is likely because the wetland surface surrounding these areas is a hummocky area where surface water was found in shallow depressions (0.5 to 1.0 in. deep) separated by emergent clumps of cattails, grasses, and saturated soils. Water collected as a surface sample might have actually flowed from the substrate at the time of collection.

Did the wetland effectively treat flow from the AMD seep? Effectiveness implies that effluent water quality meets standards set by the Surface Mining Control and Reclamation Act (SMCRA) in which Fe concentrations cannot be >7 ppm daily (or an average of 3.5 ppm monthly) and Mn limits have been set to 4 ppm daily (2 ppm for a monthly average). Effluent from the wetland met the daily Fe requirements at every sampling except for April 7, April 15, and May 1, when the inlet flow volumes were higher than average. The monthly standard was met in every month but February, April, and May. Mn concentrations remained above limitations throughout the entire period. In February, however, the mining company discontinued its downstream chemical treatment of the wetland effluent. Nevertheless, outflow from the polishing pond continued to meet Fe and Mn requirements for the remainder of the study.

Flow appeared to be an important variable affecting Fe and Mn removal: increased flow volumes coincided with drops in percentage removal of Fe and increased outlet Mn concentrations. In addition, peak inlet mass concentrations, and, consequently, peak outlet concentrations, followed the peak springtime inflow.

The importance of length in influencing wetland treatment abilities agrees with findings from other researchers. A municipal waste wetland showed that one system with a length:width ratio of 75:1 constantly outperformed another with a 4.5:1 ratio. Within the present study, the increase of Fe effluent concentrations during the spring suggests that the system was underdesigned for the maximum mass influx. The relationship between length and Fe removal suggests that a lengthwise extension of the basin would increase Fe removal. Although length was not as important in removing Mn, a longer basin would also improve Mn removal.

While outlet concentrations did not always show improvement in water quality, a few surface locations and the interstitial waters, did exhibit consistently lower Fe and Mn concentrations. Interaction with soil and/or the underlying limestone layer appears to be a common factor between these spots. A wetland designed with a more permeable substrate might provide greater contact between soil and water, provided the water remains shallow.

THE ROLE OF ALGAE IN THE TREATMENT OF ACID MINE DRAINAGE

Freshwater algae are common and widespread inhabitants of the waters associated with coal mine drainage and are recognized as effective accumulators of metals. The majority of work related to algae and the bioaccumulation of metals typically has been with metals that are generally of little concern, from a regulatory standpoint, in coal mine drainage: notably zinc (Zn), nickel (Ni), Cu, and lead (Pb).

Fe and Mn, however, are probably the two most common and regulated mineral constituents of mine drainage, accounting for millions of dollars in chemical treatment costs annually. Fe, which is relatively easy to remove from mine drainage, presents few logistical problems to chemical treatment, but Mn is a more persistent mineral and requires rather sophisticated treatment to achieve effluent standards. Additionally, concentrations of 50 to 100 mg/L of Mn are not uncommon in acidic mine drainage.

Algae require Fe and Mn as essential micronutrients. Although concentrations of Mn >1 mg/L have been shown to inhibit some blue-green and green algae, other algae are commonly found and often abundant in AMD, with high concentrations of Fe and Mn. Algae associated with these areas have been shown to accumulate Mn, with reported levels of Mn as high as 56,000 mg/kg of plant tissue (dry weight). This is of particular interest in that biological treatment systems, i.e., manmade wetlands, are proving to be a viable and cost-effective means of treating AMD.

To date, most manmade wetlands are constructed as *Sphagnum*-dominated, or more commonly, *Typha*-dominated treatment systems. Algae generally are not intentionally introduced into these wetlands but become established in the systems by way of the source of the mine drainage, the air, the soil, and most commonly, in association with the principal, transplanted wetland vegetation. Because of this secondary nature of establishment, any beneficial effects of algae on the quality of AMD often go unrecognized, or they are limited because the wetland treatment system (WTS) is not designed to optimize algal habitats and, therefore, algal biomass is limited. Little is actually known concerning the role of algae in the treatment of AMD in these WTS.

In this study, the source of the AMD that was treated by the WTS was an abandoned drift mine that was discharging since at least 1940 and was (at the time of the study) considered to be receiving recharge waters from the adjoining surface mine. The water flow rate was an average of 37 gpm. The AMD quality at the source was characterized as follows: pH, 5.0; alkalinity, 16 mg/L; mineral acidity, 450 mg L; total Fe, 130 mg/L; Mn, 71 mg/L; Al, 4 mg/L; and sulfate, 2200 mg/L.

The WTS was designed and built with an initial collection pond immediately adjacent to the drift mine source, followed by two distinct, parallel wetlands. The two wetlands combine at their respective outflow points and flow through a planted ditch to a conventional treatment pond series.

The parallel wetlands were set up as four individual units in each system: two shallow *Typha*-dominated ponds followed by two, deeper algae-dominated ponds in one series, and two algae ponds followed by two *Typha* ponds in the other series. Each pond has an approximate surface area of 600 ft^2 and is connected to the next pond in series by a section of 6 in. diameter plastic pipe. The system is graphically displayed in Figure 8.8.

The ponds were constructed with available, onsite clay and a 12-in. base of composted manure, which serves as a planting medium and a source of nutrients to the vegetation. Agricultural limestone, 10 to 15 cm, was placed beneath the compost in the final planted ditch, but no alkaline materials were used in any other portion of the wetland. *T. latifolia* was available on site and, hand dug, was planted at a density of roughly one mature plant and rhizome mass per 0.1 m^2. *Scirpus cyperinus* and *Juncus effusus* were also available onsite and were introduced in the *Typha* ponds primarily in the areas of the *Typha* ponds at the inflow/outflow pipes. *S. cyperinus* and *J. effusus* were also planted around the edges of the algae ponds to stabilize the banks and to provide possible attachment sites for algae.

Initially the study intended to concentrate on blue-green algae (Cyanobacteria) of the genus *Oscillatoria*. *Oscillatoria* spp. was noted as being prevalent in several natural and manmade wetlands that displayed reductions in Mn levels from source to outflow and was the species noted in a previous section that exhibited Mn levels of 56,000 mg/kg of plant. However, during the period between the original proposal and the actual planting of the studied wetland, the predominant algae in the previously *Oscillatoria*-dominated wetlands became the green algae of the genus *Mougeotia*. *Oscillatoria* was still present, but not dominant.

The water quality in these wetlands continued to show Mn reductions, and analysis of *Mougeotia* samples confirmed that they too were accumulating Mn.

388 ENVIRONMENTAL IMPACTS OF MINING

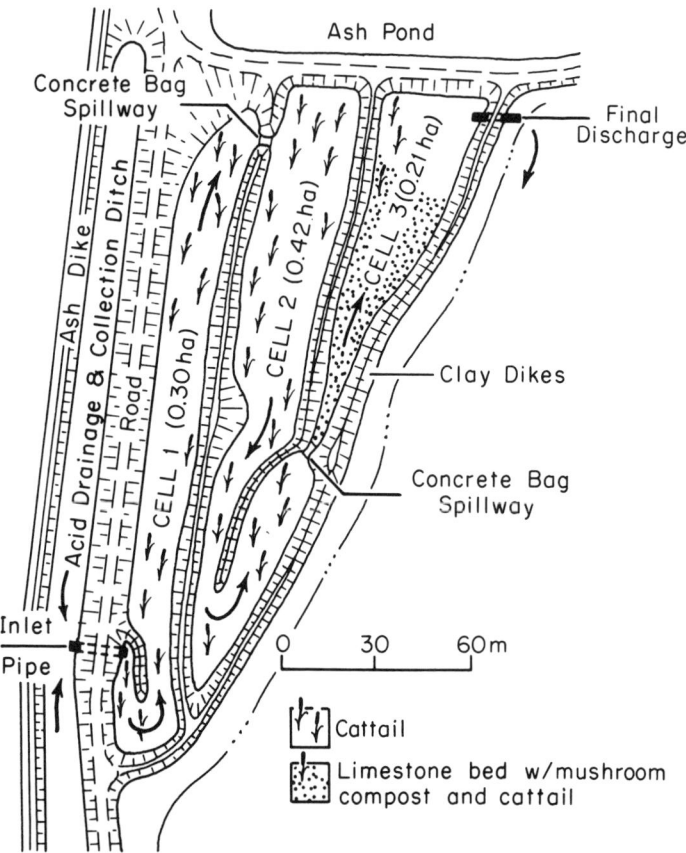

Figure 8.8. Acid drainage treatment wetland. (From Watson, J. T. and J. A. Hobson. Hydraulic design considerations and control structures, in *Constructed Wetlands for Wastewater Treatment* (Chelsea, MI: Lewis Publishers, Inc., 1990), pp. 379–391.)

Therefore, the decision was made to transplant algae from one of these sources into the study WTS, knowing that the predominant algae being "planted" was *Mougeotia*, but that *Oscillatoria* was also being introduced at the study site. Approximately one 30-gal container of *Mougeotia*-dominated algae was introduced into each of the four algae ponds at the studied wetland. Because the side of the parallel series with the *Typha* ponds downflow of the algae ponds would be immediately affected by this planting and the other side would not be affected, a container of algae was also introduced into each of the four *Typha* ponds.

Various other species of algae, primarily diatoms, were also introduced into the wetland with the principal algae planting, and a filamentous green algae was noted as being associated with the *Typha* planting. These species, while of little apparent significance regarding total numbers at the time of the planting, were identified and may be found to take on some degree of importance in treatment during the growing season.

The AMD source, the inflow/outflow for each pond, and the final outflow point of the wetland have been designed as water monitoring points in this study. Each monitoring point was sampled weekly over the 18-month study. The samples were analyzed for pH (electrometrically); specific conductance (µmhos at 25°C); alkalinity and mineral acidity (titrimetrically at pH 4.5 and 8.3, respectively); and total and ferrous Fe, total and dissolved Mn (dissolved: passing a 45-µm membrane filter), sulfate, and Al (all minerals are determined spectrophotometrically). Water temperatures (in degrees Celsius) are taken from each pond in the field at the time of collection. All analyses were conducted following the Standard Methods for the Examination of Water and Wastewater and were done in accordance with EPA guidelines.

Flows were measured weekly by a portable cutthroat flume in the channels and by bucket and stopwatch at the discharge pipes. In this manner the total water budget of the WTS may be known and any increases or decreases in the system readily noted.

Representative samples of the wetland vegetation will be collected and analyzed for Fe, Mn, and Al content beginning in the spring of 1988 and then periodically throughout the study, with principal attention given to the predominant algae species. Total metals were determined as follows: the samples are dried, ashed by means of heating at 550°C in a muffle furnace for 2 hr, acid digested by standard methods, and analyzed for mineral concentrations by atomic absorption. Results are reported as milligrams of mineral per kilogram of plant (dry weight).

To determine the site of mineral accumulation in the algae, i.e., adsorbed and intracellular concentrations, the algae samples are "washed" with an EDTA solution in a shaker bath and then filtered. The extract is analyzed by AAS to determine the concentration of adsorbed mineral associated with the algae. The "washed" algae is then ashed and analyzed as described above to determine the intracellular concentration of minerals.

The form of the minerals in and on the algae, e.g., adsorbed manganic oxide, will also be determined utilizing a form of X-ray diffraction so as to predict the long-term stability of the minerals in the wetland and the mechanisms of accumulation.

Diurnal studies were conducted during periods of accelerated algal growth to examine the DO content and pH of the ponds and, therefore, the effect, if any, the algal blooms have on pH values. Carbon dioxide removal and production in the wetland, through algal photosynthesis and respiration, can account for shifts in pH values, although this phenomenon is generally associated with alkaline waters. pH in western Pennsylvanian lakes have been noted to fluctuate 5 pH units over a 24-hr period due to the effects of blue-green algal blooms.

The *Typha* planting died back early simply because of the season, and the algae populations have not experienced favorable weather conditions to bloom. However, between the compost and the dead vegetation considerable organic matter was present in the WTS, the rhizomes in the *Typha* ponds were active, bacteria and algae were present in the system, and the WTS was designed to maximize plant contact and retention time with the flows. Therefore, even at this early date, positive water quality results can be seen.

The WTS significantly reduced the Fe and acidity load of the AMD, but Mn values were not significantly reduced at this time. To date, the quantity of water entering and exiting the system has been equal. A small amount of seepage has been noted along the coal outcrop, upslope, and immediately adjacent to the pond series ending with the *Typha* ponds, but the seepage is small enough that neither the quantity or quality of the flows in the WTS has been affected. The outcrop area has been planted in an attempt to control the seepage, as any drainage entering the WTS near its middle or end would obviously bias any results or conclusions regarding the effectiveness of the WTS.

The removal of Mn in WTS integrated with specific algae "ponds" may prove to be a successful treatment technique. Algae appear particularly well suited to this type of system in that they are widespread in AMD, are capable of accumulating relatively high concentrations of minerals, and can reproduce rapidly. The success of this single WTS and the mechanisms of the bioaccumulation of AMD minerals in algae will be determined in subsequent months.

CONSTRUCTED WETLANDS FOR ACID DRAINAGE CONTROL IN THE TENNESSEE VALLEY

The TVAs first constructed wetland for acid drainage treatment, known as Impoundment 1, is located at a reclaimed coal preparation plant site in northeast Alabama. Success of this wetland at meeting permit effluent limitations has led to the construction of 11 wetland treatment systems at TVA coal facilities and coal-fired power plants, and to an extensive research program on constructed wetlands.[20]

Site characteristics often restricted use of standardized methods; therefore, wetlands were designed for specific conditions. A generic description of predesign investigations through wetlands operation follows.

State regulators regularly met onsite with TVA officials to review alternative sites and treatment options. Wetlands systems were approved as long as all effluent discharge limitations were to be met (i.e., total Fe <3.0 mg/L, total Mn <2.0 mg/L, pH = 6.0 to 9.0 s.u., and nonfilterable residue <35.0 mg/L). Therefore, system designs occasionally included a final cell to provide for chemical treatment if necessary.

Wastewater characterization and site hydrology were the two most important predesign data needs. Preconstruction water quality sampling was conducted for all flows to be treated and any streams that would be affected by the wetlands. Analyses included pH, Eh, total and dissolved Fe and Mn, nonfilterable residues (NFR) and

TDS, sulfate, Al, DO, Mn, Zn, selenium (Se), mercury (Hg), and cadmium (Cd). Flow monitoring was incorporated into baseline site monitoring and compared to existing flow data and hydrologic modeling. Baseline population estimates of aquatic biota in receiving streams were made to provide a means of documenting stream recovery.

Often it was necessary to locate the wetland away from the immediate area or leased/owned surface; thus, negotiations with landowners for initial access and long-term future control were conducted. TVA owned or purchased many of the wetland sites, but also pursued other surface control arrangements including long-term leases and permanent easements.

Regulations neither specifically addressed nor gave guidance on constructing treatment wetlands. Permits and approvals for wetlands construction required from state and federal regulators included a National Environmental Policy Act (NEPA) review, a floodplains (Executive Order No. 11988) review, a National Pollutant Discharge Elimination System (NPDES) permit, a surface mining permit, and a water engineering report.

Topography was determined in sufficient detail to plan the number and location of cells, thereby minimizing cut and fill requirements dictated by a particular gradient. Because site regrading was usually an early step in wetlands construction, topography was completely altered, and detailed (e.g., 0.6-m contour interval) topography surveys were not warranted.

Geology was evaluated to determine if the site overlay shallow bedrock or lacked suitable growth media. If necessary, sources of borrow and adequate growth media were identified. Flow patterns and the depth of groundwater were determined to identify inflows or outflows that could affect water quality or hydrologic balance.

Preferred sources of emergent vegetation for transplantation were nearby natural wetlands developed in similar quality water. This avoided stress from abrupt changes in edaphic conditions. Cattail *(Typha)*, followed by *Scirpus, Eleocharis*, and *Carex*, were the most tolerant, readily available species for transplantation. A rush *Juncus* was used with less success. Preliminary research results[27] suggested that *Typha latifolia* and *Eleocharis quadrangulata* provided higher radial oxygen loss than other common species, thereby enhancing substrate redox conditions to bind insoluble forms of metal precipitates in the substrate.

Size of wetlands, number of cells, spillways, and dike specifications were designed for 10-year, 24-hr storm events estimated from site flow monitoring or various methodologies. Erosion and sedimentation control structure design and construction, along with best engineering estimates and practices, were used for those components having no design guidelines.

TVAs constructed wetlands ranged in size from 3.5 (38 ft^2) to 113.0 m^2 (1216 ft^2) per average flowing liter per minute, and 2.0 (22 ft^2) to 41.0 m^2 (441 ft^2) per maximum flowing liter per minute. Design sizes were dependent on water quality characteristics, storm flow hydrology, and land availability. Wetlands were designed to accommodate stormflow, then increased in size if very poor quality water had to be treated. Increasing the size of a wetland up to twice the stormflow design area only modestly increased costs and provided an adequate treatment area.

Wetlands shapes varied because of existing topography, geology, or land availability. Irregular shapes for wetlands cells enhanced natural appearance and provided hydraulic discontinuity. Configurations that increased velocities causing channelization, scouring, bank erosion, etc., were avoided.

The number of cells for constructed wetlands was determined by site topography and hydrology. Level sites were amenable to large cells hydraulically chambered with rock or earthen finger dikes, large logs, vegetated hummocks, or other baffles. Steeper gradients required more grading or a system of several cells terraced downslope.

Water depth and bottom slope depended on plant species, pollutant concentrations, freeze potential, and desired longevity of the system. *Typha latifolia* has been the preferred species in TVA-constructed wetlands. Other plants used in shallow water include *Scirpus, Juncus, Carex, Eleocharis,* and *Equisetum*. Excessive water depths not only inhibited desirable species development, but promoted anoxic, reducing conditions in the water column which seriously affected the oxidation of Fe and Mn. Shallow water depths, subject to freezing in more northern climates, are suitable in the Tennessee Valley. The primary advantage of shallow water is enhanced oxygenation and increased plant production. Potential disadvantages are reduced storage capacity and retention time. The average water depth in a TVA wetland ranged from 15 to 30 cm (6 to 12 in.), with some shallower and some deeper areas to provide for species diversification. A few deep pockets of 1 m or greater were included in many cells to provide recharge zones and aquatic fauna refuge in drought events.

Most wetlands were completed in early summer, although successful installations were completed as late as October. Wetlands construction began with site clearing, followed by grading and dike construction, and importing suitable materials as necessary to meet design specifications. Brush was burned or pushed along the site perimeter to provide wildlife habitat. Spillways were either rock lined or covered with nonbiodegradable erosion control matting and planted with *Scirpus, Carex,* or grasses. Water level control or flow monitoring devices were incorporated into the spillway design and construction.

Vegetation was dug by hand to obtain complete rootballs/rhizomes. Transplantation was completed on the same day as digging, and plants were not subjected to extreme temperatures, drying, or wind during transport. *Typha* was set into the substrate at or about nine plants per square meter and stems were broken above the water level to prevent windfall and stimulate new growth from the rhizomes. Bulrush clumps were simply placed in the desired location.

Wetlands were fertilized with a P-K (phosphorus-potassium) fertilizer such as 0-12-12 at 400 kg/ha (353 lb/acre). Mosquito fish (*Gambusia affinis*) were stocked for insect pest control.

Postconstruction activities included effluent monitoring; fertilization; maintenance of dikes, spillways, or other control structures; and pest control. Effluent monitoring was performed several meters downstream of the final spillway so that any leakage was included in the sample. Monitoring requirements included pH, total Fe and Mn, and NFR. Additional water chemistry and biological monitoring were used to quantify wetlands treatment efficiencies and wetlands habitat benefits. After the first year, fertilization was done only if vegetation showed signs of nutrient depletion. Dike repair due to muskrat burrowing was required at one wetland. Army worm (*Simyra henrici*) infestations at another wetland required control measures to prevent eradication of the cattail.

A summary of characteristics and water quality parameters for TVAs 11 constructed wetlands is presented in Table 8.27. Dates of initiated operation are based on the first effluent monitoring (i.e., usually within 1 week of initial discharge). Areas are given as surveyed inundated area. Influent water chemistry, in most cases, is based on at least 1 year of seasonal sampling of contributing seeps. Effluent monitoring, including flow, is generally based on twice-monthly discharge permit sampling results from the date the wetland system began operating.

Impoundment 1 was TVAs first constructed wetland, treating acid seepage from a coal slurry dike pond at the reclaimed Fabius Coal Preparation Plant in Jackson County, Alabama. Dominant vegetation is *Typha*, with *Scirpus, Leersia, Juncus, Eleocharis, Utricularia*, and *Sparganium* among a total of 41 species present 2 years after construction. Since construction, Impoundment 1 has produced compliance-quality effluent.

Impoundment 4, also at the Fabius plant site, was built to treat acid seepage emanating from process water recirculation ponds. These ponds (pH = 3.5 s.u.) were reclaimed in 1986 and the inflow to Impoundment 4 has been limited. Dominant plants are *Typha* and *Scirpus*. The original planting of Impoundment 4 took place in November 1986. Few plants survived in spring, and the wetland was replanted the following July. Because of the extremely low pH of the seepage, an NaOH treatment system was installed to augment the wetland treatment.

950-1 and -2 was a two-cell sedimentation basin receiving AMD from the reclaimed TVA Fabius 950 coal mine in Jackson County. A *Typha* marsh has naturally developed in the upstream cell and expanded into the lower cell. Treatment with NaOH, required from 1976 to 1984, has been discontinued. The discharge was released from NPDES permit monitoring requirements in 1987.

Impoundment 2 is a series of constructed wetlands intermediate in a 138-ha (341 acre) drainage basin receiving acid drainage from non-TVA abandoned mine land and the coarse refuse disposal area at the Fabius plant site. Effluent from the wetlands is treated with NaOH and discharged. Vegetation in the wetlands is predominantly *Typha*.

Widows Creek 018, located at TVAs Widows Creek Fossil Plant in Jackson County, receives seepage from an abandoned ash disposal area. The wetland was

Table 8.27. TVA Acid Drainage Wetlands Treatment Summary

Wetlands System	Date Initiated Operation	Area (m²)	No. of Cells	Influent Waters Parameters (mg/L)				Effluent Water Parameters (mg/L or L/m)				Flow		Treatment Area (m²/mg/min)	
				pH	Fe	Mn	NFR	pH	Fe	Mn	NFR	Ave.	Max.	Fe	Mn
W C 018	6-86	4,800	3	5.6	150.0	6.8		3.9	6.4	6.2		70	1,495	0.2	4.2
King 006	10-87	9,300	3	4.2	153.0	4.9	40.0					379	2,271	0.2	5.0
Imp 4	11-85	2,000	3	4.9	135.0	24.0	42.0	4.6	3.0	4.0	6.0	42	49	0.4	2.0
950 NE	9-87	2,500	2	6.0	11.0	9.0	19.0	6.6	0.5	0.2	49.0[a]	348	1,673	0.7	0.8
R T - 2	9-87	7,300	3	5.7	45.2	13.4		6.7	0.8	0.2	2.0	238	681	0.7	2.3
Imp 2	6-86	11,000	5	3.1	40.0	13.0	9.0[b]	3.1	3.4	14.0	0.8[b]	400	2,200	0.7	2.1
Imp 3	10-86	1,200	3	6.3	13.0	5.0	28.0	6.8	0.8	1.9	4.7	87	379	1.1	2.8
W C 019	6-86	25,000	3	5.6	17.9	6.9		4.3	3.3	5.9		492	6,360	2.8	7.4
950-1 & 2	1976	3,400	3	5.7	12.0	8.0	20.0	6.5	1.1	1.6	5.4	83	341	3.4	5.1
Imp 1	5-85	5,700	4	6.3	30.0	9.1		6.5	0.9	2.1	2.8	53	227	3.6	11.8
Col 013	10-87	9,200	5	5.7	0.7	5.3	57.0[c]	6.7	0.7	13.5		288	408	45.6	6.0

[a] One effluent sample to date.
[b] One sample, July 1987.
[c] From preconstruction instream sample.

Brodie, G. A. and D. A. Hammer. Constructed Wetlands for Acid Drainage Control in the Tennessee Valley, U.S. Bureau of Mines, IC 9183 (1988), pp. 325–331.

adjacent to a leaking coal pile runoff pond (pH = 2.8 s.u.) that has adversely affected water quality and vegetative development. Dominant vegetation is *Typha*. An NaOH treatment system was installed at this wetland for pH increase.

Widows Creek 019 also receives acid seepage from abandoned ash disposal areas. The operational needs of Widows Creek Fossil Plant at this site have resulted in plans to flood the wetland and install a facility to pump water to an existing treatment system. Effluent data may reflect additional seepage within the wetlands system and should be viewed with caution.

Impoundment 3 is located at the TVA 950 coal mine and receives AMD. It was constructed to replace a chemical treatment system which operated from 1976 to 1986 and has produced compliance-quality effluents since construction. The dominant vegetation is *Typha*.

Rocky Top 2 is located at TVAs reclaimed Fabius Rocky Top coal mine in Jackson County. The inflow is AMD, and the dominant vegetation is *Typha*.

950 NE is adjacent to the TVA 950 coal mine and receives AMD from about 32 ha (79 acre) of reclaimed area. Dominant vegetation is *Typha*; only minor discharge has occurred since construction.

Kingston 006 is located at TVAs Kingston Fossil Plant in Roane County, Tennessee. It was constructed to treat acid seepage and runoff from active ash disposal areas. Dominant vegetation is *Typha*. About 20 cm (8 in.) of high-Ca, –#16-mesh limestone covered with about 30 cm (12 in.) of spent mushroom compost for vegetative substrate was included in the final cell of the wetland system.[26]

Colbert 013 is located at TVAs Colbert Fossil Plant in Colbert County, Alabama. It receives acid drainage from an indefinite source near an active ash disposal area. This wetland is still under development and is dominated by *Typha* and *Scirpus*.

Water quality improvement has occurred at all of the operating constructed wetlands. Five systems have produced dramatic results and apparently mitigated some of TVAs most stubborn pollution problems. Where regulatory limits were not entirely achieved, costs were reduced as fewer chemicals were needed for further metals precipitation or pH adjustment.

Numerous wetlands treatment systems are planned or under construction in the coal and utility industries. Our experience suggests the following preliminary general guidelines for Fe and Mn treatment area requirements for desired discharge levels of Fe = 3 mg/L or less and Mn = 2 mg/L or less.

Fe: 2 m^2/mg < pH 5.5 > 0.75 m^2/mg
     (21 ft^2/ppm < pH 5.5 > 8 ft^2/mg)
Mn: 7 m^2/mg < pH 5.5 > 2 m^2/mg
     (75 ft^2/ppm < pH 5.5 > 21 ft^2/ppm)

For pH levels of <5.5 s.u., the suggested treatment area for Fe would be 2 m^2/mg/min (21 ft^2/ppm) and for Mn, 7 m^2/mg/min (75 ft^2/ppm). For pH levels above 5.5 s.u., the suggested treatment area for Fe would be 0.75 m^2/mg/min (8 ft^2/ppm) and Mn,

396 ENVIRONMENTAL IMPACTS OF MINING

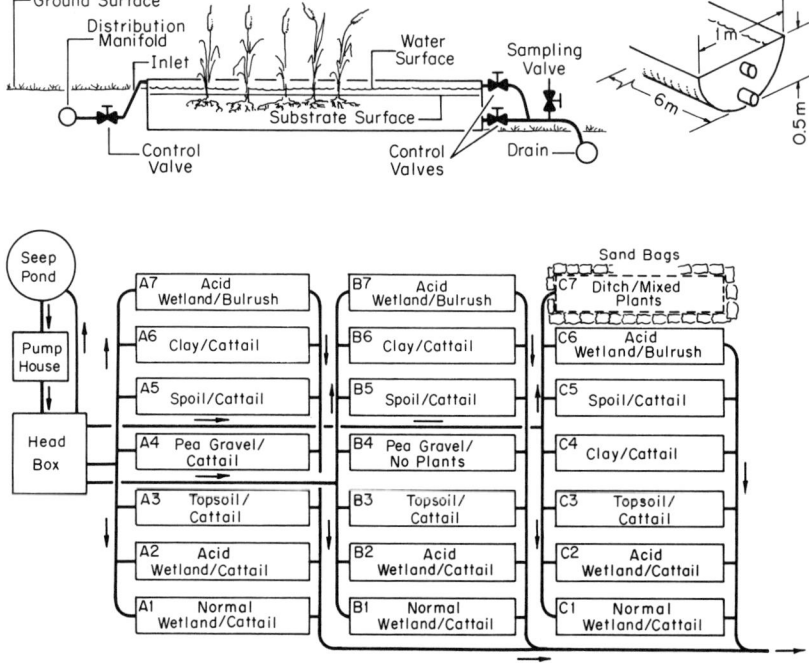

Figure 8.9. Experiment comparing treatment among substrate types. (From Brodie, A. G. et al. An Evaluation of Substrate Types in Constructed Wetland, U.S. Bureau of Mines, IC 9183 (1988), pp. 389–398.)

2 m²/mg/min (21 ft²/ppm). For example, a seep with 50 mg/L Fe, 15 mg/L Mn, pH 5.6 s.u., and an average flow of 113 L/min (30 gal/min) would require:

For Fe, the rate factor is 0.75 m²/mg/min.
Area of treatment = (0.75 m²/mg/min) (113 L/min)
(50 mg/L) = 4237.5 m²
For Mn, area = (2 m²/mg/min) (113 L/min) (15 mg/L) = 3390 m² and the wetlands treatment system area should approximate 4200 m² (45,200 ft², i.e., about 1 acre)

These treatment area estimations do not include storm flow hydrology, and the size of constructed wetlands must be increased, if necessary, to accommodate storm events and prevent dike or spillway damage.

WINDSOR COAL COMPANY WETLAND

The wetland was built on a V-shaped hollow on a hillslope. The substrate comprised a 0.5-cm limestone layer, sterile mushroom compost, and cattail plants

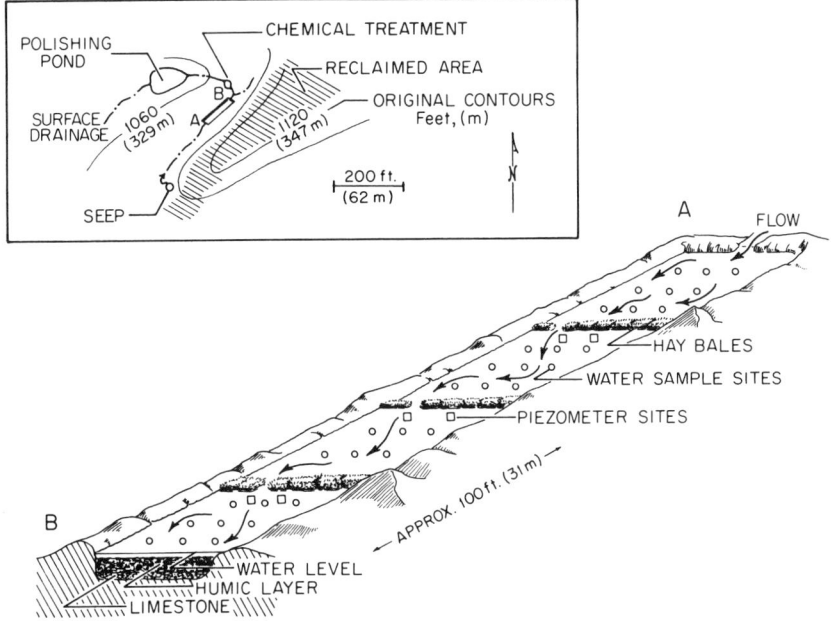

Figure 8.10. Schematic drawing of the wetland basin with sample locations. The insert locates the wetland and the site of pre-wetland chemical treatment. (From Stillings, L. L., J. J. Gryta and T. A. Ronning. Iron and Manganese Removal in a *Typha* Dominated Wetland During Ten Months Following Its Construction, U.S. Bureau of Mines, IC 9183 (1988), pp. 317–324.)

(*Typha*), as shown in Figure 8.9. The wetland was installed on the reclaimed disposal area 60 m below the hill crest on a 4:1 slope. Because of stability problems the wetland was not placed upon the bench, but installing the wetland on the slope could saturate a section of the slope face and also create a stability problem.

To protect against slope saturation, a liner installation was necessary. Because clay was not available nearby, a 36-mil Hypalon liner was used.[21]

Construction was carried out during dry weather to minimize the damage caused by equipment. Excavation was carried out by a bulldozer. Topsoil was stripped and saved to redress the area. The wetland had 2:1 side slopes, with cut materials used as fill for the downslope side. Total excavated area was 3.4 × 26 m for a total surface area of 117 m². A rock/pipe drain was installed below the liner to carry water from a 4-L/min seep at the bottom grade of the excavation. A Hypalon liner was placed and keyed into the side berm of the wetland. A small, lined sump was constructed beneath the existing seep collection pipe discharging into the existing riprap ditch, and a 4-in. PVC conveyance pipe was installed from the sump to the wetland.

Cattails were planted on 46-cm centers, and seep water was directed into the wetland. Minor weir adjustments ensured water distribution and flow across the wetland.

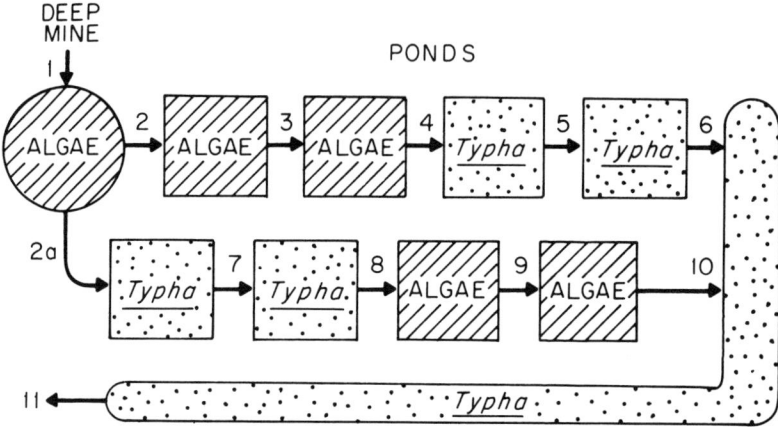

Figure 8.11. Plan view of wetland treatment system. (From Kepler, D. A., An Overview of the Role of Algae in the Treatment of Acid Mine Drainage, U.S. Bureau of Mines, IC 9183 (1988), pp. 286–290.)

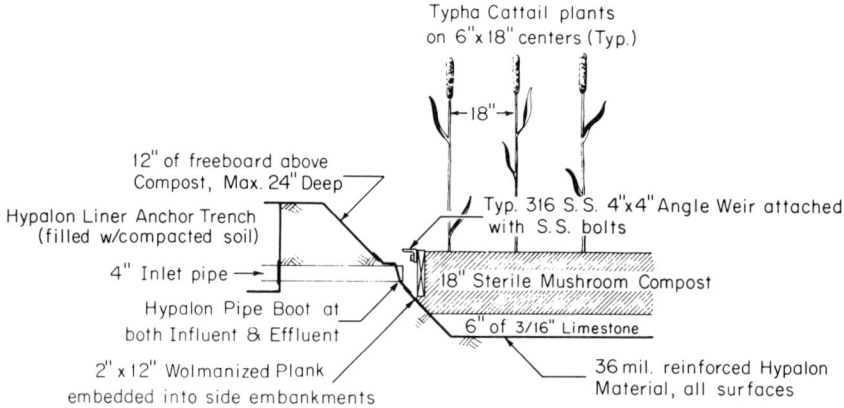

Figure 8.12. Wetland treatment area detail. (From Kolbash, R. L. and T. L. Romanos. Windsor Coal Company wetland, in *Constructed Wetlands for Wastewater Treatment* (Chelsea, MI: Lewis Publishers, Inc., 1990), pp. 788–792.)

The results are given in Figures 8.10 to 8.12. Even with less than one growing season, 50% of the incoming Fe was removed. Within-wetland water sampling indicated a gradual reduction of dissolved Fe without sharp transition zones.

Seep Mn ranged from 1.5 to 3.8 mg/L and inlet Mn increased. The pH did not change from inlet to outlet.

THE TRACY WETLAND

The Tracy Wetland consists of the Large Wetland and the Small Wetland. The Large Wetland treats a flow of 0.50 to 0.95 L/sec (11,520 to 22,000 gal/day). This flow has a pH of 2.7, and total Fe, total Al, total Mn, and sulfate concentrations of 284, 178, 1.51, and 2618 mg/L, respectively.[22]

The Large Wetland system includes a peat layer planted with cattails and sedges for metals removal, a limestone gravel channel for the neutralization of mineral acidity, and several aeration structures. A limestone-soil mix was used as a substrate throughout the system and in the construction of baffles to provide sinuosity.

The Small Wetland treats a flow of 0.38 to 0.50 L/sec (8500 to 12,000 gal/day). This flow has a pH of 3.1, and total Fe, total Al, total Mn, and sulfate concentrations of 148, 46.7, 1.2, and 1560 mg/L, respectively.

The Small Wetland includes a peat layer, a limestone gravel channel, aeration structures, and a limestone-soil substrate. It was also planted with cattails and sedges.

The area is a semiarid environment with a total precipitation of 38 cm/year (15 in.). Of the annual precipitation, 70% occurs during the months of April through September. The annual mean temperature is 7.2°C (45°F), with extremes from 37.8°C to –31.7°C (100°F to –25°F). Temperature and rainfall have a significant effect on hydrology and soil characteristics.

The Sand Coulee drainage, consisting of approximately 500 km^2 (195 mi2) in the upper Missouri River basin, is adversely impacted by AMD from 22 discharging abandoned coal mines. This is the largest concentration of abandoned discharging coal mines west of the Mississippi River. Approximately 230 km^2 (90 mi^2) of the drainage contains abandoned coal mines discharging acid mine water.

The coal in this area occurs within the upper part of the Mission Formation (Jurassic) and is exposed along outcrops in the valley of Sand Coulee Creek and its tributaries. The topography is gentle with the exception of drainages incised into the plateau, which leave a dissected and irregular terrain above the drainage areas. The Sand Coulee drainage lies in the Great Falls-Lewistown Coal Field, a large deposit of sub-bituminous coal. Unlike Eastern Montana Tertiary coal deposits, the coal in this area is higher in grade (11,118 btu/lb), and higher in sulfur (S) content (0.5 to 5.5%). The thickness of the coal seam varies from 0.3 to 3.6 m (1 to 12 ft).

This coal was the target of mining activity beginning in the 1880s. The last large-scale mine closed in 1952. During the period from 1885 to 1955 the coal mined from the Great Falls-Lewistown Coal Field exceeded 36 million tons. This amounted to about 23% of the total coal produced in Montana during that period. Coal mined from the Great Falls-Lewistown area from 1955 to 1965 was <1% of the total coal produced in Montana, and since 1965 no commercial coal production.

The area consists primarily of Cretaceous, Jurassic, and Mississippian sedimentary rocks; the coal occurs in the Jurassic rocks. The Cretaceous Kootenai Sandstone and Flood Sandstone are aquifers most consistently used for domestic wells and springs in the area. Water from these aquifers leaks into the abandoned mines in the underlying formations.

The presence of large, abandoned underground coal mines in the area has apparently produced a large change in the regional groundwater flow system. The underdraining of the basal Kootenai Sandstone aquifer by the abandoned mines has diverted the groundwater flow, which most likely discharged to the Sand Coulee drainage prior to mining.

The two mine discharges dealt with in this project are from the old Pierce Mine. Although a relatively small mine, its discharges have had a significant impact on the adjacent agricultural lands. For 40 years the discharges flowed into prime bottom land, rendering 0.08 to 0.12 km^2 (20 to 30 acres) unsuitable for agriculture. The problem persisted even after several attempts by the landowner to seal the mine opening and reroute the flow of the discharge. In the fall of 1984, the Montana Abandoned Mine Reclamation Bureau (AMRB) completed a project that installed a drainage pipeline system. This pipeline collected drainage from both mines into a main pipeline and discharged directly into Sand Coulee Creek. After the wheat field dried out, the soils were amended, and a wheat crop is growing once again. No attempt was made to treat the problem of AMD until this project in 1986.

The sites chosen for construction of the experimental wetlands offered several advantages, including:

- Two different controlled flows of <1 L/sec in proximity to each other
- Two flows having different pHs and metal concentrations
- Easy access to the sites during all seasons
- A topography at the sites that provided some freedom in the design of the wetlands

The intent of the wetland design was to provide a three-stage treatment facility: manmade peat wetlands planted with cattails and sedges, limestone-filled drainage channels, and aeration structures. The peat wetland planted with cattails and sedges provides an ion exchange facility for the removal of heavy metals. The limestone-filled channel provides a source of alkalinity to neutralize the low pH drainage. The aeration structures provide a means for the exsolution of carbon dioxide to the atmosphere, thereby reducing the concentration of carbonic acid.[21]

The primary criteria used for the design of the wetlands included:

- Sizing the wetland to allow for treatment of the quantity of water expected during all seasons and following precipitation events; a surface area of 294 m^2/L/sec (200 ft^2/gpm) was considered a minimum size
- Minimizing water velocities and maximizing retention time in the system
- Providing water depths varying from 5 to 46 cm (2 to 18 in.) in the system
- Providing optimum wetlands soils for emergent hydrophytes (such as *Typha*) composed of decomposed organic matter (peat) with some mineral soil content; the minimum depth of the peat was 0.3 m (1 ft)

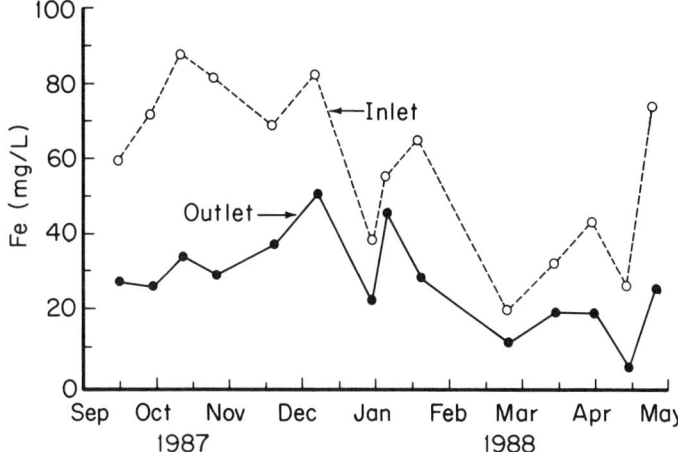

Figure 8.13. Windsor wetland, total iron. (From Kolbash, R. L. and T. L. Romanos. Windsor Coal Company wetland, in *Constructed Wetlands for Wastewater Treatment* (Chelsea, MI: Lewis Publishers, Inc., 1990), pp. 788–792.)

- Avoiding "short-circuiting" of the wetland by forming flow channels
- Providing for the placement of cattail sod mats over approximately 40% of the wetlands' surface area
- Providing a crushed limestone-filled channel downstream from the wetland to moderate pH
- Providing aeration structures along the limestone channel

The Large Wetland was designed as a rectangular impoundment with the approximate dimensions of 30 m (100 ft) × 14 m (45 ft) (Figure 8.13). This impoundment was baffled by a berm extending 28 m (93 ft) along the length of the wetland from each side. This baffling provided sinuosity for the flow through the wetlands and prevented short-circuiting of the flow across the wetlands.

The water level is controlled by a rectangular notch weir at the outlet. The flow from this outlet structure passed over 37 linear meters (124 linear feet) of limestone channel and three aeration structures before exiting the system. Each of the aeration structures had about 0.45 m (1.5 ft) of fall.

The Small Wetland was designed as two parallel linear impoundments connected at one end by an aeration structure, essentially forming a U-shaped wetland (Figure 8.14). Each of the impoundments is about 18 m (60 ft) long and 3 m (10 ft) wide. The flow from the outlet structure passed over approximately 12 linear m (40 linear feet) of limestone channel and two aeration structures before exiting the system.

The soil underlying both wetlands was amended with bentonite to prevent infiltration of the impoundments' water into the ground. The base of both wetlands and the berms used for baffling in the Large Wetland were made of a limestone-soil mix with a minimum depth of 0.45 m (1.5 ft). A minimum of 0.3 m (1 ft) of peat was

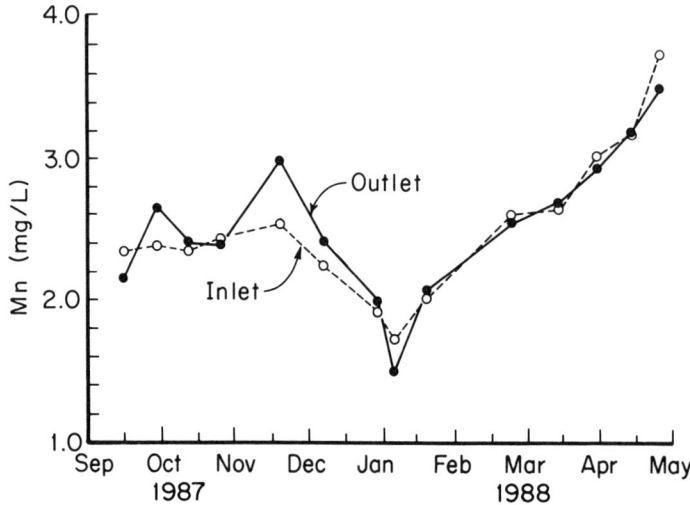

Figure 8.14. Windsor wetland, total manganese. (From Kolbash, R. L. and T. L. Romanos. Windsor Coal Company wetland, in *Constructed Wetlands for Wastewater Treatment* (Chelsea, MI: Lewis Publishers, Inc., 1990), pp. 788–792.)

placed over the base material. The peat was from an active peat extraction operation located approximately 400 km (250 mi) southwest of the project site. The species of moss, *Polytrichum piliferum* and *P. strictum*, are common to Montana wetland communities and are more often found in drier locations. Cattail sod mats measuring approximately 3.3 m^2 (36 ft^2) were placed on top of the peat, spaced evenly over approximately 40% of the wetland area. Sedge mats measuring approximately 0.09 m^2 (0.3 ft^2) were placed randomly throughout the wetlands.

The cattail and sedge vegetation sod mats were excavated from a wetland source area about 250 m (835 ft) from the constructed wetlands. The mats were excavated using a rubber-tired front end loader. The average thickness of these mats was about 15 cm (0.5 ft). Whole 2.7 × 1.2 m (9 × 4 ft) mats were placed in the wetlands.

In order to minimize the stress on the transplanted vegetation, uncontaminated water was used to partially fill the wetlands prior to inundation with mine drainage. The transplanted vegetation was allowed to adjust to the new environment for 1 week prior to coming in contact with mine drainage. It is believed that this procedure helps account for the significant "new shoot" growth shown after transplanting the *Typha*. No plant stress has been observed, and plant growth has been vigorous.

Influent and effluent water quality; metals concentrations in *Typha* roots, rhizomes, and leaves; and metals concentrations in the peat for both systems have been monitored since the end of July 1986. Initially, water sampling was performed on a biweekly basis, but after the first 2 months a monthly sampling program was adopted. Vegetation and peat sampling was performed on a monthly basis.

Two water samples were collected at the inlet and outlet of each wetland. One of the samples taken from each location was filtered and preserved with acid (HNO_3). The pair of samples from each location was analyzed for total and dissolved Al, Fe, and Mn. The sulfate concentrations, pH, and specific conductance for each pair of samples were also determined. The samples were analyzed utilizing EPA drinking water methods and Inductively Coupled Argon Plasma techniques.

The *Typha* rhizome, root, and leaf samples were collected from randomly selected plants in the wetlands. After collecting the samples, the rhizomes and roots were dried, and any residual peat was carefully removed. Each of the vegetative samples was analyzed for total Al, Fe, Mn, and S. The samples were processed using nitric acid and hydrogen peroxide digestion (EPA Method 3050). The samples were analyzed by ICP techniques.

The peat samples were collected at locations near the inlet and outlet of each wetland from the upper 15 cm (0.5 ft) of the peat layer. They were analyzed for a number of constituents, including total S, pH, and total and extracted Fe, Al, and Mn. The samples were analyzed with Montana Department of State Lands strip mining soil analysis methods.

Neither the Large Wetland nor the Small Wetland has improved the quality of water flow through the system to any degree. Table 8.28 shows a summary of the water quality data for the inlet and outlet of each wetland.

The concentrations of Fe, Al, and Mn in the rhizomes, roots, and leaves of the cattail plants in the two wetlands and the cattail source area used as a control area are summarized in Table 8.29. These data indicate that the *Typha* is concentrating metals in the system. However, because the potential uptake of metals by *Typha* is not significant when compared with the actual loading of the system, the vegetation would contribute only slightly to the total metal retention of a functioning system.

The peat has also concentrated metals in the system. Total Al, Fe, and Mn concentrations for both wetlands and the control peat are shown in Table 8.30.

Because the removal of the metals in both systems was unsuccessful, the limestone gravel in the channels downstream of each wetland became armored in <2 weeks. Although no specific tests were made, visual indications suggest that metals precipitation and pH buffering decreased after the armoring process was complete.

It was also observed that detention times in the systems were much shorter than desired. Dye tests showed that the detention times were approximately 7 hr and 3 hr for the Large and Small Wetlands, respectively. These low retention times are due primarily to the relative impermeability of the peat once it is saturated.

There seem to be two primary reasons for the ineffectiveness of these systems. First, both systems were undersized, which resulted in very low retention times and limited the contact between the AMD and the primary heavy metal removal medium, the peat moss. Second, the systems' designs did not force the AMD to flow through the peat at any location. The designs maximized the length of the flow paths through each system, but most of the flow through the systems was above the AMD-peat interface.

Table 8.28. Water Quality Summary[a]

	Large Wetland				Small Wetland			
	Inlet		Outlet		Inlet		Outlet	
	Mean	S.D.	Mean	S.D.	Mean	S.D.	Mean	S.D.
Total Al	178	17.5	180	17.9	46.7	2.1	45.7	2.1
Total Fe	284	134.9	271	110.6	148.5	37.2	94.1	52.7
Total Mn	1.51	0.14	1.67	0.23	1.2	0.1	1.3	0.3
Sulfate	2618	285	2683	201	1560	223	1551	218
pH	2.7	0.27	2.58	0.19	3.1	0.3	2.8	0.4
Specific conductivity	3349	307	3440	194	2414	246	2559	278

[a] All units mg/L except for pH (standard units) and specific conductance (µmho/cm). All samples taken between July 10, 1986 and September 30, 1987. Number of samples taken is 21.

From Hiel, M. T. and F. J. Kernis, Jr. The Tracy Wetlands: A Case Study of Two Passive Mine Drainage Treatment Systems in Montana, U.S. Bureau of Mines, IC 9183 (1988), pp. 352–358.

Table 8.29. *Typha latifolia* Analysis Summary[a]

	Total Al		Total Fe		Total Mn	
	Mean	S.D.	Mean	S.D.	Mean	S.D.
Rhizomes						
Control	558	506	590	333	42	31
Large wetland	1,398	1,268	4,237	1,735	164	74
Small wetland	1,400	1,478	4,806	2,325	134	72
Roots						
Control	5,171	2,368	8,093	7,584	475	813
Large wetland	7,444	5,006	40,884	25,437	270	151
Small wetland	4,703	2,659	34,609	29,491	241	292
Leaves						
Control	139	100	198	95	275	263
Large wetland	232	91	287	62	1,760	1,455
Small wetland	243	101	286	134	1,053	766

[a] All units are in micrograms per gram. All samples taken between December 23, 1986 and September 21, 1987. Number of samples taken is 9.

From Hiel, M. T. and F. J. Kernis, Jr. The Tracy Wetlands: A Case Study of Two Passive Mine Drainage Treatment Systems in Montana, U.S. Bureau of Mines, IC 9183 (1988), pp. 352–358.

These systems might improve if the retention time was increased and provisions were made to force the AMD to flow through the peat. Both systems were expanded during the summer of 1987. These expansions significantly increased the size of each system and provided for flow of the AMD through the peat. A report on the performance of these expansions will be made at a future date.

Table 8.30. Soil Analysis Summary[a]

	Control Peat		Large Wetland Peat		Small Wetland Peat	
	Mean	S.D.	Mean	S.D.	Mean	S.D.
Total Al	9,595	1,665	12,872	1,726.1	15,413	4,606
Total Fe	6,918	2,358	29,009	18,364	60,839	68,291
Total Mn	113	50	160.6	242.7	98.4	74.7

[a] All units are in micrograms per gram. Number of samples for control peat is 2. Number of samples Large and Small Wetland peat is 10.

From Hiel, M. T. and F. J. Kernis, Jr. The Tracy Wetlands: A Case Study of Two Passive Mine Drainage Treatment Systems in Montana, U.S. Bureau of Mines, IC 9183 (1988), pp. 352–358.

WETLAND TREATMENT IN METAL MINING

Wetlands have been effectively used for treating acid drainage problems associated with coal mining. The potential applications of wetlands in metal mining raise these questions:

- Are the effluent problems limited to deposits with abundant pyrite?
- If pyrite is present, is the problem different than AMD problems in coal mining?

For example, in Colorado, large molybdenum mines mining molybdemite ores do not have any AMD problems. In the Central City Mining District in Colorado, however, pyrite is present and Cd, Zn, and Pb concentrations are low in waters, even though the minerals of these metals are highly abundant. Fe(III) and H^+ in the groundwater catalyze the dissolution of base metal sulfides to such an extent that they become important constituents in metal mine drainage, even though the base metals may not be abundant in the deposit.

With pyrite present, almost every type of heavy metal contaminant may be present in acid drainage from metal mine operations. Contaminants such as Pb, Cd, and As are harmful at concentrations far lower than Mn; and Fe, Zn, and Cu are harmful to aquatic life at concentrations far less than drinking water standards. In a wetland, trace elements such as Cu, B, and Hg may kill plants long before the pH, Fe, or Mn would be harmful.

Acid drainage from metal mines presents a more severe problem than most coal mine drainages because priority pollutants such as As, Cd, Pb, Hg, Cu, and Zn may be present in hazardous concentrations. Designing a wetland to treat metal mine effluent should involve consideration of whether these pollutants will be toxic to flora and fauna and of whether the drainage from the wetland will meet aquatic standards. The removal process for priority pollutants should ensure that further release is not possible.

The goal of using wetlands for mine drainage treatment is low-cost, long-term immobilization of pollutants. The removal mechanisms include:

- Filtering suspended and colloidal material from water
- Uptake of contaminants into roots and leaves of live plants
- Adsorption or exchange of contaminants onto soil materials, live plant materials, dead plant materials, or algal materials
- Precipitation and neutralization through the generation of NH_3 and HCO_3^- by bacterial delay of biologic material
- Precipitation of metals in the oxidizing and reducing zones catalyzed by bacterial activity

If wetlands were buried, they would eventually become a bog deposit, coal, or black shale. Metal occurrences in these sediment types have undergone early diagenesis exhibiting long-term stability. Mineral forms of Mn, Fe, and other base metals in these sediments represent the most thermodynamically stable phases of these elements. In sediments formed by chemical precipitation, the stable Fe minerals are hematite (Fe_2O_3), pyrite, or siderite ($FeCO_3$). Stable Mn minerals are pyrolusite (MnO_2) and rhodochrosite. Trace elements such as Co, Ni, Cu, Zn, Ag, Cd, Au, Hg, and U occur as sulfides, oxides, and carbonates. Excepting Ni and V, metals are not retained by the organic fractions in organic-rich reducing sediments.[23]

Organic material in a wetland system may have only a minor role in long-term storage. Because sulfides, oxides, and carbonates are the most stable forms of trace element precipitates, immobile organic forms of these elements are intermediate products that will eventually, through diagenesis, convert to inorganic forms. Wetland system design should concentrate on forming inorganic precipitates and use organic components to promote their formation.

Microorganisms survive in nature by catalyzing chemical reactions that release energy to the organism. In aerobic zones, these bacteria promote the oxidation of Fe and Mn to more insoluble states. In anaerobic zones, sulfate-reducing bacteria promote the formation of H_2S. In a wetland, bacterial mediation of organic decay will generate NH_3 and HCO_3^-, which will raise the pH and cause hydroxide precipitation. Wetland system design should include plant species that survive and produce large amounts of biomass to support microorganism growth. Metal uptake by plants is less important. An organic substrate should:

- Have an anaerobic zone so that H_2S is produced
- Promote plant growth
- Promote growth of bacteria that increase the pH
- Conduct drainage waters to sites of bacterial activity

Precipitation of Fe and Mn oxyhydroxides in the aerobic zone not only removes primary pollutants as in coal mine drainages, but, through adsorption, these precipitates may also remove significant amounts of trace elements. Mn oxyhydroxide is far more important than the oxyhydroxide in this process, therefore, raising the pH is important.

In the aerobic zone, typical microbial-mediated reactions include:

$$4Fe^{2+} + O_2 + 10H_2O \rightarrow 4Fe(OH)_3 + 8H^+ \qquad (8.6)$$

$$2O_2 + H_2S \rightarrow 5SO_4^- + 2H^+ \qquad (8.7)$$

$$2H_2O + 2N_2 + 5O_2 \rightarrow 4NO_2^- + 4H^+ \qquad (8.8)$$

In the anaerobic zone, typical microbially mediated reactions include:

$$4Fe^{2+}(OH)_3 + CH_2O + 8H^+ \rightarrow 4Fe^{2+} + CO_2 + 11H_2O \qquad (8.9)$$

$$3CH_2O + 2N_2 + 3H_2O \rightarrow 4NH_3 + 3CO_2 \qquad (8.10)$$

$$SO_4^- + 2CH_2O \rightarrow H_2S + 2HCO_3^- \qquad (8.11)$$

CH_2O symbolizes organic material in the substrate.

It is apparent that the anaerobic reactions are approximately the reverse of the aerobic reactions. Both zones exist in a wetland. If removal involves aerobic processes, then the wetland should be constructed so the water remains on the surface. If removal involves anaerobic processes, then the wetland should be constructed so that the water courses through the substrate. In a natural wetland, the water primarily remains on the surface. The aerobic reactions generate hydrogen ions and the anaerobic reactions consume hydrogen ions.

The wetland must be constructed to maximize removal reactions and minimize competing reactions. The removal process should consume hydrogen ions, thus anaerobic processes are emphasized.

Big Five Tunnel Experimental Wetland, Colorado

A reinforced concrete structure 3.05 × 18.3 m was initially divided into three 6.1-m sections (Figure 8.15). Each compartment was fitted with two drains of 15-cm PVC pipe, one active and one reserve. Active drains consisted of standpipes initially set at a 1-m depth which deliver the overflow water to the existing pond. A 0.76-mm

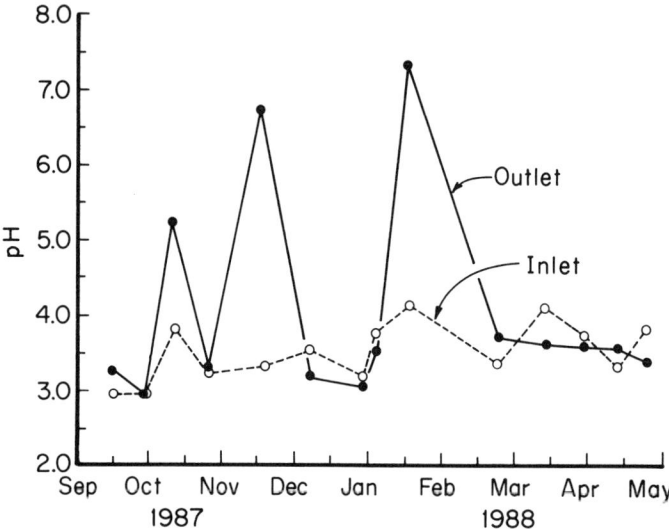

Figure 8.15. Windsor wetland, pH. (From Kolbash, R. L. and T. L. Romanos. Windsor Coal Company wetland, in *Constructed Wetlands for Wastewater Treatment* (Chelsea, MI: Lewis Publishers, Inc., 1990), pp. 788–792.)

Hypalon liner was used to separate one compartment from another and to prevent chemical reactions between construction materials, organic substrates, and mine drainage. Six sampling wells of 15-cm PVC in each compartment were used to allow water to enter from the lowest 30 cm, the middle 30 cm, and the upper 30 cm of the organic substrates. A small concrete dam inside the tunnel provided head to distribute water to each compartment through insulated PVC lines fitted with ball valves for flow control. A perforated drain pipe, 10-cm PVC, across the width of each compartment distributed water to each compartment, and excess water was drained into the pump.

Compartment sections were filled with 1 m of organic substrates. The first compartment contained mushroom compost; the second had equal parts of peat, aged steer manure, and decomposed wood shavings and sawdust; the third had 10 to 15 cm of 5- to 8-cm limestone rock below the same mixture as the second compartment. Substrates were initially saturated with municipal water at 3.78 L/min.

Cattails (*Typha latifolia*) and sedges (*Carex utriculata*) were transplanted from a lake at a similar elevation in Grand County, Colorado, into 25 to 30% of each treatment bed. Sedges (*C. aquatilis*) and rushes (*Juncus arcticus*) transplanted from a fen 6.5 km from the demonstration site covered 25% of the treatment beds. Cattails from a foothills wetland and sedges from an adjacent wetland covered an additional 35% of the treatment beds.

Initially, cell A with mushroom compost was the most effective in raising pH and removing contaminants. Selected values for the mine drainage input and the cell

WETLANDS 409

Table 8.31. Concentrations (mg/L) of Metals, Percent Reduction of Metals, pH, and Flow Rates (gpm) in the Big Five Mine Drainage and Wetland Cell Output Waters[a]

Water Sample	Mn	% Red.	Fe	% Red.	Zn	% Red.	Cu	% Red.	pH	Flow Rate
Initial operation										
December 11, 1987										
Mine drainage	34		32		10.6		1.02		2.8	1.0
Cell A	27	21	18	45	7.8	27	0.44	57	4.6	1.0
Cell B	33	1	24	26	9.8	12	0.89	12	3.1	1.0
Cell C	34	0	22	32	9.6	9	0.91	10	3.3	1.0
August 19, 1988										
Mine drainage	26		37		8.1		0.91		2.9	
Cell A	25	4	20	56	<0.1	100	0.17	81	5.5	0.51
Cell B	26	0	15	59	6.1	24	0.55	40	3.2	0.24
Cell C	25	4	11	70	5.8	28	0.38	58	3.5	0.34
Cell A Modification										
February 21, 1989										
Mine drainage	27		32		9.3		0.56		3.0	
Cell A	22	19	12	63	4.5	52	<0.05	100	5.1	0.28
Cell B	27	0	28	13	6.1	34	0.82	0	3.4	0.31
Cell C	25	7	31	3	7.2	23	0.26	53	3.5	0.32
May 6, 1989										
Mine drainage	30		42		10.4		0.76		3.0	
Cell A	33	−10	28	33	7.8	25	<0.05	100	3.5	0.92
Cell B	27	10	10	76	6.2	40	0.36	53	3.0	0.83
Cell C	29	3	9	79	6.4	38	0.46	39	3.2	0.81

Table 8.31 (continued). Concentrations (mg/L) of Metals, Percent Reduction of Metals, pH, and Flow Rates (gpm) in the Big Five Mine Drainage and Wetland Cell Output Waters[a]

Water Sample	Mn	% Red.	Fe	% Red.	Zn	% Red.	Cu	% Red.	pH	Flow Rate
August 1, 1989										
Mine drainage	32		43		9.4		0.75		2.9	0.30
Cell A	31	3	39	9	5.2	45	<0.05	100	4.1	0.48
Cell B	26	19	24	44	6.6	30	0.46	39	3.1	0.43
Cell C	32	0	18	58	4.8	48	0.09	88	3.7	
Cell B Modification										
October 3, 1989										
Mine drainage	35		46		9.9		0.66		3.2	
Cell B-Up	34	3	39	15	9.3	6	0.59	11	3.5	0.29
Cell B-Down	23	34	8	83	0.8	92	<0.05	100	6.5	0.16
Cell E	24	31	<1	100	<0.1	100	<0.05	100	6.3	0.062
November 5, 1989										
Mine drainage	32		38		8.7		0.61		2.9	
Cell B-Up	31	3	17	55	8.4	3	0.48	21	3.6	0.22
Cell B-Down	20	38	<1	100	6.0	31	<0.05	100	5.9	0.24
Cell E	20	38	<1	100	<.1	100	<0.05	100	6.5	0.11
January 13, 1990										
Mine drainage	31		33		9.0		0.59		2.9	
Cell B-Up	30	3	33	0	8.9	1	0.49	17	3.2	0.21
Cell B-Down	27	13	12	64	8.1	10	0.44	25	3.2	0.20
Cell E	28	10	10	70	<.1	100	<0.05	100	6.0	0.10

[a] The area of cells A, B, and C is 200 ft^2; the area of cells B-Up, B-Down, and E is 100 ft^2.

Wildeman, T. R. and S. L. Lauden. Use of wetlands for treatment of environmental problems in mining: non-coal-mining applications, in *Constructed Wetlands for Wastewater Treatment* (Chelsea, MI: Lewis Publishers, Inc., 1990), pp. 221–231.

Table 8.32. Bacteria per Gram in Initial Substrate Components; Substrate Samples in the Top 15 cm and 0.9 m from the Surface in Wetland Cells After 2 Months of Mine Drainage Flow at the Big Five Tunnel, Idaho Springs, CO

Initial Component	Iron Oxidizers		Sulfate Reducers	
	MPN/g $\times 10^3$	Factor of Confidence	MPN/g $\times 10^4$	Factor of Confidence
Aged manure	0		9	3.3
Wood products	0		0.3	3.3
Mushroom compost	0		50	3.3
Peat	0.2	3.3	0.3	3.3
Peat/manure/wood	0		2	3.3
Two-month samples from 15 cm				
Cell A Well 3	20	3.8	1000	3.3
Cell A Well 5	900	3.3	>3000	3.3
Cell B Well 3	5	3.3	>1000	3.3
Cell B Well 5	8	3.3	>1000	3.3
Cell C Well 3	40	3.3	>1000	3.3
Cell C Well 4	5	3.3	>1000	3.3
Two-month samples from 0.9 m				
Cell A Well 3	30	3.3	1000	3.3
Cell A Well 4	600	3.3	1000	3.3
Cell B Well 3	10	3.3	500	3.3
Cell B Well 5	2	3.3	200	3.3
Cell C Well 3	4	3.3	>800	3.3
Cell C Well 4	18	3.3	900	3.3

Wildeman, T. R. and S. L. Lauden. Use of wetlands for treatment of environmental problems in mining: non-coal-mining applications, in *Constructed Wetlands for Wastewater Treatment* (Chelsea, MI: Lewis Publishers, Inc., 1990), pp. 221–231.

outputs are shown in Table 8.31. Mn, Fe, Cu, and Zn are the primary heavy metal contaminants. The removal strongly depended on the loading rate (expressed in square feet per gallon per minute). The size of the cell was fixed at 18.6 m² (200 ft²). The flow is inversely proportional to the loading rate. For cell A, the best results were 100% removal of Cu and Zn, 63% removal of Fe, and an increase in pH from 3.0 to 6.2 at a loading rate of 600 ft²/gpm. In the first stage, Mn was not removed. The experimental results gave convincing evidence that the important removal process was bacterial reduction of sulfate in the mine drainage to hydrogen sulfide and subsequent precipitation of metals as sulfides.

Microbial tests were conducted for both autotrophic Fe oxidizers and sulfate reducers. One gram of soil components and actual soil samples were taken to determine initial bacterial content. Two months after mine drainage flowed in the system, samples were collected from the top 15 cm and 0.9 m below the surface at two sites in each cell (Table 8.32).

Table 8.33. Colony Count for Heterotrophic Bacteria (10^4 Colonies/g)

Cell	Sample	Total Heterotrophic Count		Heterotrophic Count for	
		Plates with $FeSO_4$	Plates with $MnSO_4$	Manganese Oxidizers	Iron Oxidizers
A	(a)	>2000	10	0	200
A	Duplicate	>2000	30	50	7
A	(b)[a]	>300	50	3	100
A	(c)[a]	>800	>800	>800	>800
A	(d)[a]	>700	30	20	200
A	(e)[a]	>700	60	10	300
C	(f)[a]	>400	>400	20	200

[a] Average of two samples.

Wildeman, T. R. and S. L. Lauden. Use of wetlands for treatment of environmental problems in mining: non-coal-mining applications, in *Constructed Wetlands for Wastewater Treatment* (Chelsea, MI: Lewis Publishers, Inc., 1990), pp. 221–231.

The results indicated that different soil components and soils were good sources of sulfate-reducing bacteria. Autotrophic Fe oxidizers were initially present only in peat, and the population was only 200 bacteria per gram. After 2 months of mine drainage flow, Fe-oxidizing autotrophs were present on the surface and at depths in all substrates. In mine drainage, the bacterial population was 50/mL and the number in cell A output was not significantly different.

The Fe- and Mn-oxidizing heterotrophs existed on the surface environment in cells A and C (Table 8.33). *Clothrix* did not promote the growth of Fe-oxidizing heterotrophs, but did severely depress growth of the Mn oxidizers.

It was found that removal strongly depended on the loading rate expressed in square feet per gallon per minute. Because the size of the cell is fixed at 18.6 m^2 (200 ft^2), the flow is inversely proportional to the loading rate. The best results for cell A during this period were 100% removal of Cu and Zn, 63% removal of Fe, and an increase in pH from 3.0 to 6.2 at a loading rate of 600 ft^2/gpm. In the first stage of operations, Mn was not removed. The removal patterns and results from other experiments performed on the substrates gave convincing evidence that the important removal process was bacterial reduction of sulfate in the mine drainage to hydrogen sulfide and subsequent precipitation of the metals as sulfides.

A possible toxic metal removal mechanism may be through accumulation in leaves, stems, and roots of wetland plants. Should this occur, it could create problems by concentrating toxic metals in the biologically very active wetland surface. However, at the Big Five site, careful collection and analysis of wetland plant samples has revealed only minor accumulation of toxic metals after 2 years of operation. This result is consistent with studies of wetland removal systems used for coal mine drainage that show that uptake by plants accounts for at most 5% of the metal removal.

Gradually over considerable time, toxic metals will accumulate in the anaerobic and aerobic zones of a wetland. At some point, these metal sulfides and hydroxides mixed with organic substrate will have to be removed from the wetland and treated for disposal or be recycled for metal recovery. However, the quantity of sludge to be handled thusly represents a small fraction of that which would be produced by more conventional processes and this sludge may even have economic potential for metal recovery.

Using constructed wetlands for wastewater treatment is still a developing treatment. However, the results from the Big Five Pilot Wetland show promising removal of heavy metals and an increase of pH for AMD. Conclusions from the project include:

- Toxic metals such as Cu and Zn can be removed and the pH of mine drainage can be increased on a long-term basis
- The major removal process is sulfate reduction and subsequent precipitation of the metals as sulfides; exchange of metals onto organic matter can be important during the initial period of operation
- A trickling filter type of configuration achieves the best contact of the water with the substrate
- Removal efficiency depends strongly on loading factors; in the Big Five wetland, factors above 1000 ft^2/gpm are needed for reasonable removal
- Permeability of the substrate is a critical design variable for successful operation; using laboratory and bench-scale tests, a good indication of the soil permeability in a constructed wetland can be determined
- Solutions to problems such as plugging or plumbing by ferric hydroxides and freezing of discharge lines during winter have to be designed and constructed into the passive nature of wetlands to achieve long-term operation

Nickel and Copper Removal by a Natural Wetland

The initial phase of the study was conducted at the LTV Steel Mining Company's Dunka Mine in northeastern Minnesota. The Dunka Mine is a large open pit taconite operation, covering approximately 160 ha. The pit is 4 km long, 0.4 km wide, and has a maximum depth of 110 m. At this location, the Duluth complex, an igneous intrusion overlies the taconite ore and must be removed and stockpiled. The material has been separated based on copper content and has been stockpiled along the east side of the open pit. Drainage from all stockpiles and mine dewatering discharges (011, 012) flow to Unnamed Creek (Figure 8.16).

Drainage from a stockpile containing Duluth complex material with an average grade of 0.30% copper and 0.09% nickel flowed through a white cedar (*Thuja occidentalis*) peatland into Unnamed Creek (Figure 8.16). The stockpile covers 0.12 km^2; flow rates varied from 0 during the winter to a maximum of 16 L/sec. Mean trace metal concentrations were 17.9 mg/L nickel and 0.62 mg/L copper; the mean pH of the drainage was 7.2 (Table 8.34).

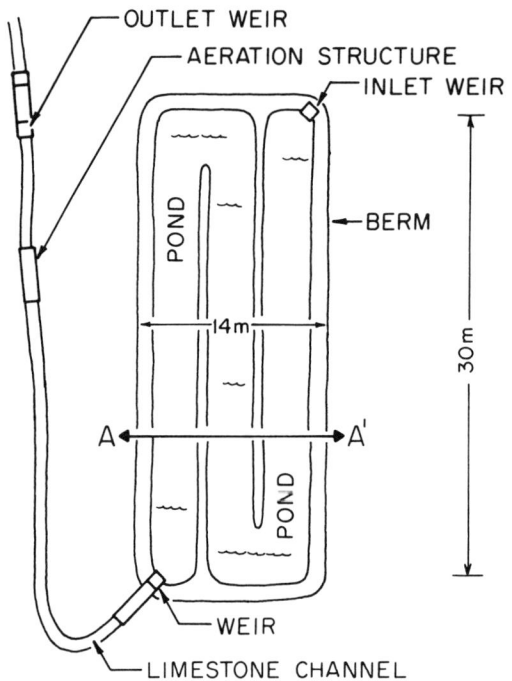
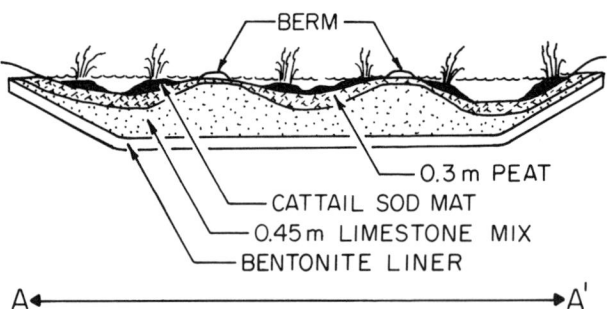

Figure 8.16. Large wetland. (From Hiel, M. T. and F. J. Kernis, Jr. The Tracy Wetlands: A Case Study of Two Passive Mine Drainage Treatment Systems in Montana, U.S. Bureau of Mines, IC 9183 (1988), pp. 352–358.)

The peatland covers 0.04 km²; the depth of the peat ranges from 1.5 to 1.8 m, and the area is covered by 3 to 5 cm of standing water. The peat is generally well decomposed, with decomposition increasing with depth, and is underlain by a layer of silty blue clay. Another white cedar peatland in the same watershed, but remote from trace metal sources, was selected as a control.

Table 8.34. Stockpile Drainage Water Quality

	Mean	Range	n
pH	7.2	6.68-7.79	24
Alkalinity (mg/L as CaCO$_3$)	113	47-206	24
Specific conductance (μsec/cm)	2540	890-3550	25
Calcium (mg/L)	285	93-388	17
Magnesium (mg/L)	225	52-288	12
Sulfate (mg/L)	1300	370-2600	17
Copper (mg/L)	0.62	0.04-1.7	24
Nickel (mg/L)	17.9	0.4-39.8	24
Cobalt (mg/L)	1.16	0.4-2.4	8
Zinc (mg/L)	0.38	0.07-0.65	12
Iron (mg/L)	1.44	0.4-5.4	24

From Eger, P. and K. Lapakko. Nickel and Copper Removed from Mine Drainage by a Natural Wetland, U.S. Bureau of Mines, IC 9183 (1988), pp. 301–309.

Sampling stations were established at six sites along the stream at each of the major stockpile flows (Em-8, Seep 1, Seep 3), at the two dewatering discharges (011, 012), and at 26 sites in the peatland. Continuous flow data were collected at three sites along the stream (Em-1, Em-3, Em-5), and at the largest volume of stockpile seepage (Em-8); pumping records were available for the mine dewatering discharges. Flow measurements at all other stream and stockpile sites were collected every 2 weeks.

Water quality samples were collected at each site twice monthly during open water and approximately monthly during the winter. There was no drainage from any of the stockpiles during the winter.

In addition to the twice-monthly samples, special sampling programs were conducted to better quantify the source of metal input and the amount of metal transport in various parts of the watershed. Automatic samplers were used to collect composite samples along the stream, low flow samples were collected under the ice, and dye studies with Rhodamine WT were conducted to determine the travel times between the different sampling stations on the streams. Water quality samples were collected sequentially (from upstream to downstream) at the specific time intervals measured in the dye study so that the same parcel of water was sampled as it moved downstream.

In the peatland (Figure 8.17), at eight of the sites (3-1 to 3-8), shallow wells and piezometers were installed. The shallow wells consisted of a perforated 10-cm diameter, 46-cm long PVC pipe, taped at the top to minimize surface water infiltration. The piezometers (1.5-m depth) were constructed of PVC, with a flange and a bentonite seal to prevent leakage. Specific conductance was sampled in the surface water at all 26 sites, and in the piezometers and wells. Specific conductivity surveys of the surface water were conducted monthly during the summer to establish the path of stockpile seepage through the peatland. All 26 sites and additional sites perpendicular to the well line were sampled. At sites 3-1 to 3-8, the water at the surface and in the wells and piezometers was filtered and analyzed for trace metals by AAS.

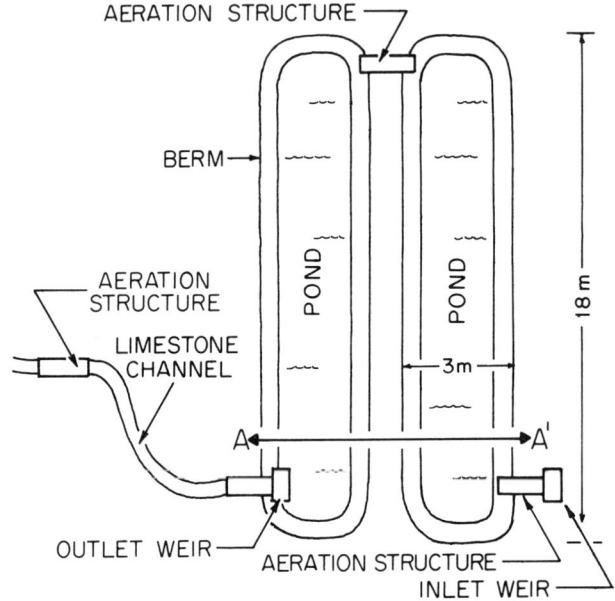

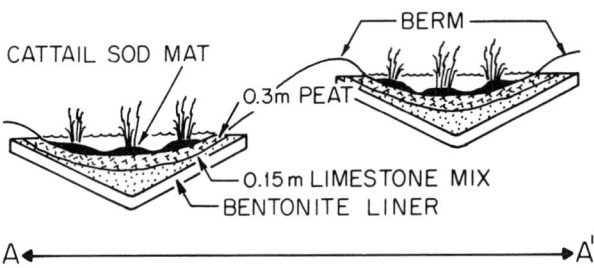

Figure 8.17. Small wetland. (From Hiel, M. T. and F. J. Kernis, Jr. The Tracy Wetlands: A Case Study of Two Passive Mine Drainage Treatment Systems in Montana, U.S. Bureau of Mines, IC 9183 (1988), pp. 352–358.)

At the odd-numbered T sites and at the control peatland within the same watershed, relevés were used to describe the plant community, and visual estimates of plant damage were made. Samples of leaf tissue were collected at each T site for each of three species: white cedar (*Thuja occidentalis*), alder (*Alnus rugosa*), and *Carex* spp. These samples were wet-washed using an $HNO_3/HClO_3$ digestion and analyzed for trace metal content using inductively coupled plasmospectroscopy. Digestions and analyses were performed by Barringer Laboratories in Toronto, Canada.

At each site a composite of three peat samples was collected from the top 20 cm and analyzed for trace metals by total acid extraction (HF, HCl, HNO_3). At four stations (3-1, 3-2, 3-5, 3-8) and the control site, trace metal concentrations were determined at approximately 25-cm depth intervals for the entire depth of the peat.

Stockpile drainage typically began in late March and ended around the middle of November, while mine dewatering and groundwater inputs maintained flow in the stream for the entire year. The mean nickel concentration at the mouth of the stream (site Em-1) was 0.1 mg/L and ranged from 0.03 to 0.22 mg/L. Concentrations throughout the streams exceeded the natural background levels (0.001 to 0.005 mg/L) for streams in the area, even during the winter when stockpile input ceased. Copper concentrations ranged from 0.001 to 0.006 mg/L and were similar to the values measured in unimpacted streams (0.001 to 0.005 mg/L).

Water quality in the peatland varied with distance from the drainage input, depth in the peatland, and time. The specific conductance values in the surface water were used to define the path of stockpile drainage through the peatland. The boundary on the west side was particularly dramatic; specific conductance decreased from over 2000 to 400 within several meters. Specific conductance decreased as distance from the seep and depth in the peatland increased. Conductivity values decreased from 3000 to 3500 µsec/cm at the seep to 600 to 2000 µsec/cm near the stream, with the majority of the reduction occurring over the final 200 m. Specific conductance values were highest in the surface water, an order of magnitude greater than those measured in the piezometers (Table 8.35).

The reduction of copper and nickel concentrations with increasing distance from the seep in surface samples was greater than that of specific conductance. In both June and August the copper concentration was reduced by about two orders of magnitude. The greatest reduction occurred within the initial 100 m of flow, with concentration remaining relatively constant: essentially at or slightly above background levels (≤ 0.005 mg/L) over the final 200 m. In June, nickel was reduced by about two thirds within the first 150 m, but in August little reduction occurred in this portion of the peatland. In August, nickel did decrease by about two thirds over the final 200 m of the peatland, but the concentration near the stream (8.4 mg/L) was still about three orders of magnitude greater than background values. Surface concentrations of both metals were two to three orders of magnitude greater than the values from the wells and the piezometers.

The predominant vegetation in both the impacted and control peatlands was white cedar, alder, and *Carex* (Table 8.36). Damage levels were determined by the presence of chlorosis in the leaves which gave only a qualitative indication of plant stress. High levels of damage to vegetation were observed at two of the ten T sites (T_1, T_{13}), but no definitive pattern relating damage to distance from the seep was observed.

Leaf tissue nickel concentrations were about one to two orders of magnitude higher in the impacted area than the values measured in the control peatland, where nickel was not detectable (≤ 1 mg/kg). The highest nickel concentrations were generally found within the zone of high conductance water, and the lowest values were measured outside this zone in the northeast section of the peatland. Elevated nickel values were, however, found at sites $T1_6$ to T_{19}, which were outside the zone of high specific conductivity. Copper concentrations were about an order of magni-

Table 8.35. Water Quality Summary at White Cedar Peatland

	Surface Samples		Shallow Wells		Piezometers	
	Mean	Range	Mean	Range	Mean	Range
Specific conductance (μsec/cm)	3100	2250–3500	2600	2100–3000	270	213–325
Copper (mg/L)	0.13	0.002–0.46	0.001	0.001–0.002	0.001	0.001–0.002
Nickel (mg/L)	20	8.4–26.0	0.6	0.18–0.93	0.03	0.01–0.06

From Eger, P. and K. Lapakko. Nickel and Copper Removed from Mine Drainage by a Natural Wetland, U.S. Bureau of Mines, IC 9183 (1988), pp. 301–309.

Table 8.36. Trace Metal Concentrations in Vegetation

Plant Species	Peatland Receiving Stockpile Drainage (sites T1 - T19)				Control Peatland (mean, composite sample)		Regional Values		
	Copper mean	(mg/kg) range	Nickel mean	(mg/kg) range	Copper (mg/kg)	Nickel (mg/kg)	Copper (mg/kg)	Nickel (mg/kg)	No. of Sites
White Cedar (*Thuja occidentalis*)	2.6	2.1–3.8	37	8–62	4.1	1.0	No Data	No Data	1
Alder (*Alnus Rugosa*)	7.8	4.8–10.1	82	10–239	12.1	1.0	3.9–12.1	1.0	3
Carex spp.	6.1	3.3–9.2	76	7–233	2.6	1.0	0.9–8.6	1.0–8.0	10

From Eger, P. and K. Lapakko. Nickel and Copper Removed from Mine Drainage by a Natural Wetland, U.S. Bureau of Mines, IC 9183 (1988), pp. 301–309.

Table 8.37. Overall Mass Balance for Watershed

	Vol (L × 10⁶)	Mass (kg)		
		Nickel	Copper	Sulfate
Inputs				
Stockpile seepage				
Seep 3	90	1,810	72	103,000
Em-8	184	150	4	128,000
Seep 1	29	110	16	72,000
Mine dewatering pumps				
011	2,530	14	16	205,000
012	255	0.8	2	78,000
Natural runoff[a]	426	2	2	13,000
Total input	3,514	2,087	112	599,000
Outflow				
Em-1	3,514	340	9	563,000
Overall removal[b]		84	92	6

[a] Computed by difference. Volume at Em-1 minus the sum of all input volumes.
[b] (input - output)/(input) × 100%.

From Eger, P. and K. Lapakko. Nickel and Copper Removed from Mine Drainage by a Natural Wetland, U.S. Bureau of Mines, IC 9183 (1988), pp. 301–309.

tude lower than the nickel values, even at sites near the seep, and there was no relationship between the copper values and location in the peatland. Concentrations in the leaves of cedar and alder were slightly higher at the control than the values measured at the impacted area.

Nickel and copper concentrations in the peat decreased as distance from the seep and depth in the peat column increased. The maximum copper concentration of 3600 mg/kg was found in the top 20-cm sample near the seep (site 3-1). The concentration decreased with depth, and the maximum value was about six times the value at the base of the peat. Samples collected at sites farther from the seep had a more uniform concentration vs depth profile, and concentrations near the creek were only 200 to 400 mg/kg. Surface nickel concentrations ranged from 6400 mg/kg near the seep to 1116 mg/kg near the creek and decreased with depth at all sites in the impacted area. Isopleths of the concentrations in the top 20 cm and the specific conductance in the surface water were used to define the area of drainage impact.

Generally the zone of high specific conductance corresponded to the area with the highest nickel concentrations in the peat. However, concentrations increased along the line from $T1_6$ to T_{19}, and the nickel values at T_{19} were comparable to the values within the zone of influence. Concentrations of the trace metals at the control peatland (25 mg/kg average copper, 65 mg/kg average nickel) and across the stream (site 3-8) were one to two orders of magnitude lower than the concentrations near the stockpile and were relatively constant with depth at both sites.

To quantify the overall removal of copper and nickel occurring in the peatland, a mass balance was calculated for the period of study. An overall watershed balance and a balance for the peatland alone were computed. Flow and concentration mea-

Table 8.38. Nickel Mass Input from Stockpile Drainage and Release from the Peatland to Stream

	Nickel Mass Loading	
Time Period	Seep (mg/sec)	Input to Stream from Peatland (mg/sec)
7-15 to 10-25-76	45.4	3.6
12-07 to 12-10-76	0.0	1.2
5-05 to 5-13-76	1.1	5.4
8-16 to 8-17-77	111.0	28.0

From Eger, P. and K. Lapakko. Nickel and Copper Removed from Mine Drainage by a Natural Wetland, U.S. Bureau of Mines, IC 9183 (1988), pp. 301–309.

surements were combined to compute overall mass inputs and outputs. Individual time periods were analyzed and error estimates for the nickel mass were made for seep 3, Em-8, and Em-1. The overall mass values were estimated to be within ±15%.

For the watershed, the major source of nickel and copper was the stockpile drainage, with seep 3 contributing about 87% of the total nickel input and 78% of the total copper input (Table 8.37). The total input from stockpiles comprised 99.2% and 82% of the nickel and copper inputs, respectively. The total nickel input was 2087 kg and total output, as measured at the mouth of the stream (Em-1), was 340 kg. The overall nickel removal was 84%. Total copper input was 112 kg and the output was 9 kg, resulting in an overall copper removal of 92%.

The only significant input of nickel and copper to the white cedar peatland was stockpile drainage from seep 3. Nickel input during the 13-month study period was 1800 kg, and copper input was about 70 kg. The output from the peatland was much more difficult to quantify because no single point could be found at which the drainage from the peatland entered the creek. The drainage was diffuse along the stream for about 250 m (determined by specific conductivity measurements). The output of nickel was determined through the overall balance for the watershed by comparing upstream and downstream nickel loads in the stream and through low flow sampling along the stream. During the study period, discharge rates of nickel from seep 3 ranged from 0 during the winter to 111 mg/sec, while contributions to the stream ranged from 1.2 to 28 mg/sec during the same period (Table 8.38). The total estimated load from the peatland to the stream was about 300 kg; about 80% of the nickel had been removed in the peatland. The copper output was not calculated because water quality data in the peatland and in the stream indicated that (1) copper concentrations were reduced to background levels within 200 m from the seep, and (2) no measurable input of copper was made from the peatland to the stream.

Metal accumulation was estimated in the water, vegetation, and peat, using the specific conductance values to determine the boundary of the impacted area. As the upper portion of the peat is about 90% water and field observations indicated about 3 to 5 cm of standing water in the peatland, a substantial amount of metal could be stored in the water. Using the concentrations measured at the well sites and dividing

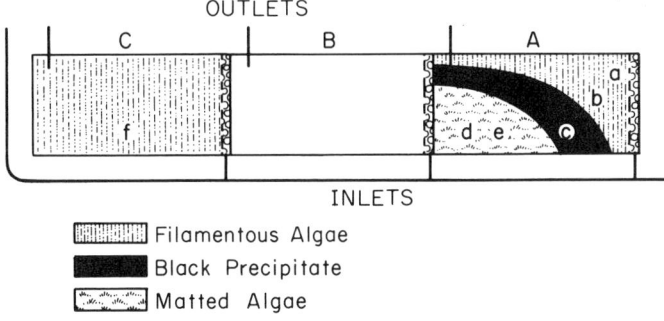

Figure 8.18. Big Five tunnel experimental wetland.[23]

the area between the sites, estimates were made for the mass of metal stored in the water. For nickel the total was 42 kg, with 76% of the metal being stored in the top 20 cm of the peatland.

Biomass estimates were made for cedar, *Carex*, and alder, and the average concentrations for each species were used to estimate the mass of nickel in each species. The total mass was <6 kg with cedar containing about 70% of the total.

The total mass of nickel contained in the top 20 cm of peat was calculated by using the area between the isopleths and applying the total mean concentration to that volume of peat. A certain percentage of the metals was contained in the peat prior to the stockpile drainage, therefore, a background correction was applied. The concentrations at depths >75 cm at sites 3-2 and 3-5 were assumed to be representative of natural conditions. Estimates were made for the 20 to 75 cm depth based on the measured depth profiles and the assumption that the isopleths at depth would be similar to those in the top 20 cm. Using these assumptions the nickel mass was estimated to be about 1150 kg in the top 20 cm and 350 kg between 20 to 75 cm. The mass of copper was calculated in a similar fashion and estimated to be 77 kg.

In conclusion, it is evident that significant removal of nickel and copper has occurred in this watershed. The overall watershed balance indicated that about 1750 kg of nickel and 100 kg of copper were removed. The largest source of metal input was seep 3, but analyses of upstream and downstream loads in the stream revealed that only a portion of the nickel and none of the copper were transported through the peatland. Most of the metal removal in the watershed was occurring in the white cedar peatland.

Both the input-output calculations and the accumulation estimates indicated that about 1500 kg of nickel and about 70 kg of copper were removed and stored in the peatland during the period of study. Based on the overall mass balance and water quality results, essentially 100% of the copper and about 80% of the nickel were removed. Analysis of the individual compartments demonstrated that >90% of the metals were associated with the peat. Although the nickel and copper concentrations in the peat are on the order of several tenths of a percent, laboratory experiments with stockpile drainage have demonstrated that peat can accumulate up to 2% nickel

422 ENVIRONMENTAL IMPACTS OF MINING

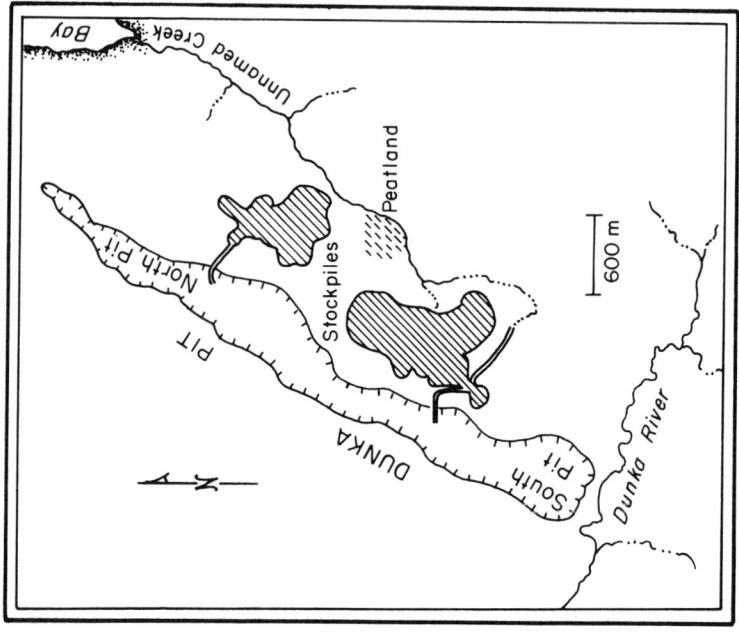

Figure 8.20. White cedar peatland.[24]

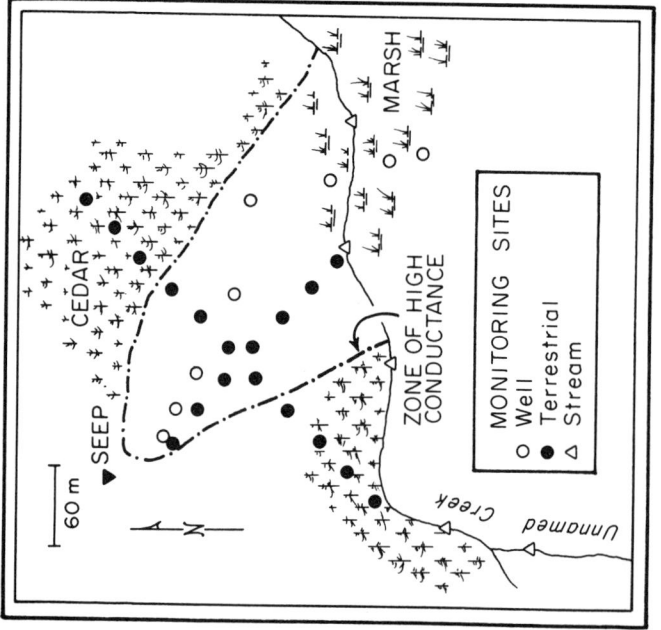

Figure 8.19. Stockpile and sampling locations.[24]

during a continuous column removal experiment, and peat from the Tantramar Swamp in New Brunswick was found to contain as much as 10% copper. Typical metal concentrations for peat in Minnesota are about two orders of magnitude less than the values measured in the impacted peatland. Metal values above background have been found near zones of mineralization, but were generally an order of magnitude less than the concentration in the study area.

Specific conductance and peat metal concentrations indicate that the contact zone between the stockpile drainage and the peatland is confined to the upper portion of the peat, a zone referred to as the acrotelm, or the zone of active water movement. Peat decomposition generally increases and hydraulic conductivity decreases with depth, thus, only a small amount of the stockpile drainage will contact the lower levels of the peat. Treatment is, therefore, restricted to the upper, more permeable zone.

Even though the nickel values in the vegetation were elevated, the removal of nickel by plant uptake accounted for <1% of total nickel removal. Increasing the biomass in a wetland might provide greater metal removal, but for the metals in this study the increased removal would not appear to be significant.

REFERENCES

1. Donald, A. H. and R. K. Bastian. Wetland ecosystems: natural water purifier, in *Constructed Wetlands for Wastewater Treatment*, Hammer, D. A., Ed. (Chelsea, MI: Lewis Publishers, Inc., 1990), pp. 5–19.
2. Kadlec, R. H. Hydrologic factors in wetland water treatment, in *Constructed Wetlands for Wastewater Treatment*, Hammer, D. A., Ed. (Chelsea, MI: Lewis Publishers, Inc., 1990), pp. 21–39.
3. Faulkner, S. P. and J. L. Richardson. Physical and chemical characteristics of freshwater wetlands soils, in *Constructed Wetlands for Wastewater Treatment*, Hammer, D. A., Ed. (Chelsea, MI: Lewis Publishers, Inc., 1990), pp. 41–72.
4. Guntenspergen, G. R., F. Stearns and J. A. Kadlec. Wetland vegetation, in *Constructed Wetlands for Wastewater Treatment*, Hammer, D. A., Ed. (Chelsea, MI: Lewis Publishers, Inc., 1990), pp. 73–88.
5. Portier, R. J. and J. P. Stephen. Wetland microbiology: form, function, processes, in *Constructed Wetlands for Wastewater Treatment*, Hammer, D. A., Ed. (Chelsea, MI: Lewis Publishers, Inc., 1990), pp. 89–105.
6. Girts, M. A. and R. I. P. Kleinmann. Constructed Wetlands for Treatment of Acid Mine Drainage, National Symposium on Mining, Hydrology, Sedimentology and Reclamation, Lexington, KY (1986), 165.
7. Emerick, J. C. and E. A. Howard. Using wetland for the control of western acid mine drainage, in *SME Annual Meeting Preprint 88-173* (Denver, CO: Society of Mining Engineers, 1988).
8. Brodie, A. G. Selection and evaluation of sites for constructed wastewater treatment wetlands, in *Constructed Wetlands for Wastewater Treatment*, Hammer, D. A., Ed. (Chelsea, MI: Lewis Publishers, Inc., 1990), pp. 307–317.

9. Watson, J. T., C. R. Sherwood, H. R. Kadlec and R. L. Knight. Performance expectations and loading rates for constructed wetlands, in *Constructed Wetlands for Wastewater Treatment*, Hammer, D. A., Ed. (Chelsea, MI: Lewis Publishers, Inc., 1990), pp. 319–351.
10. Steiner, G. R. and R. J. Freeman. Configuration and substrate design considerations, in *Constructed Wetlands for Wastewater Treatment*, Hammer, D. A., Ed. (Chelsea, MI: Lewis Publishers, Inc., 1990), pp. 363–377.
11. Watson, J. T. and J. A. Hobson. Hydraulic design considerations and control structures, in *Constructed Wetlands for Wastewater Treatment*, Hammer, D. A., Ed. (Chelsea, MI: Lewis Publishers, Inc., 1990), pp. 379–391.
12. Brodie, A. G. et al. An Evaluation of Substrate Types in Constructed Wetland, U.S. Bureau of Mines, IC 9183 (1988), pp. 389–398.
13. Allen, H. H., G. J. Pierce and V. R. Wormer. Considerations and techniques for vegetation establishment, in *Constructed Wetlands for Wastewater Treatment*, Hammer, D. A., Ed. (Chelsea, MI: Lewis Publishers, Inc., 1990), pp. 405–415.
14. Samuel, D. E., J. C. Sencindiver and H. W. Rauch. Water and Soil Parameters Affecting Growth of Cattails, U.S. Bureau of Mines, IC 9183 (1988), pp. 369–374.
15. Sencindiver, J. G. and D. K. Khumbla. Effects of Cattails (*Typha*) on Metal Removal from Mine Drainage, U.S. Bureau of Mines, IC 9183 (1988), pp. 359–368.
16. Wieder, R. K. Determining Capacity of Metal Retention in Man-Made Wetlands Constructed for Treatment of Coal Mine Drainage, U.S. Bureau of Mines, IC 9183 (1988), pp. 375–381.
17. Spratt, A. K. and R. K. Wieder. Growth Responses and Iron Uptake in *Sphagnum* Plants and Their Relation to Acid Mine Drainage Treatment, U.S. Bureau of Mines, IC 9183 (1988), pp. 279–285.
18. Stillings, L. L., J. J. Gryta and T. A. Ronning. Iron and Manganese Removal in a *Typha* Dominated Wetland During Ten Months Following Its Construction, U.S. Bureau of Mines, IC 9183 (1988), pp. 317–324.
19. Kepler, D. A., An Overview of the Role of Algae in the Treatment of Acid Mine Drainage, U.S. Bureau of Mines, IC 9183 (1988), pp. 286–290.
20. Brodie, G. A. and D. A. Hammer. Constructed Wetlands for Acid Drainage Control in the Tennessee Valley, U.S. Bureau of Mines, IC 9183 (1988), pp. 325–331.
21. Kolbash, R. L. and T. L. Romanos. Windsor Coal Company wetland, in *Constructed Wetlands for Wastewater Treatment*, Hammer, D. A., Ed. (Chelsea, MI: Lewis Publishers, Inc., 1990), pp. 788–792.
22. Hiel, M. T. and F. J. Kernis, Jr. The Tracy Wetlands: A Case Study of Two Passive Mine Drainage Treatment Systems in Montana, U.S. Bureau of Mines, IC 9183 (1988), pp. 352–358.
23. Wildeman, T. R. and S. L. Lauden. Use of wetlands for treatment of environmental problems in mining: non-coal-mining applications, in *Constructed Wetlands for Wastewater Treatment*, Hammer, D. A., Ed. (Chelsea, MI: Lewis Publishers, Inc., 1990), pp. 221–231.
24. Eger, P. and K. Lapakko. Nickel and Copper Removed from Mine Drainage by a Natural Wetland, U.S. Bureau of Mines, IC 9183 (1988), pp. 301–309.
25. Lang, G. E. Personal communication.
26. Pasavento, B. Personal communication.
27. Copeland, S. R. Unpublished data.

CHAPTER 9

Blasting

The goal of blasting is to obtain maximum fragmentation of the consolidated material in the overburden with optimum drilling and blasting cost. The amount of fragmentation required is determined by the stripping unit to be used in overburden removal. Many coal seams by surface coal mines must also be broken by blasting; this is conducted before coal removal. Environmental factors as well as due regard for public safety, health, and welfare must be considered in choosing the blasting plan.[1]

The blasting plan should be made during preplanning and is based on data from the overburden cores. Analysis of the data will help determine the kind of drilling equipment and bit types that will be needed for overburden preparation.

A variety of complaints has always been received by industry pertaining to blasting. The population explosion and urban sprawl have acted in concert to bring industry and the public into closer physical contact. In many cases, structures were built on property adjacent to surface mining operations. As a result, the number of complaints increased drastically and presently constitute a major problem.

Some complaints registered are legitimate claims of damage from blasting vibrations. The advances in blasting technology and a more knowledgeable explosives profession have minimized real structural damage. However, vibration levels that are completely safe for structures may be annoying and unpleasant for humans. Though no actual damage is done, air blast pressures may cause windows to rattle and the loud noise may be intolerable. Repeated vibrations, such as those from a nearby quarry, may eventually cause damage.[2]

Control of vibrational damage to natural scenic formations is a very important environmental consideration in surface mining. The wind-eroded formations are very fragile, and damage as far as one fourth of a mile (P:402 km) from the operation has been noted.

Where a conventional detonating cord is used to link blastholes, most airborne noise results from the connecting trunk lines. A new, low-energy detonating cord has been developed that can be substituted for the conventional cord. A 150-ft (45.6 m) length of this cord makes about as much noise as one electric blasting cap or 2 in. (50.8 mm) of the conventional cord.

If detonating cord is used on the surface, noise can be reduced by covering the trunk lines with up to 10 in. (254 mm) of dirt. When detonating cord is used only in the holes to fire the primers, a shovelful of dirt at each hole will effectively cover the exposed cord and cap.

Millisecond delays can be used to decrease the vibration level from blasting, because it is the maximum charge weight per delay interval rather than the total charge that determines the resulting amplitude. Also, many mines limit the number of holes per shot, using millisecond delays in series to minimize concussion and noise, especially near population centers, natural scenic formations, wells, water impoundments, and stream channels.[3-5]

Weather conditions can increase airborne noise. When temperature inversions prevail, blasting should be avoided. This condition exists frequently in the early dawn and after sundown. Foggy, hazy, or smoky days are unfavorable for blasting. When the wind is directed toward residential areas, blasting should be postponed.

When blasting is performed in congested areas or close to a structure, stream, highway, or other installation, the blast should be covered with a mat to prevent fragments from being thrown by the blast. The possibility of dust problems resulting from blasting is very remote.

The possibility does exist, however, and precautions must be taken to control dust pollution if the operation is close to high-use areas. During periods of dry weather, dust from explosions has been carried by air currents for many miles, and in certain isolated instances, it has been a public nuisance.

Several states, including West Virginia, Tennessee, Ohio, Montana, and Kentucky, have established guidelines for preventing or holding vibrational damage to a minimum. Most state laws concerning blasting pertain only to safety, storage, handling, and transportation of explosives.

When a blast is detonated, the bulk of energy is consumed by fragmentation and some permanent displacement of the rock close to the location of the drilled holes containing the explosive. This activity normally occurs within a few tens of feet (meters) of the blast hole. The leftover energy is dissipated in the form of waves traveling outward from the blast, either through the ground or through the atmosphere. The ground waves produce oscillations in the soil or rock through which they pass, with the intensity of these oscillations decreasing as distance from the blast increases.

One measurable quantity of interest that is caused by seismic waves or oscillations is particle velocity. This quantity defines how fast a particle (or structure) is moved by passing seismic waves, measured in inches (millimeters) per second. The results of a 10-year study program in blasting seismology by the U.S. Bureau of Mines concluded that particle velocity is more directly related to structural damage than particle displacement or particle acceleration. It is not how much but how fast the ground under a structure is moved by the passing seismic waves that determines the likelihood of damage. Particle velocity, therefore, becomes the vibration quantity of greatest concern to those engaged in blasting activities. They also concluded that a safe blasting limit of 2.0 in./sec (50.8 mm) peak particle velocity as measured from any of three mutually perpendicular directions in the ground adjacent to a structure should not be exceeded if the probability of damage to the structure is to be small (<5%). Kentucky is the only coal-producing state that has passed a law based on seismographic measurements. They limit vibrations adjacent to any structure to levels producing a particle velocity of 2.0 in./sec (50.8 mm) or less.[6,7]

Where instrumentation is not used or is unavailable, the U.S. Bureau of Mines found that a scaled distance of 50 feet per square root of pounds (22.62 m per square root of kilograms) can be used as a control limit with a reasonable margin of safety, and the probability is small or finding a site that produces a vibrational level that exceeds the safe blasting limit of 2.0 in./sec. For cases in which a scaled distance of 50 feet per square root of pounds (22.62 m per square root of kilograms) appears to be too restrictive, a controlled experiment with instrumentation should be conducted to determine what scaled distance can be used to ensure that vibrational levels do not exceed the particle velocity of 2.0 in./sec (50.8 mm).[8]

West Virginia uses the scaled distance formula, $W = (D/50)^2$, for control of vibrational damage. W equals the weight in pounds (kilograms) of explosives detonated at any one instant and D equals the distance in feet (meters) from the nearest structure, provided that explosive charges are considered to be detonated at one time if their detonation occurs within 8 msec or less of each other (see Table 9.1) for maximum explosive charges. A blasting plan for each method for a typical blast must be submitted with the permit application.

Citizen complaints concerning blasting on surface mining operations have been drastically reduced since the West Virginia law became effective. This success can be attributed to the conscientious efforts by the operators in using the scaled distance formula and guidelines for blasting issued by the State of West Virginia Department of Natural Resources.

Ammonium nitrate-fuel oil (AN/FO) blasting agents and slurries, used as breaking mediums for overburden, have greatly improved the efficiency of surface mine blasting operations and have reduced the cost of explosives considerably. AN/FO is an excellent heterogeneous fertilizer, because it contains readily available ammonia nitrogen and nitrate nitrogen and does not leave unfavorable residues in the soil. As a constituent of various types of explosives, it functions as an oxidizer and an explosive modifier. AN/FO mixes lead all other types of explosives in bank prepa-

Table 9.1. Maximum Explosive Charges[a] Using Scaled Distance Formula, $W = (D/50)^2$

Distance to Nearest Residence Building or Other Structure (ft)	Max. Explosive Charge (to be detonated) (lb)	Distance to Nearest Residence Building or Other Structure (ft)	Max. Explosive Charge (to be detonated) (lb)
100	4	2,100	1,764
150	9	2,200	1,936
200	16	2,300	2,116
250	25	2,400	2,304
300	36	2,500	2,500
350	49	2,600	2,704
400	64	2,700	2,916
450	81	2,800	3,136
500	100	2,900	3,364
550	121	3,000	3,600
600	144		
650	169	3,100	3,844
700	196	3,200	4,096
750	225	3,300	4,356
800	256	3,400	4,624
850	289	3,500	4,900
900	324	3,600	5,184
950	361	3,700	5,476
1,000	400	3,800	5,776
		3,900	6,084
1,050	441	4,000	6,400
1,100	484		
1,150	529	4,100	6,724
1,200	576	4,200	7,056
1,250	625	4,300	7,396
1,300	676	4,400	7,744
1,350	729	4,500	8,100
1,400	784	4,600	8,464
1,450	841	4,700	8,836
1,500	900	4,800	9,216
1,550	961	4,900	9,604
1,600	1,024	5,000	10,000
1,700	1,156		
1,750	1,225		
1,800	1,296		
1,850	1,369		
1,900	1,444		
1,950	1,521		
2,000	1,600		

Metric unit conversion: foot = 0.304 m; pound = 0.453 kg.

[a] Where blast sizes would exceed the limits of the scaled distance formula, blasts shall be denoted by the use of delay detonators (either electric or nonelectric) to provide detonation times separated by 9 msec or more for each section of the blast complying with the scaled distance of the formula. Explosive charges shall be considered to be detonated at one time if their detonation occurs within 8 msec or less of each other.

ration. Several types are available and can be obtained in prilled, granular, crystalline, or grained forms.

A new line of metalized blasting agents has become commercially available. These products are reported to be three times more powerful by weight and five times more powerful by volume than AN/FO combinations. Based on ammonium nitrate in combination with aluminum chips, the blasting agents vary in aluminum content between a low of 5% and a high of 30%. They are soft, silvery gels and maintain that softness even at 0°F (−18°C).

A trend is developing for casting overburden with explosives. The goal is to cast as much overburden as possible into the parallel cut with blasting techniques. With proper loading, spacing, and detonation delays, a good portion of overburden can be moved, thus reducing backfilling costs. This method also minimizes the need for recasting. Some mines report that 30 to 50% of their overburden is moved with explosives. This method works very well in deep, narrow pits by casting overburden into the pit away from the highwall and up on the spoil pile on the low wall side.

REFERENCES

1. Nicholls, H. R., C. F. Johnson and W. J. Duvall. Blasting Vibrations and Their Effects on Structures, U.S. Department of Interior, Bureau of Mines Bull. 656 (1971).
2. Perkins, B., Jr. and W. F. Jackson. Handbook for Prediction of Air Blast Focusing, Ballistic Research Laboratories Report No. 1240, U.S. Army, Aberdeen, MD (1964).
3. Perkins, B., Jr., P.H. Lorrain and W. H. Townsend. Forecasting the Focus of Air Blast Due to Meteorological Conditions in the Lower Atmosphere, Ballistic Research Laboratories Report No. 1118, U.S. Army, Aberdeen, MD (1960).
4. Berger, P. R. Blasting controls and regulations, *Min. Congr. J.* 59(II):21-25 (1973).
5. West Virginia Code, Chapter 20: Surface Mining and Reclamation: Section 6-11a, Blasting Restriction; Formula; Filing Preplan; Penalties; Notice, 1971.
6. Siskind, D. E. Damage to residential structures from surface mine blasting, in *SME-AIME Preprint 80-336* (Denver, CO: Society of Mining Engineers, 1980).
7. Linehan, P. and F. J. Wiss. Vibration and air blast noise from surface coal mine blasting, *SME-AIME Preprint 80-336* (Denver, CO: Society of Mining Engineers, 1980).
8. Siskind, D. E. and M. S. Stagg. Structure Response and Damage Produced by Ground Vibration from Surface Mine Blasting, U.S. Bureau of Mines Report of Investigations, RI 8507 (1980).

CHAPTER 10

Mining Subsidence

INTRODUCTION

Underground mining creates a void system that disturbs the existing stress field established by natural processes. The overburden responds with the first cut into the coal and continues until the applied stress reaches a new equilibrium within the overburden. These stress changes result in deformation and displacement of the surrounding overburden material. The magnitude of the change that takes place is controlled by the size of the cavity and the strength characteristics of the affected strata. Figure 10.1 is a simplified illustration of the forces acting on a stratigraphic section as a result of a mine void. The force vectors above the void are deflected toward the void proportionate to the distance from the center of the void. Below the coal seam an upward component of force also acts toward the center of the void. As the cavity increases in size, fractures develop in the overburden which may result in subsidence of the surface and sometimes heaving of the mine floor.

In general, mine subsidence problems develop where postmining pillar support systems and coal barriers ultimately fail. Many interrelated factors control when, where, and how failure will occur, including:

- Thickness of coal mined
- Size, shape, and distribution of pillars and rooms
- Percent extraction of coal
- Thickness and physical characteristics of the overburden
- Method of mining (e.g., longwall, shortwall, room and pillar, room and pillar with full or partial retreat)

432 ENVIRONMENTAL IMPACTS OF MINING

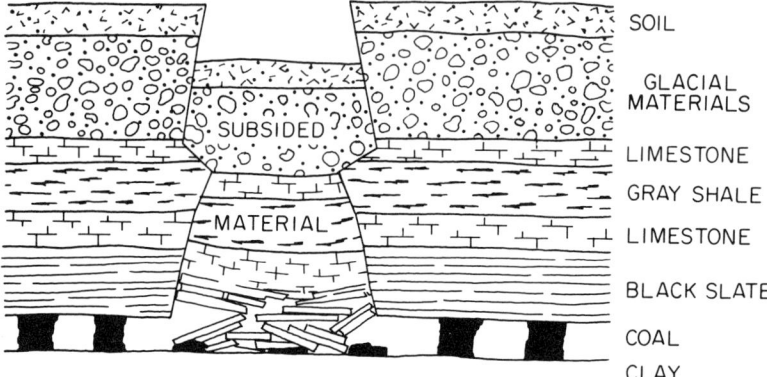

Figure 10.1. Pit subsidence. (From Elbert, J. and L. Guernsey. An Analysis of Extreme Danger Problems Associated with Subsidence of Abandoned Coal Mine Lands in Southwestern Indiana, U.S. Bureau of Mines, IC 9184 (1988), pp. 383–389.)

- Dry or flooded conditions in the mine
- Actual or potential level and degree of fracturing in overburden
- Mineralogy of overburden (e.g., clay minerals that swell when water is added, sulfide minerals that chemically and physically change in the presence of oxygen and moisture, and minerals that react with water to form new minerals)

Three main types of subsidence are seen: pit, trough, and shaft. Pit subsidence is responsible for more of the extreme danger problems than trough or shaft subsidence. Each of these three types of subsidence have specific movements that separate them from each other. Eventual erosion is a type of movement that accentuates each of these subsidence problems.

Typically, pit subsidence occurs from the failure of the roof (Figure 10.2). The result is an opening from 2 to 2.5 m deep and 0.6 to 12 m in diameter. The main movement, at least initially, is vertical. In the eastern United States, pit subsidence usually occurs where mines are <30 m deep; in Pennsylvania, most pit subsidence is found where the mine overburden is <15 m. Areas in which the mine is deeper than 45 to 50 m experience decreased frequencies of subsidence occurrence. Pit subsidence usually occurs over rooms and entries resulting in a surface reflection of the mining pattern.

The second type of subsidence is trough, which is usually a gradual disintegration of the pillars or squeezing of floor materials into the mine voids. The disintegration and squeezing may also occur simultaneously. Troughs (Figure 10.3) may occur in areas of mining at any depth, but most frequently associated with deeper mines than those in which pit subsidence occurs. Troughs are usually 0.6 to 1.3 m deep and cover larger areas than pit subsidence. The movements are both vertical and horizontal as the whole area is depressed. The horizontal areas may extend over a broader area than is undermined. For a deep mine, the area affected may be 20 ft beyond the

MINING SUBSIDENCE   433

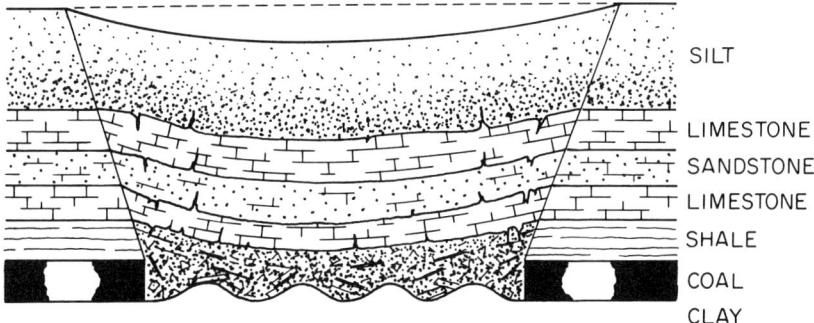

Figure 10.2. Trough subsidence. (From Elbert, J. and L. Guernsey. An Analysis of Extreme Danger Problems Associated with Subsidence of Abandoned Coal Mine Lands in Southwestern Indiana, U.S. Bureau of Mines, IC 9184 (1988), pp. 383–389.)

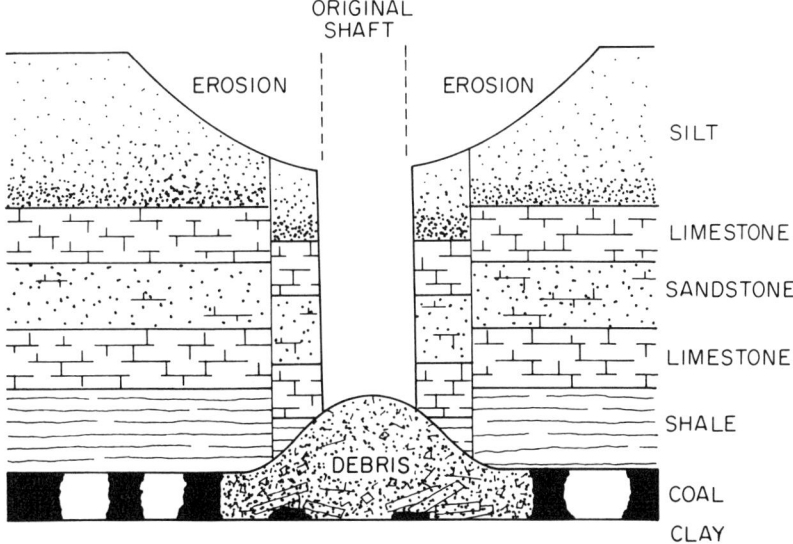

Figure 10.3. Shaft subsidence. (From Elbert, J. and L. Guernsey. An Analysis of Extreme Danger Problems Associated with Subsidence of Abandoned Coal Mine Lands in Southwestern Indiana, U.S. Bureau of Mines, IC 9184 (1988), pp. 383–389.)

mined-out areas. This is an important point as many individual homeowners feel that they and their property are safe if the mine does not extend beneath their house.

The third type of subsidence is associated with the shaft itself. This is commonly associated with older mines. The older shafts were not well supported when active mining was ongoing. Frequently the added weight of the buildings and equipment near the shaft was not considered in the required support. Decades later, the supports fail and erosion widens the original shaft opening (Figure 10.4). New subsidence

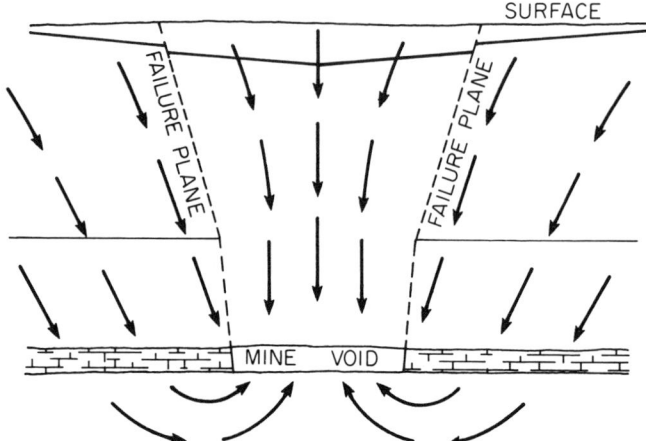

Figure 10.4. Ground movement due to subsidence. (From Elbert, J. and L. Guernsey. An Analysis of Extreme Danger Problems Associated with Subsidence of Abandoned Coal Mine Lands in Southwestern Indiana, U.S. Bureau of Mines, IC 9184 (1988), pp. 383–389.)

sites pose continuous problems because they are a result of collapsing shafts and tunnels. In the western United States, subsidence occurs around mine shafts. This subsidence is similar in appearance to pit subsidence, except that the pits are much deeper.

SUBSIDENCE INVESTIGATIONS

The approach and procedures for the study of subsidence over abandoned mine lands have been established. Time is essential because the damaged conditions may be disturbed and the exposed features may disappear. Therefore, a site investigation is always the first step for subsidence research because most of the information must be collected from the subsidence site, especially during the first time reconnaissance. For most cases the causes, damage pattern, and the status of ground movement can be determined by the obtained information. The information mainly consist of six categories such as: (1) general information, (2) structure information, (3) surface condition, (4) geological information, (5) geometry and characteristics of the subsidence, and (6) mining activity. Each category consists of several subcategories. In order to ensure that complete information can be collected a site investigation checklist has been developed by this research group.

With the data collected from a site investigation, a hypothesis should be established to clear up the logic and narrow down the number of possible causes. In order to determine the cause of the damage, the cause identification system with an identification list have also been developed by this group to evaluate the most possible cause. If the cause cannot be determined due to insufficient data, an in-depth exploration should be adopted.

An in-depth exploration mainly consists of surface and subsurface measurements and instrumentations, including surface survey, borehole TV camera examination, crackmeter, inclinometer, and Sondex settlement measurements. The results can be used to determine the cause(s) of the damage, the direction and magnitude of movements, and can assist in the design of a better remedial program.

When subsidence occurs several types of ground damage will be induced. The most common damage is depressions, cave-in pits, ground cracks, and compression ridges. The ground depression is associated with trough or sag subsidence. Its mining-related causes are pillar and floor failures; non-mining-related causes are differential settlement on the backfilled area, fluid withdrawal and organic soil drainage. A ground depression is a gentle settlement over a broad area which may range from a few tens to hundreds of centimeters in diameter, while the depth of depression ranges from 0.1 to 1 m for mine subsidence. The depth of the depressions due to non-mining-related causes is relatively unknown, but a 4-m depression has been reported to be due to fluid withdrawal.

The cave-in pits are most likely associated with potholes, sinkholes, or chimney subsidences. Their mining-related cause is roof failure, while the non-mining-related causes are limestone solution and soil piping. The newly developed cave-in pits are likely to be steep-sided with straight or bell-shaped walls depending on soil type. The diameter of cave-in pits ranges from 0.6 to 12 m, but most of them are <16 ft in mine subsidence. The depth of cave-in pits is about 2 to 2.5 m for mining-related cases and the coal seams are <55 m deep. For the non-mining-related cave-in pits, the diameter ranges from a few meters to 3000 m, and the depth ranges from a few meters to 20 m. All the pothole subsidence cases investigated in this research showed that the diameter of the holes ranged from 0.0 to 6 m; the depth was about 0.6 to 5 m; and the coal seam was about 6 to 18 m deep.

The ground cracks that occur at the margin of the depression and cave-in pits are the result of convex bending and are associated with the stretching of the ground surface. Generally, the ground cracks at the margin of the depression, and cave-in pits indicate the location of the maximum tensile strain. Beyond this maximum tensile strain zone, structure and ground damage is usually minor or nonexistent. Therefore, a subsidence damaged area can be determined by locating the ground cracks. In general, ground cracks are wider, more abundant, and more extensive near cliffs and steep terrain than they are in flat or gently rolling topography. Moreover, ground cracks are more often located at the asphaltic pavement than that at soil zone because asphalt is more brittle than soil and can absorb less stress.

Compression ridges occur in the depression where the ground surface is subjected to concave bending and associated with shortening of the ground surface. Generally, compression ridges are located in the central part of the depression. The ground cracks and compression ridges sometimes are not visible when they are located in the soil zone, especially when the surface is covered by vegetation. The crack does not always reach the surface. The stress at the tip of the crack in propagation can be less than the soil cohesion and root binding forces. Crack propagation is either stopped or changes its direction to the weaken part of the soil. In addition, a ground crack may have been developed in the subsurface and cannot reach the ground surface. Therefore, if the ground crack cannot be seen in the area it does not necessarily mean that a ground crack has not developed.

STRUCTURAL DAMAGE

Structural damage due to subsidence ranges from slight to very severe. Some structures may have a few hairline cracks while others may have cracks a few inches wide. The damage severity is dependent upon (1) the types of ground surface deformation, (2) the type of stresses, (3) the type and size of structures, (4) the location of the structure with reference to the mine openings, (5) the material of the structures, and (6) the type of foundation of structure attached to the ground. For instance, structures attached firmly into the ground are likely affected more by ground movement.

Damage to buildings can be divided into functional (structural) and cosmetic (architecture) damage. The major functional damage consists of misalignment of windows, doors, and walls; slanting floor; leakage in the roof; and breakage of utility lines. This type of damage may impair the function and usage of the structure. The major cosmetic damage consists of cracks along the mortar lines on the exterior and interior walls, cracks on the stone facing and plaster, separation of siding and frames, and cracks on the floor. This type of damage is more annoying than dangerous. Both functional and cosmetic damage can occur separately or simultaneously, based on the level of the damage source. For instance, a low level of vibration due to traffic or industry may cause cosmetic damage; however, a high level of vibration due to blasting or earthquake will cause not only cosmetic damage, but also functional damage and sometimes even cause structural collapse. If a building has a basement, the deformation of the building due to subsidence most likely will be initiated from the foundation or basement and propagate upward to the upper structure. The foundation or the basement is likely to suffer more severe damage than the rest of the building because the basement is directly subjected to the stresses. The amount of damage is referred to the degree of structural deformation, not the repair cost.

Generally, structural damage is associated with ground movements, the structures assumed to move with the ground. However, a resistance or friction exists, which is due to the weight of the structure, between the structure and the ground. Therefore, the ground movement cannot be entirely transmitted to the surface structures. Accordingly, several types of stresses such as axial, shear, and bending are induced against the foundation of the structures or the structures directly. The axial stresses, either extension or compression, are caused by horizontal movement. This type of stress will induce tensile and compressive damage to the structures. The shear stresses are caused by the friction between structure and ground. The bending stresses are caused by the vertical movement of the ground and will induce tilt and curvature damage to the structures.

As discussed earlier, linear deformation is due mainly to nonuniform horizontal movement, which will induce tensile and compressive damage to the structures. If the structure is subjected to tension the structure will be lengthened and open cracks will be induced along the joints and weak points, such as the contact line between the windows, door frames, and the walls. All the tensile cracks are uniformly developed along the brick faces and mortar lines because these are the weakest places on the brick wall. If a structure is subjected to compression, the foundation walls buckle inward and induce horizontal cracks perpendicular to the direction of the compression along the mortar on the brick faces. Tensile cracks and buckled structures are the specific characteristics of tensile and compressive damage, respectively. Those characteristics can be used to identify the types of stresses that cause damage to the structure.

When the ground slope changes, tilting is induced on the structures. Tall structures with a small base, e.g., towers and chimneys, are the most sensitive to tilting. Natural ground slope will also affect the stability of the ground surface and structures when subsidence occurs. If the dip of a slope is consistent with the ground movement direction, ground movement is enhanced. On the other hand, when the dip of a slope is in the opposite direction from that of the ground movement, ground movement is reduced. In addition, a slope area may be a location prone to instability or landslide.

Depending on the location of the structure in the depression, the structure may be subjected to sagging or hogging damage. Sagging and hogging are two different types of ground curvatures which are caused by the differential settlement. Sagging is a negative or concave curvature. For structures located on the concave area, the lower portions of the structures are subjected to tension and the upper portions are subjected to compression. As a result, vertical cracks with wider gaps at the bottom occur at the window and door frames, on the walls, and at the wall joints.

Hogging is a positive or convex curvature. For structures located in the convex area, the upper portions of structures are subjected to more severe stretching than the lower portions. As a result, vertical cracks with wider gaps are found at the top of the window and door frame, on the wall, and at the wall joints. This crack pattern indicates that the structure has been subjected to hogging. In general, structures located at the edges or the inflection point of the depression will be subjected to hogging damage. The most specific damage characteristics on the structure due to hogging are the convex bending and v-shaped crack.

DAMAGE CRITERIA

There are two types of damage criteria for buildings affected by mining subsidence. One type is based on the appearance of the damage and the other on the magnitude of the ground movement parameters. The damage criteria discussed here belong to the second category and their development involves three steps. The first

Table 10.1. Building Characteristics and Their Ratings

Characteristics	Rating
Foundation	
Isolated footing	1
Continuous footing	4
Raft foundation	8
Buoyancy foundation	16
Superstructure materials	
Brick, stone, and concrete	2
Reinforced concrete	4
Timber	6
Steel	8
L:R ratio	
<0.1	4
0.10–0.25	3
0.26–0.15	2
>0.50	1
H:L ratio	
<1.0	4
1.0–2.5	3
2.6–5.0	2
>5.0	1

step serves to classify buildings based on their structural characteristics, the second to specify the damage levels, and the third to determine the critical subsidence indices for each building category.

The building classification method used in this section is similar to that used in Poland. Several building characteristics are taken into account and each of these characteristics is rated into different classes, according to its response to subsidence. The final ratings is the sum of all the ratings of the various building characteristics, which in this case include building foundation, superstructure material and L:R and H:L ratios, where L is the length of the building, H the height, and R the radius of influence.

Several methods are available to specify damage levels. For example, under the system proposed by the British National Coal Board, the change in structure length combined with a physical description of damage is used to categorize damage. In a classification proposed seven parameters were considered and rated; damage level was determined by adding the ratings of the individual parameters.

Several subsidence indices have been used when developing building damage criteria. The most commonly used indices are horizontal strain, deflection ratio, curvature, tilt, and angular distortion. It is not necessary to use all these indices in establishing damage criteria as some of them are interrelated.

REMEDIAL MEASURES

Currently, four types of remedial measures are available for abandoned mine subsidence: point support, local backfilling, areal backfilling, and strata consolida-

Table 10.2. Building Classes

Total Rating	Class
4–10	I
11–17	II
18–25	III
26–32	IV

tion. In the point support method, a large number of grouting holes are drilled and a relatively small quantity of grouting materials are injected to form the grouting piles in the mine voids or openings and to achieve firm contact with the main roof. By doing so, the amount of subsidence potential and subsidence severity can be reduced. The advantages of these methods are that they are simple and use a small amount of grouting materials. The disadvantages are that a large number of grouting holes are needed to be drilled, the distribution of grouting materials cannot be controlled, and its support capacity is uncertain. Based on different procedures and approaches the point support methods can be subdivided into gravel column method, grout column method, fly ash grout injection method, and fabric formed concrete method.

The local backfilling method is designed to fill the small and shallow potholes or surface cracks without drilling the grouting holes. The backfill materials include gravel, refuse, and dirt. Generally, the backfill materials are directly dumped to fill the cracks and holes.

Areal backfilling methods are designed to fill a large area of underground openings in order to provide general protection to urban areas that are measured in terms of hundreds of acres. These methods involve a large quantity of grouting materials such as coal refuse, fly ash, and gravel which are injected into underground openings under high pressure.

Strata consolidation methods are designed to bind or grout the shallow strata beneath the damaged structure into a single rigid unit. If subsidence continues, the unit will be moved as a rigid body without being damaged. Based on different grouting materials and approaches, the strata consolidation methods can be divided into the polyurethane binder method, the cement grout pad method, and the rock anchor method.

Stress redistribution due to abandoned mine subsidence was initially conducted on a model represented a typical retreated abandoned room and pillar mine by. At first stage, the pillars at the central area were subjected to higher stress than the surrounding area. Multiple pillar failure and subsidence were initiated. At second stage, more pillars failed and subsidence increased. Simultaneously, destressing occurred at the failure zone while stress concentration occurred in the vicinity. At third stage, pillar failure and subsidence were continuously developed. Meanwhile, because of the compaction of gob materials stress again built up at the failure zone, but stress concentration still remained at the outward area. In summary, the magnitude of subsidence and surface affected area increase as stress concentration continues to cause pillar failure.

REFERENCES

1. Craft, J. L. and T. M. Crandall. Mine Configuration and Its Relationship to Surface Subsidence, U.S. Bureau of Mines, IC 9184 (1988), pp. 373–381.
2. Elbert, J. and L. Guernsey. An Analysis of Extreme Danger Problems Associated with Subsidence of Abandoned Coal Mine Lands in Southwestern Indiana, U.S. Bureau of Mines, IC 9184 (1988), pp. 383–389.
3. Hsiung, S. M., P. M. Lin and S. S. Peng, Structures and Ground Damages Over Abandoned Mine Lands, U.S. Bureau of Mines, IC 9184 (1988), pp. 362–368.

CHAPTER 11

Postmining Land Use

INTRODUCTION

During surface coal mining activities, all other uses of land are precluded. This results in a public perception that a conflict exists between surface mining and other established land use demands. However, surface mining is only a temporary use of land that can be accommodated in a sequence of future land uses. This can be achieved by successful reclamation efforts in restoring the land to its premining use capabilities.

Land use planning is conducted at two levels: macro- and microscale. Macroscale planning, sometimes called comprehensive planning, is directed toward guiding growth, development, and land use for an entire region through the formulation of a comprehensive plan. Microscale planning, generally referred to as site planning, is concerned with obtaining a site plan that satisfies given performance standards and an overall comprehensive plan for the area. Macroscale planning is generally conducted by public planners at the county, state, and federal levels. Site planning is generally carried out by the private sector, by land developers, or mine planners.[1]

The general process of land use planning is shown in Figure 11.1. The mined land use planning process is illustrated in Figures 11.2 and 11.3.

The processes used for both types of land use planning are essentially the same, but the level of effort may vary significantly. Comprehensive planning is conducted before site planning. However, it is not uncommon that site planning be conducted in an area in which no comprehensive plan exists. This places the burden on the site planner to evaluate regional objectives. Once a mining company has defined the

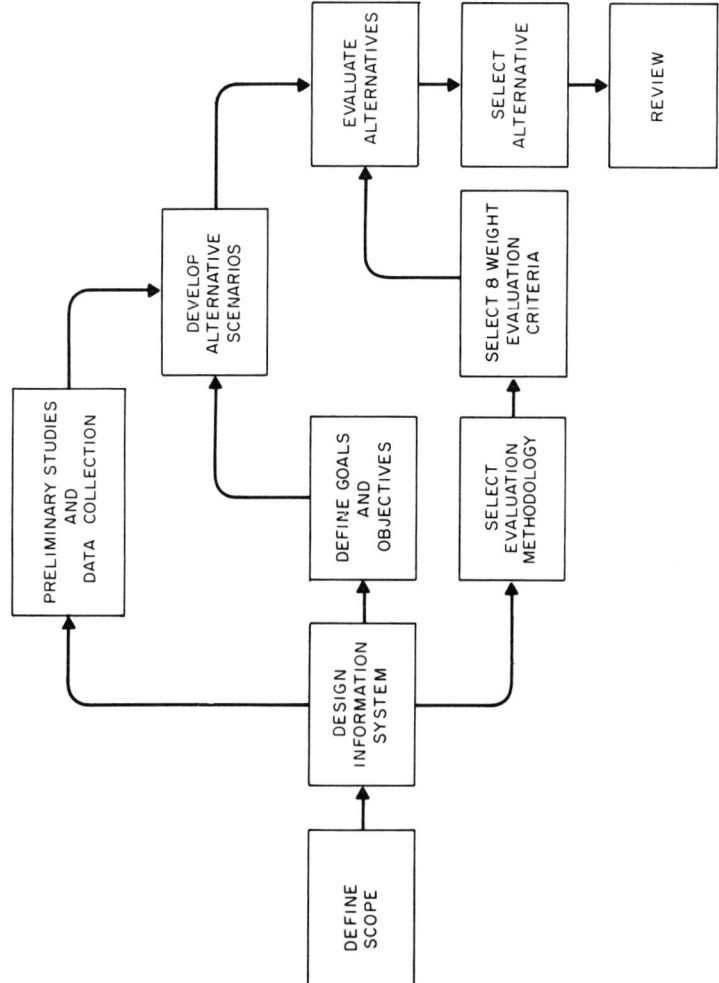

Figure 11.1. Land use planning model. (From Sweigard, R.J. and R.V. Ramani. Regional comparison of post mining land use practices, in *SME Preprint 83-105* (Littleton, CO: Society of Mining, Metallurgy and Exploration), 1983.)

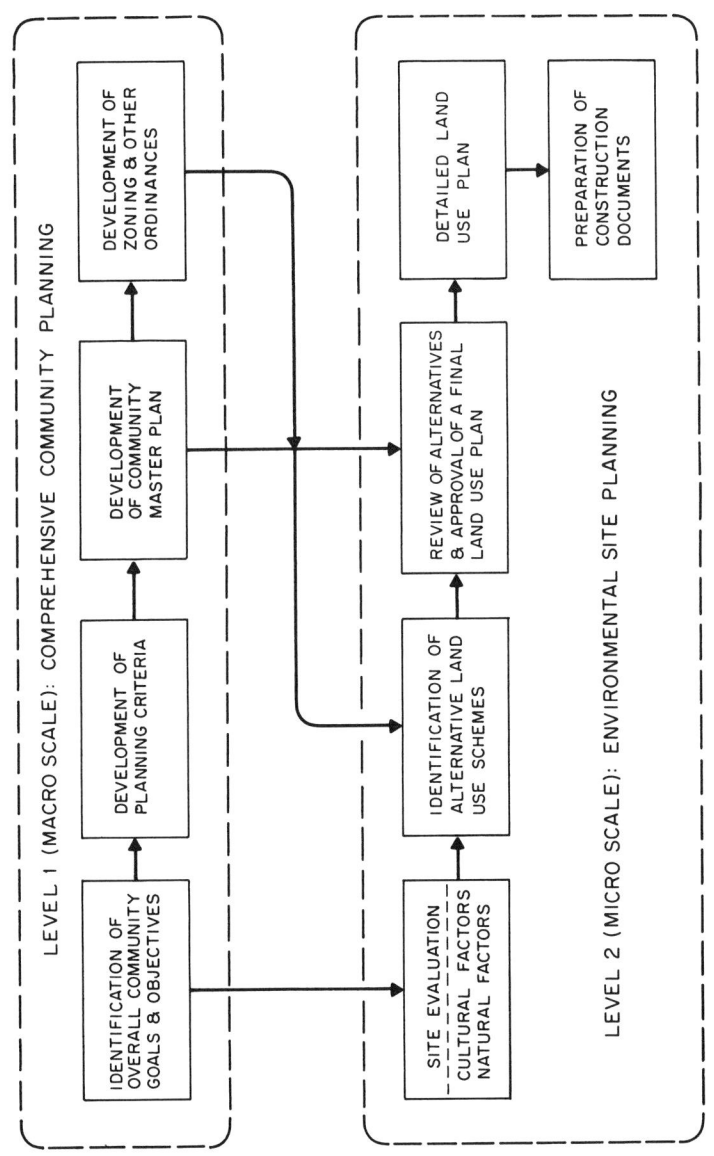

Figure 11.2. Macro- and micro-level land use planning. (From Sweigard, R.J. and R.V. Ramani. Regional comparison of post mining land use practices, in *SME Preprint 83-105* (Littleton, CO: Society of Mining, Metallurgy and Exploration), 1983.)

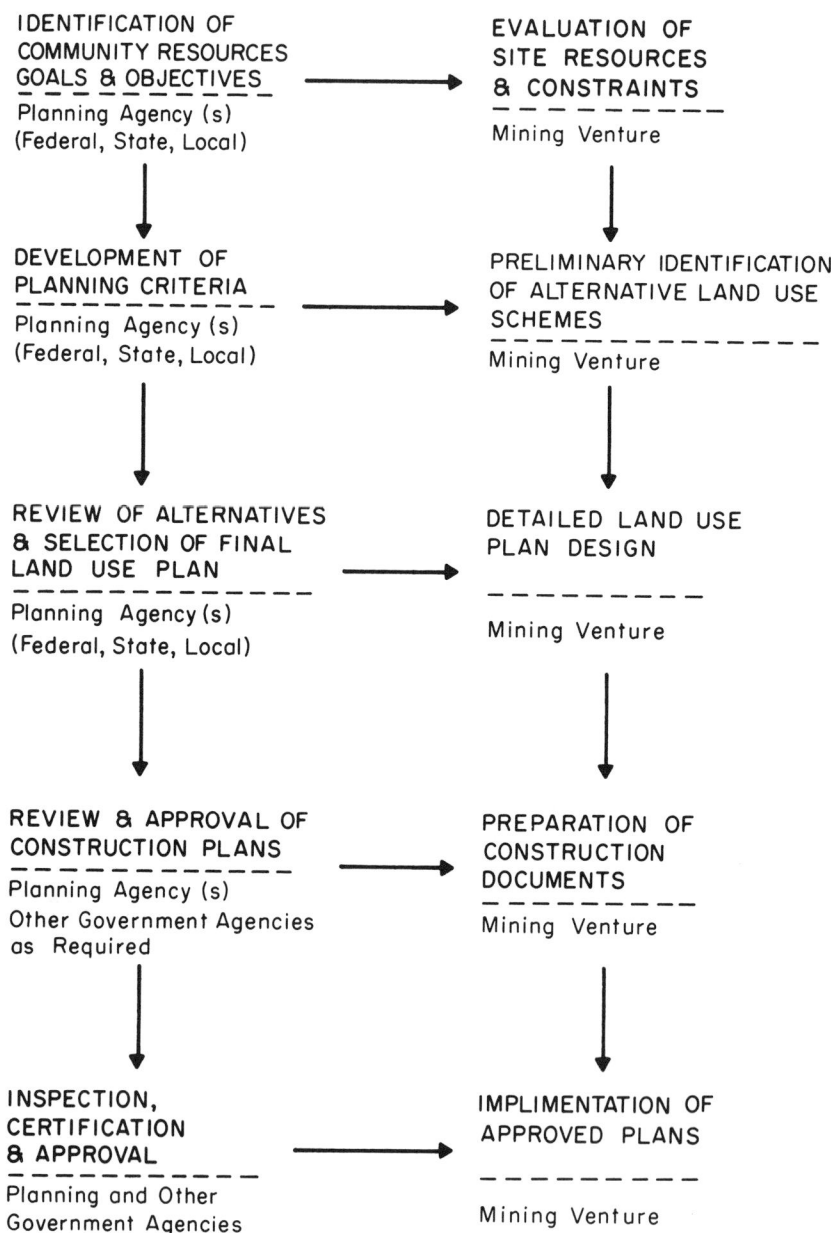

Figure 11.3. Mine-land use planning. (From Ramani, R.V. and R.J. Sweigard. Impacts of land use planning on mineral resources, in *SME Preprint 82-418* (Littleton, CO: Society of Mining, Metallurgy and Exploration), 1982.)

scope of its reclamation and land use planning, this scope is unlikely to change for different mining operations. Scope includes specifying the type of output required from the planning process; e.g., final design, conceptual plan, and organizational structure of the planning process.

The local goals and objectives are determined by public planners. Mine planners interact with local and regional planners so that the postmining land use plan is compatible with the overall plan for the area. Besides satisfying local goals and objectives, the mining company may wish to establish other goals and objectives such as improving the value of the land, promoting good public relations, or developing other company goals for postmining land use.

The environmental differences between the various regions impact the postmining land use potential of surface mined lands. The socioeconomic conditions such as population trends, employment trends, land values, the influence of public planners, regulatory constraints, and the availability of cultural resources may have an impact on the actual use of reclaimed land.

Two case studies of postmining land use are discussed below.

APPALACHIAN REGION CASE STUDY[2]

The Spingola No. 1 mine is located in Clearfield County, Pennsylvania, and is worked by a small operator, Simca Mining Inc. This mine is typical of the region in that the most recent operation has had a relatively short life (about 3 years), but a large portion of the site has been disturbed by past surface and underground operations. All of the land is privately owned and the parcel has been assembled through lease agreements with six different landowners. Land uses that existed before recent mining include forest land, open fields, and disturbed land that has either been reclaimed to forests, permanent grasses, or left unreclaimed.

Soils in this area are relatively thin, particularly on the steeper slopes. In many instances, both at this site and over the entire region, previous mining operations have resulted in a complete loss of topsoil. Two soil types at this mine have been identified as prime farmland. However, they have been exempted from prime farmland standards due to previous stripping, forest cover, or a lack of agricultural activity. Agriculture does not play a significant role in this area, with only about 8% of the land in Clearfield County devoted to crops or pasture. In addition to the soil limitations, this area has a fairly short average growing season.

It was estimated that 29% of the population lived in urban areas, while the remaining 71% lived in rural areas of the county. Based on a countywide average, the population density was about 70 persons per square mile. Jordan Township, which includes the Spingola No. 1 mine, had a 1980 population of 580 inhabitants. The township, therefore, had a population density of only 25 persons per square mile. The entire county experienced a small population growth in the period from 1970 to 1980 of 12%.

Distances to major transportation arteries, public utilities, places of employment, shopping centers, public transportation terminals, parks, and other recreational areas play a large part in determining the suitability of a site for various users. The Spingola No. 1 mine, like many eastern surface mines, is located in a remote rural area and is accessible only by two-lane secondary roads. Higher intensity land uses in this area are inhibited by its inaccessibility.

The primary land use planning agency for this area is the Clearfield County Planning Commission. Jordan Township has no planning board and no zoning ordinances. The County Planning Commission is adopting a countywide subdivision ordinance. General land use goals and objectives have been outlined by the Commission.

The mining company does not own the land on which the mine is located. The site planning process was applied to develop three plans without regard for ownership and one plan to reflect current ownership patterns.

Initial phases of the land use planning process have indicated that economically feasible postmining land use alternatives for this site are basically limited to low-intensity uses such as pasture and wildlife habitat. Many times land use plans evolve through commonsense and reason during data collection and preliminary studies. In this case, it is obvious that a large portion of the 1.7-km^2 (430 acres) site will be preserved as open space.

The major features of each land use alternative that was considered can be summarized as follows:

- **Alternative 1A:** recreational development — an 0.2-km^2 (56 acres) campground along with 1 km^2 (253 acres) of forest land and 0.5 km^2 (121 acres) of open fields designed to provide a desirable habitat for wildlife.
- **Alternative 1B:** large acreage residential development — forest land and open fields combined in such a manner as to encourage wildlife habitation. The unique aspect of this proposal is that it provides for nine large building lots (at least 40,470 m^2, or 10 acres) that are accessible from existing roads.
- **Alternative 1C:** new community residential development — a conceptual design for a residential community of 100 new homes. The 0.4-km^2 (90 acres) development is intended to fit well into the surroundings by having the homes built off the main access road and providing for extensive open space on at least one side of each property.
- **Alternative 1D:** existing ownership land use plan — consists mainly of revegetating the land to a combination of open fields (permanent grasses and legumes) and forest land. In addition to the vegetation and wildlife benefits, at least two building sites could be developed on prime developable land.

Evaluation of Alternatives

Each of the postmining land use alternatives was evaluated from an economic, environmental, and social point of view. Quantitative analyses were performed for the economic and environmental considerations, and the social implications for each plan were discussed.

Economic Evaluation

Estimates of the total reclaimed land values based on the various land use plans were made. The unit prices for each land use were determined with the aid of experienced realtors operating in Clearfield County.

Alternatives 1A and 1C are presumed to be infeasible due to the market conditions. Alternative 1B would be selected based on economic analysis and the assumption that present ownership presents no barrier to such development. Alternative 1D, based on the existing ownership pattern, demonstrates that initial ownership can act to constrain higher land use realization.

Environmental Evaluation

Alternative 1D is the most acceptable plan environmentally because it involves the least amount of change from premining conditions. Alternative 1B also represents a very small change from premining conditions, and the only foreseen difficulty may be some stability problems with building foundations. The negative impacts of Alternative 1A are estimated to be small, but some decrease in infiltration and a slight increase in erosion may occur. Alternative 1C would have the greatest negative impact due to the high level of development. Many of the negative impacts would be mitigated through construction measures engineered to recharge or improve the properties of the soil.

Social Impact Evaluation

Alternative 1A is consistent with the county's objective of providing outdoor recreational facilities. Possible negative social impacts could be experienced by adjacent landowners and the residents in the nearby village of McCartney. These inconveniences may be offset by local economic benefits. Aesthetically, a campground could be designed in such a way so as not to detract from the rural atmosphere.

Alternative 1B is consistent with the local desire for rural living. This plan could be aesthetically pleasing if the homes are properly designed and oriented on the lots. No negative social impact should occur to the adjacent landowners or the village of McCartney.

Alternative 1C would result in the greatest social impact due to the addition of 250 to 300 new residents. This increase would place a strain on community facilities such as schools, medical centers, and police and fire protection.

Alternative 1D is viewed as having little social impact because it calls for no change in the existing social conditions.

Selected Alternatives

Based on the economic analysis, Alternative 1B is considered to be the most feasible of the three alternatives that are not influenced by the existing land ownership pattern. Although the calculated land value is not as high as those estimated for

Alternatives 1A and 1C, it is believed that these values are unattainable in light of the local real estate market. If Alternative 1D is disregarded, as stated at the outset, the environmental analysis also indicates that Alternative 1B is the most desirable. Finally, the social impact analysis does not indicate that Alternative 1B poses any threat to the social conditions.

MIDWEST CASE STUDY

The Chinook Mine of the AMAX Coal Co. is located in west-central Indiana, about 9 km (6 mi) east of Terre Haute. Although the area immediately adjacent to the mine is rural in nature, the mine is situated in the heart of the Midwest industrial belt. This mine has been in operation since 1928 and has the distinction of being the oldest continually operating surface coal mine in Indiana.

Meadowlark Farms Inc., a subsidiary of AMAX Inc., performs land holding and land management services for AMAX Coal. Meadowlark Farms carries out all revegetation operations at AMAX Coal Co. mines, operates four corporate farms, and conducts a cropshare-lease program on company land. One of these corporate farms is located at Chinook Mine and uses reclaimed land for agricultural purposes. In 1980, Meadowlark Farms produced from all its Indiana operations a total of 656,258 bushels of corn, 196,226 bushels of soybeans, 26,177 bushels of wheat, and 309,502 pounds of marketed livestock. Approximately 62% of the 12-km^2 (2976 acres) West Field is owned by Meadowlark Farms. The remainder is leased from private landowners.

Locally, the overburden consists mainly of dark gray to black shales and gray sandstones. A brown, oxidized sandstone is also present in some locations. Overburden analyses have shown the consolidated overburden to have high total sulfur values. This is particularly true for the deeper overburden and the interburden between some of the coal seams. Much of this material has the potential for acid production.

High-capacity soils are abundant in the area, as evidenced by the percentage of agricultural land. The climate is also conducive to agricultural production with an average growing season of 207 days and average annual precipitation of about 40 in. Besides cultivated fields, plant communities characteristic of fallow fields, old fields (not cultivated for 3 to 10 years), and woodlands are found in the vicinity of the mine.

With a few local exceptions, this portion of Indiana has lagged behind the remainder of the state in population growth and economic development. Over a period of 50 years, from 1930 to 1980, the population of Indiana has increased 69.53%. For the same period, the population of the two counties surrounding the mine increased only 9.5%.

Although this region has been one of high unemployment in the past, changes made during the 1980s have helped to reverse this trend. One of these changes has been the success of Terre Haute in attracting new industries and encouraging the expansion of existing industries. A second factor has been a diversification of industry in the area. Vigo County has experienced relatively low unemployment during that period. However, the most drastic improvement was seen in Clay County, which went from nearly 4% above the state average in 1971 to 1.4% below the state average in 1975.

Local planning in the project area is conducted at the county level. Of the two counties affected by Chinook Mine, only Vigo County presently has a planning commission. A visit was made to the Area Planning Department of Vigo County during the "site visit" stage of the project. At that time, the Area Planning Department was in the process of formulating a comprehensive county plan, nonexistent prior to this period. Vigo County has no zoning ordinances, but a subdivision ordinance is in effect.

The West Central Indiana Economic Development District Inc. (WCIEDD) serves a six-county region that includes the case study site. Although the major objective of this organization is to promote the economic growth of the region, the WCIEDD has prepared a District Land Use Element. The agency, however, has no authority to implement any of its recommendations. Therefore, the land use plans serve mainly as a source document for future planning activities.

A need has been identified in this area for improvements to the existing county and municipal parks and the acquisition of additional park land, particularly in and around Terre Haute. The per capita acres of outdoor recreational facilities for Clay County and Vigo County, 0.097 and 0.040, respectively, are considerably lower than the state average of 0.168 acres per capita.

Row crops accounted for 54% of the premining land use in the West Field. Pasture land and forests also occupied significant portions of the premining landscape. About 20% of the field has been disturbed by earlier mining operations and had been either reclaimed to pasture land or was still unreclaimed due to ongoing operations. Minor amounts of land were used for water impoundments, roads, and utilities.

Evaluation of Alternatives

The three alternatives developed for this case study were

- **Alternative 2A:** agricultural use
- **Alternative 2B:** low-density residential development
- **Alternative 2C:** residential development with integrated open space and recreational uses

These alternatives were subjected to the same evaluation procedure that was developed for the Appalachian case study.

Economic Evaluation

The economic evaluation attempts to estimate the resale value of the reclaimed land for a variety of alternative uses. For each alternative, the largest percentage of the total land value is derived from the land's usefulness in agricultural production. Alternative 2C has a slight advantage over both 2A and 2B because it makes better use of land having marginal agricultural activities. By using some of the residential land for multifamily dwellings, its value is increased. Also, private recreational and commercial uses occupy more land in this alternative than in the other two. Based on this evaluation procedure, it is estimated that Alternative 2C produces the highest potential total land value. However, the increase over the other two alternatives is rather small.

Environmental Evaluation

The environmental impact assessment matrix method was used to rank the alternatives according to the seriousness of the environmental impact that would be carried by each alternative. In reviewing the total environmental impact of the three land use plans, it can be concluded that Alternative 2A has the least impact because it differs the least from the premining uses. Although the impact of Alternative 2B and Alternative 2C varies slightly on certain points, the overall impact points of the plans are essentially identical. It should be noted that while Alternatives 2B and 2C do represent larger negative environmental impacts than Alternative 2A, impact is in no way disproportionate than any other development project and can be mitigated, in many instances, through proper design and construction methods.

Social Impact Evaluation

Alternative 2A would cause the least social impact for the same reason that it would cause the least environmental impact: the great similarity between this land use and the premining land uses. Alternative 2C would result in much of the same social impact as Alternative 2B. The magnitude of the impact would be increased slightly, however. Because this alternative would provide more housing units, it would more adequately meet the need for geographically suitable housing. It also would help alleviate the recognized shortage of rental housing. The premining study phase of the investigation has pointed to a deficiency in outdoor recreation facilities. Alternative 2C would provide a variety of recreational opportunities. Aesthetically, a properly designed residential community is often much more satisfactory than the haphazard development that is likely to occur without any planning. Alternative 2C is essentially the same as Alternative 2B in preserving prime agricultural land and in providing buffers between developed and rural areas. As Alternative 2B would have some negative impact on utilities, public facilities, and highways, Alternative 2C would have slightly larger negative impact. It appears that the magnitude and importance of the positive social impact generated by Alternative 2C would outweigh any negative impact.

Selected Alternative

Upon reviewing the three different analyses, it becomes apparent that Alternative 2B is the least desirable. The choice, then, is between Alternative 2A, which is similar to the premining land use, and Alternative 2C. Based strictly on the maximization of land value, Alternative 2C would be selected. Although this alternative could result in a slightly larger negative environmental impact than Alternative 2A, it would also provide the largest and most diverse social benefits by helping to meet several regional objectives. For these reasons, Alternative 2C is the recommended alternative.

It must be reemphasized that this decision was reached strictly from a land use potential perspective and was not influenced by regulatory requirements. Another point that requires clarification is the time frame for implementing the selected alternative. Because the field will be actively mined for more than 10 years, it is necessary to have an interim land use plan that productively utilizes the land until the entire area is mined. Such an interim plan would likely be similar to Alternative 2A, as the land could be used for agricultural production without preempting or diminishing its potential for later development.

REFERENCES

1. Ramani, R.V. and R.J. Sweigard. "Development of a Procedure for Land Use Potential Evaluation for Surface Mined Land," National Technical Information Service, PB-83-253401 (1983).
2. Sweigard, R.J. and R.V. Ramani. "Regional comparison of post mining land use practices," in *SME Preprint 83-105* (Littleton, CO: Society of Mining, Metallurgy and Exploration, 1983).
3. Ramani, R.V. and R.J. Sweigard. Impacts of land use planning on mineral resources, in *SME Preprint 82-418* (Littleton, CO: Society of Mining, Metallurgy and Exploration, 1982).

CHAPTER 12

Environmental Effects of Gold Heap-Leaching Operations

INTRODUCTION

Two of the issues raised in the Life-of-Mine Environmental Assessment (EA) relate to long-term conditions within reclaimed heaps. The amount of seepage generated and the concentration of metals and cyanide in seepage are of concern in establishing closure criteria for the facility. In addition, the fate of metals and cyanide in released solutions (e.g., attenuation, dissociation of complexes, etc.) is also an issue in the EA. These issues are appropriately addressed by an environmental risk assessment.

A risk assessment quantifies the perceived risk posed to defined "receptors" by a proposed action. A risk assessment normally consists of four distinct steps:

- Hazard identification
- Exposure assessment
- Toxicity assessment (receptor dose-response)
- Risk characterization

Hazard identification involves the characterization of potential contaminants, their relative mobility, and relative toxicity, and identifies target contaminants of concern. The exposure assessment consists of a conceptual model of contaminant fate and transport, and modeling of potential exposure for a specific event.[1] The toxicity assessment identifies the dosage of target contaminants at a point of exposure to potential receptors and compares this exposure to known toxicological information. Finally, the risk characterization summarizes the overall environmental risk.

A case study on the Landusky operation in Montana is described below.

HAZARD IDENTIFICATION

Several potential contaminants may exist within solutions retained within the heaps at closure. Metal-complexed (WAD) cyanide, copper, zinc, and arsenic are present in active heaps at levels that could affect environmental or human receptors if released to the environment at significant levels. In order to meet closure criteria the Montana Water Quality Bureau has determined that effluent from a heap must be less than 0.22 mg/L WAD cyanide. Data from the Landusky heap suggest that when 0.22 mg/L WAD cyanide concentrations are achieved, the levels of copper, zinc, arsenic, and nitrate will have decreased significantly. When cyanide levels have reached compliance (e.g., ≤0.22 mg/L WAD cyanide), then copper is only 4% of the drinking water standard, zinc 1% of the standard, and arsenic 4% of the drinking water standard. Metal concentrations follow the trend in cyanide levels because their primary form within the heap solution is as a cyanide-complex. Zinc, copper, and arsenic are readily attenuated during unsaturated or groundwater flow. Although cyanide has been shown to be strongly attenuated during unsaturated flow (less so during saturated flow), its attenuation is less than that for most metals. Hence, at closure cyanide is the constituent of greatest concern. If the environmental and human health risk of cyanide exposure is acceptable, then the inferred risk from other contaminants can also be assumed to be low.

EXPOSURE ASSESSMENT

In an exposure assessment, the concentration of target compounds within a "contaminant source" and the mobility of target contaminants are identified. Specific "migration routes" are defined depending on waste and environmental characteristics. Typically, one or more media (e.g., surface water, groundwater, air, or soil) may be affected by migrating contaminants. A schematic of the Landusky facility is shown in Figure 12.1, which indicates the primary source, migration routes, and potential receptors used in the risk assessment.

Solution contained within the rinsed and reclaimed heaps is considered to be the primary source of potential contaminants. Movement of solution containing concentrations of metals and cyanide may occur in response to infiltration events. A limited number of observations on existing reclaimed heaps suggests that only intermittent flow will occur from heaps after they are reclaimed. Water use by plants and surface runoff from reclaimed sideslopes will decrease infiltration into the heaps.

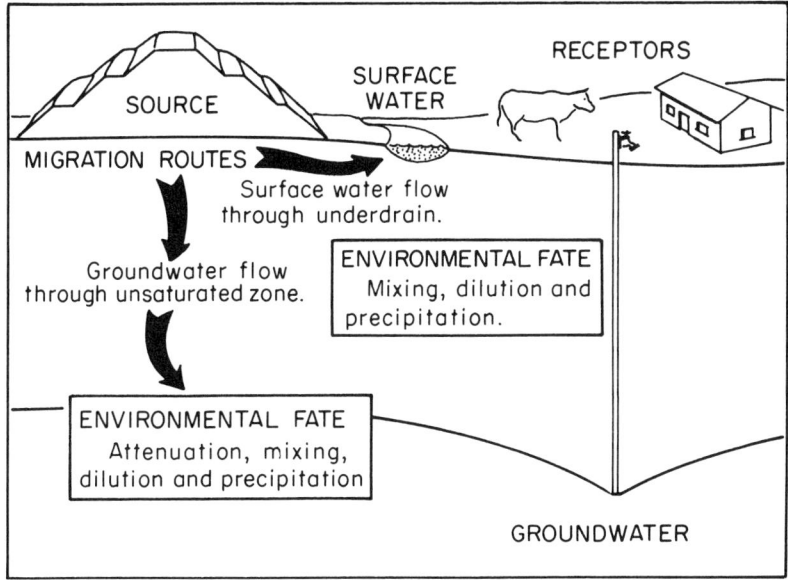

Figure 12.1. Conceptual model of migration routes and environmental fate of cyanide and metals in heaps after facility closure. (From Schafer, M.W. *Cyanide Degradation and Rinsing of Spent Heap Leach Gold Facilities* (Bozeman, MT: Schafer and Associates, 1990).)

Water in the form of rainfall or snowmelt which falls on heaps will either run off, be stored in the root zone, and subsequently transpired by plants, or will percolate through the heap. At closure the liner of all valley-leach pads will be perforated. Hence, percolating water will drain out of the bottom of the heap, potentially carrying low concentrations of residual levels of cyanide and metals.

Percolating water exiting the base of the heap will enter the finger drain system beneath the liner. Water within the finger drain system will flow out of the drainage system only if saturated conditions develop. Saturation will occur if the rate of water flowing into the drain is greater than the permeability of the foundation material below the drain. If saturated conditions do not occur, then solution will flow downward under unsaturated conditions until it encounters groundwater. Although a combination of surface water and groundwater flow will likely occur, two separate exposure assessments were performed, assuming all flow would go independently to groundwater or surface water. These individual exposure assessments represent "worst-case" conditions.

For the surface water exposure assessment, water percolating from the heap will exit the drain and then mix with surface water at the discharge point for the underdrain system. Potential receptors include all users of surface water and are thought to include humans, livestock, and wildlife using water for consumption. The point of compliance used for the surface water system was the confluence of Mill Gulch and Rock Creek. Actual points of beneficial use of surface and groundwater by humans are near this point, so this choice of a point of compliance is appropriate.

Because water movement within the heap is under unsaturated conditions, water may not enter the drain system (saturated conditions are required for gravel drains to accumulate water). If unsaturated conditions prevail during drainage, then percolation will continue downward until the groundwater system is reached. Percolating water will mix with existing groundwater within the local aquifer system. For purposes of the risk assessment, the contaminant concentrations at the point of beneficial use are of concern. Groundwater resources within the Mill Gulch/Rock Creek watershed upgradient of Landusky are not currently utilized for domestic purposes. The fractured bedrock system is not expected to be appropriate for utilization as a domestic water supply source in the future. As a result, the shallow alluvial groundwater system at the confluence of Mill Gulch and Rock Creek was used as the point of compliance.

The hydrologic event used for the risk assessment is the 100-year return interval maximum 24-hr rainfall. This event should provide an upper estimate of flow quantities and contaminant concentrations from the reclaimed heaps. The 100-year event was calculated to be 3.94 in. However, because two storms in the last 10 years have exceeded this event, a storm that dumped 6 in. of precipitation was used as the design storm, which exceeds the 100-year event by 50%.

The exposure assessment model consists of three distinct analyses: (1) generation of seepage and transport of contaminants from the heap, (2) calculation of the watershed hydrograph, and (3) identification of groundwater location, flow direction, and flow rate. A simple "mass balance" model was used to predict cyanide concentrations in surface water and groundwater. No attenuation or precipitation of cyanide and no dispersion within the groundwater system were assumed. These processes are significant for both the surface water (due to volatilization of HCN from surface water at pH of 6.5 to 7.0) and groundwater pathways (due to oxidation, volatilization, and adsorption) and will greatly reduce the magnitude of potential environmental risk.

An unsaturated flow model was used to predict water movement in the heap. The percolation rate into the heap was calculated using the SCS curve number method, which was used to calculate runoff from the heap sideslopes and heap surface. For the 6-in. designed storm event, the calculated weighted average runoff was 2.226 in., so 3.774 in. of infiltration occurred. The antecedent soil moisture conditions were assumed to equal field capacity, therefore, all infiltration into the heap percolated through the applied soil cover as well. The timing of the infiltration was determined by the difference between the rainfall hydrograph and the calculated soil infiltration rate.

PC Seep was initialized for a heap with an average thickness of 200 ft (8 to 25 ft lifts), which corresponds closely to the average thickness of the 1987 and 1991 heaps (195 ft average thickness for the 1991 pad). The pressure distribution used to initialize the model corresponded to an equilibrium outflow rate of 0.1 in./day, representative of fully drained conditions. The peak outflow (or leading edge of the wetting front) drained from the heap about 1.1 days after the peak infiltration. The maximum rate of outflow observed was about 0.14 in./day, while the peak infiltration rate was greater than 1 in./hr. The long-term rate of solution outflow declined to just <0.1 in./day. Based on the model results, the drainage from the heap would last for 20 to 30 days at this outflow rate.

The movement of solution within the heaps was explained by a "two-region" flow model. A region of rapid flow developed during leaching, while a region of slow flow also occurred. The "slow flow" region may be thought to correspond to the water-filled pore space after draindown (about 7.4% of the total volume). The mobile flow region is the additional water-filled pore space which exists during leaching (about 0.6% of the total volume).

A mass balance solute mixing and dispersion model was developed to simulate solute movement within a two-region flow system. Initially, both regions are thought to have equal cyanide concentrations. Once infiltration into the heap occurs, the mobile flow region must fill up with water before percolation can occur. This phenomenon leads to a lag period between infiltration and percolation out of the heap. Movement of solutes out of the mobile flow region was described by hydrodynamic dispersion. Dispersion explains the nature of solute movement in porous media. When solution with a low concentration of a particular constituent is added to a system with a higher concentration, the chemistry of the drainage will follow the curve for "ideal flow." Concentration changes gradually after passage of the approximately 1 pore volume of solution. For the two-region flow model an "effective pore volume" is defined as the pore space occupied by the mobile solution phase, or about 0.6%.

$$\frac{C}{C_o} = \frac{1}{2}\left[\text{erfc}\left(\frac{1-Vt}{\left(2\sqrt{D_l t}\right)}\right)\right] \quad (12.1)$$

where l   = distance traveled
       V   = average velocity
       t   = time of travel
       D_l = dispersion coefficient
       erfc = a function as described by Freeze and Cheery

The dispersion coefficient for unsaturated soils was estimated to be 0.2 to 2.0 times the effective velocity, or higher for gravelly soils. We used 100 times the effective velocity due to the extremely coarse texture of Landusky heap material (median grain size of 10 to 75 mm). Effective velocity was defined as the Darcy velocity divided by the effective pore volume (0.006).

Solutes within the immobile or slow flow region would tend to slowly mix with the rapid flow region, especially as the concentration difference between the two regions develops. A diffusion term was used in the numerical model of solute movement to account for mixing between the two regions, according to Equation 12.2:

$$C_{out} = C_{mobile} + (C_{heap} - C_{mobile})Jt \qquad (12.2)$$

where C_{out} = concentration in outflow from the heap
C_{heap} = average cyanide concentration in an immobile region
C_{mobile} = average concentration in mobile flow region
J = a diffusion term set by the user
t = elapsed time

The solute transport and solution flow models were validated by simulating the rinsing event on the Landusky 1986 heap to see if the models simulate observed cyanide concentrations measured in lysimeters during the field investigation. The model indicated that when initial levels of cyanide within the heap are 400 mg/L, initial leachate is also at 400 mg/L. Dispersion within the mobile solution zone causes a rapid decline in effluent levels of cyanide to about one half of the average concentration found in the heap (found by averaging concentrations in the mobile and immobile regions). With continued leaching, levels of both average and effluent cyanide continued to decline. At the end of a 10-day rinse period, effluent and average cyanide levels are 20 and 40 mg/L. These results are very similar to lysimeter nest 03 on the 1986 pad, which was leached and rinsed in a similar manner. The initially rapid decline in effluent WAD cyanide (after addition of 6 to 8 in. H2O) is dispersion-controlled within the mobile solution region, while the slower decline is controlled by the rate of solution mixing between the mobile and immobile regions.

If leaching was temporarily discontinued at any point, the cyanide concentrations in both the mobile and immobile regions equalize. When leaching is begun again, effluent levels exhibit a "spike." The two-region flow and solute transport model appears to explain the somewhat erratic decline in cyanide levels observed in heap effluent. The model predicts that the concentration level in "spikes" will average twice the typical concentration in the effluent measured during rinsing. As a result, if a heap is rinsed until effluent WAD cyanide reaches 0.22 mg/L, the average concentration in the heap could be as high as 0.44 mg/L.

The two-region solute model was run again for a 200-ft-high heap with an initial cyanide level of 0.44 mg/L. The average cyanide concentration predicted in the 3.77 in. of drainage is shown in Figure 12.2. Very little change in effluent cyanide level was predicted because the amount of water added by the 6-in. precipitation storm is much less than an effective pore volume (15.4 in.), which is the amount of water necessary to change the composition of effluent. On shallower heaps an effective

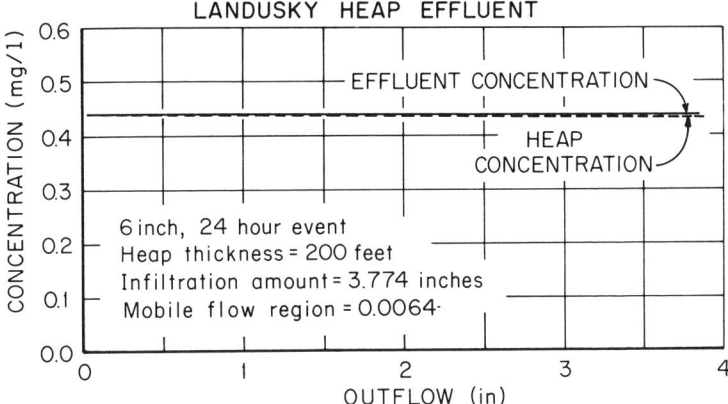

Figure 12.2. Predicted concentration of WAD cyanide in the heap and in effluent during drainage after the design 6-in. storm on a 200-ft heap.[1]

pore volume is only 2 in., so a small infiltration event can cause a change in effluent chemistry. The outflow cyanide concentration used for heap exposure assessment was 0.44 mg/L. The timing and rate of solution released from the heaps are shown in Figure 12.2.

Groundwater Pathway

Interpretation of the potential groundwater pathways in the Little Rocky Mountains requires a brief review of the physiology and geology of the area. The Little Rocky Mountains consist of a 15 × 30 mi region domed up about 3000 ft higher than the surrounding plains. The average annual precipitation is in the range of 15 to 20 in./year, favoring recharge to local aquifers.

The geology of the Little Rocky Mountains area was mapped. This work indicated the interior of the mountains is composed of Tertiary intrusives, primarily syvenite and trachyte porphyry and Precambrian metamorphics (gneiss and schist). A band of upturned Paleozoic and Mesozoic sedimentary rocks flanks the core and forms a more or less continuous band, except on the southern boundary where the strata have been disrupted by faulting. Generally flat-lying strata of the upper Cretaceous (Campanion and Maestrichtian) Bearpaw and Judith River formations surround the uplift.

The Madison Group discharges in the immediate vicinity of the Little Rocky Mountains, with several springs emanating from the Madison at various locations. Lower stratigraphic sedimentary units, notably the Cambrian-Devonian section, also apparently discharge in the area. The Judith River formation, exposed at the surface in the plains surrounding the mountains, is apparently recharged in this region, and it is possible that other Cretaceous units are similarly recharged. Locally, the groundwater regime is characterized by a three-component system consisting of a bedrock

Table 12.1. Pump Test Data Summary for the Landusky Mine Area

Well Identification	Drainage	Aquifer Type	Hydraulic Conductivity
136	Mill Gulch at Landusky	Bedrock (sandstone)	2.3×10^{-3} cm/sec
138	Mill Gulch at Landusky	Alluvium	1.69×10^{-3} cm/sec
139	King Creek	Bedrock (granitic)	2.24×10^{-3} cm/sec
141	Bull Creek	Bedrock (shale)	2.26×10^{-6} cm/sec
141A	Bull Creek	Alluvium	8.26×10^{-4} cm/sec
142	Ruby Creek	Bedrock (shale)	1.78×10^{-5} cm/sec
147	Goslin Creek	Alluvium	1.63×10^{-3} cm/sec
148	Goslin Creek	Bedrock (shale)	5.9×10^{-6} cm/sec
152	Goslin Creek	Bedrock (shale)	9.9×10^{-6} cm/sec

Source: Hydrometrics Inc.

fracture flow system in the core area, shallow alluvial systems leading away from the core, and flow in the upturned sedimentary rocks surrounding the core. Limited pump test data from the site suggest that the hydraulic conductivity (K_s) of alluvial materials ranges from 1.6×10^{-3} to 8.3×10^{-4} cm/sec. A single well tested in the hardrock core material ("granitic") had an apparent K_s of 2.2×10^{-3} cm/sec. The remaining pump tests were conducted in sedimentary units, with K_s values ranging from 2.3×10^{-3} cm/sec for a sandstone unit in Mill Gulch to 2.3×10^{-6} cm/sec for a shale unit in the Bull Creek drainage (Table 12.1).

Although no local potentiometric surface maps are available, the data available indicate that the hydraulic gradient in the shallow alluvial aquifers closely parallels the topographic gradient of the valley bottom. The data also suggest the fracture flow system in the core area is discontinuous (very steep apparent gradients in some areas). The relationship of the shallow alluvial aquifers to the local Paleozoic and Mesozoic strata is uncertain. It is apparent that in certain specific locations, the Madison Group discharges to overlying alluvium or to the surface in the form of springs; however, the quantity of flow is not known. For the purpose of evaluating the fate of cyanide from the heaps, discharge of the sedimentary aquifers to the shallow alluvium has been ignored because data are inadequate to quantify these flows. The effect of ignoring the recharge of shallow alluvium from the Madison is to overestimate actual cyanide levels in groundwater because further dilution is neglected. The existence of artesian heads in underlying sedimentary rocks makes significant infiltration into these units unlikely.

In summary, flow within the Mill Gulch/Rock Creek groundwater system is as follows. Recharge to the fractured bedrock system occurs at higher elevations in the Little Rocky Mountains. This system discharges into the shallow alluvial systems which are also recharged through runoff infiltration. The alluvial systems extend from the mountains onto the adjacent plains, where they are underlain by the Bearpaw Shale or, to a lesser extent, the Judith River formation. After exiting the mountains, most of the surface water systems are ephemeral, with flow occurring only during snowmelt or large storm events. Consequently, during typical stream

Table 12.2. Calculated WAD Cyanide Concentration in Shallow Alluvium After a 6-in. Storm Event Received Immediately After Facility Closure[1]

Cyanide Conc. in Shallow Alluvium			
Watershed Conc.		**Heap Contribution**	
Mill Gulch area = 579 ac		Heap area = 170 ac	
Heap area = <92> ac		Infiltration = 3.77 in..	
Total area = 487 ac		= 53 ac-ft	
Infiltration = 3.08 in.			
= 125 ac-ft			
W. Fork Rock Creek = 212 ac			
Heap area = <78> ac			
Total area = 134 ac			
Infiltration = 3.24 in.			
= 36 ac-ft			
Rock Creek area = 734 ac			
Infiltration = 3.48 in.			
= 213 ac-ft			
Total infiltration = 374 ac-ft			

Cyanide Mass Balance
$[(374*0) + (53*0.44)]/(374+53) = 0.055$
Average WAD cyanide
in groundwater = 0.055 mg/L

flow, surface water will tend to infiltrate into the shallow groundwater system. When the alluvial groundwater is recharged by an infiltration event, groundwater levels may rise sufficiently that groundwater discharges to surface water at least temporarily until the groundwater system again subsides. Water in the alluvial system will remain in the aquifer until depleted by evapotranspiration, incorporated into the alluvial system of a larger drainage, is lost to infiltration to deeper aquifers, or is returned to surface flow at the termination of the alluvial system.

The concentrations of cyanide within the shallow alluvial system (Table 12.2) are determined by the quantity of water received from the heaps (containing 0.44 mg/L WAD cyanide) and the remainder of the watershed (containing WAD below detection). The quantity of water entering the shallow groundwater system is the infiltration from the design storm event minus the infiltration which is stored in the soil (soil water deficit assumed to equal 1 in. in watershed and 0 in. in heap). The calculated WAD cyanide level is 0.055 mg/L. This assessment is extremely conservative, as no groundwater mixing or attenuation of cyanide was considered in the assessment. As a result, 0.55 mg/L is a maximum expected level in response to a storm in excess of the 100-year storm.

A similar mass balance assessment could be completed for the groundwater system immediately below the heap. The fractured bedrock groundwater system lies below the heap-leach pad. The porosity of the fractured system is assumed to be 10%, while the aquifer depth is thought to be confined in the upper weathered zone (6 ft deep). After mixing percolation from the heap with existing groundwater, the WAD cyanide concentration immediately below the heap would be the weighted average

Table 12.3. Analysis of Groundwater-Surface Water Interaction Following a 6-in. Design Event in the Mill Gulch/Rock Creek Watershed[1]

Groundwater Flow in Shallow Alluvium and Groundwater Discharge to Surface Water	
Total groundwater flux	= 427 ac-ft
Alluvium cross-section area	= 30 × 100 ft
Hydraulic conductivity	= 5.67 ft/day
Gradient	= 0.04 ft/ft
Maximum groundwater flow	
$Q = AK(dh/dx)$	= (30)(100)(5.67)(0.04)
	= 680 ft^3/day

Groundwater flow would continue for 75 years at this rate of flow. Hence, discharge of shallow alluvium to surface water is evidently occurring. The average surface water discharge expected over a 20-day base-flow period is

Surface water discharge (cfs)	= Volume/Time
	= (427 ac-ft)/(20 days)
	= 10.76 cfs

of the heap effluent (3.774 in.) and the groundwater below the heap (6 ft at 10% porosity = 7.2 inches) [3.774"/(7.2" + 3.774")]*0.44 mg/L = 0.15 mg/L.

The quantity of groundwater which is expected to discharge through the shallow alluvial system is 427 ac-ft (Table 12.2, 374 + 53 ac-ft). The groundwater discharge following a large event typically lasts roughly 10 days for small watersheds (SCS-National Engineering Handbook, Section 4, Hydrology; SCS-NEH4). The characteristics of the shallow alluvial system suggest that it does not have the capacity to accommodate this quantity of flow (Table 12.3). Discharge of the groundwater system to surface water is expected until the excess groundwater moves out of the system. The calculated amount of surface flow due to groundwater discharge would average 10.76 cfs over a 20-day period. This long-term flow represents the surface water "base flow" contributed by groundwater.

Surface Water Pathway

The upper reaches of Rock Creek, including Mill Gulch, are typical of the postmine landscape after facility closure and contain reclaimed heap-leach pads and other reclaimed mine areas, as well as undisturbed areas. Mill Gulch is a tributary to Rock Creek, which is a tributary of the Missouri River, located 22 miles southeast of the Landusky townsite. The Rock Creek drainage was selected to perform runoff calculations in order to determine typical runoff timing and volume from postmining reclamation topography.

The upper Rock Creek drainage area is generally steeply sloping with moderate to heavy cover of lodgepole pine in undisturbed areas and reseeded native grasses in

Table 12.4. Surface Water Flow Measurements for Selected Sites in the Rock Creek Watershed Area[1]

Date	Surface Water Monitoring Station (flow in gpm[a])			
	L-4	L-7	L-22	L-23
6/85	Dry	2–3	—	—
10/85	3100	450	30	—
6/86	500	2–3	10	250
10/86	Dry	75	30	150–200
6/87	100	2–3	5	600–900
10/87	Dry	15–20	—	100–150
5/88	1500	2–3	5	900
10/88	Dry	Dry	2–3	Dry

[a] Flow in cfs = gpm*0.002228.

the reclaimed areas, tending to eventual reforestation. Maximum topographic relief is about 1500 ft (4000 ft near the confluence of Mill Gulch and Rock Creek to 5500 ft at the upper reaches). Average channel slopes are 5 to 15% with sideslopes ranging from 10 to 55%.

A soil survey completed in 1985 shows soils in the Mill Gulch area to be mixed typic cryochrepts, ranging from coarse loamy to loamy skeletal. Soils in the Rock Creek drainage area include both mixed typic cryochrepts and mixed typic cryoborolls (loamy, coarse loamy, and loamy skeletal).

Average rainfall is 15 to 20 in./year with a predicted 100-year, 24-hr precipitation event of 3.94. Storm events producing approximately 6.5 in. (May 1988) and 9 to 13.25 in. (September 1986) were recorded by Zortman Mining, and within Landusky townsite. A 6-in. design storm was used in the runoff calculations.

Surface flows in the extreme upper reaches of the drainage area are ephemeral, with the exception of several small seeps or springs. Flows within the lower channel reaches are seasonal. Generally, all runoff is in response to spring snowmelt or intense rainfall events. Recorded surface water flows (Table 12.4) show a large variation in flows from spring to autumn and from year to year. In addition, stream flow data indicate surface flows above the confluence of Mill Gulch and Rock Creek appear to be a losing reach near the confluence and a gaining reach immediately downstream of the confluence near Landusky.

A runoff model was developed to simulate the runoff from a 6-in., 24-hr precipitation event. The model was based on the SCS-NEH4. The SCS method finds a watershed flow hydrograph using the curve number method. A complete description of the background, methods, and procedures is given in NEH4. A brief description of the modeling system is provided below.

The SCS curve number method was developed for areas having little rainfall data, particularly for storm duration and intensity. Runoff does not begin until after some period of "initial abstraction" (Ia) where interception, infiltration, and surface stor-

Table 12.5. Hydrologic Input Parameters Used for Simulating Surface Water Hydrographs for a 6-in. 24-hr Storm Using SCS Type II Rainfall Distribution Typical of a Large Convective Thunderstorm[1]

Watershed Name	Area (acres)	Weighted Curve No.	Av. Slope (%)	Channel Length (ft)	Runoff (in.)	Peak Flow (cfs)	Time of Peak (hr)
Surface water hydrograph[a]							
Mill Gulch	579	60	12.1	10,540	1.922	341	13.5
Rock Creek	734	55	10.1	13,450	1.518	233	14.1
West Fork Rock Creek	212	58	17.9	6,020	1.768	182	12.9
Heap infiltration hydrograph[b]							
Heap side-slope	40[c]	66.2	15.7	2,382	2.458	64	12.4
Heap top	40[c]	56	2.0	1,593	1.596	21	13.1

[a] Simulation of surface water hydrograph for a design storm event.
[b] Simulation of runoff and infiltration on heap to compute Inflow hydrograph for input to PC Seep.
[c] Typical hydrograph for 40-acre heap was computed. Inflow volumes were adjusted for actual areas of heap sideslopes (170 acres * 75% = 127.5 acres) and heap surfaces (42.5 acres).

Table 12.6. Boundary Conditions Used for the SCS Flow Hydrograph Model[1]

- The Rock Creek/Mill Gulch drainage area containing Sullivan Park and other leach pad areas will be typical of postmining topography within the permit boundary
- A 6-in., 24-hr design storm to simulate a 100-year event; this is larger than the predicted NOAA (1973) 100-year event, and within the range of recent major events
- Soils within the drainage classify as SCS type "B" soil group; the soils contain average in situ antecedent moisture
- Fair to good soil cover, majority being lodgepole pine and native grasses and shrubs on undisturbed areas, and fair cover of reseeded native grasses on reclaimed areas
- Curve numbers of 50 for level areas (top) of reclaimed heaps, 66 for sideslopes of heaps and all other reclaimed areas, and 55–58 for undisturbed areas within the drainage
- Of the surface area of the heaps 23% are level areas and 73% are sideslopes
- The influence of the West Fork of Rock Creek on the total hydrograph (confluence of Mill Gulch and Rock Creek) was delayed about 15 min (based on Mannings equation) to allow the runoff time to travel from the mouth of West Fork to the confluence area
- The total hydrograph at the Rock Creek/Mill Gulch confluence is the summation of the individual watershed hydrographs with a coincidental time sequence

age occur. The Ia is estimated to be 20% of the maximum potential runoff. Rainfall-runoff relations, based on SCS curve numbers, are then developed to estimate the runoff volume and timing from precipitation events.

Curve numbers are selected based on land use, soil type, cover, hydrologic condition, and antecedent moisture. Other necessary information includes average slope, drainage area, and longest runoff length from maps, areas of disturbance/reclamation from end-of-mine plans, and rainfall distributed as an SCS Type II convective thunderstorm event. Lag time, time to concentration, time to peak, etc.,

Table 12.7. Calculated WAD Cyanide Level in Surface
Water After a 6-inch Design Storm Event[1]

Cyanide Conc. in Surface Water	
Surface water base flow	= 10.76 cfs
Heap flow rate	= 0.08 to 0.14 in./day
	= 0.57 to 1.00 cfs
WAD cyanide level (mg/L)	= [1.00/(10.76 + 1)]*0.44
	= 0.037 mg/L

are calculated from the curve numbers. A series of elemental hydrographs, based on peak flows and the values of the dimensionless unit hydrograph (SCS), are developed for each duration. The elemental hydrographs are summed to produce a runoff hydrograph for each drainage area. The hydrographs for each drainage area are then totaled to determine the runoff at the confluence of Mill Gulch and Rock Creek. Table 12.5 shows input parameters for the simulated hydrographs. Table 12.6 shows the boundary conditions used for the flow hydrograph model.

Weighted curve numbers are used for the drainages that contain more than one soil cover complex, such as reclaimed areas within undisturbed forest areas. These curve numbers are arrived at by "weighting" the curve numbers for each soil cover complex within the drainage area to get a watershed-wide curve number. This takes into account the different runoff rates from differing soil-cover complexes within the drainage area.

The peak runoff for Mill Gulch (341 cfs) occurred 13.5 hr after the beginning of the storm event. The peak rainfall intensity was timed at 12 hr after the beginning of the storm. Peak runoff for Rock Creek (233 cfs) occurred at 14 hr, and for West Fork of Rock Creek (182 cfs) at about 13.25 hr. The total hydrograph for the Rock Creek/Mill Gulch drainage area was obtained for summing the three individual hydrographs. The West Fork of Rock Creek hydrograph was delayed by 15 min to allow for flow from its mouth to the confluence of Mill Gulch and Rock Creek. The peak flow was 724 cfs, which occurred 13.5 hr after the beginning of the event.

The total runoff for the drainage area was 217 ac-ft with 92.7 ac-ft (1.92 in.) from Mill Gulch, 92.9 ac-ft (1.52 in.) from Rock Creek (main), and 31.2 ac-ft (1.77 in.) from West Fork of Rock Creek.

The SCS storm runoff hydrograph does not take into account the interaction of the surface water with the groundwater, i.e., the groundwater contribution to the base streamflow (surface) following a major storm. The groundwater will recharge the base streamflow at about 11 cfs (Table 12.3) for a period of time (20 days) after all surface flow caused by direct runoff has ceased.

The timing of solution discharge from the heaps is delayed after the runoff hydrograph has peaked. As a result, the heap effluent collected by the underdrain system mixes with the stream baseflow contributed by the groundwater system. The quantity of baseflow was calculated to be 10.76 cfs for a 20-day period. The resulting WAD cyanide level in surface water by the surface water exposure route is 0.037 mg/L (Table 12.7).

Table 12.8. Summary of Pertinent Water Quality Criteria, and Nonlethal and Lethal Cyanide Dose-Response Characteristics for Specific Receptors[1]

Standard	Conc.[a] (mg/L)	Duration	Frequency
Water Quality Standards			
Freshwater-chronic	0.0052	4 days	1/3 years
Freshwater-acute	0.022	1 hr	1/3 years
Drinking water advisory	0.200	—	—

Receptor	Effect	Dosage	Equivalent[a] Conc.
Chronic Dose-Response Data			
Humans	Nontoxic dose	10 mg/day	5 mg/L[b]
	None observed	5 mg/day	2.5 mg/L[b]
	Noninjurious	<10% of drinking water standard	2.0 mg/L
Chickens	Nontoxic	6 mg/kg	34 mg/L[c]
Acute Dose-Response			
Humans	LD_{50}	1–3 mg/kg	35–105 mg/L[b]
Mice	LD_{50}	3 mg/kg	105 mg/L[d]
Birds	LD_{50}	0.1 mg/kg	3.5 mg/L[d]
Cattle	LD_{50}	0.9 mg/kg/hr	315 mg/L[d]
Chickens	LD_{50}	21 mg/kg	120 mg/L[c]
Bluegill	LD_{50}	—	0.126 mg/L
Perch	LD_{50}	—	0.107 mg/L

[a] Concentrations are based on free cyanide levels. The EPA suggests the use of total cyanide, however, due to the lack of an acceptable method for free cyanide determination.
[b] Based on 2 L of water consumption for every 50 to 70 kg in adults.
[c] Based on a 2-kg adult weight and 0.35 L water consumption.
[d] Assuming an equivalent water consumption to that in adult humans.
[e] Assuming water intake equivalent to humans over a 10-hr waking period and a 250-kg adult weight.

TOXICITY ASSESSMENT

Toxicological data from the literature have been summarized indicating the dose-response characteristics of humans, cattle, and other receptors to cyanide (Table 12.8).

Cyanide can be acutely toxic (lethal after a short-term exposure) to most organisms. An important characteristic of the dose-response behavior of these organisms is that it is difficult to identify the effects of chronic exposure. For example, cyanide does not accumulate in the human body; it is regularly excreted by the urine. For this reason, the dose-response data for chronic effects are uncertain and are therefore conservative.

ENVIRONMENTAL EFFECTS OF GOLD HEAP-LEACHING   467

The acute dose-response values given in Table 12.8 range from 0.1 mg/kg of body weight for birds to 21 mg/kg for chickens. A value of 1 to 3 mg/kg of body weight has been found for humans. The value of 0.1 mg/kg (3.5 mg/L in water) appears to be too low, as experience with migratory waterfowl in Nevada indicates that birds survive in tailings ponds with solution levels up to 30 mg/L.

Consistent data on dose-response characteristics for chronic cyanide exposure are difficult to obtain. Table 12.8 lists typical values obtained from the literature. In a 1985 *Federal Register* the Environmental Protection Agency (EPA) reported that the World Health Organization (WHO) guideline for drinking water is based on consumption of 4.7 mg/day of cyanide, which is recorded as having no effect. Assuming consumption of 2 L of water per day, WHO concluded that cyanide concentrations of 2.35 mg/L could be consumed in water. The WHO allowed for an additional safety factor in adopting their drinking water standard. The values given for human dose-response are similar to the WHO values.

RISK CHARACTERIZATION

The exposure assessment model indicated that WAD cyanide levels would be 0.037 mg/L for surface water and 0.055 mg/L for groundwater. These concentrations would only occur as a result of a hydrologic event, with a very low probability of occurrence. In addition, each of the assumptions made within the exposure assessment are conservative; hence, the WAD cyanide concentrations predicted represent maximum predicted levels from a worst-case event. Actual levels are likely to be much lower than those predicted. Some of the assumptions used for the exposure assessment are

- No cyanide attenuation was considered; however, WAD cyanide attenuation has been measured in these materials.
- No volatilization of cyanide in surface water was included, though at the pH of area runoff, considerable volatilization would be expected.
- No degradation of cyanide; however, long-term cyanide degradation rates have been predicted using the formula $C = 0.00124\, Co e^{-0.203T}$, for a cyanide half-life of 3.4 years.
- The event was modeled to occur immediately after rinsing to 0.22 mg/L WAD cyanide in effluent. As time after closure elapsed, the WAD cyanide levels would decline substantially.
- All heaps were assumed to be decommissioned simultaneously, though the Mill Gulch heap will likely be rinsed 3 to 5 years before the Sullivan Park heap, resulting in lower cyanide levels for at least one half of the area covered by heaps.
- The event modeled was 50% in excess of the 100-year storm.
- No soil moisture deficit was assumed to be present within the heap, while 1 in. of soil water deficit was assumed for the surrounding watershed. The heap would be expected to have a higher soil water deficit than the watershed because of the dominance of grass vegetation on the heaps (after closure).

The toxicity values summarized in Table 12.8 clearly show that these concentrations do not harm human or other environmental receptors, even if they are exposed for an extended period.

The nearest sustainable fishery downstream of this site is at the Missouri River, over 20 mi to the south. These minute cyanide concentrations would be diluted to below detection after a travel distance of 20 mi. Therefore, no impact to fisheries can occur from runoff or groundwater discharges containing low levels of WAD cyanide.

It is therefore concluded that using a conservative exposure model for cyanide, and considering surface and groundwater as two independent media, full exposure to human and other receptors does not lead to any risk of impact to human health or the environment.

Decommissioning of Heap-Leach Facilities — Industry Experience

Only a few heap-leach facilities have been completely rinsed and decommissioned to date. Those involved in a number of heap-leach projects are in the process of developing closure methods and designs and, therefore, have been running extensive column rinse studies, as well as pilot scale tests. Examples of natural degradation, freshwater rinsing, and rinsing with treated barren solutions were found. Water rinse has been used for cyanide removal in five mines: Carson Hill, California; Roundmountain, Nevada; Maggie Creek, Nevada; Standard Hill, California; and Illapah, Nevada.

This section reviews some of the experiences at a number of projects in which rinsing of heap-leach facilities is proposed or has been completed on full- or pilot-scale levels.

Borealis Mine, Echo Bay Minerals Company, Hawthorne, NV

The Borealis Mine decommissioned their first run-of-mine heap. The pad area was 345,000 ft and contained an estimated 800,000 tons of ore stacked to a height of about 120 ft at its zenith. Cyanidation started in August 1985 and was suspended in August 1987. The pad remained idle until June 1989, at which time the decommissioning program was started. This rest period allowed for the natural degradation of cyanide. Freshwater rinsing began in June 1989 and continued for several months. The pad was rinsed in sections of about 20,000 ft^2 at a flow rate of about 300 gpm (i.e., an application rate of about 0.005 gpm/ft^2). The water was applied with Rainbird sprinklers.

Each section of the pad was rinsed for approximately 10 to 20 days, or until WAD cyanide levels were below 0.2 mg/L. The effluent was monitored on a daily basis for WAD cyanide and pH. Effluent pH levels for 5 months following the rinsing period show that the pH remained below 8.5 for most of the time throughout the monitoring period. Over the same period, the cyanide levels remained at <0.2 mg/L. Solid samples removed from three large backhoe test pits had cyanide levels far below the 10 mg/kg limit set by NDEP for agglomerated ores.

A pad closure plan was submitted to the NDEP by the Borealis Mine on March 26, 1990, for their remaining heaps. The mine will be terminating their gold mining operations and is, therefore, actively looking for an approach to neutralize the leach pads.

The general requirements for leach pad closure in Nevada are

- WAD cyanide levels of the effluent rinse water must be <0.2 mg/L
- The pH level of any effluent rinse water must be between 6.0 and 9.0
- Contaminants in any effluent from the process ore that result from meteoric events must not degrade state waters

Variances can be granted if these conditions cannot be reached. In such cases, the owner must demonstrate by representative sampling of the spent leach ore that:

- It will not contain levels of contaminants that are likely to become mobile and degrade the waters of the state under the conditions that exist at the site
- The spent leach material is stabilized in such a fashion as to inhibit waters from meteoric events from migrating through the material transporting contaminants with the potential of degrading state waters

The initial permit for the Borealis leach facilities states that spent heaps will be rinsed with freshwater until solution return has a pH value of <8.5 for 3 consecutive days. The rinse water will be contained in lined ponds for evaporation or reuse in another leach circuit. After the Borealis permit was granted, the State of Nevada revised their rinsing guidance document (dated October 17, 1986). New guidelines provide more alternatives for the acceptable closure of agglomerated heaps by rinsing. Echo Bay Mines has indicated its concern that the existing closure criteria may not be achievable and has, therefore, proposed alternate closure criteria.

The objectives of the alternate heap leach and pond closure program will be to utilize the known natural degradation processes of cyanide in conjunction with an initial rinsing to accelerate the degradation of contained cyanide to nontoxic levels prior to regrading and reclaiming the heaps. The strategy recommended for the Borealis heaps includes:

- Use column tests to select the most appropriate rinsing technique to satisfy the criteria above
- Rinse the entire surface of the heap until the pH of the rinsate is <8.5 and the concentration of WAD cyanide is reduced by 50% from the initial rinsate concentrations
- Rinse edges of the heaps that will be regraded until the rinsate meets NDEP standards
- Evaporate excess rinsate
- Fill ditches and ponds with gravel
- Topsoil and revegetate heaps

In developing a preferred treatment process for rinse waters, Echo Bay Mines evaluated:

- Freshwater rinsing
- Ferrous sulfate addition
- Alkaline chlorination
- Copper-catalyzed hydrogen peroxide

In laboratory column rinse studies, freshwater was used to rinse the ore, but was not recirculated. This leaching cycle is typical of an operational mine in which return water from rinsed pads would be used for makeup. It was found that the free cyanide levels in crushed ore from the initial heaps could be reduced from about 1.2 to 3.7 mg/L to <0.2 mg/L in 2 to 10 days. In a full-scale sequential rinse conducted by Echo Bay on one of their heaps, the effluent WAD cyanide levels were reduced to <0.20 mg/L and the entire pad was detoxified in about 60 days.

Preliminary laboratory studies conducted for Borealis indicate that hydrogen peroxide is the preferred treatment process for the detoxification of the heaps. Rinse water collected from columns was treated with hydrogen peroxide and was recirculated into the columns. Tests performed by the Degussa Corporation showed that barren water containing 37 mg/L WAD cyanide was successfully treated to 0.1 mg/L WAD cyanide. Borealis decided to design a mobile treatment plant of 500 to 1500 gpm treatment capacity that can be moved from heap to heap.

Alkaline chlorination was examined in a series of bench scale experiments using calcium hypochlorite. The results indicated that hypochlorite could reduce the initial free cyanide from 1.2 mg/L to <0.2 mg/L in about 2 to 3 days for one ore and from 3.6 mg/L to <0.2 mg/L in about 5 days for another. A cost comparison indicated that calcium hypochlorite would cost from 1.2 to 2.8 times that of hydrogen peroxide for treating spent ore.

Ferrous sulfate additions were not successful in lowering the WAD cyanide levels sufficiently. The poor treatment performance was directly related to the form of cyanide in the parent solutions. The solutions contained elevated levels of copper due to the copper content of the ores. The copper forms stable complexes with cyanide, which is not degraded by the addition of ferrous sulfate.

Barrick Goldstrike Mines, Inc., Carlin, NV

On August 15, 1990, Barrick Goldstrike Mines, Inc. submitted a report on the decommissioning of their first heap leach, Post Pad No. 1. This pad was leached between autumn 1986 and July 1989. At that time, Barrick began rinsing the pad by adding hydrogen peroxide to the barren solution pond. This solution was recirculated to the heap. Additional rinsing of the pad using sodium hypochlorite was conducted in August and September of 1989. From December 1989 through the end of 1990, Barrick added additional hydrogen peroxide to the circulating rinse solution on a continuous basis. The reason that so much treatment was conducted is because the

Table 12.9. Fondaway Mine Spent Ore Rinse Tests (Samples Taken After 50 gal/ton Applied)[1]

	Spent Ore Solution Water Treatment Method				
Parameter	Raw Barren	Water Rinse	Cl_2	H_2O_2	Well #3 Water
pH	10.5	7.9	7.2	7.8	6.7
Electrical conductivity	31.0	7.2	9.5	8.8	4.94
F (mg/L)	4.25	2.5	2.2	3.1	2.15
NO_3-N (mg/L)	36.5	4.8	6.5	7.7	1.0
Free Cl^2 (mg/L)	—	0.07	74	0.10	—
Chloride (mg/L)	—	1910	1930	2040	656
Free cyanide (mg/L)	497	1.15	0.067	1.12	0.001
WAD cyanide (mg/L)	478	1.08	0.012	1.20	0.011
Total cyanide (mg/L)	690	2.19	0.03	1.62	0.015
SO_4 (mg/L)	6987	22.0	20.7	22.3	1330
Arsenic (mg/L)	3.64	0.538	0.287	0.332	0.032
Cadmium (mg/L)	<0.005	<0.005	<0.005	<0.005	<0.005
Copper (mg/L)	0.223	0.196	0.026	0.207	0.116
Lead (mg/L)	0.252	0.120	0.15	0.138	0.12
Mercury (mg/L)	0.003	0.0013	0.0082	0.0015	0.001
Magnesium (mg/L)	3.2	46.0	55.6	31.2	112
Selenium (mg/L)	0.10	<0.005	<0.005	<0.005	<0.005
Zinc (mg/L)	260	0.290	0.677	0.48	1.70

WAD cyanide analyses remained higher than the 0.2 mg/L values throughout the period. After careful evaluation of the test results and initiating a multiple-laboratory quality assessment (QA) evaluation, it was found that the laboratory that performed the original WAD analyses tested for something closer to total cyanide than WAD cyanide. The analytical laboratory used a UV-pretreatment step before distillation, which degraded some of the strongly complexed cyanides such as ferrocyanide and cobalt-cyanide complex. It was subsequently found that at the end of the rinsing, the WAD cyanide levels were between 0.1 and 0.7 mg/L. Because of the relatively long rinsing time, natural degradation of cyanide could have played an important role in lowering WAD cyanide levels.

The final closure plan that was proposed included collection of all the effluent and analyzing it until the amount of effluent discharge was <1 gpm, at which point it is released to the environment. Furthermore, the heap is surrounded and covered by 300 ft of waste rock. The waste rock surfaces will be sloped in such a way that further infiltration of meteoric waters is limited.

Fondaway Mine, Tenneco Minerals

The Fondaway Mine started leaching in early 1990, and due to poor economic conditions, has closed temporarily. A temporary closure plan was submitted by Fondaway Mine to the Nevada DEP in July 1990. This closure plan involves rinsing the heaps with treated pregnant solution or with freshwater makeup. At the time that the plan was submitted, column tests were underway to evaluate the rinsing methodologies. Table 12.9 gives the results from the column test reported at that time. It is expected that natural degradation of cyanide did not play a significant role in the column testing. Unfortunately, no details are presented on the rinsing behavior or the types of ores that have been rinsed at any of these mines.

Gilt Edge Mine, Deadwood, SD

Brohm Mining operates a reusable pad heap-leach operation in the Black Hills of South Dakota. The process cycle consists of leaching followed by neutralization. After leaching with cyanide, a heap is rinsed and the cyanide is detoxified using hydrogen peroxide, with copper sulfate as a catalyst for the reaction. In deciding on the offload scheduling, discrepancies were observed between the WAD cyanide in the effluent and that found in pore-water sampling, with effluent giving higher values than the pore water samples. A study was performed to investigate these discrepancies. The study program included size fraction analyses and WAD cyanide analyses for various sections of the spent ore from the actual leach pad. Similar test work was also done on samples from four 29-ft-high laboratory columns.

Leach solution and neutralization solution were applied continuously at a rate of about 0.004 gpm/ft^2 with no rest (draindown) period provided. WAD cyanide was determined in effluent as well as in solution extracted from spent ore through bottle-roll tests. The results indicated that under the continuous rinsing scenario, an amount of rinsate that is approximately equal to the amount of ore (expressed in tons of rinsate per ton of ore) is necessary to reduce the WAD cyanide level from 200 to 0.5 mg/L.

It was concluded that solid sampling is more indicative of heap neutralization than effluent sampling at Brohm. The differences observed between the effluent WAD cyanide and the solid sampling were ascribed to the presence of a layer of finer bedding material placed at the bottom of the heap to protect the liner. The mechanism is described as follows. The bottom layer of ore, or in the case of the columns, the geotextile filter, is continuously saturated with high concentrations of cyanide-bearing solution. It appears that fines collect in this bottom layer. Having continuous contact with the cyanide solution, the fines may indeed adsorb cyanide on their surface area. The cyanide slowly diffuses from this saturated zone, never being thoroughly flushed. The cyanide-enriched layer causes elevated cyanide levels in the effluent which are not indicative of the pore water in the neutralized spent ore above this layer. The enhanced cyanide levels during draindown appear to be coming from these cyanide-enriched fines. The cyanide is in the form of WAD copper-cyanide complex.

Annie Creek Mine, Lead, SD

The Annie Creek Mine was not initially designed as a reusable pad system; however, as the project developed, it became necessary to unload some of the heaps and dispose of the spent ore in nearby Ross Valley. In South Dakota, spent ore is considered suitable for offloading when either the effluent or the pore water from the heap meets the following designated treatment standards:

- Effluent concentrations of <0.5 mg/L WAD CN
- pH of 6.5 to 8.5
- For all other parameters, South Dakota drinking water standards, surface water quality standards, groundwater quality standards, or the ambient concentrations of waters potentially affected, whichever is more appropriate based on a pathway and fate analysis
- Other parameters of concern, as determined by the Department of Water and Natural Resources (DWNR), which are not specified in the South Dakota drinking water standards, or water quality standards must not exceed ambient concentrations as determined by baseline monitoring for potential receiving ground or surface waters

The Wharf heaps are on the order of 0.75 to 1 million tons each, and they have not experienced problems neutralizing cyanide in a timely manner. The heaps undergo a 10- to 12-month leach cycle, followed by a 60-day neutralization cycle. Treating the heaps to the required offload criteria appears to be more a function of neutralization solution volume than contact time with the spent ore. It is noted that a 2.5-pore volume rinse with neutralization solution is generally required to achieve the treatment criteria in Wharf's heaps. The neutralization solution that Wharf is using is barren solution treated by hydrogen peroxide. This approach does not allow for significant natural degradation of cyanide in the heaps and, therefore, requires the high volumes of rinsate.

In some typical leachate qualities over a period of 6 months for the Wharf spent ore dump, WAD cyanide levels varied from 0.07 to 1.03 mg/L. It is concluded that current neutralization requirements in South Dakota have been sufficient to avoid significant environmental impact to date.

Golden Maple Gilt-Edge Mine

The Golden Maple Gilt-Edge Mine was a small heap-leach gold operation near Lewistown, MT. The mine, originally permitted for 100,000 tons/year of ore, was attempting to leach the Old Jig tailings and unprocessed ore previously mined from the Lookout and Cuba Pits. Ore mined underground was also being leached. No agglomeration was utilized and the material had a significant clay content.

Heavy rains during 1985, generally poor leaching performance of the heap, and use of the emergency pond for process-water storage led to a number of small discharges from the process ponds. Though a number of discharges had a total cyanide of 10 to 30 mg/L within the mine site, the low levels found in surface water offsite (below detection to 0.013 mg/L total cyanide) could not be traced to the active mine. Runoff from the historic tailings upgradient from the mine had detectable levels of cyanide as well. Groundwater contamination was noted in the production well of the mine (1.3 mg/L total cyanide in April 1985 after discharge from the emergency pond), which was located just downgradient of the emergency pond. Stock wells (and springs) located downgradient of the mine had levels of total cyanide from 0.05 to 0.15 mg/L in 1985 and 1986 and chloride-amenable cyanide generally <0.3 mg/L. These cyanide levels were thought to be caused by the mine, though due to the lack of historical data, mining waste could not be ruled out as a potential source.

The mine was closed by regulatory enforcement agencies in late 1985, and closure of the site was begun using bond money. A contractor was employed to treat process solution with calcium hypochlorite and to land-apply the treated solution. Preliminary testing was conducted to see if onsite soils would attenuate the low levels of residual metals in the treated solution. Cyanide destruction by hypochlorite, followed by land application, appeared to be a suitable treatment method at Golden Maple.

During excavation of the heap, which was rinsed with an unknown quantity of treated solution, a number of zones were observed that appeared to have high levels of residual cyanide and which were not rinsed effectively. The apparent poor rinsing performance at the Golden Maple site as well as at a few other small heap-leach sites that have undergone closure has reinforced the premise that heaps cannot be uniformly rinsed; hence, cyanide removal during decommissioning will be incomplete. It must be emphasized that the apparent poor rinsing performance at Golden Maple as well as at other small facilities was exacerbated by a number of factors, including

- Poor overall leaching and rinsing performance resulted in slow or poor gold recovery leading to economic hardship
- Closure activities were accomplished at an accelerated rate due to the approach of winter
- The amount of heap rinsing was unknown and may not have been adequate to remove cyanide under the best of circumstances
- Facility closure was brought about by inadequate environmental safeguards employed by mine operators during operations (e.g., use of the emergency pond for process water storage) rather than by the inherent environmental risk posed by the heap-leaching process itself

Previous Experience at ZMI

ZMI has leached and rinsed a number of heaps on the Zortman, as well as Landusky, properties. The total cyanide and WAD cyanide contents for these various facilities are summarized in Tables 12.10 to 12.14. Notes are included in each of the tables explaining the input conditions. Note that a 2-in. precipitation event took place on May 23, 1990. WAD cyanide levels (Tables 12.10 to 12.12) in the effluent from the heaps increased directly after the precipitation (Table 12.10). Increases in WAD cyanide levels were also observed after the start of the rinse cycles.

Table 12.10. Zortman 1979 Heap Historical Effluent Cyanide Concentrations[1]

Date	Total Cyanide (mg/L)	WAD Cyanide (mg/L)	Remarks
3/29/89	0.03	0.02	
8/5/89	1.11	0.12	Freshwater rinse at
8/6/89	1.45	0.12	0.005 gpm/ft^2 from 4:30
8/7/89	0.98	0.08	p.m. on 8/5 until 8 a.m.
8/8/89	0.97	0.13	on 8/8; effluent solution pH ranged from 7.0 to 8.6
8/8/89	0.96	0.12	Regrading and topsoiling
8/9/89	0.92	0.13	heap
8/10/89	0.84	0.08	
8/11/89	0.73	0.08	
8/14/89	0.72	0.07	
8/21/89	0.52	0.06	
3/10/89	—	0.08	Effluent appeared after
5/7/90	0.43	0.10	spring thaw and 2 in.
5/25/90	0.96	0.06	of precipitation
6/14/90	0.44	0.04	

Leaching and Rinsing History

1. Leach pad last sprayed in the summer of 1982.
2. No process solution or rinsing other than precipitation, including a 500-year precipitation event in 1986 (13.5 in.) and a 100-year precipitation event in 1988 (6.5 in.).
3. No effluent return since the 6/14/90 sample collection.

Table 12.11. Zortman 1980/81 Heap Effluent Cyanide Concentrations[1]

Date	Total Cyanide (mg/L)	WAD Cyanide (mg/L)	Remarks
3/1/90	1.04	0.36	
5/23/90	1.35	0.23	2 in. precipitation
6/5/90	1.22	0.25	On-off rinse cycle with
6/22/90	0.53	0.15	freshwater on north
8/29/90	0.35	0.10	side
9/6/90	0.42	0.11	
9/12/90	1.42	0.13	On-off rinse cycle with
9/14/90	0.83	0.11	freshwater on south
9/17/90	2.41	0.25	side
10/12/90	1.65	0.20	

Leaching and Rinsing History

1. Last sprayed with cyanide in the summer of 1983.
2. On-off rinse cycle: 2 weeks on, 2 weeks off, 1 week on.

476 ENVIRONMENTAL IMPACTS OF MINING

Table 12.12. Zortman 1982 Heap Effluent Cyanide Concentrations[1]

Date	Total Cyanide (mg/L)	WAD Cyanide (mg/L)	Remarks
5/23/90	14.9	6.0	
5/25/90	7.02	2.61	After 2 in. rain (?)
6/15/90	12.20	10.32	
6/17/90	13.50	9.92	

Leaching and Rinsing History

1. Last sprayed with cyanide in the autumn of 1988.

Table 12.13. Landusky 1979 Heap Effluent Cyanide Concentrations[1]

Date	Total Cyanide (mg/L)	WAD Cyanide (mg/L)	Remarks
5/23/90	1.28	0.09	
5/25/90	0.14	0.06	After 2 in. precipitation
6/1/90	2.72	0.30	
6/2/90	2.36	0.40	
6/3/90	2.80	0.48	
6/5/90	2.40	0.27	
6/14/90	2.80	0.63	Spray rinse with freshwater, 3 days on, 3 days off, water off for 2 hr per shift
6/20/90	0.96	0.11	
6/21/90	2.60	0.28	
7/5/90	0.81	0.11	
7/26/90	0.81	0.11	Rinsing as above continued
7/27/90	0.78	0.11	
7/28/90	0.80	0.08	
7/29/90	1.29	0.54	
7/30/90	0.70	0.11	
7/31/90	0.56	0.08	
8/1/90	1.38	0.19	
8/3/90	0.32	0.18	Rest
8/6/90	0.82	0.19	
8/7/90	0.75	0.19	
8/15/90	0.65	0.18	
8/18/90	0.33	0.11	
8/20/90	0.54	0.11	
8/29/90	0.59	0.10	

Leaching and Rinsing History

1. Last sprayed with cyanide in the summer of 1982.
2. Sample on 7/29 taken by Security is of questionable quality.

Table 12.14. Landusky 1980/81/82 Heap Effluent Cyanide Concentrations[1]

Date	Total Cyanide (mg/L)	WAD Cyanide (mg/L)	Remarks
5/23/90	5.71	0.54	
6/14/90	5.37	0.40	
8/17/90	5.10	1.17	Rinse with freshwater
8/18/90	3.40	1.50	during pilot study
8/20/90	3.90	1.13	
8/29/90	5.10	1.05	

Leaching and Rinsing History

1. Last sprayed with cyanide in the summer of 1986.

REFERENCE

1. Schafer, M.W. *Cyanide Degradation and Rinsing of Spent Heap Leach Gold Facilities* (Bozeman, MT: Schafer and Associates, 1990).

Index

Abandoned mines
  acid rock drainage, 224–227
  subsidence, 431–432
Abiotic oxidation of iron, 125
Absorption, wetlands treatment system, 345
Acid generation
  chemical and biological reactions, 171–182
    kinetic tests, 180–182
    metal leaching and migration processes, 175–176
    prediction of acid drainage, 176–179
    static tests, 178–180
  control of, 182–184
  impacts of mining, 9, 11
Acid-generation potential, 177
Acid mine drainage, 18–21
  bactericides in control of, 162–163
  chemical treatment, 132–136
    caustic soda, 135, 136
    lime, 129–132
    limestone, 132–135
  chemistry of formation, 122–126
  conventional neutralization process using lime, 126–129
    high-density sludge process, 128
    other treatment processes, 128–129
  determination of acid-generating potential, 163–164
  Equity Silver closure plan, 247–249
  ion exchange, 155–162

  chemical softening, 160–162
  modified desal process, 158, 159
  sul-bisul process, 156–158
  two-resin process, 158–160
iron oxidation, 135–143
  aeration systems, 137–140
  biological oxidation, 140–143
  oxidation rate, 142–143
from metal mines, 405
reverse osmosis, 151–155
role of bacteria, 124–126
sludge dewatering and disposal, 143–151, 152
wetlands for control of
  algae, role of, 386–390
  metal retention capacity, 374–376
  in Tennessee Valley, 390–396
Acid potential (AP), 163–164, 178–180
Acid removal, wetland processes, 332–333
Acid rock drainage and metal migration
  acid-generation process, 168–169
  bactericides, 186
  base additives, 186–189
  chemical and biological reactions related to acid generation, 171–182
    kinetic tests, 180–182
    metal leaching and migration processes, 175–176
    prediction of acid drainage, 176–179
    static tests, 178–180

conditioning of tailings/waste rock, 184–185
control of acid generation, 182–184
covers and seals to control acid generation, 190–192
  saturated soil or bog, 192
  soil covers, 190–191
  synthetic covers, 191–192
  water cover, 192
designing closure of open pit mine in Canada, 246–249
environmental control measures after the closure of lead-zinc mine in Greenland, 232–238
hydrologic solution, 227–228
impact of abandoned mine on water quality, 224–227
metal contents and treatment of mine water, 249–258
mine environmental rehabilitation, 238–246
monitoring programs for each mine component, 215–224
  haul roads, 224
  open pits, 218–219
  quarries, 223–224
  tailings impoundments, 222–223
  underground workings, 219–221
  waste rock dumps, stockpiles, and heap-leach sites, 221–222
subaqueous deposition, 192–215
  covers and seals to control infiltration, 200–214
  disposal into flooded mine workings, 194
  disposal into manmade impoundments, 193–194
  groundwater interception, 198–200
  lake disposal, 194–195
  marine disposal, 195
  migration control of ARD, 195–198
  waste rock and tailings placement methods, 214–215
sulfide minerals, 169–171
waste segregation and blending, 185
water resource problems in a lead belt, 228–232
Acid wetland substrate comparison, 357–359
Adsorption, wetlands metal removal mechanisms, 332, 345, 349, 354, 406
Aeration systems, acid mine water treatment, 137–140
Aerial photographs, wetlands siting, 349
Aggregation, reclamation processes, 116

Agricultural area land classification, 285
Air quality
  impacts of mining, 5
  liability issues, 29
Alder, 416–418, 421
Alfalfa, 111
Algae
  acid rock drainage from lead mines and, 228
  wetland, 350, 386–390
Alnus rugosa, 416, 417, 418, 421
Altitude, and wetland composition, 343
Alumina-lime-soda process, acid mine water treatment, 161–162
Aluminum, 114
  and cattail growth, 365
  spoil banks, 113
AMAX Coal Co., 448–451
Ammonia nitrogen, wetlands construction, 360
Ammonium nitrate-fuel oil (AN/FO) blasting agents, 427, 429
Anaerobic zones, wetlands, 331, 342, 406, 407
Annie Creek Mine, 472–473
Appalachian region
  hydrologic impact, 274–276
  postmining land use, 445–448
Aquifers, see also Hydrologic impact
  coal, 261–263
  cyanide migration from gold heap-leach operations, 456
  impacts of mining, 10
Areal backfilling methods, 439
Area mining, 12, 15
Argo Tunnel, 224–227, 232
Army worm, 393
Arsenic, 405, 454
Asphaltic membrane, 210–211
Audits, environmental, 29–32
Australia
  reclamation procedures, 105–111
  Rum Jungle site, 190, 206–208

Backfill, 439
Bacteria, see also Microorganisms
  acid generation, 124–126, 174, 256
  control considerations, 182–183, see also Bactericide treatment
Bactericide treatment
  acid mine water, 162–163
  acid rock drainage, 186
Baffled pond, sludge dewatering, 145

INDEX 481

Bank protection structures, erosion and sediment control, 291, 292
Barrel medic, 111
Barren areas, land classification for site evaluation, 285
Barrick Goldstrike Mines, Inc., 470–471
Base additives, acid rock drainage, 186–189, see also Lime; Limestone
Basins, sediment, 296
Bed drying, sludge disposal, 148, 151, 152
Bedrock, wetland siting, 349
Benches, erosion and sediment control, 294
Beneficiation process, 266–267
Benzoate, 186
Berkeley Pit, 249–258
Best management practices (BMPs), 317
Bicarbonates, 350
Big Five Tunnel Experimental Wetland, 407–413
Bioaccumulation of toxic chemicals in wetlands, 332
Biochemical oxygen demand (BOD)
    Nalco 634 and, 303–304
    wetlands performance expectation, 349
Biological oxidation
    acid mine water treatment, 140–143
    microbial, see Microorganisms
Biological transformation
    microbial, see Microorganisms
    wetland treatment processes, 332
Biomass
    of cattails, 369
    and metal removal, 374
    white cedar peatland, 421
Black Angel Mine, 232–238
Black spruce, 117
Blasting, 4, 425–429
Blending, waste rock, 185
Blue-green algae, 387–390
Bluejoint, 117
Blue mussel (*Mytilus edulis*), 234, 235
Bog iron, 327
Bogs, 326, see also *Sphagnum*
    acid generation, 169
    acid rock drainage, 192
    definitions, 325–326
Bone Valley formation, 271
Boom length, 61
Boreal forest zone, 117
Borealis Mine, 468–470
Borrow pit, box pit, 49

Box pits, coal, 45–51
Bucket capacity, 61
Building classification method, 438
Bulk density, reclamation processes, 104
Bulrush (*Scirpus*)
    altitude and, 343
    substrate evaluation, 355, 357–359

Cadmium, 242, 405
Calcite, 173–174
Calcium, cattail growth studies, 362
Calcium carbonate equivalent, 188
Calcium sulfate, reverse osmosis, 154
Canada, acid rock drainage, 246–249
Capacity factor, 370
Carbonates, 350, 406
Carex, see Sedge
Catalysis factor, 124
Cation exchange capacity (CEC)
    and plant growth, 104
    wetlands, 343, 346
Cations
    cattail growth studies, 362
    reclamation processes
        overburden, 107
        soil profile, 109
Cattail (*Typha*), 338, 408
    algae and, 387–390
    altitude and, 343
    Big Five Tunnel wetland, 408
    performance expectation, 349–350
    range of, 363–367
    removal efficiency, 342
    substrate evaluation, 355–361
    survey, 336
    Tennessee Valley wetlands project, 391–395
    Tracy wetland, 401–404
    wetland construction, 361–367, 368–374, 381–386
    Windsor Coal Company Wetland, 396–397
Caustic soda, acid mine water treatment, 128, 135, 136
Cave-in pits, 435
Centrifugation of sludge, 148
Cetraria nivalis, 234
Channel dimensions, erosion and sediment control, 315
Check dams, erosion and sediment control, 314
Chelation, peat, 375
Chemical pollution, impacts of mining, 9

Chemical reactions, wetlands, 327
Chemical treatment, acid mine water, 129–136, 160–162
  caustic soda, 135, 136
  lime, 129–132
  limestone, 132–135
Chinook Mine, 448–451
Chopping, overhand, 74
Clay
  erosion and sediment control, 288, 289, 294, 300
  phosphate mining, 266
  wetlands construction, 354, 355, 357–359
Clean Air Act, 30
Cleanup costs, 29
Clean Water Act, 31
Climate
  control of acid generation, 183
  and impacts of mining, 12, 13
  and wetland system removal efficiencies, 344
Closure
  acid rock drainage after, 232–238, 246–249
    environmental control measures, 232–238
    environmental rehabilitation, 238–240
    open pit mine, design for, 246–249
    water quality impacts, 224–227
  gold heap-leach operations, 455, 468–477
Clothrix, 412
Coal mining, 12, 13
  acid mine drainage, see Acid mine drainage
  hydrologic impact
    Appalachian region, 274–276
    Decker mine, 261–263
    Longwall mining, 276–281
  reclamation, 113–115
  reclamation processes
    in Montana, 112–113
    in Ohio, 113–115
  sdimentation control, 300–307
  subsidence, 431–439
  surface, see Surface coal mining with reclamation
Colonization, reclamation processes, 116
Colorado
  Argo Tunnel, 224–227
  wetland treatment, 342–346, 405–413
Commercial interests, 26
Compaction
  erosion and sediment control, 294
  soil, 105
Complexation, wetlands performance expectation, 349
Compliance, 32
Comprehensive Environmental Response, Compensation and Liability Act (CERCLA), 29–30
Compression ridges, 435–436
Conditioning of tailings/waste rock, acid rock drainage, 184–185
Conductance, specific, 415, 417–420, 422
Conductivity, electrical
  cattail growth effects, 362
  iron removal, 373
Conductivity, hydraulic, see Hydraulic conductivity
Consolidation, 213
Construction equipment, 12, 16
Contour mining, 12, 15, 96
Contract hauling, 155
Control structures, wetland construction, 351–354
Copper
  Big Five Tunnel wetland, 409–411
  gold heap-leach operations, 454
  metal mining and, 405
  wetlands removal, 349
    algae and, 386–390
    Dunka Mine, 413–422
Copper mines
  acidic leaching processes, 124
  environmental rehabilitation, 238–246
Covering operation, Falun copper mine rehabilitation, 244–245
Covers and seals, acid rock drainage management, 190–192, 200–214
Crimping, 296
Curve numbers, 464–465
Cyanide, see Gold heap-leaching operations
Cyanobacteria, 387–390

Darcy velocity, 457
Decker Mine, 261–263
Decommissioning, see Closure
Decomposition, in wetlands treatment system, 345
Deep bench coal mining, 57–58
Deep mine sludge disposal, 148, 151
Deflectors, erosion and sediment control, 291, 292
Depressions, 435
Desal process, 158, 159

INDEX   483

Design process
  acid rock drainage management, closure of open pit mine, 246–249
  erosion and sediment control, 287–296
  reclamation process, landforms, 107–108
  wetlands, 339, 341, 351–354, 361
Dikes
  erosion and sediment control, 294
  wetlands, 348, 353
Dispersion coefficient, 457
Dispersion terraces, 319
Dozer-dragline system, 81, 82
Dragline bench, extending, 61
Dragline operation, coal, 35–39
Draglines, 12
Dragline stripping, 39–66
  box pits, 45–51
  deep overburden, multiple lifts, one pass, 58–62
  deep overburden, split bench mining, 54–58
  moderately deep overburden, two lifts, one pass, 53–54
  moderate overburden depth, single lift, single pass, 51–53
  pullback, 63, 65–65
  single machine subsystems, 39–45
  single seam stripping with nonselective spoil placement, 39
  two-pass extended bench method, 62–64
Drainage
  surface mine, 317–322
  impacts of mining, 9
  erosion and sediment control, 308
  soil, 105
  wetland site selection, 348
Dump slopes, 213
Dunka Mine, 413–422

Echo Bay Minerals Company, 468–470
Economic evaluation, postmining land use, 447, 449–450
Effective pore volume, 457
Effective velocity, 457
Electrical conductivity, 362, 373
Eleocharis, 391, 393, 394
Elevated bench method, 78–79
Elliot Lake project, 205
End cut box pits, 46–47, 49–51
End-cut method, material rehandle, 46–47
End loaders, 12

Engineering characteristics, erosion and sediment control, 284
Environment, mining and
  effects of surface mining, 4–17
  environmental audits, 29–32
  land use, 18, 22–26
  subsidence of mined land, 26–28
  uniqueness of mining, 2–4
  water pollution, 17–21
Environmental audits, 29–32
Environmental control measures, acid rock drainage, 232–238
Environmental effects of surface mining, 4–17
Environmental evaluation, postmining land use, 447, 450
Environmental Protection Agency (EPA), 30
Environmental rehabilitation, see Reclamation; Reclamation and revegetation; Rehabilitation
Equipment, 12, 15–16
Equisetum, 336, 338, 392
Equity Silver Mines Ltd, 246–249
Erosion
  dump slopes, 213
  impacts of mining, 8, 10
Erosion and sediment control
  formulation of plan, 296–300
    maintenance, 298–300
    operation, 297–298
  planning, 287–296
    final design, 289–296
    preliminary design, 287
    preliminary site investigation, 287
    subsurface investigations, 288–289
  preliminary site evaluation, 284–287
    floodplains, 286
    groundwater conditions, 286–287
    impoundments, 286
    land type, 285
    soil and rock, 285
    streams, 285–286
    vegetative cover, 287
  surface mine, 300–307
    drainage control, 317–322
    sedimentation control, 300–316
Evaporation, reverse osmosis waste stream, 155
Evapotranspiration
  hydrologic impact of mining, 265, 266
  phosphate mining effects, 269
  wetland system iron and manganese removal, 381

484   ENVIRONMENTAL IMPACTS OF MINING

Exposure assessment, gold heap-leach operations, 454–466
Extended bench method, 38–39, 57
  tandem machine systems, 58–62, 71, 73–76
  two-pass, 62
  two seams, 73–76

Fabriform mats, 291, 292
Falun Mine, 238–246
Faults, wetland siting, 349
Ferric hydroxide sludges, 148
Fiberglass mulch, 295–296
Filter berm, 299
Filtration
  mine drainage sludge, 151
  sludge disposal, 148, 151
  in wetlands system, 327, 332, 345
    metal removal mechanisms, 406
    performance expectation, 349
Finniss River, 206, 208
Fish
  lead and zinc contamination, 234
  wetlands, 327
Flooded mine workings, acid rock drainage, 194
Floodplains, 286, 391
Florida, phospate mining, 263–274
Flow channels, wetland, 401
Flow patterns
  Tennessee Valley wetlands project, 391
  wetlands configuration, 352–353
Flow rate
  erosion and sediment control, 315
  wetland
    Big Five Tunnel, 409, 410, 411
    substrate selection, 356
Fondaway Mine, 471–472
Framboids, 169
Freezing, and surface metal concentrations, 255, 258
French drains, 338
Freshwater wetlands, 326
Fucus, 234, 236
Fulvic substances, 354

Gabions, 291, 292
Gambusia affinis, 392
Geochemical removal, wetland treatment processes, 332

Geologic hazard maps, wetland siting, 349
Geology
  erosion and sediment control, 284
  wetlands, 326
    site selection, 347, 348, 349
    Tennessee Valley project, 391
Geopolymer, 212
Geotechnical monitoring, 221
Gilt Edge Mine, 472, 473–474
Gold heap-leaching operations
  exposure assessment, 454–466
  decommissioning, industry experience 468–477
    Annie Creek Mine, Lead, SD, 472–473
    Barrick Goldstrike Mines, Inc., Carlin, NV, 470–471
    Borealis Mine, Echo Bay Minerals Company, Hawthorne, NV, 468–470
    decommissioning, industry experience 468–477
    Fondaway Mine, Tenneco Minerals, 471–472
    Gilt Edge Mine, Deadwood, SD, 472
    Golden Maple Gilt-Edge Mine, 473–474
    previous experience at ZMI, 474–477
  hazard identification, 454
  risk characterization, 467–477
  toxicity assessment, 466–467
Gold tailings, conditioning of, 184–185
Grafton Coal Company, 162
Grasses, 118
  erosion and sediment control, 293, 315
  reclamation process, 110–111
Grasslands, 117
Gravel
  tailings area, 204–206
  wetlands, constructed
    function of, 330, 332
    substrate, 359
Gravitational settling, sedimentation control, 300
Great Falls-Lewistown Coal Field, 399
Greenland, 232–238
Ground cracks, 435
Ground vibration, 4
Groundwater, 4, see also Hydrology
  acid rain drainage, 196
  control of acid generation, 183
  cyanide migration from gold heap-leach operations, 456
  erosion and sediment control, 284, 286–290

gold heap-leach operations, 459–462
impacts of mining, 6, 8, 10, 11
liability issues, 29
wetland site selection, 348
Groundwater interception, acid rock drainage, 198–200
Groundwater monitoring, see Monitoring
Gypsum, 154, 173, 256
Gypsum disposal areas, hydrologic impact, 269–274

Hanaford Creek, 300–307
Haulback mining, 12, 15, 96–101
Haul roads, 224
Hawthorn Formation, 271, 272
Hazard identification, gold heap-leach operations, 454
Headcutting, 8
Health, 26
Heap-leach sites
  acid rock drainage monitoring, 221–222
  gold, see Gold heap-leaching operations
High-density sludge process, 151, 153
  acid mine drainage, 128
  Equity Silver closure plan, 247
Horseshoe mining sequencem coal, 89–91
Horsetail (*Equisetum*), 336, 338, 392
Humic content, wetlands construction, 354
Humic substrate
  iron and manganese removal study, 381
  performance expectation, 349
Humidity cells, 181
Hydraulic conductivity
  peat versus mineral soils, 329
  wetlands, constructed, 328
Hydraulic residence time, wetlands, 351
Hydrogen peroxide, iron oxidation, 140
Hydrologic solution, acid rock drainage, 227–228
Hydrology
  control of acid generation, 183
  Falun copper mine rehabilitation, 244–245
  gold heap-leach operations, 454–456
  impacts of mining, 12–14
    longwall, 276–281
    phosphate gypsum disposal areas, 269–274
    phosphate, 263–269
    subsurface, 224–226
  site evaluation for sediment control, 285–287

wetlands, 326, see also Wetlands
  design, 339, 341
  site selection, 347, 348
  Tracy wetland, 399

IMC Fertilizer (IMCF), 272
Impoundments
  acid rock drainage, 193–194
  effects on water, 18, 19–21
  erosion and sediment control, 286
Individual Degree of Saturation (IDS), 252–253, 254
Infiltration, 293
  acid rain drainage, 196
  cyanide migration from gold heap-leach operations, 456
  erosion and sediment control, 293
  covers and seals, 200–214
  placement of waste rock and tailings, 214
Insurance, 29
Intensity factor, 370
Interceptor dike, 299
Ion exchange
  acid mine water treatment, 129, 155–162
    chemical softening, 160–162
    modified desal process, 158, 159
    sul-bisul process, 156–158
    two-resin process, 158–160
  and plant growth, 104
  wetlands, 343, 346, 349, 354
Iron ions and minerals
  acid generation, 169
  acid mine drainage chemistry, 122–124
  acid mine water treatment, 135–143
  oxidation of, 171–172
  solubility of, pH and, 135, 136
  wetlands, 327–328, 376–380, 381–386
    Big Five Tunnel, 409–411
    cattail effects on, 364–367
    cattail growth studies, 362
    efficiency of, 340
    precipitation of, 342
    removal efficiency, 350
    removal efficiency of *Sphagnum*, 341–342
    site selection, 348
    Sphagnum, 374–380
    stability of, in sediments, 406–407
    substrate comparison, 357–359
    vegetation effects on removal of, 362–363
    Windsor wetland, 401

Juncus, 387, 393, 394
Jute netting, waterway stabilization, 293

Kellog Tunnel, 227–228
Kentucky, blasting guidelines in, 426, 427
Keycut, 37–39, 55–57
Kinetic tests, acid rock drainage, 180–182

Lagooning
  mine and mill separation, 232
  sludge disposal, 148, 150–151
Lake disposal, acid rock drainage, 194–195
Landform design, reclamation procedures, 107–108
Land type, erosion and sediment control, 285
Land use, 441–451, see also Postmining land use
  impacts of mining, 6, 12, 14, 18, 22–26
  wetland site selection, 347
Land-use conflicts, 26
Landusky heap, 454
Layered cutting, coal, 37–39
Layered tailings placement, 184
Leaching, acid rock drainage, 175–176
Lead
  acid rock drainage, 228–232
  wetlands treatment, algae and, 386–390
Lead mine reclamation, 115–116
Lead-zinc mine, acid rock drainage, 232–238
Legislation, 29–31, 426
Level spreaders, 300, 338
Liabilities, 29, 246
Lichen, 234
Life-of-Mine Environmental Assessment (EA), 453
Lime
  acid drainage control, 187
  acid mine water treatment, 126–132
  Berkeley Pit water treatment, 255–258
  reverse osmosis waste stream, 155
Lime-soda process, acid mine water treatment, 160–161
Limestone, 182
  acid drainage control, 187
  acid mine water treatment, 128, 132–135
  constructed wetland systems, 339
  cover material, 203
  erosion and sediment control, 289
  sludge management, 147

and wetland treatment removal efficiencies, 342
Limonite, 327
Loading rates, wetlands, 350
Local backfilling method, 439
Longwall mining, hydrologic impact, 276–281
Lysimeters, 242

Magnesium, 362
Maintenance, erosion and sediment control, 298–300
Manganese, wetland treatment systems, 327–328, 381–386
  Big Five Tunnel, 409–411
  cattail effects on, 364–367
  cattail growth studies, 362
  site selection, 348
  substrate comparison, 357–359
  stability of, in sediments, 406–407
  vegetation effects on removal of, 362–363
  Windsor wetland, 402
Marcasite, acid generation, 169
Marine disposal, acid rock drainage, 195
Marshes, definitions, 325–326
Mass balance model, 456, 457
Membrane covers, see Synthetic covers
Mercury, 405
Metal mining, see also specific metals
  acid mine drainage, 405
  acid rock drainage, 232–238
  reclamation, 115–116
  acid mine drainage, 405
Metals, see also Acid rock drainage and metal migration
  acid rock drainage, 249–258
  wetland processes, 327–328, 332–333, 342–347, see also Iron; Manganese
  cattail effects on, 364–367
  retention of, 374–376
Microorganisms
  bioelement sources and function, 331
  sulfate-reducing bacteria, 412
  wetlands, 328–329, 332
    bacterial morphology, 355
    metabolism, 345
    metal removal by, 343
    metal removal mechanisms, 406
  wetland system construction
    design and development of, 361
    performance expectation, 349

INDEX 487

reactions of, 406–407
sources of, 412
substrate composition, Big Five Tunnel wetland, 411
Migration, see Acid rock drainage and metal migration
Mineral soils, 328, 329
Mine shafts, wetland siting, 349
Mine spoil, see Spoil material
Mining and environment, see Environment, mining and
Mining methods, 12, 15
Minnesota, natural wetland, 413–422
Missouri, water resource problems, 228–232
Modified desal process, acid mine water treatment, 158, 159
Molybdenum mining, 405–413
Monitoring, 32
  haul roads, 224
  open pits, 218–219
  quarries, 223–224
  tailings impoundments, 222–223
  underground workings, 219–221
  waste rock dumps, stockpiles, and heap-leach sites, 221–222
Montana
  blasting guidelines in, 426
  gold heap-leach operations, 473
  reclamation procedures, 112–113
  Tracy wetland, 399–405
Mosquito fish, 392
Moss, 482, see also Peat; *Sphagnum*
Mougeotia, 387–390
Mountaintop removal, 12, 15
Mulches, erosion and sediment control, 295–296
Multiple seam systems, 17, 71, 85–88
Mytilus edulis, 234, 235

Nalco 634, 303–307
NAPP technique, 164
National Environmental Policy Act (NEPA), 391
National Pollutant Discharge Elimination System (NPDES), 31
National Pollutant Discharge Elimination System (NPDES) permit, 391
Natural wetland, 188–189
  nickel and copper removal by, 413–422
  substrate comparison, 357–359

Net neutralization potential (NETNP), 164
Netting, waterway stabilization, 293
Neutralization
  acid mine drainage, 126–129
  waste water, base additives, 186–189
  wetlands metal removal mechanisms, 406
Neutralization potential (NP), 164, 178, 188–189
Nevada, gold heap-leach operations, 468–472
New Brunswick, 421
Nickel, wetlands removal, 349, 386–390, 413–422
Nitrogen, wetlands construction, 360
Nonpoint sources, 31
Nutrients
  and plant growth, 104
  wetlands, 326, 327, 349

Ohio
  blasting guidelines in, 426
  reclamation procedures, 113–115
One-pass extended bench method, 74–76
Open pits, 12, 15
  acid rock drainage, 167
    designing mine closure, 246–249
    groundwater interception, 199
    metal migration control, 197
    monitoring, 218–219
  equipment, 12, 16
  water treatment, 249–258
Operation, erosion and sediment control, 297–298
Organic loading, in wetlands, 352–353
Organic material, wetlands
  long-term storage role, 406
  performance expectation, 349
Organic soils
  comparison of physical and chemical attributes of, 328
  wetlands, 332, 354
Organic sulfur, 164
Oscillatoria, 387–390
Overburden
  characterization for reclamation, 106, 107
  Equity Silver closure plan, 247
  impacts of mining, 9, 12, 13
  unstable, stripping of, 83–84
Overhand chopping, 74
Oxidation
  acid generation process, 168–169, 171

acid mine water chemistry, 122–126, 135, 137–143
  bacteria, 124–126
  oxidation rate, 142–143
  impacts of mining, 9, 11
  iron, 256
  and precipitation of metals, 342
  wetland processes, 332, 335
Oxidation rate, acid mine water, 142–143
Oxides, stability of, 406
Oxygen
  and acid generation, 172, 183
  pore gas, 206
  wetlands, 327, 348, 361
Oxygen measurements, Falun copper mine rehabilitation, 245
Oxygen transfer efficiencies (E), 138
Ozone, acid mine water treatment, 139

Panel (pit) width, 37–38
Paper birch, 117
Particle size, reclamation processes, 107, 109
Pasture, 111
PC Seep, 456
Pea gravel, 359
Peak rate of runoff, erosion and sediment control, 315
Peat, see also *Sphagnum*
  altitude and, 344
  cation exchange capacity, 343
  physical characteristics for, 329
  Tracy wetland, 399
Peatland, white cedar, 413–422
Percolation, gold heap-leach operations, 455
Permeability
  erosion and sediment control, 284
  tailings and waste rock, 184
Permit performance audits, 32
Permit process, wetlands construction, 391
pH, 340, 346, 406
  and acid generation, 172, 182–183
  erosion and sediment control, 284, 289, 295
  and iron solubility, 135, 136
  and plant growth, 104
  and precipitation of metals, 342, 343
  reverse osmosis, 153
  wetlands
    Big Five Tunnel, 409–411
    cattail effects on, 364–367
    cattail growth effects, 362

    sediments, 370
  site selection, 348
  substrate comparison, 357–359
  substrate selection and, 356
  variability of, 350
Phosphate compounds, 186
Phosphate mining
  gypsum disposal areas, 269–274
  hydrologic impact, 263–269
Phosphorus, wetlands, 360
Photooxidation, iron, 256
Pillar failure, 439
Pit subsidence, 432
Plants, see also Vegetation
  reclamation process, factors affecting growth, 104
  seeded mine spoil, 112
  toxic metal removal mechanism, 412
  wetlands, 328, 329, 376–380
    bioaccumulation of toxic chemicals, 332
    definitions, 325–326
    metabolism of, 345
    performance expectations, 349
    species, 329–330
    and system removal efficiencies, 344
Plug-flow kinetics, 351–352
Polychlorinated biphenyls (PCB), 31
Polyelectrolytes
  Nalco 634, 303–307
  sedimentation control, 300, 301
  sludge settling, 147
Polymer netting, waterway stabilization, 293
Polytrichum, 402
Ponds, erosion control structure, 296
Pore gas, oxygen in, 206
Porosity
  peat versus mineral soils, 329
  size distribution, reclamation considerations, 104
  wetlands configuration, 352
Porous bed drying, mine drainage sludge, 151
Postmining land use, 441–451, see also Reclamation; Reclamation and revegetation
  Appalachian region case study, 445–448
  Midwest case study, 448–451
Potential acidity, 178
Powelson Wildlife area, 335
Prairie Hill Mine, 309–317
Precipitation (chemical)
  Berkeley Pit water treatment, 255–258

INDEX 489

wetlands, 345
  metal removal mechanisms, 406
  performance expectation, 349
Precipitation (rainfall), see Rainfall
Pressure filtration, mine drainage sludge, 151
Priority pollutants, metal mining and, 405
Productive capacity, postmining, 103
ProMac system, 162–163
Public access to information, 26
Pullback method, 55, 63, 65, 67
Pyrite, see also Acid mine drainage; Iron ions and minerals
  acid generation, 169
  oxidation chemistry, 122–126, 171–172

Quarries, acid rock drainage monitoring, 223–224
Quicklime
  Equity Silver closure plan, 247
  types of, 129–130

Radiation, 26
Rainfall, 9
  Appalachian region, 274, 276, 277
  and erosion and sediment control, 300–302
  gold heap-leach operations, 455
  and phosphate mine sites, 269
  wetland systems
    iron and manganese removal study, 381
    Tracy wetland, 399
Receiving-environment surface monitoring stations, 221
Reclamation, see also Postmining land use; Surface coal mining with reclamation
  factors affecting, 12
  phosphate mine, 263–269
Reclamation and revegetation
  in Australia, 105–111
    landform design, 107–108
    material characterization, 106, 107
    pasture management, 111
    reclamation procedures, 109–110
    revegetation methods, 110–111
    tree planting, 111
    use of topsoil, 108–109
  factors affecting natural revegetation of coal mine spoil banks in Ohio, 113–115
  revegetation at the Usibelli Coal Mine, Alaska, 117–119

revegetation of a surface mined land in Montana, 112–113
soil, 103–105
vegetation development on old and abandoned lead and zinc mine 113–116
Redox conditions, wetland, 332, 333, 342, 343, 346, 350, 373–374
Regulation, 26
  audits, 28–32
  wetlands construction, 391
Rehabilitation, see also Reclamation; Reclamation and revegetation
  acid rock drainage, 238–246
  covers and seals
    acid generation control, 190–197
    infiltration control, 200–214
  Falun copper mine, 241–245
Rehandle method
  borrow pit, 49, 51
  for high and deep benches, 59
Remedial investigation and feasibility study (RI-FS), 29
Remedial works, Falun copper mine, 241–242
Removal efficiencies, wetlands, 341–342, 344, 350, 352, 358
Resource Conservation and Recovery Act (RCRA), 29
Revegetation
  impacts of mining, 10–12
  reclamation, see Reclamation and revegetation
Reverse osmosis
  acid mine drainage treatment, 129
  acid mine water treatment, 151–155
Revetments, erosion and sediment control, 291, 292
Rhizobium, 111
Rhizome systems, 361
Rhodes grass, 111
Riprap, 291, 292, 307, 308, 310–311, 314–316
Risk characterization, gold heap-leach operations, 467–477
Risk management audits, 32
Road construction material, acid generation, 224
Roads, erosion and sediment control, 284
Robinia pseudo-acacia, 115
Rock check dams, erosion and sediment control, 314
Root systems
  growth in wetland cells, 360–361

metal-rich, 343
Root zone
   revegetation process, 103–105
   wetland processes, 332, 361
Rotating biological contractors (RBC), 140–142
Rum Jungle site, 190, 206–208
Runoff
   erosion and sediment control, 315
   impacts of mining, 9, 265
   phosphate mining effects, 269
   reclamation processes, 115
Runoff model, 463
Rushes (*Juncus*), 393, 394, 408
Ryegrass, 111

Safe Drinking Water Act (SDWA), 31
Sag subsidence, 435
Saltwater marshes, 326
Sampling program, prediction of acid generation, 177–179
Sand, wetlands, 330, 332
Sand Coulee drainage, 399
Sand tailings, phosphate mining, 267
Saturated soil, acid rock drainage, 192
Saturation Index (SI), 252–254
Scirpus, 343, see also Bulrush
   algae and, 387
   Tennessee Valley wetlands project, 391–394, 395
   substrate evaluation, 355
Scirpus cyperinus, 355
Scouring, erosion and sediment control, 315
Scrapers, 12
SCS curve number method, 456
Sealing, copper mine, 241–244
Seams mined per pit, 17
Seasonal cycles
   erosion and sediment control, 295
   wetlands systems
      and metal removal, 371–372, 381–383
      performance, 357–359
Seaweed, lead and zinc contamination, 234, 236
Sedge (*Carex*), 343, 344
   Tennessee Valley wetlands project, 391, 392
   white cedar peatlands, 416, 417
Sediment, impacts of mining, 8
Sedimentation
   acid mine drainage, 126
   impacts of mining, 8, 9
   wetlands, 327, 345, 349
Sediment control, see Erosion and sediment control
Sediment pond, 296
Seepage
   erosion and sediment control, 289
   wetland site selection, 348
Seepage models, 269
Settlement, 213
Settling ponds
   clarification with Nalco 634, 303–307
   mine drainage treatment, 150
   sedimentation control, 300–301
Shaft subsidence, 433–434
Shellfish, wetlands, 327
Shoe Basin Mine, 346
Short-circuiting, wetland, 353, 401
Shotcrete, 212
Sidecast method, 8, 10, 11
Side cut box pits, 46, 48
Simca Mining Inc., 445–448
Simyra henrici, 393
Single dragline, 17
Single-machine subsystems, 17, 39–45
Single-seam stripping, 39, 70
Single-stripping shovel subsystems, 17
Sinkholes, wetland siting, 349
Site assessment audits, 32
Site investigation, erosion and sediment control, 287
Slope
   dump, 213
   erosion and sediment control, 288, 289, 294
Sludge
   acid mine water treatment, 143–152
   Berkeley Pit, 255
Sludge recirculation process, 128
Slug test method, 279
Slurry feed system, acid mine water treatment, 131–134
Snowmelt, gold heap-leach operations, 455
Social impact evaluation, postmining land use, 447, 450
Sodium, wetland, 362
Sodium hydroxide, acid drainage control, 187
Sodium lauryl sulfate, 186
Soil
   cyanide migration from gold heap-leach operations, 456
   erosion and sediment control, 284, 285, 288, 289, 294, 295
   liability issues, 29

INDEX 491

reclamation processes, 104, 115
  as sources of bacteria, 412
  wetlands, 361–367
    comparison of, 328
    definitions, 325–326
    function of, 330, 332
    site selection, 347, 349
Soil binders, erosion and sediment control, 295
Soil covers, acid rock drainage, 190–191, 200–209
Soil temperature, and reclamation, 114, 115
Soil texture
  impacts of mining, 12
  reclamation processes, 104, 107
Sorbate, 186
South African gold tailings, conditioning of, 184–185
South Dakota, gold heap-leach operations, 472–473
Sparganium, 393
Special purpose audit, 32
Specific capacities, coal mining and, 277–279
Specific conductances, 415, 417–420, 422
Spent heap-leach piles, acid rock drainage, 198, 200
Sphagnum, 338, 376–380
  adverse effects studies, 335
  maximum removal efficiencies, 341
  metal retention capacity of, 374–376
  wetlands performance expectation, 349
Sphagnum recurvum, 335
Spingola No. 1 mine, 445–448
Split bench mining, 54–58
Spoil materials
  acid rock drainage, 198, 200
  surface coal mining, see Surface coal mining with reclamation
  impacts of mining, 8, 9
  natural revegetation of, 113–115
  wetlands substrate comparison, 357–359
Static tests, acid rock drainage, 178–180
Statutes, 29–31
Steep-slope mining methods, 12, 15, 91–101
  general sequence of mining and reclamation operations, 92–96
  haulback mining methods, 96–101
Stockpiles, acid rock drainage, 198, 200–202
Strata consolidation methods, 439
Streams, see also Erosion and sediment control
  erosion and sediment control, 285–286
  wetlands construction, 353

Strip mining, see Surface coal mining with reclamation; Surface mining
Strip mining spoil banks, see Spoil materials
Stripping equipment, types of, 12, 15–16
Subaqueous deposition, acid rock drainage, 192–215
Subsidence, 431–439
  damage criteria, 437–438
  impacts of mining, 26–28
  investigations, 434–436
  remedial measures, 438–439
  structural damage, 436–437
  wetland siting, 349
Substrates, wetlands, 328, 354
  and cattail biomass, 369–370
  definitions, 325–326
  function of, 330, 332
  site selection, 348
Subsurface coal mining, hydrologic impact, 274–276
Subsurface drainage, erosion and sediment control, 294
Subsurface investigations
  erosion and sediment control, 288–289
  wetland site selection, 347
Sul-bisul process, acid mine water treatment, 156–158
Sulfate, wetlands water treatment, 340–342, 348
Sulfides, see also Acid rock drainage and metal migration
  acid generation process, 168–169
  acid rock drainage, 169–171
  stability of, 406
Super Floc 330, 307
Superfund, 29
Superfund Amendments and Reauthorization Act (SARA), 30
Superlien, 29
Surface area, wetland system development, 361
Surface coal mining with reclamation
  dragline operation, 35–39
  dragline stripping, 39–66
    box pits, 45–51
    deep overburden, multiple lifts, one pass, 58–62
    deep overburden, split bench mining, 54–58
    moderately deep overburden, two lifts, one pass, 53–54

moderate overburden depth, single lift,
single pass, 51–53
pullback, 63, 65–65
single machine subsystems, 39–45
single seam stripping with nonselective
spoil placement, 39
two-pass extended bench method, 62–64
horsehoe mining sequence, 89–91
multiple seam systems, 85–88
steep slope mining, 91–101
general sequence of mining and reclamation operations, 92–96
haulback mining methods, 96–101
other conventional contour mining
situations, 96
tandem machine systems, 67–84
deep overburden, two lifts, one pass, 71
dozer-dragline system, 81, 82
elevated bench method, 78–79
extended bench method for two seams, 73–74
moderate overburden depth, one lift, single
pass, 70–71
moderate overburden depth, two lifts, one
pass, 71
multiple seam stripping with draglines, 71
one-pass extended bench method for two-
seam stripping, 74–76
selective spoil placement in two-seam
condition, 81–83
single dragline, two seams, nonselective
spoil placement, 71–73
single-seam stripping with selective spoil
placement, 70
stripping of unstable burden material, 83–84
tandem dozer/dragline stripping, 67–70
tandem machine, two-seam condition,
nonselective spoil placement, 80
tandem machine-multiple dragline, 81, 83
tandem shovel-dragline system, 80–81
two-pass spoil side method, 77–78
terrace mining, 66, 67
Surface flows
acid rain drainage, 196
wetlands configuration, 352
Surface mining
drainage control, 317–322
environmental effects, 4–17
erosion and sediment control, 300–307, 316–322

hydrologic impact, 262–263
reclamation processes
in Australia, 105–112
in Montana, 112–113
in Ohio, 113–115
sedimentation control, 307–316
drainage control, 317–322
erosion and sediment control, 307–316
Surface runoff, see Runoff
Surface texture, impact of mining, 7
Surface water, 4, 332, see also Hydrology
acid rock drainage, 197
to Berkeley Pit, 251
control of acid generation, 183
diversion from acid-generating rock, 197–198
gold heap-leach operations, 455–456, 462–466
impacts of mining, 5, 8
liability issues, 29
site evaluation for sediment control, 285–287
Suspended solids (SS), wetlands
performance expectation, 349
site selection, 348
substrate comparison, 357–359
Swamps
acid generation, 169
definitions, 325–326
zinc removal, 249
Sweden, environmental rehabilitation in, 238–246
Synthetic covers
control of acid generation, 191–192
control of infiltration, 209–212

Tailings
acid rock drainage
conditioning of, 184–185
groundwater interception, 200
migration control, 198
monitoring, 222–223
placement methods, 214–215
seals and covers, 204–206
Equity Silver, 249
phosphate mine, 267
Tandem machine subsystem, 17
Tandem machine systems, 17, 67–84
deep overburden, two lifts, one pass, 71
dozer-dragline system, 67–70, 81, 82
elevated bench method, 78–79

extended bench method for two seams, 73–74
moderate overburden depth, one lift, single pass, 70–71
moderate overburden depth, two lifts, one pass, 71
multiple dragline, 81, 83
multiple seam stripping with draglines, 71
one-pass extended bench method for two-seam stripping, 74–76
selective spoil placement in two-seam condition, 81–83
shovel-dragline system, 80–81
single dragline, two seams, nonselective spoil placement, 71–73
single-seam stripping with selective spoil placement, 70
stripping of unstable burden material, 83–84
two-seam condition, nonselective spoil placement 80
two-pass spoil side method, 77–78
Tantramar Swamp, 421
Technical processes-practices audits, 32
Temperature, wetland, 399
and acid generation, 172, 182–183
and cattail growth, 362
Tenneco Minerals, 471–472
Tennessee
blasting guidelines in, 426
wetland treatment, 390–396
Tennessee Valley Authority (TVA) wetlands guidelines, 350
Terrace mining, coal, 66, 67
Terraces, erosion and sediment control, 294, 319
Thiobacillus ferrooxidans, 124–126, 140, 173–175, 186
Thuja occidentalis, 413, 416–418, 421
Thunder Mountain Mine, 317–322
Topdressing, erosion and sediment control, 295
Topography, 12, 15
control of acid generation, 183
erosion and sediment control, 284
impacts of mining, 12, 13
wetland siting, 349
Topsoil
erosion and sediment control, 284, 289
reclamation procedures, 108–109
wetlands, see Soil, wetlands
Toxic compounds
bioaccumulation on wetlands, 332

and plant growth, 104
Toxicity assessment, gold heap-leach operations, 466–467
Toxic Substances Control Act (TSCA), 31
Trace elements, 107
Tracy Wetlands, 399–405, 416
Transmissivity values, 277—281
Transpiration, 293, see also Evapotranspiration
Treatment-storage-disposal (TSD) facilities, 30
Trough subsidence, 432–433, 435
Tub Run Bog, 337
Two-pass extended bench method, coal, 62–64
Two-pass spoil side method, coal, 77–78
Two-resin process, acid mine water treatment, 158–160
Two-seam mining
extended bench method, 73–74
one-pass extended bench method, 74–76
selective spoil placement, 81–83
single dragline nonselective spoil placement, 71–73
tandem machine, nonselective spoil placement, 77–81
Two-seam stripping, coal, 74–76
Typha, see Cattail
Typha latifolia, 392

Underdrains, erosion and sediment control, 293
Underground disposal of sludge, 148, 151
Underground mines
acid rock drainage
groundwater interception, 199
migration control, 197
monitoring, 219–222
subsidence, 431–439
Unsaturated flow model, cyanide migration from gold heap-leach operations, 456
Uranium Mill Tailings Remedial Action Project (UMTRAP), 201
Uranium mine soil cover, 206
Usibelli Coal Mine, Alaska, 117–119
Utricularia, 393

Vacuum filtration, mine drainage sludge, 151
Vegetation, see also Reclamation and revegetation; Revegetation
erosion and sediment control, 287–291, 295, 315
tailings site, 203

wetlands, 332, 355–356, 361, see also specific plants
Vibrational damage, blasting, 426

Wad, 327–328
WAD cyanide, 454, 461, see also Gold heap-leaching operations
Washington State, erosion and sediment control, 300–307
Waste management issues, 26
Waste rock, acid rock drainage
 conditioning, 184–185
 covers, 214–215
 groundwater interception, 200
 migration control, 198
 monitoring, 221–222
 segregation and blending, 185
Wastewater, wetland, 332, 348
Water
 acid rock drainage, 250–258
 control of acid generation, 183
 reclamation processes, 115
 wetland construction, 361–367
 wetland definitions, 325–326
Water balance model, 265, 269
Water budget, and wetland function, 381–383
Water cover, acid rock drainage, 192
Water monitoring, see Monitoring
Water quality
 acid rock drainage, 224–227
 impacts of mining, 17–21
 wetlands, 327
  site selection, 348
  Tracy wetland, 404
Water resources
 acid rock drainage, 228–232
 coal mining and, 277–279
 impacts of mining, 10
Waterways, erosion and sediment control, 291, 292
Weathering trials, 181
West Virginia
 blasting guidelines in, 426, 427
 cattail wetlands, 368–374
 membrane barrier, 210
Wetlands
 algae, treatment of acid mine drainage, 386–390
 cattails, metal removal, 368–374
 constructed, 328–335

definitions, 325–326
functions of, 326–327
iron and manganese removal in *Typha*-dominated wetland, 381–386
metal retention capacity for treatment of acid mine drainage, 374–376
mine drainage treatment, 335–354
 hydraulic design and control structures, 351–354
 metal removal, 342–347
 performance expectations, 349–350
 site selection, 347–349
Sphagnum, iron uptake, 376–380
substrate, 354
 evaluation for AMD systems, 354–361
 vegetation, 361
 water and soil parameters affecting growth of cattails, 361–367
in Tennessee Valley, 390–396
Tracy wetland, 399–405
wetland treatment in metal mining, 405
 Big Five Tunnel experimental wetland, Colorado, 407–413
 nickel and copper removal by a natural wetland, 413–422
Windsor Coal Company wetland, 396–398
White cedar, 413, 416–418, 421
White cedar peatland, 413–422
White spruce, 117
Widows Creek Fossil Plant, 393, 395
Windsor Coal Company wetland, 396–398, 401, 402
Wisconsin, reclamation of abandoned mines, 115–116
Woodlands land classification, 285

Yazoo gully formation, 314
Yellowboy, 121, 342

Zinc
 in acid rock drainage, 241
 Big Five Tunnel wetland, 409, 410, 411
 Equity Silver mine, 249
 gold heap-leach operations, 454
 wetlands treatment, algae and, 386–390
Zinc mine
 acid rock drainage, 232–238
 reclamation of, 115–116